Advanced Manufacturing

Springer
London
Berlin
Heidelberg
New York
Barcelona
Budapest
Hong Kong
Milan
Paris
Santa Clara
Singapore
Tokyo

Other titles published in this series:

A Systems Approach to AMT Deployment
D.R. Towill and J. Cherrington (Eds)

Human-Intelligence-Based Manufacturing
Y. Ito (Ed.)

Intelligent Manufacturing: Programming Environments for CIM
W.A. Gruver and J.C. Boudreaux (Eds)

Automatic Supervision in Manufacturing
M. Szafarczyk (Ed.)

Modern Manufacturing
M.B. Zaremba and B. Prasad (Eds)

Advanced Fixture Design for FMS
A.Y.C. Nee, K. Whybrew and A. Senthil kumar

Intelligent Quality Systems
D.T. Pham and E. Oztemel

Spyros G. Tzafestas (Ed.)

Computer-Assisted Management and Control of Manufacturing Systems

With 178 Figures

Springer

Spyros G. Tzafestas, DSc, PhD, MSc, CEng, FIEE, FIEEE
National Technical University of Athens,
Department of Electrical and Computer Engineering,
Intelligent Robotics and Automation Laboratory,
15773 Zographou, Athens, Greece

Series Editor

Professor Duc Truong Pham, PhD, DEng, CEng, FIEE
University of Wales Cardiff
School of Engineering, Systems Division,
P.O. Box 917, Cardiff CF2 1XH, UK

ISBN-13: 978-1-4471-1242-6 e-ISBN-13: 978-1-4471-0959-4
DOI: 10.1007/ 978-1-4471-0959-4

British Library Cataloguing in Publication Data
A catalogue record for this book is available from the British Library

Softcover reprint of the hardcover 1st edition 1997

Typesetting: Camera ready by contributors

69/3830-543210 Printed on acid-free paper

Preface

Modern manufacturing systems involve many processes and operations that can be monitored and controlled at several levels of intelligence. At the highest level there is a computer that supervises the various manufacturing functions, whereas at the lowest level there are stand alone computer controlled systems of manufacturing processes and robotic cells. Until recenty computer-aided manufacturing systems constituted isolated "islands" of automation, each oriented to a particular application, but present day systems offer integrated approaches to manufacturing and enterprise operations. These modern systems, known as computer-integrated manufacturing (CIM) systems, can easily meet the current performance and manufacturing competitiveness requirements under strong environmental changes. CIM systems are much of a challenge, and imply a systemic approach to the design and operation of a manufacturing enterprise. Actualy, a CIM system must take into account in a unified way the following three views : the *user view*, the *technology view*, and the *enterprise view*. This means that CIM includes both the engineering and enterprise planning and control activities, as well as the information flow activities across all the stages of the system.

In contrast to early automation efforts which were focusing on mechanical gear-and-cam technologies and hardwired electrical circuits, present day automation efforts give emphasis to the information technology, and management and control issues of CIM. Two new concepts emerged from these latest efforts are the concept of *concurrent (or simultaneous) engineering* (CE), and the concept of *intelligent information system* (IIS). CE is one of the primary concepts that enables enterprises to reach a world-class level, and IIS is the synergy of CE and CIM. Today, artificial and computational intelligence are embedded in many different ways to assist the operation of CIM systems and ensure best performance under strong internal and external variations.

This book involves sixteen timely contributed chapters that reflect very well the above tendencies in the design and operation of modern manufacturing systems. A brief outline of these chapters is as follows.

Chapter 1 (by S. G. Tzafestas) provides a short account of a number of information technology (IT) issues applied to manufacturing systems. First, the basic engineering CIM functions are discussed , and then it is shown how and why IT enhances the competitive advantage of manufacturing companies. The core of the chapter outlines two important CIM models, namely the CIM-OSA and CMSO models, that cover intra- and inter-organizational managerial, communication and IT operations. This chapter ends with a discussion of three principal areas of artificial intelligence that are extensively used in CIM systems, and a look at the economic evaluation and justification of automated manufacturing.

Chapter 2 (by Moravčik and Mišút) deals with the coupling of conventional and artificial intelligence-based methods for the design of intelligent "decision support systems" (DSS) in the manufacturing systems management area. First, some general information about DSS along with an explanation of the knowledge structure and knowledge representation for intelligent DSS is introduced. Then, special issues of manufacturing problems are discussed. Finally, the topic is demonstrated by two examples of DSS. The first example was chosen from a very popular problem area, namely operation scheduling ; the second example is also related to the problems of manufacturing management and presents DSS for process planning.

Chapter 3 (by Kovács) is devoted to the application of artificial intelligence to flexible manufacturing systems (FMS) simulation, scheduling and control. An overview of the field is first given covering all important issues. Then, the chapter presents some expert systems' applications to assist in simulation, evaluation, scheduling, quality control, and control of FMS in the form of a case study discussing research and development efforts in the author's laboratory. Well known simulation (SIMAN/Cinema) and networking (MAP) packages are combined with high performance expert system shells (ALL-EX, G2) to lead to acceptable solutions.

Chapter 4 (by Kratz and Rose) first identifies perspectives on process modelling. It starts with general observations drawn from the field of business process modelling in order to set the context. These observations provide evidence for the need of a holistic framework of process modelling that covers different phases and approaches for specification, analysis and implementation of business processes. While refining the functional and non-functional requirements of the proposed framework, an architecture is designed that supports these perspectives and in particular various kinds of performance analysis with respect to quantitative and qualitative measures. These performance analysis tools- called advisors- are each founded in a very specific performance perspective of a process -be it cost, time, quality, flexibility etc. Two applications in the field of administration and production are employed in order to illustrate the approach and the pertaining aspects of process modelling and reengineering.

Chapter 5 (by Bajic and Richard) presents how quality management activities must be conducted and based on both functional and informational modelling approaches, according to pertinent and standardized modelling frameworks like CIM-OSA, and also product data representation models and methodologies like STEP. Particular attention is given to three topics, namely quality modelling in information systems, quality management in production control, and communication support for quality assurance. A promising concept, named "nomadic manufacturing object", and an associated methodological support and application framework are presented as a new sight for quality management and relationship within a CIM architecture.

Chapter 6 (by Little) is based on a research project that reviewed planning and scheduling procedures in some thirty five industrial companies to establish best practice in short term scheduling and to identify the key factors for scheduling process. The dominant position held by MRP and MRPII within manufacturing industry today is examined and the objectives of the research undertaken given. This is followed by an explanation of the method developed to provide a framework for the analysis of scheduling performance both by industrial sector and by scheduling approach. It presents the main findings of the research and then attempts to account for the comparatively low penetration of finite scheduling tools within the UK when compared to the evident current success of MRPII and MRPII hybrid models. An examination of a number of identified weaknesses of the MRPII paradigm in meating the challenges of agile manufacture is made that leads to the generation of some new models which show how different business sectors are adapting the paradigm to provide a more dynamic scheduling capability.

Chapter 7 (by Fenchel and Chen) considers a generic FMS where a set of n input part types (PTs) is processed subject to processing and system constraints. Both constraint categories are considered for a set of m parallel servers. All servers provide identical operating conditions for a certain PT, but nonidentical processing capabilities. Thus it is assumed that there is no inherent priority between between servers for processing a PT, if a PT can be processed at the considered server. A routing policy for a generic Discrete Event Dynamic System (DEDS) of the following charateristics is developed : (1) real-time capable, (2) maintains or returns to prescribed upper and lower buffer bounds, and (3) balances assigned workload between eligible servers. It is assumed that the process times, buffer bounds, and routing conditions are known in advance, processing is non-preemptive, machines allow for a class of operating breakdowns, and no precedence relationships between PTs of one class exist. A simulation experiment using 2^k factorial design illustrates attributes of the developed routing policy.

Chapter 8 (by Michelini, Acaccia, Callegari, Molfino and Razzoli) starts with a review of efficiency in intelligent manufacturing which is related to the *economy of scope* issues, and some basic requirements (such as : piece wise continuous betterment for flexible specialization, co-operative knowledge processing with simultaneous engineering, company wide quality control by trend monitoring maintenance, lean engineering verifications with material and information redundancies removal). Then, it deals with typical example applications, where the judgmental blocks are developed to be included for on-process operations, directly interfaced to real plants. The three case discussions cover : (i) the adaptive scheduling at emergencies, by integrated control-and-management; (ii) the design and robotised (unattended) assembly of contractors' driven manufacts; (iii) the expert-govern of an assembly facility for customers' adapted product mixes. Details on the material set-ups and on the information out-fits are summarized, with sample results. Each case distinguishes for the software requirements. The first is based on an effective, general purpose AI language, to make easy the inclusion of plant reconfigurability; the second uses a dedicated package, with programming extensions ruling the unattended operations; the third addresses to a sophisticated shell, to support the flexible specialization abilities for broadening product mixes options.

Chapter 9 (by Vervoort) is dealing with the design issues of CIM systems through the Yourdon methodology as described by Ward and Mellor. This methodology, which is based on the separation of analysis and implementation, was used in an ESPRIT (European Strategic Planning for Research in Information Technology) project to design and implement a part of a computer integrated manufacturing system. This method was chosen in order to compensate for the different backgrounds of the European partners in the project. It also guided and simplified the discussions with the management of the testbed where a part of the CIM was implemented. The method is introduced, its applicability investigated and it is related to other design methods.

Chapter 10 (by Szafarczyk) presents a study of the monitoring and automatic supervision processes in manufacturing systems. Monitoring and automatic supervision (ASM) are crucial at the present state of art in manufacturing. ASM system is a condition for full automation of production and is very important for a new manufacturing processes and new requirements of product quality. Comprehensive automatic supervision makes manufacturing system "intelligent" - the system can deal by itself - with disturbances that occur during its operation. The classification of disturbances into three groups, according to their consequences, is proposed as a main arrangement approach. Sensorial methods of monitoring, and methods of influencing manufacturing systems are presented as well as typical ASM systems. Thanks to last developments in integrated circuits, computers and sensor technology, and also thanks to the deeper knowledge of manufacturing processes, there are on the market relatively simple ane reliable systems directed at specific disturbances. The need for an holistic approach to automatic supervision in manufacturing is discussed.

Chapter 11 (by Colombo and Carelli) presents the main concepts and definitions related to Petri nets (PN) and their interpretation when applied to manufacturing systems. Methods and concepts used for modelling FMS by means of PN and for validating PN models are overviewed. An introduction to coloured Petri nets for designing FMS is also presented. Quantitative analysis is considered using model simulation by means of a temporized Petri net and by defining performance indexes. Finally, a methodology for designing the PN based model of the FMS control system is proposed and applied to a comprehensive example of a flexible manufacturing system. As a mathematical and graphical method developed for modelling and analyzing discrete-event systems, Petri nets have reached a well founded theory that enables to formally represent parallelism and synchronization of events as well as to investigate the structural properties of modelled FMS.

Chapter 12 (by Tzafestas and Čapkovič) shows how Petri nets are employed for the synthesis of control of discrete event dynamic systems (DEDS), like FMS, transport systems and communication systems. It deals both with the mathematical model of the DEDS under control and the mathematical model of the knowledge base via PNs. The former is used for the simulation of the DEDS dynamics (which helps in the construction of the knowledge base rules), whereas the latter facilitates the control synthesis procedure. The concepts and techniques of the chapter are illustrated by a set of simple but representative FMS examples.

Chapter 13 (by Bernhardt) is primarily based on the results of the so-called "Realistic Robot Simulation (RRS)" project, and presents novel techniques and tools for commissioning manufacturing lines with robots. It starts with an exposition of the state-of-art of the field and then discusses the concepts of realistic robot simulation, robot calibration and optimal trajectory planning (OTP). A suitable interface called the RRS-interface,which was implemented on several available SOP (simulation and off-line programming) systems, is described. For OTP a software package was developed that allows a minimization of cycle times or mechanical stress. This OTP package which has open interfaces and thus can be easily linked to commercially available SOP systems is briefly discussed. Of course much work remains to be done for reaching the overall goal of efficient off-line generation of robot programms and their fault-tolerant operation at the shop floor.

Chapter 14 (by Camarinha-Matos, Rabelo and Osório) investigates the concept of balanced automation which addresses an industrial environment with appropiate level of technical automation. In contradiction to both the pure human-based and the total automation environments, the emphasis is put on the combination of anthropocentric, technocentric, and economic approaches, with the right weights , to satisfy the local environment requirements and criteria. Balanced automation stresses the value of the economic and socio-organizational aspects, while taking the legacy assets of the enterprise and its region into account. Motivation, historic factors and current difficulties are presented. In a second part of the chapter, comprehensive examples of the opportunities, and first attempts to implement a balanced automation approach in the areas of shop floor, scheduling, concurrent engineering and virtual enterprises are discussed.

Chapter 15 (by Thompson) describes a major commercial systems integration project involving the co-ordination of shop-floor robot control systems at the highest layer in the factory automation system hierarcly. Although relating to an analytical chemistry laboratory, the principles described apply equally well to the supervision of manufacturing systems. The importance of integrating the factory control system with the bussiness, in addition to the components of automation, is stressed. Specific application requirements in discrete areas of the plant are examined, followed by a detailed analysis of the management system required to support laboratory operation, including the less evident processes necessary to sustain automated production. The chapter then describes the integration process in terms of the application build life-cycle and the commissioning of the plant. Finally, based on the experiences gained in this innovative automation project, the chapter provides advice to those contemplating an undertaking magnitude.

Finally, **Chapter 16** (by Tzafestas and Kapsiotis) presents an overview of the generalized network (GN) algorithms (Jewell's algorithm, dual algorithm, relaxation algorithm) and discusses their application to manufacturing operations (scheduling, planning, logistics, optimal lot sizing, machine loading, and strategic distribution planning). The chapter also describes a new expert decision support tool based on the GN model and includes a practical numerical example.

Taken together the contributions of the book provide a well-balanced and representative picture of the current status of the modelling, optimization, information technology, decision support, and intelligence methodologies of modern computer-assisted manufacturing systems. Each chapter is self-contained and involves both well established and fresh results. The editor is indebted to the Series editor Professor D. T. Pham for his encouragement, and to the contributors for their enthusiastic support in this project and their experience offered to the book. Many thanks also go to Mr. Nicholas Pinfield (Scientific Editor), Miss Imke Mowbray and Mr. Christopher Greenwell of Springer Verlag for their care throughout the book production.

The book is suitable for the researcher and practitioner, as well as for the educator and student in related postgraduate research courses.

May 1996 Spyros G. Tzafestas

Contents

Chapter 3

AI in Manufacturing : Application to FMS Simulation, Scheduling and Control

G. L. Kovács

Chapter 4

Modelling and Analyzing Processes in Production and Administration

N. Kratz and T. Rose

Chapter 5

Quality Management in CIM

E. Bajic and J. Richard

Chapter 6
Best Practice in Shop Floor Scheduling
D. Little

Chapter 7
A Stable, Distributed Routing Policy for Flexible Manufacturing Systems
J. Fenchel and Y. H. Chen

Chapter 8
Shop Controllers-Managers for Intelligent Manufacturing
R. C. Michelini, G. M. Acaccia, M. Callegari, R. M. Molfino and R. P. Razzoli

Chapter 9

A CIM Designed According to Ward and Mellor

W. A. Vervoort

Chapter 10

Monitoring and Automatic Supervision in Manufacturing Systems

M. Szafarczyk

Chapter 11

Petri Nets for Designing Manufacturing Systems

A. W. Colombo and R. Carelli

Chapter 12

Petri Net-Based Approach to Synthesis of Intelligent Control Systems for DEDS

S. G. Tzafestas and F. Čapkovič

Chapter 13

New Methods and Tools for Commissioning of Manufacturing Lines with Robots

R. Bernhardt

Chapter 14

Balanced Automation

L. M. Camarinha-Matos, R. Rabelo and L. Osório

Chapter 15
Factory Principles Applied to Analytical Chemistry : An Integrated Laboratory Management System
J. N. Thompson

Chapter 16
The Generalized Network Model : Algorithms and Application to Manufacturing Operations
S. G. Tzafestas and G. Kapsiotis

Contributors

G. M. Acaccia
Industrial Robot Design Research Group
University of Genova, Via all'Opera Pia 15/A
I-16145 Genova, Italy

E. Bajic
Manufacturing Systems Research Group
Research Center for Automatic Control of Nancy (CRAN)
Faculty of Sciences, Henri Poincaré University, Nancy 1
B.P. 239, 54506 Vandoeuvre, Cedex, France

R. Bernhardt
IPK : Fraunhofer-Institut für Produktionsanlagen und Konstruktionstechnik
Pascalstrasse 8-9, D-1000 Berlin 10, Germany

M. Callegari
Industrial Robot Design Research Group
University of Genova, Via all'Opera Pia 15/A
I-16145 Genova, Italy

L. M. Camarinha-Matos
Center of Intelligent Robotics
Department of Electrical Engineering
New University of Lisbon (UNINOVA), Quinta da Torre
P-2825 Monte Caparica, Portugal

F. Čapkovič
Institute of Control and Robotics
Slovak Academy of Sciences
Dúbravská cesta 9
84237 Bratislava, Slovak Republic

R. Carelli
Institute of Automation, Faculty of Engineering
National University of San Juan
Av. San Martin 1109 (oeste)
5400 San Juan, Argentina

Y. H. Chen
School of Mechanical Engineering
Georgia Institute of Technology
Atlanta, Georgia 30332-0405, U.S.A.

A. W. Colombo
Institute of Automation, Faculty of Engineering
National University of San Juan
Av. San Martin 1109 (oeste)
5400 San Juan, Argentina

J. Fenchel
School of Mechanical Engineering
Georgia Institute of Technology
Atlanta, Georgia 30332-0405, U.S.A.

G. Kapsiotis
Intelligent Robotics and Automation Laboratory
Department of Electrical and Computer Engineering
National Technical University of Athens
15773 Zographou, Athens, Greece

G. L. Kovács
CIM Research Laboratory
Computer and Automation Research Institute
Hungarian Academy of Sciences
XI. Kende u. 13-17, POB 63
H-1518 Budapest, Hungary

N. Kratz
FAW Ulm
Bereichsleiter Geschäftsprozessmanagment
University of Ulm
Postfach 2060,D-89 010 Ulm, Germany

D. Little
Manufacturing Systems Research Group
School of Engineering
The University of Huddersfield, Queensgate
Huddersfield HD1 3H, U.K.

R. C. Michelini
Industrial Robot Design Research Group
University of Genova, Via all 'Opera Pia 15/A
I-16145 Genova, Italy

M. Mišút
Department of Applied Informatics and Automation
Faculty of Materials Science and Technology
Slovak Technical University, Hajdóczyho 1
91724 Trnava, Slovak Republic

R. M. Molfino
Industrial Robot Design Research Group
University of Genova, Via all 'Opera Pia 15/A
I-16145 Genova, Italy

O. Moravčik
Department of Applied Informatics and Automation
Faculty of Materials Science and Technology
Slovak Technical University, Hajdóczyho 1
91724 Trnava, Slovak Republic

L. Osório
Center of Intelligent Robotics
Department of Electrical Engineering
New University of Lisbon (UNINOVA), Quinta da Torre
P-2825 Monte Caparica, Portugal

R. Rabelo
Center of Intelligent Robotics
Department of Electrical Engineering
New University of Lisbon (UNINOVA), Quinta da Torre
P-2825 Monte Caparica, Portugal

R. P. Razzoli
Industrial Robot Design Research Group
University of Genova, Via all 'Opera Pia 15/A
I-16145 Genova, Italy

R. Richard
Manufacturing Systems Research Group
Research Center for Automatic Control of Nancy (CRAN)
Faculty of Sciences, Henri Poincaré University, Nancy 1
B.P. 239, 54506 Vandoeuvre, Cedex, France

T. Rose
FAW Ulm
Bereichsleiter Geschäftsprozessmanagment
University of Ulm, Postfach 2060,
D-89 010 Ulm, Germany

M. Szafarczyk
Warsaw University of Technology
Narbutta 86
02-524 Warsaw, Poland

J. N. Thompson
SEMA Group UK Ltd
75 Berwick Road, Buxton
Derbyshire SK179PD, UK

S. G. Tzafestas
Intelligent Robotics and Automation Laboratory
Department of Electrical and Computer Engineering
National Technical University of Athens
15773 Zographou, Athens, Greece

W. A. Vervoort
Department of Computer Science
Twente University, P.O. Box 217
7500AE Enschede
The Netherlands

1 Modern Manufacturing Systems : An Information Technology Perspective

Spyros G. Tzafestas

1 Introduction

Manufacturing systems perform many activities and operations that can be monitored and controlled at several levels of abstractions [1-3]. A modern manufacturing system has to be able to adapt to quick internal and external changes. To this end, a variety of successful models and control techniques have been developed during the last two decades, which are based on the principles and tools of *information technology* (IT) and *management science*. At the technological level the current developments in manufacturing systems stem from the advances in machine tools, robots and controllers. Both direct numerical control (DNC) and computer numerical control (CNC) use adaptive techniques to optimize machine operations, while robotic systems provide new solutions for manufacturing tasks such as material handling, assembly, welding, and spray painting.

Modern computer-integrated manufacturing (CIM) systems have a hierarchical structure with three principal hierarchical levels [4]. At the highest level a powerful computer supervises the manufacturing functions (activities or operations) of the various phases of the enterprise (product design, planning of product manufacture, automatic production of parts, automatic assembly, automatic quality control, control of the flow of materials and parts through the plant, and so on). All these functions must be accomodated into a computer network, supervised by the CIM central computer, which controls the various interrelated tasks according to overall management policy. At the lowest level there are stand-alone computer-controlled machines and robots, the operation of which is controlled by the intermediate (coordinator) level of the CIM system. Throughout the operation of the system there is feedback from a lower level to its superior level(s).

Therefore, CIM implies a systemic approach to the operation of a manufacturing enterprise, i.e. it involves:
- research engineering
- plant functions
- production
- business functions
- administrative functions

Each of the above areas is comprised of a large collection of functional activities which must be organized and performed effectively by the company's personnel. The interfaces to the CIM system are workstations or interactive terminals for people, and instrumentation for equipment. The variety of processes to be handled makes it necessary to use a repertory of computing, control and instrumentation equipment for the implementation of a CIM system. Clearly, communication among the distributed components of the system is essential, while for some critical components the use of fault-tolerant devices or facilities is required.

The purpose of this chapter is to provide an introduction to some key information technology, managerial and communication issues that must be addressed for designing and implementing integrated CIM systems. We start in Section 2 with an outline of some basic CIM functions namely ; product design and design for assembly ; process planning, scheduling and control ; dynamic simulation of FMS ; equipment selection ; quality assurance, and facility layout. Then, in Section 3 we discuss how and why IT enhances the competitive advantage of manufacturing (and other) companies. Sections 4 and 5, which form the core of the chapter, present two comprehensive CIM models derived under the auspices of the "European Strategic Programme for Research in Information Technology" (ESPRIT). The first model which is called CIM-OSA (Open System Architecture for CIM) deals mainly with intra-organizational operations, whereas the second model which is called CMSO (CIM for Multi-Supplier Operations) deals with the inter-organizational operations. Section 6 gives a short account of some major issues about the application of artificial intelligence (AI) in manufacturing systems. Especially three basic AI areas which are applicable to CIM (automatic planning, automatic learning, and qualitative modelling and simulation) are discussed, and a number of knowledge-based (expert systems) designed for manufacturing operations are reviewed. The chapter closes with a look at the economic evaluation of automated manufacturing systems, and some remarks on the so-called "informed choice" among management and control strategies.

2 Basic Engineering CIM Functions

In this section we provide a brief discussion of some basic intra-organizational engineering CIM functions that provide a challenge for IT techniques and tools. These functions are the following [1, 5-7] :

- Product design and design for assembly
- Process planning, scheduling, and control
- Dynamic simulation of FMS
- Equipment selection
- Quality assurance
- Facility layout

Specific examples of expert and knowledge-based systems developed for some of these functions will be presented in Section 6.3.

2.1 Product Design and Design for Assembly

Product design is a complex task that requires intelligence and experience and is performed in several stages : definition of the product's functional structure, definition of physical principles, drawing of product (or part) shape, and specification of manufacturing data (detailing). The design of a product involves a preliminary design process, a refinement process, an analysis procedure, and finally the implementation process.

Computer-aided design (CAD) is a well-defined set of hardware and software methods and tools by which the product design is computer-automated. Very often, commercial CAD systems are turnkey systems that provide an integrated combination of both CAD hardware and CAD software for carrying out a specific design process. Many CAD systems consist purely of computer-independent CAD software that can work on hardware systems supported by appropriate workstations. A CAD system has to support the designer via a graphic-interactive communication technique together with appropiate processing tools such as geometry processing and graphics processing. The data that describe the technical solution as well as the results of the processing are stored in a database and form the computer *internal product model*. The purpose of the computer internal model is to provide data for the optimization and control of the product (tolerance analysis, control of correctness, completeness, etc.) as well as for other manufacturing functions such as the generation of production schedules, generation of numerical control programs for machine tools and robots, and simulation of manufacture and assembly operations.

Design for assembly (DFA) belongs to a more general field of *design for manufacturing* (DFM) [6, 8]. DFM deals with understanding how product design interacts with other parts of the manufacturing system. It also involves the problems of specifying product design alternatives that help to globally optimize the manufacturing system as a whole. DFA is concerned with the design of products with ease of assembly in mind. By using DFA, a product can be systematically designed to minimize the technological and financial efforts required for assembly and simultaneously satisfy all constraints on the product functionality. To facilitate the designers in this job, a *decision support system* (DSS) is needed that can take into account the knowledge of assembly requirements and help the product engineers in their task (i.e., in designing for easy manual and automated assembly, designing for assembly process planning, and reducing reworking). Such a system should, among other things, be able to :

- Analyze all decision relevant to assembly and point out if there is some fault
- Indicate the critical features in the design
- Propose possible changes of inappropiate design charateristics
- Suggest examples of design changes and check whether a design change will affect product functionality

In general there are three fundamental approaches to designing a product for assembly [9] :

1. Simplify the assembly task
2. Improve the assembly organization
3. Facilitate the assembly execution

To simplify the assembly task, one has to minimize the number of parts, reduce the variety of parts, and select simple assembly methods. To improve planning performance in assembly through specific design features, one has the following alternatives :

- Allow arbitary decomposition of the product in subassemblies
- Use standard parts and subassemblies
- Avoid compulsory assembly sequences in order to be able to solve easily the line balancing problem.

2.2 Process Planning, Scheduling, and Control

Computer-aided process planning (CAPP) is the computer-aided activity that determines the appropiate procedure to transform raw material (usually in some prespecified form) into a final product. It defines what has to be produced, how it should be manufactured, and what means should be used to meet competition and economic goals. *Process scheduling* determines how many items have to be produced, when and where they should be manufactured, and by whom. The correction of the observed deviations between actual product and planned output is achieved through appropiate *feedback control* [10, 11].

Process planning can be categorized as *variant* or *generative*. In variant process planning a plan is selected from among a set of standard plans. In generative process planning one uses information on product or part features (coming, for example, from a CAD system) for creating a process plan. Examples of available CAPP systems are AUTAP (Eversheim et al., [12]), CAPP (Link, [13]), CAPSY (Spur and Hein, [14]), CPP (Dunn and Mann, [15]), MIPLAN (Schaffer, [16]), and GENPLAN (Tulkoff, [17]). The real integration of CAD and CAM becomes possible with the introduction of a CAPP system in the chain through a connective information flow and a central database. CAPP systems reduce the deficiencies of traditional process planning systems realized by skilled planners ; that is, they improve planning time and promptly prepare the information for decision-making and planning objectives. Knowledge-based CAPP systems can further improve the capabilities of CAPP systems by utilizing human experience and providing increased flexibility.

Process (production) scheduling determines a sequence (a schedule) of part lots to be machined in the FMS so as to meet the due dates of lots and minimize machine idle times, queues at machines, and so on. Process scheduling involves two basic tasks :

1. Production scheduling
2. Real-time rescheduling

The production scheduling is based on a medium-term horizon (e.g., 2-3 weeks) and determines the estimated starting times of lots to allocate auxiliary resources. The production scheduling must take into account several constraints such as planned maintenance periods for machines and raw materials availability times. Real-time rescheduling is needed when the planned schedule must be modified because of the occurence of unexpected events (e.g., when raw material is not available when required or when a machine breaks down). Examples of conventional production scheduling systems can be found in Rolstadas [18] and Alexander and Jagannathan [19].

Process control involves control structures and algorithms that support planning and supervision tasks at various levels of the manufacturing system (production scheduling, materials flow, maintenance control, quality control, etc.). In general, CIM possesses three distinct control levels :

1. Highest-level control (management control)
2. Intermediate-level control (shop-floor control)
3. Lowest-level control (control of machines)

At the highest level, the processing of management data (production scheduling, operational management, etc.) takes place. Also, at this level, horizontal communication to other factories or plants and vertical communication to local production and control units are coordinated. Intermediate-level control relates to plant supervision and coordination, the disposition of materials flow, production process control, quality assurance, and the supervision of the lower level controllers. Horizontal and vertical communication is again involved. The lowest level of control involves specific control algorithms (e.g., DDC algorithms and optimal control algorithms). Future systems will be characterized by increased intelligence located at different hierarchical control levels (adaptive control, rule-based control, intelligent control, and so on)(Dillman [20], Williams [21], Conterno, [22]).

2.3 Dynamic Simulation of a Flexible Manufacturing System

Dynamic FMS simulation imitates the operation of the system through the use of some dynamic model. The objective of the simulation is to study the system behavior and evaluate alternative decision, design, and operation rules. The simulation must be dynamic because the system behavior and the results obtained are time-dependent and deterministic or stochastic. Appropriate models for the simulation are queuing models for the buffer stores or the automated storage and retrieval system. Of increasing use for manufacturing system simulation and performance evaluation are the various types of Petri nets [130].

According to Ranky [23-25] the FMS simulation models are classified as

- Graphical models
- Mathematical models
- Scaled-down physical models

Most present-day graphic systems and CAD packages provide not only three-dimensional solid graphics representations of models and animation techniques, but also high-level task description languages.

Mathematical models use functional expressions and appropriate simulation languages such as GPSS (general-purpose simulation system), DYNAMO, SIMSCRIPT, or SLAM. Simulation of a flexible manufacturing system needs appropriate data such as manufacturing time distributions, inspection time distributions, transportation times, fixturing time distributions, and unloading time distributions.

The scaled-down physical models are controlled by computers or microprocessors and are used in plant engineering for getting information about the layout and the behavior of the system. These models are static and are expensive to build.

2.4 Equipment Selection

The equipment selection problem is very important to the design of a manufacturing system, since it is a dominant factor in overall capital cost. A successful solution to this problem leads to (Heragu and Kusiak [26,29]) :

- Lower capital cost
- Lower operating and maintenance cost
- Increased utilization rate
- Improved layout facilities
- Increased total production efficiency

A survey and comparison of the available approaches to solve the equipment selection problem is provided by Miller and Davis [27]. Many of the assumptions made in this technique are unrealistic. As Kusiak explains [28], it is very difficult to precisely formulate the equipment selection problem to reflect practice accurately. To this end, it is necessary to use engineering-based expert system techniques (including quantitative and qualitative issues).

2.5 Quality Assurance

Quality control (QC) is a crucial operation in a manufacturing system that determines the amounts by which the quality of actual products deviates from the specified product quality. The quality standards are set by the market (customer standards, industrial acceptance), by legislation (producer liability, authorities, safety guards for workers), and by the enterprise (failure rates, costs, profits, image). Quality control has to be performed at many different levels of a manufacturing process. *Quality assurance* (QA) is the procedure that includes all phases of the life cycle of a product. It integrates all QC activities into one system. Quality assurance is performed in a hierarchical way according to a local network of operational data acquisition and the equipment configuration conceived for the various QC levels.

Quality control must be applied during every production cycle of the product and involves some or all of the following :

- Design quality (planning, development, layout design)
- Supply quality (purchasing, material reception)
- Manufacturing quality (manufacture, assembly)
- Delivery quality (acceptance, final test)
- User quality (sale, user installation, use of the products)

2.6 Facility Layout

The optimal location of each piece of equipment or facility to its corresponding site is known as the facility layout problem [7, 30]. The optimality criterion is the minimization of the total material handling cost. The facility layout problem can be cast in the quadratic aasignment task, or the set covering problem form or linear integer programming form, and so on. When the system involves more than ten facilities it is very difficult, if not impossible, to find optimal algorithms for the solution. A survey of heuristic algorithms which can in some cases give a solution relatively quickly is provided in [30]. However these solutions require further examinations and can not be applied directly. This is because (a) the configuration for some facilities provided may not be acceptable, and (b) the algorithms do not consider some constraints such as the space needed for materials-handling equipment and clearance between machines. Here the application of expert systems may be of considerable help.

3 Information Technology Enhances the Competitive Advantage

Here we discuss in very general terms how IT helps in the enhancement of a CIM company's competitive advantage [31-34]

Very broadly IT is influencing competition in three major ways :

- By changing industry's structure and thus creating new rules of competition.
- By creating competitive advantage through the development of new ways of outperforming the company's competitors.
- By creating entirely new business, often from within a company's existing operations.

IT influences the whole process by which CIM companies create their products including the reshaping of the product itself. The role of IT in competition can be best understood through the concept of *value chain* which decomposes the operations and activities of a company into the technologically and economically distinct activities it performs to do business [33]. These activities are called "value activities". Actually, the value created by a CIM company is measured by the amount that customers are prepared to pay for a product or service. To achieve competitive advantage over its competitors a company must perform its value activities in a cheaper way or such that to produce more value.

The value chain involves *primary activities* (such as inbound logistics, operations, outbound logistics, marketing / sales, service activities) and *support activities,* such as human resource management, technology development and procurement. The activities within the value chain are connected by *linkages* that exist whenever the way in which one activity is performed influences the effectiveness or cost of other activities. The value chain of a company in the CIM industry is embedded in a broader network of activities called "value system". The value system involves the value chains *of suppliers* (who provide raw material, components and purchased services) and *distributors.*

An important tool for creating competitive advantage is "competitive scope". *Broad scope* gives a company the capability to exploit interrelationships among the value chains of different industry sectors, geographic areas or other related industries. *Narrow scope* enables a company to shape the value chain to a desired target segment such that to achieve lower cost or higher benefit.

IT enters the value chain at all points and transforms the way in which the value activities are carried out including the type of the linkages among them. Moreover, IT influences the competitive scope and modifies the ways products meet customer demands. The above simple effects make clear why IT has achieved strategic significance for the effective operation of CIM systems.

Every value activity is characterized by a *physical* and an *information-processing* element. For example logistics activity involves information like transportation rates, scheduling requirements and production plans. A service activity is based on information concerning service requests or part orders, and produces information on company's corrective actions for revising product designs and manufacturing methods such that to avoid observed product failures.

The competitive advantage achieved through changes of the physical component of what a CIM company does reaches very quickly a saturation level. Here is exactly where IT contributes very much, since IT is advancing faster than technologies of physical components. IT affects all categories of value activity from computer-aided design to automated operations. In nowdays IT is used over the whole value chain and helps in performing computerized optimization, decision and control functions. Also, IT modifies the physical processing component of activities. For example CNC machine tools are faster, more flexible and more precise than manually operated machines. Through the new information flows generated, IT increases the capability of a company to beneficially use existing linkages between activities within the company itself or within its related companies.

4 An Intra-Organizational CIM Model

In this section the CIM-OSA (Open System Architecture for CIM) model is reviewed [35-38]. This model covers all intra-organizational functions in a manufacturing company, and supports all levels of management, namely the strategic, the tactical and the operational planning, including the direct operation of the shop floor.

The purpose of the CIM-OSA model is to enable the CIM company to perform its business in an adaptive and real-time way. To this end, the model supports operational flexibility and multi-disciplinary knowledge, as well as system integration. Using the architectural constructs of CIM-OSA the structured description of business requirements and CIM implementation are facilitated . To satisfy the desires of different users, CIM-OSA provides the required constructs for multiple representations of a CIM company.

CIM-OSA actually provides a reference framework involving two major elements:

- The CIM-OSA reference architecture
- The CIM-OSA particular architecture

The former provides building blocks and guidelines, whereas the latter contains the instantiated (particularized) building blocks for a specific company.

4.1 The Business Enterprise

CIM-OSA is based on a clear separation between a 'business enterprise' and concepts like data processing, computer applications and manufacturing technology. The manufacturing system can then be designed, modified and run under the guidance and control of the business enterprise. In other words the technology (which is logically different and separate from the business) must follow and satisfy business objectives.

The business objectives for the design of a CIM system may be dynamic or static. In the dynamic case the market requirements are applied to the manufacturing operation for generating the approppriate actions (changes) in response within short periods of time. As market requirements vary from day to day, the manufacturing operations have to be adapted within the same periods. This type of CIM design is known as 'pulling' design.

On the other hand, if the current business objectives are (or are assumed to be) stable they are employed for forecasting future requirements. Therefore, the resulting design is static and cannot be changed as market pressures vary. This design can lead to high short-term efficiency but cannot maintain its effectiveness as the objectives change with the time. This type of CIM design is known as 'pushing' design.

4.2 The CIM-OSA Reference Framework

The CIM-OSA Reference Framework possesses three levels of architectural generecity (generic level, partial level, particular level) three modelling levels (enterprise modelling level, intermediate modelling level, implementation modelling level) and four different views (organization view, resource view, information view, function view) as shown in Fig. 1.

The *generic* and *partial* levels provide all constructs needed to gather the user requirements for the operation of this system and to transform these requirements into a suitable description and implementation. The constructs (building blocks) of the generic level which deal with components, constraint rules, terms, service functions and protocols, have the widest application in CIM. The partial level deals with sets of partial models applicable to a particular class of manufacturing enteprises (e.g. automotive, electronics, aerospace, etc.) and thus provides a realistic and usable aid for developing a particular architecture. The *particular* level involves the specific requirements for the particular company operations, and all determined, selected and implemented components of the system that meet these requirements.

The three CIM-OSA modelling levels provide the means for modelling both the user and CIM issues of the manufacturing company. The *enterprise* modelling level allows the collection of the business requirements of a given enterprise. From these requirements one can isolate different *views* and optimize them according to user specified criteria (e.g. simulation). The *intermediate* modelling level helps to optimize and logically restructure the representation of the requirements collected at the enterprise modelling level. This is a system oriented description (non-redundant) of the business requirements. Actually, the intermediate modelling level acts as an isolation between the user representation and the system representation, on the levels below and above the intermediate level, to limit the effect of changes. Finally, the *implementation* modelling level allows the translation of the content of the intermediate modelling level into a complete system description. The content of the implementation modelling level consists of the descriptions of the specified, selected (from the CIM render lists) and implemended system components, along with other appropriate information such as flow of actions and responsibilities.

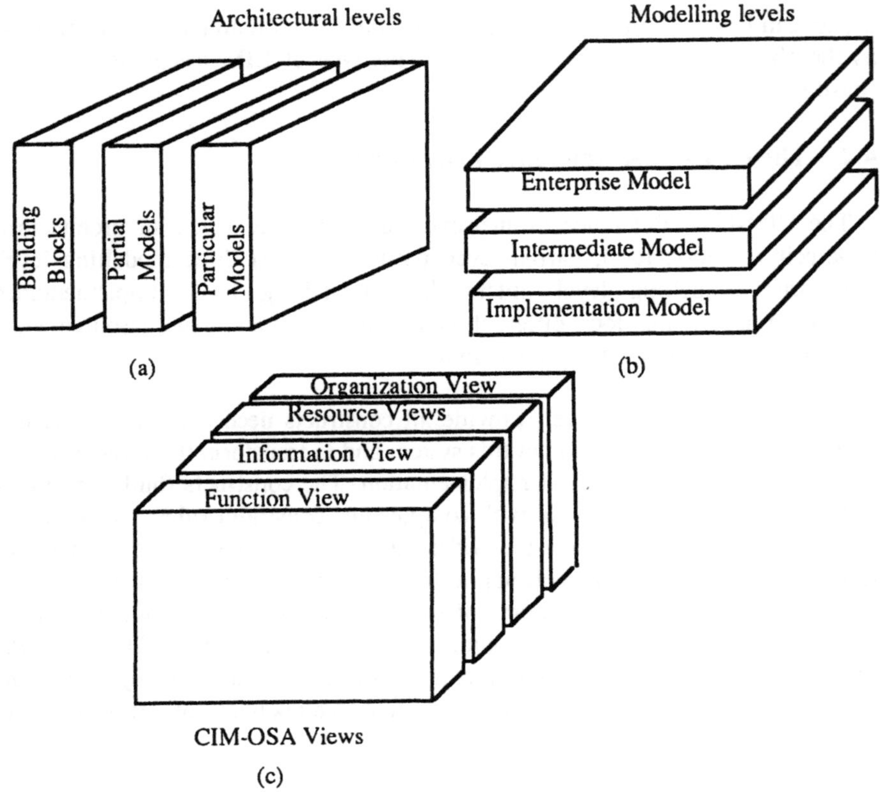

Fig. 1 Architectural levels, Modelling levels, and Views of CIM-OSA.

The four different views allow the independent modelling and optimization of particular user and specific CIM system issues of the company. The *organization view* and *resource view* involve all relevant knowledge regarding the enterprise responsibilities and resources in the company. The *information view* involves all information defined and contained in the company. Finally, the *function view* involves the representation of the company operation through a set of hierarchically structured business processes. Each business process is defined by its triggering events, by the result it generates, and by its procedural rule set (control flow description) that acts on either lower level business processes or directly on executable enterprise activities.

Figure 2 depicts an overview of the CIM-OSA framework which is the result of overlaying all architectural and modelling levels with the views.

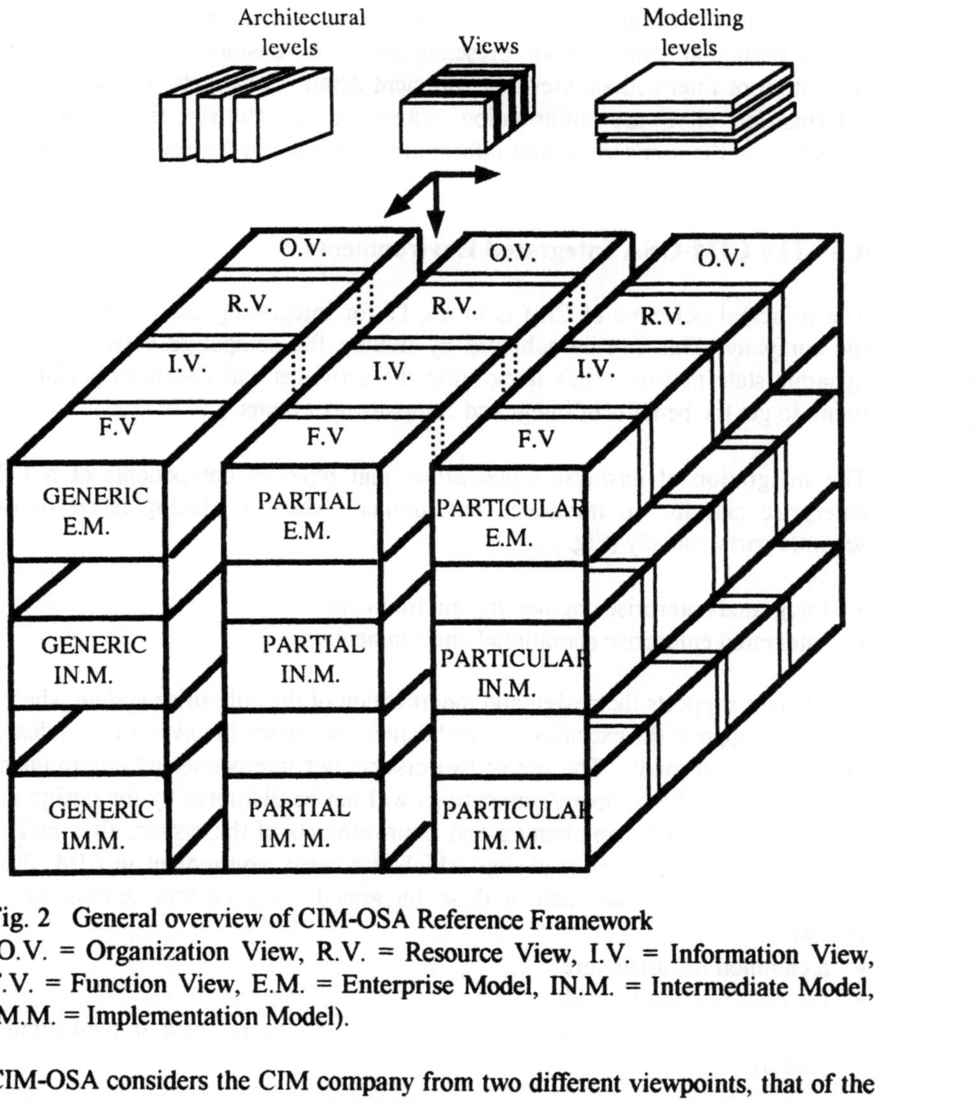

Fig. 2 General overview of CIM-OSA Reference Framework
(O.V. = Organization View, R.V. = Resource View, I.V. = Information View, F.V. = Function View, E.M. = Enterprise Model, IN.M. = Intermediate Model, IM.M. = Implementation Model).

CIM-OSA considers the CIM company from two different viewpoints, that of the *business environment* as seen by the business user, and that of the equivalent *physical environment* (i.e. manufacturing engineering and technology) These views are represented by the *enterprise* modelling level and the *implementation modelling level,* which are connected by the *intermediate* modelling level. The creation of the *functional view* of the enterprise model starts with the overall description of the entire company (using the business process concept), then proceeds in building the enterprise system from the business requirements using the concept of *procedural rule sets,* and finally is completed by the definition of the functionality of the business process using the content of the enterprise activities it employs (inputs, transfer function, outputs). The purpose of the *information view* of the *enterprise level* is to collect the information requirements of the particular company for presentation to the business user, and for the derivation of intermediate views of sufficient detail. Constructs provided for the information view are information categories structuring the information according to the user's view, and information items as the lowest elements in any information category.

4.3 The CIM-OSA Integrated Environments

The principal objective of CIM is to use IT for integrating the functions within the company. This can be achieved by linking the company's activities into a dynamic state network, thus integrating the activities and resources performing them, to get the benefits of integrated company operations.

The integration of *business, application,* and *physical* components of a CIM enterprise call for an integrated environment which is distinguished in two separate parts, namely (Fig. 3) :

- Integrated enterprise engineering environment
- Integrated enterprise operational environment

The former supports the design and modification of the enterprise system,whereas the latter supports the execution of the business processes involved in the released implementation model. The use of two distinct but interconnected environments guarantees that the company's operations will not be disrupted by the design and implementation of future changes and improvements of the system, thus making possible *the management of change* which is a basic requirement in CIM. From Fig. 3 one can see that each of these integrated environments consists of the following :

- a common infrastructure
- the real (physical) world
- a common dynamic definition of the system (the released implementation model).

The common infrastructure is provided by the CIM-OSA infrastructure which provides a common and specified set of services and protocols. The real world is expressed in the form of a set of implemented functional objects that provide the required functionality. The released implementation model involves the CIM-OSA defined model of the business processes to implement the adopted guidelines for the specification design, implementations and release requirement of the enterprise system. This model also contains the relevant information, resource and organization views. The CIM-OSA integrating infrastructure is common to both the *engineering* and *operational* environments.

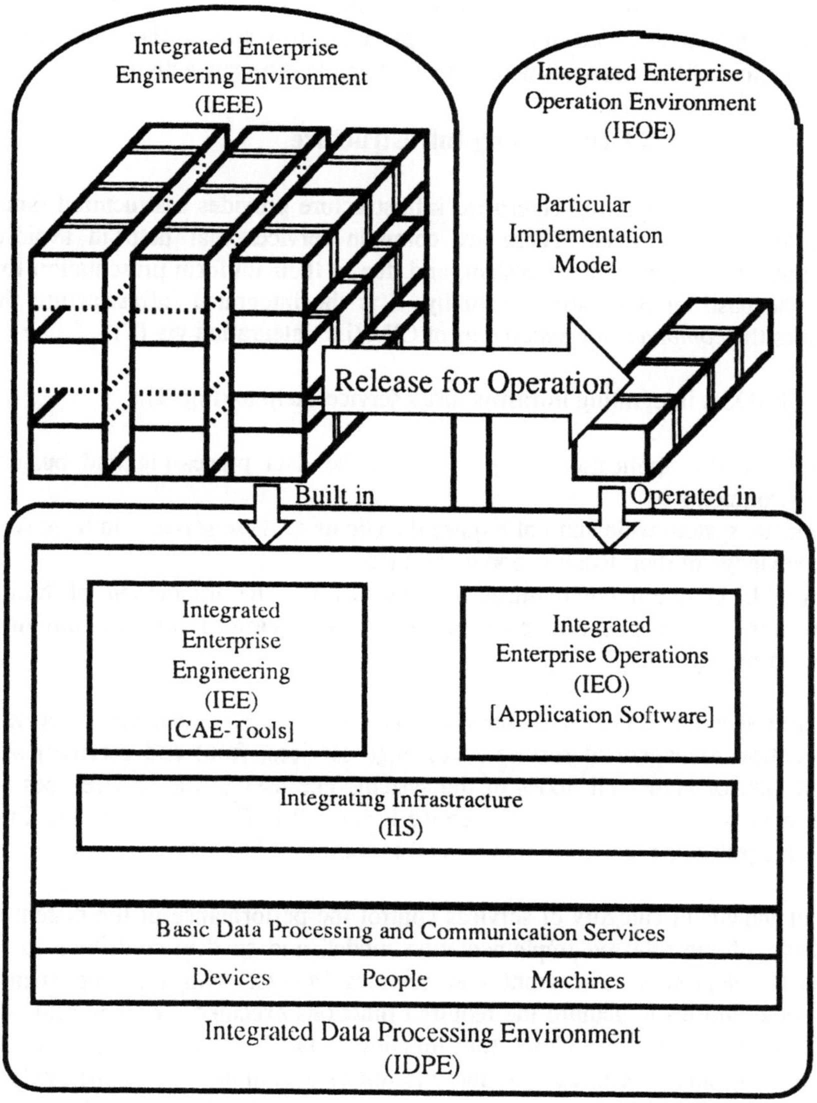

Fig. 3 Integrated environments of CIM-OSA

The physical integration is achieved by the integration of the entities that perform the enterprise activities, and the communication needed between all entities. Communications constitute a common service where both *intra* and *inter* system communication appear identical to the user. The application integration is achieved by integrating the information needed to carry out the activities. Finally, the business integration is achieved by the integration of the management or control of activities joined into business processes (oriented to the results of the company)and of the resources needed to execute the company activities.

4.4 The CIM-OSA Integrating Infrastructure

The CIM-OSA integrating enterprise infrastructure provides a structured family of *system wide services*. These are common services that help in avoiding functional redundancy in the system, and due to their uniform presentation they form the basis of integration. Actually, it is the integrated infrastructure that provides the *computer integrated part* of CIM (i.e. integration via IT).

The CIM-OSA integrating infrastructure's services help to (Fig. 4) :

- separate the application programs from the data processing and business environments.
- operate system wide and not require the clients of these services to have some knowledge of their location and ditribution.
- provide common communication protocols for the interaction of human machine and application programs with the integrated infrastructure in a uniform way.

The user sees the *system wide services* (business process management services, information management services, exchange services, front end services) as a simple service across all nodes of the system. The user of the service does not need to know how and where the service is provided. The services of the CIM-OSA integrating structure are categorized as shown by Table 1.

The management category of services control the performance of the system on the basis of the released implementation model, and so they contribute to the business integration. The front end services interact with the implemented functional entities to acquire the required functions executed, i.e. they deal with *application integration*. Now, the function related services deal with the enterprise (management, control and execution) operations and so they help to achieve the *integration of function*. Information related services deal with the enteprise information, i.e. with the location, access, storage and consistency of data and thus they contribute to the *integration of information*. Finally, communication related services involve all services for the management of all communications (*intra* and *inter*), i.e. they provide the *integration of communication*.

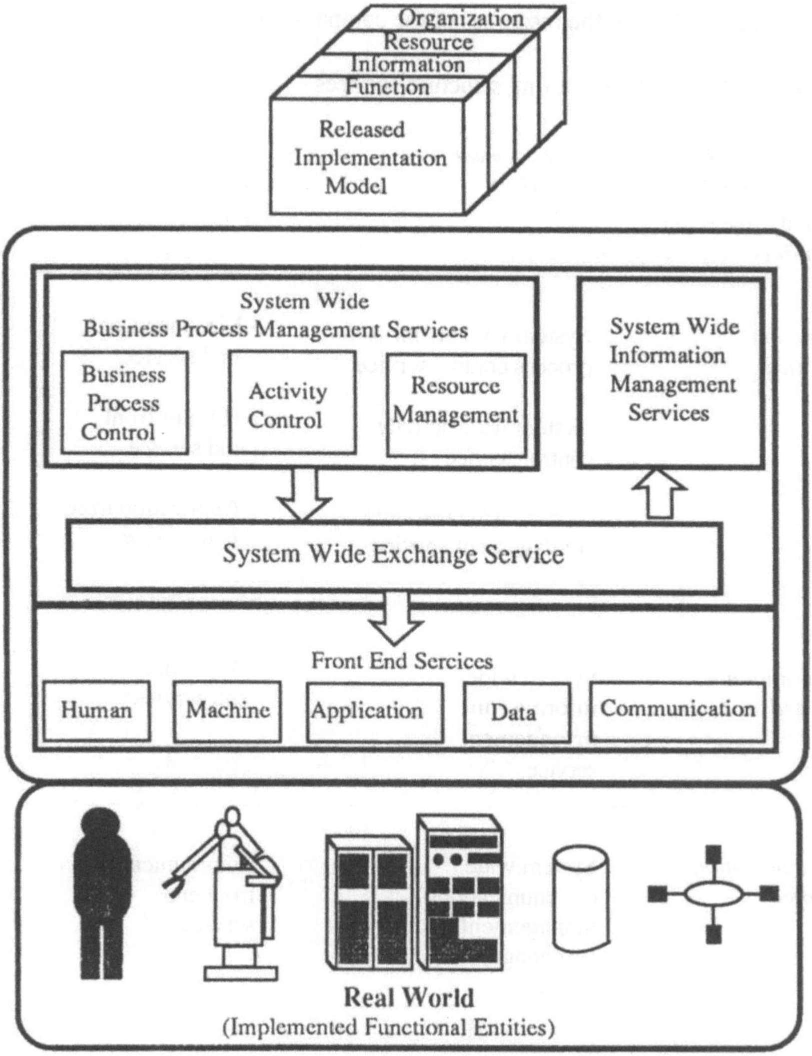

Fig. 4 The CIM-OSA integrating infrastructure services.

The business process control services use the function view of the released particular impementation model to control the business processes of the company system wide. The activity control services use the function view of the released particular implementation model to control the execution of the company activities system wide. Finally, the resource management services use the resource view of the released particular implementation model and the business process occurences to schedule the resources of the company worldwide.

Table 1 : CIM-OSA integrating structure services

Set of Services \ Category of Services	*Management Services*	*Front End Services*
Function Related	- System wild business process control service - System wide activity control service - System wide resource management service	- Medicine front end service - Human front end service - Application front end service
Information Related	System wide information management service	Data front end service
Communication Related	System wide communications management (exchange) service	Communication front end service

The status of integration has really been promoted by the CIM-OSA model. Previously, the most advanced level of integration was the *physical system integration* with available solutions the *open system interconnection* (OSI) and the *manufacturing automation protocol* (MAP) models [39, 42]. The *application integration* (integration of enterprise functions, integration of information) and the *business integration* (integration of enterprise control) was a task of the user. As we have seen from the above discussion CIM-OSA has been concentrated on the system wide integration of CIM specific application processes including human-implemented and computer-implemented processes. CIM-OSA can beneficially use the commercially available interfaces designed for the MAP-application layer as the main platform for implementing the various communication functions required [39].

5 An Inter-Organizational CIM Model

The model which is described here has been developed under the CMSO (CIM for Multi-Supplier Operations) European project [43-48]. The CMSO addresses the issues of inter-organizational structures, market requirements, and improved effectiveness of multi-supplier / multi-distributor (MS/MD) chains in terms of a generalized "customer service" performance measure that includes : factors such as quality, delivery, price, innovation, and product range. Each of these factors can be partitioned in a group of sub-factors, e.g. quality includes performance, reliability, conformability and serviceability. Finally, each of these sub-factors can be further exploded until controllable factors are determined. For example, lead time is not a controllable factor itself, but it contains elements such as review period or order transmission time that are controllable factors.

5.1 The CMSO Reference Model

The main body of the CMSO MS/MD partial reference model is formed by a combination of the various organizational units which manage the business of the automotive supply industry. It integrates in a conceptual and operational way the following three types of chains :

- Manufacturing (logistics) chain (MC)
- Distribution (logistics) chain (DC)
- Product development and support chain (PDSC)

which cover all areas and problems involved.

A pictorial representation of this model is shown in Fig. 5, where each chain consists of a set of elements that represent particular entities of the automotive industry (e.g. VMs, supplier companies, part distributor companies, etc). The manufacturing chain starts with sub-supplier(s) at the lowest level, goes to the supplier level and then to the VM level, and ends up with the dealer who sells the vehicle to the end customer. The connection of the individual elements is performed by appropriate EDI (Electronic Data Interchange) communication functions. Similarly, all logistics operations concerning the processing / handling of spare parts are accomodated in the distribution chain branch of Fig. 5. The DC starts again at the sub-supplier level and ends-up at the installer with intermediate levels the supplier, prime distributor, area distributor, and local distributor. Supplier and sub-supplier companies are (usually) identical for the part production for VMs or aftermarket operations (spare parts). So in Fig. 5 no distinction between them is made. The PDSC handles technical information which is involved in both the MCs and DCs, and so it is not explicitly shown in Fig. 5 as a separate chain.

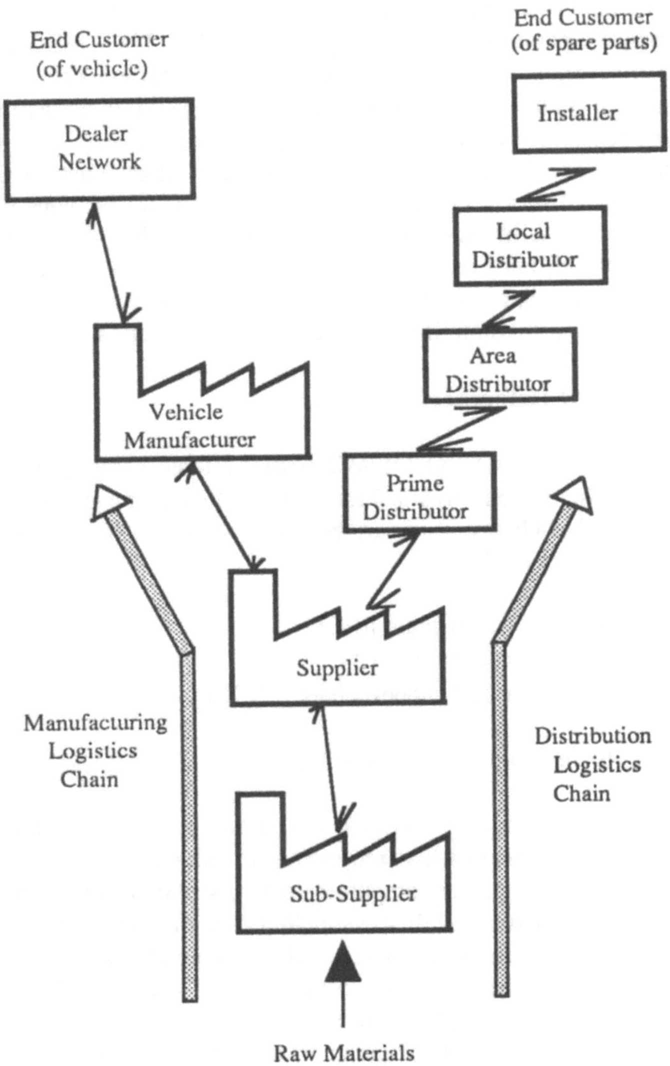

Fig. 5 CMSO model of the manufacturing logistics and distribution logistics chain.

It is actually hidden within the MCs and DCs. The product development part of the PDSC chain deals with the commercial information for one-of-a-kind and mass production for distribution logistics, while the product support part comprises information required at the installer sites of the DC such as manuals, part lists and fault diagnosis.

The individual chain elements (organizations), independently of their type, are treated by the CMSO model in a similar way. The management of each organization is implemented in three layers : *strategic, tactical* and *operational*. Each of these layers controls logistics and technical functions ; in trading terms these are classified as the *initial information phase*, *order phase, delivery phase*, and *settlement phase*. The single technical function is product development and support. Each of the functions is divided into three primary business processes, namely *buy, produce/store* and *sell*.

To incorporate the three managerial levels and the technical/business functions of the MS/MD chain an overall model, called the "CMSO Toblerone" model, has been developed as shown in Fig. 6. In this model the *logistics* and *process chains* are combined in a clear and usable way. Each typical organization/company is represented as a *pyramid* in the toblerone model. The shaded areas between the companies indicate the inter-organizational operations considered. Here lies the main contribution of the CMSO model for the European automotive industry. Through business procedures, which are improved by EDI communications, the chain elements grow together such that the industry can overcome the current differentiation with all known disadvantages.

A key element of the CMSO MS/MD model is an intelligent EDI box developed within the project [47]. This communication box is called *CMSO box*. Actually many reference controllable factors can be connected to the CMSO box (and to other CIM technologies). For example, EDI can assist in the reduction of order transmission time and hence in the reduction of the lead time.

5.2 CMSO Model Integration Issues

The integration achieved at all levels, i.e. strategic, tactical and operational (see Fig. 6). Very broadly, the integration is twofold. Among the three levels the integration is mainly done in a conceptual manner. This is supported by the CMSO toblerone model and the EDI reference model. Experiences gained at one

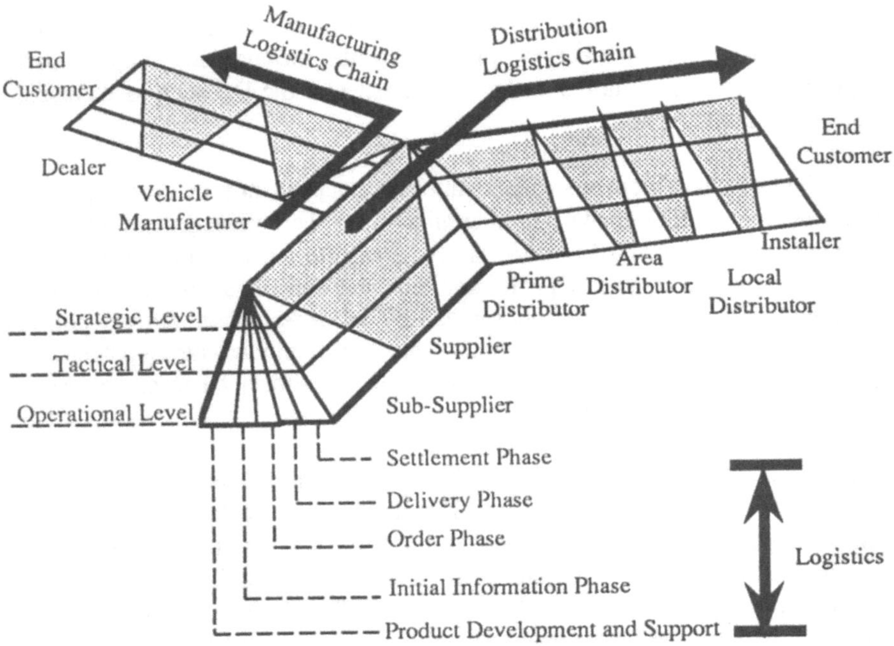

Fig. 6 The CMSO toblerone model.

level influence the development of solutions at another level. The feedback action works in both directions. At the two higher levels either methods can be applied or simulations can be performed. On the other hand, a strong physical integration is achieved within the levels through the CMSO-box and the integrated EDI architecture developed.

The type of integration offered by the CMSO model is shown in Fig. 7.

For the strategic level an integrated methodology is offered by CMSO which provides a set of procedures (user guides) and support tools (expert systems). Application of this methodology enables users or managers to analyze business problems and identify the key processes at particular nodes in a chain where improvement is required if end customer expectations are to be met and the competitive position of the chain maintained or improved. The definition of the processes at the strategic level is of a generic type.

For the tactical level, *a logistics chain simulator* was developed within CMSO which has the form shown in Fig. 8. Here, the process definitions are more detailed than those of the strategic level, but less detailed than those of the operational level which also require user and communication interfaces.

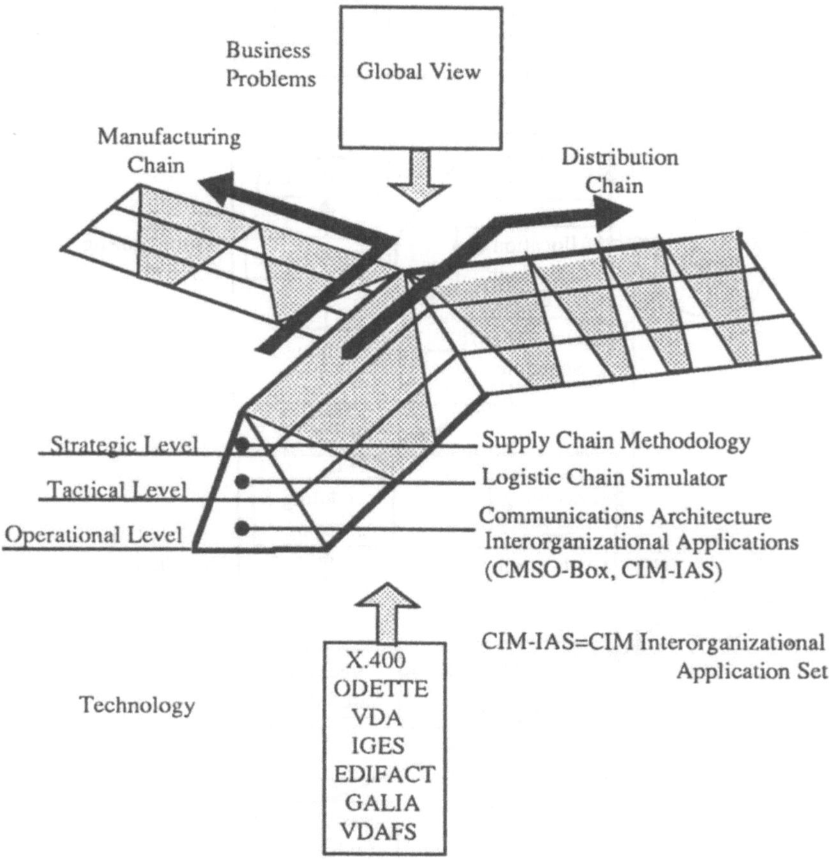

Fig. 7 Integration levels based on the toblerone model.

In the simulator example of Fig. 8, the important processes such as forecasting and scheduling are treated in detail (actually a number of options for each of them are provided) whereas other processes such as order processing are in skeletal form only. At this level the communication processes (such as EDI) are completely absent. The impact of these processes is simulated through change of parameters such as interorganizational lead time factors. The simulator permits the simulation of chains involving multiple manufacturing and distribution nodes and data are exchanged among nodes through conventional data sets.

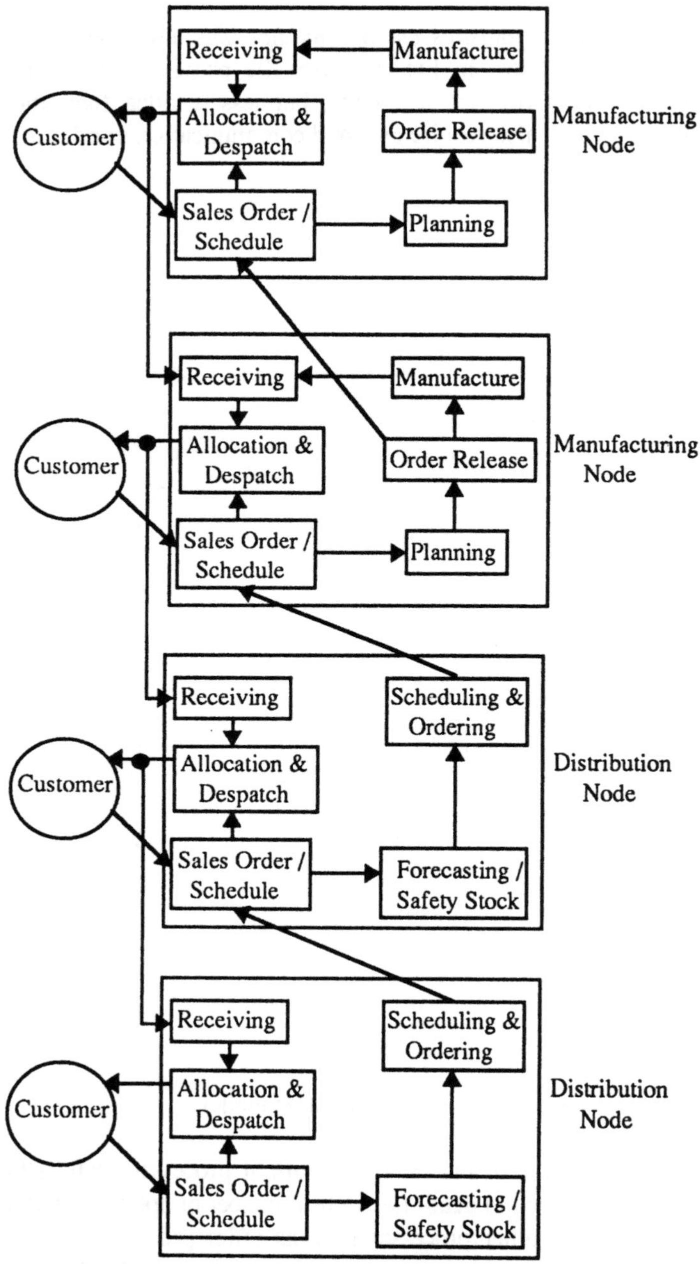

Fig. 8 CMSO logistics chain simulator.

At the operational level, the CMSO model involves operational information systems for manufacturing chain logistics, product development and product support. In manufacturing logistics CMSO has been directed towards interorganizational applications during the phase of serial and mass production. It has therefore been focused on international open EDI and Just-in-Time solutions dealing with interface of manufacturing companies to their business partners in the chain and with graphical planning tools for interorganizational processes.

5.3 The CMSO EDI Reference Model

The CMSO EDI Reference Model (mainly located within OSI layer 7) offers a conceptual framework for services dealing with the exchange of managerial and technical EDI messages. This model actually structures the wide spectrum of EDI functions located in the CMSO-Box. The setting up of dedicated CMSO-Boxes is performed within a powerful environment developed by CMSO which is called the *Layer Shell System* [46].

The services of the CMSO EDI reference model are structured into five sub-layers ranging from communication support functions (such as OSI application services FTAM and X.400) to CIM applications (such as PPC front end applications or development coordination) at the highest layer (Fig. 9).

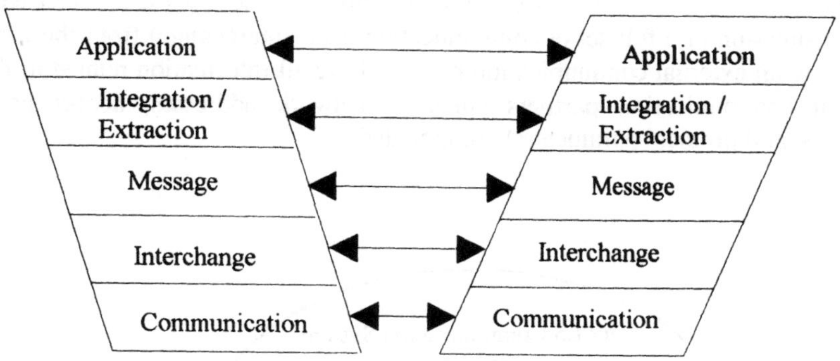

Fig. 9 The CMSO EDI multilayer reference model.

This EDI reference model does not impose any limitations on the system configuration in the communication partner side (e.g. PC based front end EDI server, host based system, distributed system).

A brief description of the principal features of the five layers of Fig. 9 is as follows.

Application layer : This sub-layer deals with business partners and business data (e.g. orders, invoices and credit notes in the commercial area, or drawings and bills of material in the technical area). These elements cover all applications which are involved in interorganizational operations.

Integration/Extraction layer : This sub-layer involves functions that deal with the semantics of EDI messages that are structured according to a unique data format syntax. Here, services such as consistency and integrity check against the application database are located. These services make message data accessible for the application, commitment handling, message management and control, and error/emergencies handling.

Message layer : This sub-layer deals with single messages, logical communication partners and the relation between message type/data format standard and communication partner. Here, a full knowledge about the syntax rules of various data format standards and the relation of standards (to some extent) are located.

Interchange layer : Here the generation and analysis of "EDI-interchanges", and security services are located. An EDI-interchange is built up by one or more EDI messages which are embedded in a data format specific envelope (e.g. ODETTE or EDIFACT service segments).

Communication layer : This lowest sub-layer offers all services concerned with the transmission of a full set of communication data (interchange) from the local domain to an external communication partner. Here all information related to the physical communication partners (such as network addresses, access path, protocols, and protocol parameters) are included.

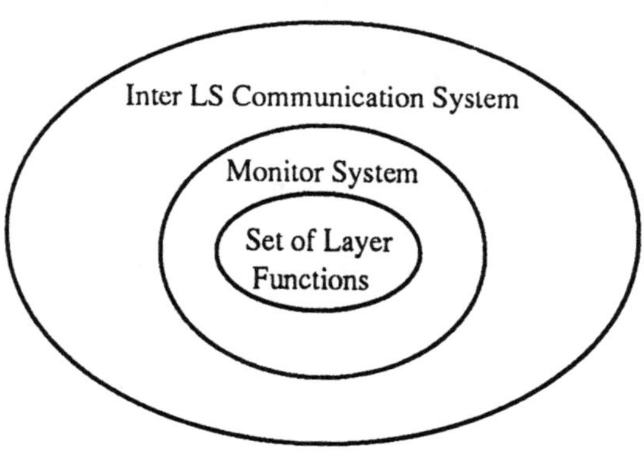

Fig. 10 The layer shell architecture.

We now outline the operational charateristics of the *layer shell system* mentioned at the beginning of this subsection. This is a generalized data and control flow management system that provides the following (Fig. 10) :

- normalized and uniform interfaces for the embedded layer functions (any kind of executable program or command language procedure).
- generalized protocols for data/control flow within one *layer shell* instance (monitor system).
- generalized protocols for the data and control flow between *layer shell* instances (inter-layer-shell communication).

In general, the layer shell concept allows one to build distributed and layered system architectures. Within CMSO this concept was used to build a specific EDI communication system architecture which involves the CMSO box and the CIM-IAS (CIM Interorganizational Application Set) architecture (i.e. the CMSO EDI reference system).

5.4 The CMSO-Box and CIM-IAS Concepts

The CMSO-Box and CIM-IAS concepts for manufacturing logistics consist of applications for manufacturing companies handling their logistics problems in the chain from the vehicle manufacturers to suppliers right down to sub-suppliers. With regard to the EDI reference model, these applications are specific layer functions. The coordination among functions situated at the five layers of the model (Fig. 9) is managed by the layer shell system.

The CMSO-Box for manufacturing logistics is able to manage commercial business procedures. An example of such a procedure is the handling of delivery calls in a manufacturing chain. A delivery instruction is actually a time-phased set of quantity-date pairs for one or more items that belong to log-term orders previously contracted. It involves the call-offs of the items which the customer is aiming for. For this business procedure an application set, called "Front-End Sales Logistics", was developed within CMSO. This CIM-IAS application has access to the classical PPC systems located in CIM-INTRAS (CIM-Intra Organizational Application Set). The functional overview of the prototype CMSO-Box manufacturing logistics is shown in Fig. 11.

In addition to the "Front-End Sales Logistics", the CMSO-Box for manufacturing logistics contains a second application set called "Integral Logistics Planning Workstation" (ILPW). The ILPW is a group of decision support systems for specific logistics planning problems (e.g. short term production planning, supplier rating, transpotration and goods receiving planning).

To see how the CMSO-Box operates, let us consider an incoming order based on ODETTE. The communication layer provides (among others) an ODETTE File Transfer Protocol (OFTP). The order is unpacked in the ODETTE Segment Handler, lying at the interchange layer. The unpacked order is transformed, at the message layer, to a data structure called "multi-format" (a combination of EDIFACT, ODETTE, GALIA and VDA). The order is now treated by the EDI data integrator and distibutor of the integration/extraction layer. It is transformed into a form appropriate for the "Front-End Sales Logistics" application set, which then feeds the required data to the intra-organizational PPC system. The overall process is carried out in a similar way in the other direction to send delivery notes and invoices back.

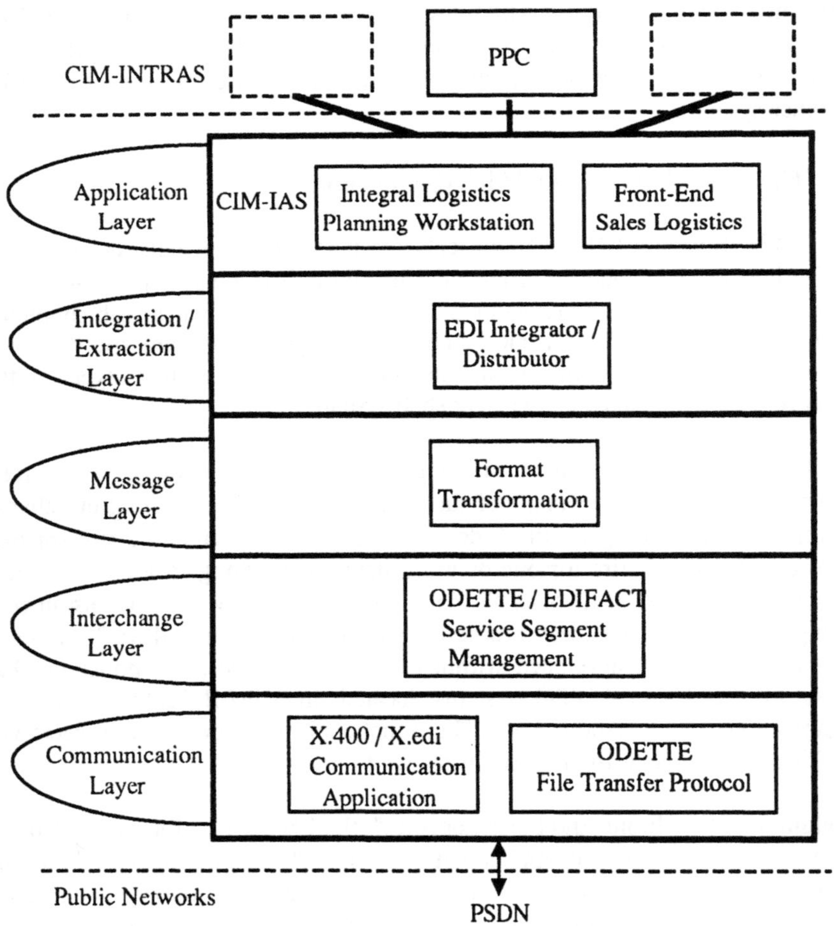

Fig. 11 General functional overview of the CMSO-Box for manufacturing logistics.

The CMSO-Box for the product development operations acts in a similar way and has the structure shown in Fig. 12.

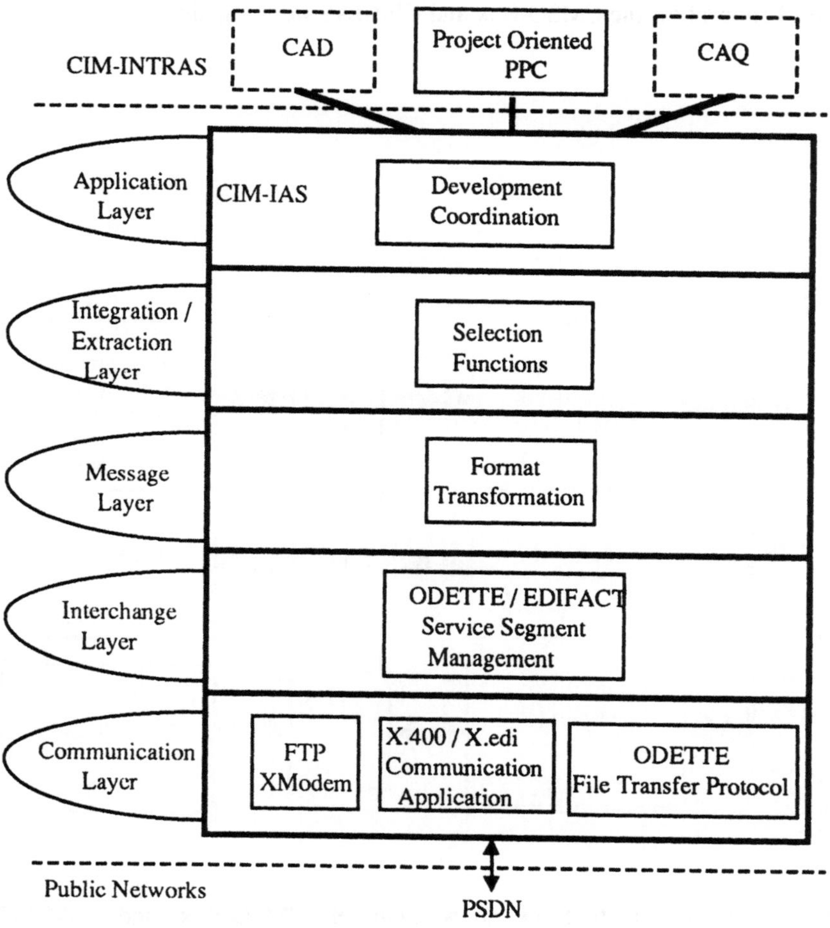

Fig. 12 General functional overview of the CMSO-Box for product development.

The communication and interchange layers are almost the same as in Fig. 11. The messages provided by the interchange layer are represented in a standardized format (ENGENV messages). These messages are transformed at the message layer into an internal representation which is a database for commercial data or a set of technical data files (VDAFS, IGES). At the integration/extraction layer these data are compared with the available interfaces (pre/postprocessors) of the originating and receiving CAD systems. The result is a suggestion for the best combination of the available CAD systems. Finally, the application layer has the development coordinator which controls the entire business process that deals with the development of a one-of-a-kind product.

Figure 13 shows how the CMSO-Box and CIM-IAS work together.

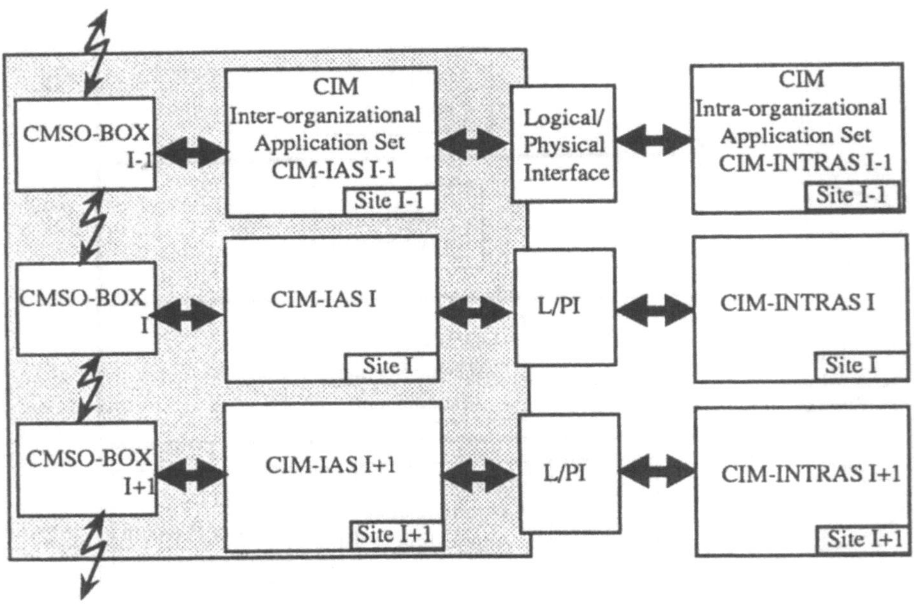

Fig. 13 General structure of cooperation of CMSO-Box and CIM-IAS (Information System Level)

The CIM-IAS in the manufacturing logistics application layer consists of the following subsystems (Fig. 14) :

- Front-end sales logistics (FESL).
- Front-end purchasing logistics (FEPL).
- Integral logistics planning workstation (ILPW).

- Cumulative figure manipulative system (CUMS).
- Advisory forecaster (ADFOR).
- Optimized transportation planning (GENET -Optimiser).
- Planning of distributed production (DP-PSS).

At the integration/extraction layer the CMSO-Box for the manufacturing logistics involves the following subsystems :

- Integration of EDI delivery instructions and JIT-call-offs/Extraction of EDI despatch advices and invoices.
- EDI data integrator and distributor.

Figure 14(a) concerns the case of single company plant, while Fig. 14(b) shows how the CMSO-Box, the CIM-IAS and CIM-INTRAS are combined in a two-company plant.

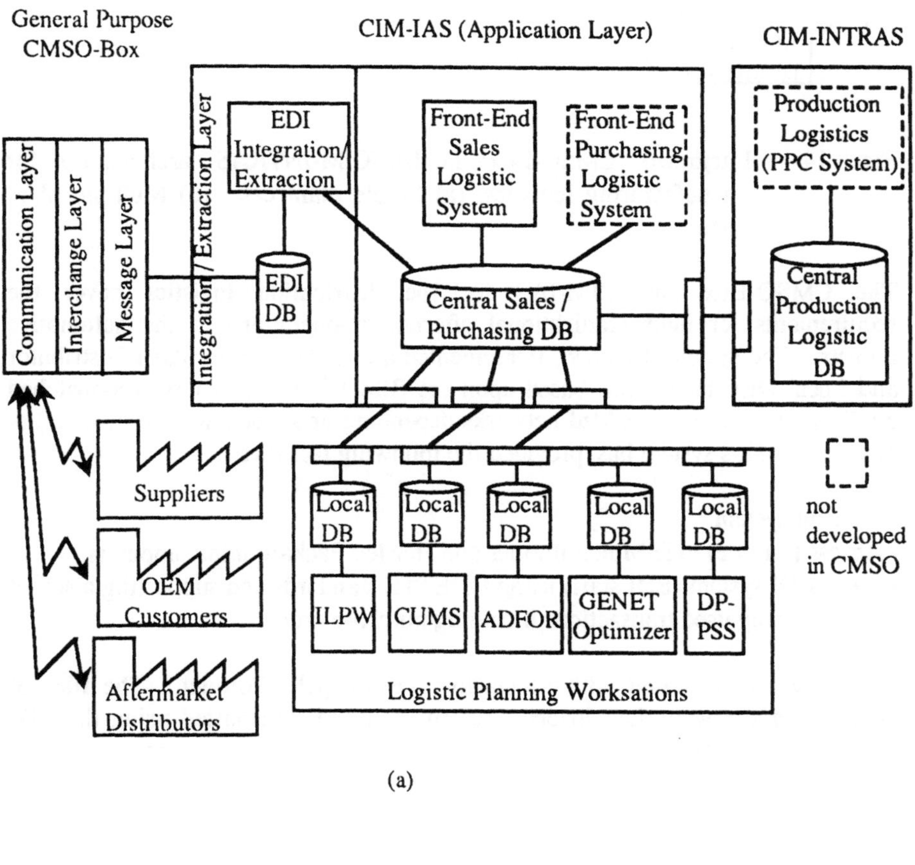

(a)

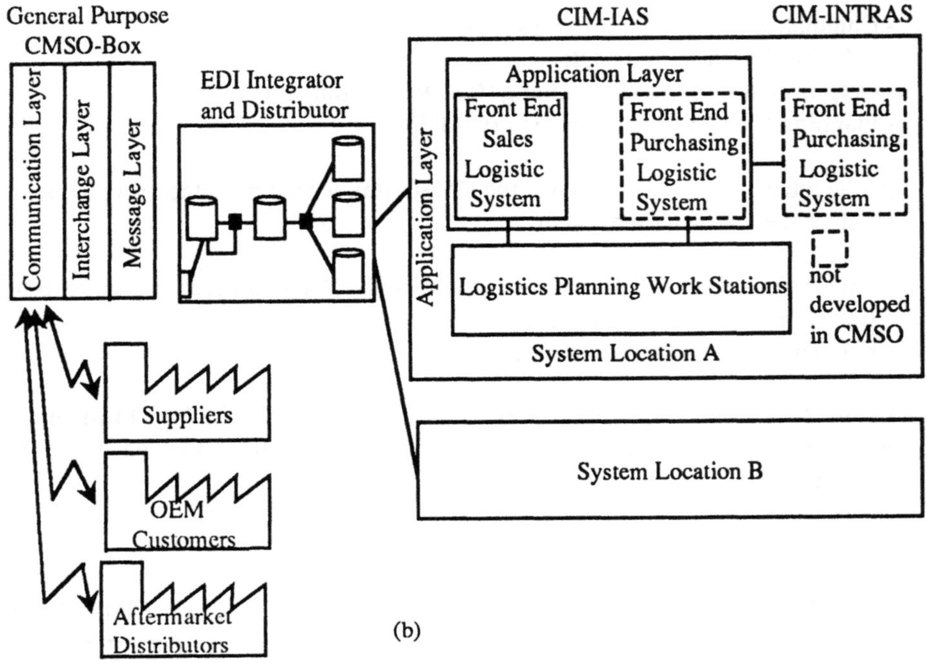

Fig. 14 Integrated CMSO-Box/CIM-IAS/CIM-INTRAS architecture for manufacturing logistics. (a) Single plant case, (b) Multiple plant case.

The CMSO-Box and CIM-IAS for the distribution logistics cover the requirements for interorganizational aftermarket-operations in the automotive industry. The system designed, implemented and tested for the Area Distributor and Local Distributors was based upon the IBM PC and either packet-switched public services or value added networks depending upon location and availability of services. This system has provided the following facilities

- A menu system.
- Access (on line) to IBM mainframes and Hewlett Packard minicomputers.
- Basic EDI system (EDI start) using ODETTE standards and supporting a subset of commonly used transactions including purchase orders.

The system was tested with satisfactory results (UK, Portugal). The files of purchase orders were downloaded, reformatted and transmitted using the EDI start facilities. The achieved business benefits include : (i) a decrease in order processing costs for the supplier, (ii) lead-time reductions (of 1-2 days), (iii) a substantial improvement in delivery reliability (25% to 85%), and (iv) reductions in inventory (between 1 and 1.5 weeks) as a direct consequence of the reduced times and improved delivery reliability.

For the product development process a prototype CMSO-Box was designed and implemented dealing with technical data exchange among independent companies of the logistics chain. This was named "Technical Information Management System" (TIMS) and has the architecture shown in Fig. 15. Actually, TIMS acts as a gateway (or agent) between internal organizational applications and standard data available for transmission on public networks. TIMS provides the information exchanged in such a way that the applications have the right data at the right time. This is achieved via manipulation and routine coordination of orders, products, progress data, etc., performed by the TIMS components : (i) routine data administration, (ii) order data administration, (iii) coordination, (iv) management and temporary data, (v) interface selection, and (vi) conversion of information.

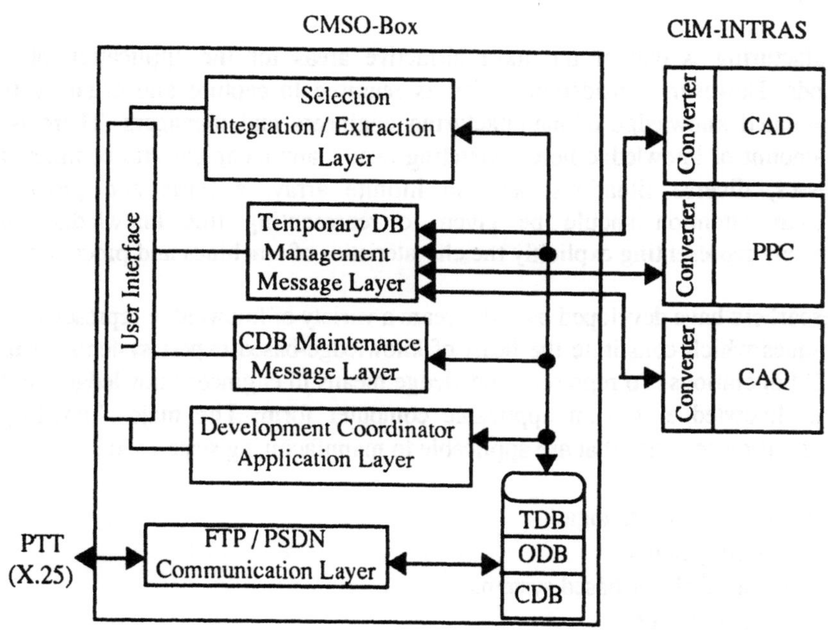

Fig. 15 General architecture of TIMS.

Several tests on the use of TIMS between a vehicle manufacturer and a supplier (in Germany) have shown a reduction of lead time from 100% to 25%. The integration of high speed communication techniques has reduced the exchange time to about 10% in comparison with transfer via magnetic tape.

In the area of product support, CMSO has integrated a natural language input/output handler, and a diagnostic expert system shell with a CD-ROM facility. The natural language system consists of a parser and semantic mapper, an object/entity model, and a dialogue manager [49-50].

6 Artificial Intelligence in CIM Systems

6.1 General Issues

Artificial intelligence (AI) has now arrived at a point where many of its techniques have seen important applications in CIM systems or more generally in a variety of engineering systems [51-53]. AI technology offers the tools for capturing and retaining expertise that was gained over many years of engineering practices, for amplifying this expertise, and for designing systems that reason intelligently about necessary actions to take in real time.

Manufacturing is one of the most attractive areas for the application of AI methods. However, considerable effort is required to capture and organise the accumulated knowledge of manufacturing engineers and managers. There is a vast amount of knowledge here, extending over many technical and commercial processes, diverse situations, and an infinite array of parts and products. Particular attention should be given to representing this knowledge and adequately representing explicitly the charateristics of machines and processes.

AI researhers have developed over the years a variety of knowledge representation techniques which constitute the heart of knowledge-based expert systems design for CIM operations. To represent knowledge means to engineer knowledge, i.e. to convert knowledge into an applicable computer form. The major knowledge representation schemes that are applicable to manufacturing systems are :

- First-order predicate logic
- Rule-based systems
- Frame-based/object-based systems
- Semantic networks

Predicate logic (or predicate calculus) constitutes the fundamental logic-based knowledge representation technique. Knowledge is represented by simple and complex predicate statements or propositions. It uses the so called general inference rules, i.e. i) modus ponens rule, ii) modus tollens rule, iii) double negation rule, iv) AND introduction rule, v) reductio ad absurdum rule, and vi) universal specialization rule [51]. The unified proving and problem solving method employed in first-order predicate logic is called *resolution refutation* and is based on the resolution principle of Robinson [51].

Rule based systems originated from a framework called production systems [54]. A rule based system may be regarded as the assembly of three major component i) a *working memory* that holds the facts, the goal and the intermediate results, ii) a *rule memory* that holds the system rules, and iii) a *rule interpreter* that decides what rules are applicable and in which order they must be executed.

Semantic (or associative) networks are based on the very simple and ancient idea that memory consists of associations between concepts. In semantic networks, the *concepts* are represented as nodes, and *relations* between pairs of concepts are represented by labeled arrows (directed arcs). Semantic networks (like frame-based systems) organize knowledge into hierarchical structures that inherit properties to their ancestors [55].

In *frame-based systems*, knowledge is represented in data structures called "frames". The concept of frame is a result of organizing the properties of some object or event (or a class of objects) to form a prototype (structure). A frame, for example, can be used to represent a part of a manufacturing operation (e.g. drilling) or a controller or a class of vehicles. The power of frame representation is due to the fact that those elements that are conventionally present in a description of an object or event are grouped together and can be accessed and processed as a unit.

Object-oriented systems are based on *"objects"* which are data structures representing descriptive knowledge and armed with private procedures. The concept of a private procedure is that it can only be activated by the objects it has been designed for and its scope is restricted entirely by those objects. Objects communicate with each other by messages which can be regarded as triggering mechanisms for the execution of a private procedure. By sending a message to an object, the sender specifies *what* the receiver has to do, not *how* it must be done. In general, the object-oriented knowledge representation is based on three basic features, namely inheritance hierarchies, private (or attached) procedures, and default values. The relationships in object-oriented systems are classified in hierarchical, topological and conceptual relationships. Object oriented systems [56-57] are very appropriate for building intelligent systems in the manufacturing area.

6.2 Three Basic AI Areas for CIM

Here we give a brief outline of three basic AI areas that are used in CIM functions. These are the areas of automatic planning, automatic learning, and qualitative simulation.

6.2.1 Automatic Planning

Automatic planning can be used for robot and manufacturing task planning. Planning is an important area of AI and is connected to general problem solving. A typical example of AI planning is the well-known "blocks world" problem. The planning consists of finding a path (or subgraph) in a graph in which the nodes represent situations (e.g. states in the block world) and the links between nodes represent the possible moves (e.g. the robot's elementary operations).

The design of a state feedback controller can be formulated as an AI planning problem. The initial state of the AI planning system consists of the open-loop system and the desired closed-loop charateristics. The final state consists mainly of the control law that leads the system to the desired performance or sufficiently near to it. Each node of the tree (or graph) represents a unique model of the process and/or a unique control law. Lines of reasoning about the design method are represented by branches of the tree [75].

Daniel [58] reviews two planning systems, STRIPS and NOAH. STRIPS can also be found in Fikes and Nillson [59]. Other examples of AI-based planners are GPS (General Problem Solver) [60], WARPLAN [61], AL 3 [62] and SIPE [63]. STRIPS is actually a Lisp implementation of GPS [64], and WARPLAN is an improved version of STRIPS in Prolog. STRIPS uses formal logic for representing domain knowledge, and so the current state is represented by a logical statement involving terms of relations among objects. STRIPS is appropriate for robot planning systems. WARPLAN allows nonlinear constuction of plans and is somewhat superior to STRIPS, but it possesses the drawbacks of the depth-first search strategy it employs. NOAH is a planner that can examine the nature of incomplete ordering for actions in plans. SIPE has the capability of replanning whenever some plan fails to achieve the desired goal. During the execution of a plan by a robot, some action in the plan may not lead to the desired goal. This can be detected by the robot sensors (e.g. visual or tactile), and the robot has to make a new plan. SIPE modifies the original plan so as to work well with the new, unforeseen conditions. Planning in CIM is neededd not only for generating robot actions on the basis of sensory feedback, but also for planning the procedures of collecting information and process planning.

6.2.2 Automatic Learning

Automatic learning can be accomplished using the following approaches :

- Learning from instruction
- Learning by analogy
- Learning from examples (inductive learning)
- Learning by discovery

Of course, there are other styles of learning, such as learning through visual imagery or through tactile sensing. So far, machine learning has mainly been connected with symbolic forms of learning.
Generally the learning problem is the following :

Given a set of objects specified in some description language and the classes to which the objects belong.

Find a general classification rule that "explains" the learning set of objects and can also be used for the classification of new objects.

Explaining a set of objects means classifying them correctly into classes as specified in the initial specification. Learning algorithms differ in their use of particular generalizing and specializing rules, rule selection criteria,whether negative instances are included in the training set, and whether a bottom-up or top-down learning sraegy is used.

A useful collection of chapters on various approaches of learning is provided in [65]. A discussion on the relation of machine learning techniques and expert systems is presented in [66]. General examples of learning systems are given in [67-70].

6.2.3 Qualitative Modeling and Simulation

Qualitative modeling and simulation is based on symbolic processing techniques, in contrast with traditional modeling, which is based on differential equation models and numerical techniques. Qualitative models are more appropriate for performing the structural synthesis of a process and carrying out fault diagnosis.

Qualitative modeling and reasoning techniques :

- Can explain the past and predict the future
- Can be used to interpret measurement results

- Can be applied to detect analogies between different domains
- Are suitable for use as teachware

Qualitative reasoning systems use abstract models of their domain, employing the model and causal relationships between the model components to draw conclusions. The domain models are abstract in the sense that they do not use any numerical values to represent the objects, their properties, and their relationships. Examples of qualitative models are given in [71-72]. A survey of qualitative modeling techniques of dynamic systems is given in [73], where the techniques for qualitative modeling of time, space, and motion are described. A combination of qualitative and quantitative modeling and simulation is also provided in [73] for performance-reliability modeling (see also [74]).

6.3 Some Knowledge-Based Systems for CIM

This section provides a brief survey of a number of knowledge-based expert systems (KBES) for manufacturing functions that are available in the open literature. It is not intended to be exhaustive, but it is hoped that the reader will obtain a good picture of the present status of this area.

6.3.1 KBES for Product Computer-Aided Design

The product design task in manufacturing depends on whether the desired product (or part) is a typical one for which standard well-established design procedures exist or a new or a rarely used part that needs the development of a new design procedure. If the product is a modification of an existing product, then one can accordingly design it by suitably modifying available design procedures.

PROPLAN (PROcess PLANning) [76-77] is a knowledge-based environment that integrates the design and planning phases for mechanical parts. Its main advantage is the reduction of human intervention between design and planning. To this end CAD data are automatically transformed into a symbolic form that permits automatic geometric feature extraction.

AIFIX [78] is an expert system suitable for designing fixtures for parts being produced on milling machines. It is an if-then rule-based system and involves Fortran routines that are called by Lisp functions. To design a fixture, all appropriate orientations for the part are at first generated and evaluated, and a number of them are retained. Then the fixture design around the part is carried out in three steps : fixture configuration proposal, fixture configuration realization, and final evaluation (including cost issues).

DOMINIC I [79] is a rather domain-independent expert system (coded in Common Lisp) for iretative design of mechanical parts. Design and performance data are introduced through a knowledge acquisition module. The system asks the user to provide an initial design or generates such a design automatically. This design, which may be pure, is then improved iteratively by the expert system until an acceptable design is found. Otherwise the redesign process is terminated. The refinement process in DOMINIC I is based on only one performance parameter at a time, and in some situations it exhibits serious difficulties. But it is still under improvement.

Other expert systems available for product design can be found in [80-85]. In Brown and Chandrasekaran [85] the design problem of mechanical parts is solved in a top-down way using a hierarchy of specialists. A prototype is described for the design of a small table that consists of a cylindrical support and a circular top. The expert system AIR-CYL presented in [85] is suitable for the design of air cylinders. It is based on the same principles as that of [84], that is, it follows a redesign/refinement process. It is coded in DSPL (Design Specialists and Plans Language), which is a task-level language.

6.3.2 KBES for Process Planning and Scheduling

This is the area for which many expert systems with important properties and features are available.

Fellenstein et al. [86] describe a prototype knowledge base for manufacturing planning that is implemented using the expert system shell Syllog. The knowledge base is built with a combination of a relational database and an expert system shell. The Syllog system is writen in Prolog, and its inference engine can perform many sets of recursive syllogisms that are outside the scope of pure Prolog [87]. Knowledge in the Syllog shell is written as facts and syllogisms. Facts are similar to types in a relational table (although in Syllog a table has an English-like heading), and syllogisms are English-like rules for the use of facts. In [86], the *tester capacity planning and yield analysis task* , which are used in the Syllog shell, are described.

Descotte and Latombe [83] describe an if-then rule-based expert system called GARI, which employs a general problem solver (GPS) and appropriate knowledge for planning the sequence of machining cuts for mechanical parts. The approach of iterative refinement is followed, and the part geometry is represented in terms of features. GARI together with other manufacturing expert systems such as TOM [89], PROPLAN [76-77] and HI-MAPP [90] formulate the process planning problem as a sequence of actions by which the goal state (i.e., the finished parts) is obtained from the initial state (i.e., the raw material), a set of allowable actions (operations), and a group of available resources (robots, machines, tools, etc.).

TOM (Technostructure of Machining) is written in Pascal and generates a plan for a given finished geometry in the hole-making domain. It is again an if-then rule-based system with backward inference (alpha-beta) strategy. TOM provides a process plan with minimum machining time that must be used with care because sometimes its heuristic conflict resolution strategy leads to wrong results.

PROPLAN (Process Planning) involves a parts design subsystem and a process planning subsystem. The process planning subsystem was designed to be generative in the prototype phase, but enhancements are under development to allow both variant and generative planning whenever appropriate. Presently, PROPLAN is implemented in Interlisp-D on a Xerox 1108 machine or in muLisp on an IBM personal computer. One of the potential applications of the variant approach in a hybrid generative/variant environment is the retrieval of existing plans, previously produced by generative planning methods, and then the editing of machining operation descriptions in order to tune the plans to specific shop-floor installations.

Hi-MAPP (Hierarchical Intelligent Manufacturing Automated Process Planner) is similar to GARI with regard to part presentation but differs in two respects : (1) it produces hierarchical plans and (2) it produces initially an abstract of a correct plan, whereas GARI generates initially a loosely constrained plan and uses a time-consuming refinement process. Other examples of expert process planning systems are EXCAP [91], OPEX [92], SIPP [93], SAPT [94] and those described in [95-98].

Regarding the process scheduling problem, the systems described by Szenes [99], Fox and Smith [100], Bruno et al. [101], Mill and Spraggett [102], and Litt et al. [103], are worth mentioning. Szenes [99] presents a prototype tool for modeling planning decision support. This is more closely related to manufacturing than to computer algorithms. Fox and Smith [100] describe the system ISIS,which is suitable for process scheduling in a job-shop facility. Bruno et al. [101], propose a production scheduling system that combines the expert systems methodology with queueing network analysis for fast performance evaluation. The scheduler follows a simple guideline given by the priority of lots and was implemented in the rule-based domain-independent production system language OPS5. In [102] a rule-based expert scheduling system is described that includes the design of a database of part geometry. Finally, Litt te al. [103] developed a rule-based forward-chaining expert system called FSAS (Furnace Scheduling Advisory System) to schedule a multipass glassing and furnacing operation for glass-lined vessels. This system captures the expertise of the shop-floor supervisor, attempting to satisfy all resource constraints on the basis of heuristics. The system is written in

OPS5 and is implemented on a microVAX II. The resulting heuristic scheduler ensures a best solution that meets delivery date requirements, optimizes utilization of multiple furnaces, and minimizes energy consumption. The resource constraints include availability and capacity of the furnaces, and availability of firing tools. Operational constraints include part mix, firing temperature, and allowed thickness difference.

6.3.3 KBES for Assembly and Equipment Selection

A large database for assembly operations (feeding, handling, orienting, and insertion times) is provided in Rembold and Dillmann [1], and Boothroyd and Dewhurst [104-105]. Other works in the assembly area using AI and expert system tools are those of De Winter and Van Brussel [106], Jozefowicz and Urbanski [107], Lee and Gossard [108], Ambler et al. [109], Kak et al. [110], and chang and Wee [111]. In [110] the automated assembly cell developed at the Robot Vision Laboratory of Purdue University is described. It consists of five modules : supervisor, global knowledge base, current world model,motion controller, and sensory unit. The object representation is made using the slot-filler technique which works well for objects with distinctive landmarks. For solid objects of high symmetry, the extended Gaussian image concept representation is used. Experimental results are included in parts mating using three-dimensional vision feedback.

In [111] use is made of a new approach to attack the problems of planning and mechanical assembly. The authors describe system knowledge types in detail and examine thoroughly the system's control structure, which functions in two separate phases (analysis and generation). A versatile system for computer-controlled assembly, using television cameras, a moving table, and an intelligent robot is proposed in [109]. The operator places a number of parts in a heap on the table, and the machine has to separate them, recognize them, and then assemble them into a predetermined configuration. The system occupies about 50K of 24-bit words of POP-2 code in a 128K time-shared ICL 4130 machine. The program is written as two distinct subprograms, layout and assembly, which rely on common conventions. The layout subprogram uses tactile and visual information and has internal descriptions of parts and tabletop. It can face failures, errors, and accidents. The assembly subprogram employs tactile data only and has no internal descriptions of parts. It can cope with small positioning errors but cannot recover from accidents.

The application of AI in the domain of equipment selection is limited. A rule-based expert system for the selection of the materials handling equipment, called MATHES (MATerial Handling Equipment Selection), is decribed in [112]. The domain of application is the transportation of material between facilities in a manufacturing environment. The conclusions are derived on the basis of four main parameters (path, volume of flow, size of load, and distance between facilities) and several minor ones concerning specific situations.

In [113] a Prolog-based expert system for truck selection is proposed. Truck selection is very important in dock operations, unit load storage, order picking, in-process handling, yard operations, and elsewhere. The selection process is based on truck features, including type of engine, type of loading, type of tires, and capacity, as well as on the specific activities taking place in the application areas. The system provides a user-system interaction facility by means of which the user can ask for alternative truck selection if she or he is not satisfied with the current selection.

Two expert systems for industrial robot selection are presented in McGlennon et al. [114] and Tzafestas and Tsihrintzis [115]. The first, called ROBOSPEC, has a knowledge base that includes information on several application areas (handling, manufacturing, assembly). The system is written in OPS5 and provides as a final result the specifications on the selected robot for the application area specified by the user. The second robot-selecting system, ROBBAS, was realized with the expert system shell M1., which runs on the IBM PC. Sesssions with M1. are divided into several cycles of dialogue, so that the user is able to modify the value of a robot charateristic without having to go through the entire session from the beginning. As usual, ROBBAS consists of two main components, one containing pure expertise (rules, facts and metafacts) and one containing a database of commercially available industrial robots and their charateristics.

Another expert system for the specification and selection of robots is presented in [116], which discusses the relative advantages of the expert system and integer programming approaches. The expert system identifies by forward inference the required technology alternatives that are needed to carry out the desired tasks, and selects by backward inference the best available robots that satisfy the technological requirements, Finally, ROKON 1, an expert system for selecting a suitable sensor and gripper configuration written in the Franzlisp version of OPS5 on a VAX 750 under UNIX, is presented in [117]. More on expert systems in control, robotics and manufacturing can be found in [125-128].

7 Economic Evaluation of Automated Manufacturing

7.1 General Issues

The three primary approaches to industrial production systems are :

- hard automation
- robotics/flexible automation
- manual labour

Hard (fixed) automation involves the use of special-purpose equipment, which needs very high capital investment and is appropriate only for very high volumes of products. Less special equipment involves the use of numerical control (NC) or computer numerical control (CNC). Industrial robots form a kind of automated machines that combine some features of hard and manual labour. The principal question here is whether the overall cost of automation is justified by the benefits obtained.

Table 2 : Charateristics of production systems

Charateristics	*Hard (Fixed) Automation*	*Robotics*	*Manual*
(I) Capabilities (what can the system do)			
Flexibility	None	Moderate	High
Sensing/ Intelligence	None	Limited	High
Manipulation	Neglitible	Simple	Complex
Transport	Excellent	Good	Limited
(II) Manufacturing charateristics			
Type of production	Mass	Mass/Batch	Batch/Job shop
Number of shifts	2 ~ 3	2 ~ 3	1
Task complexity	Low	Medium	High
Degree of structure	High	High	Low
Production volume	Longrun	Medium run	Short run
Production rate	High	Slow	Slow
Degree of hazard	No limit	No limit	Non-hazardus
(III) Performance			
Productivity	High	Medium	Low
Quality	High	Medium	Low

Table 2 summarizes the main features, i.e. capabilities, manufacturing charateristics, and performance of industrial production systems classified by degree versus type of automation [118-119].

The decision about what type of system must be used for the production of a particular product is taken in the following three steps :

Step 1 : Selection of a small number of canditate systems (commercially available) on the basis of technical hardware/software specifications (including issues of safety, devices, maintenance, tooling, and spare parts).

Step 2 : Estimation and evaluation of impacts or consequences of the adoption of each candidate system including economic issues such as directed labour, savings, inventory savings, as well as important non-economic consequences such as corporate prestige and employee motivation.

Step 3 : Final selection of the production system to be adopted employing multi-attribute decision making theory. This can be done by appropriately aggregating the particular utility (or preference) functions of the various factors involved.

In the following , the emphasis will be given on the economic aspects along the lines of [118-120].

7.2 Economic Evaluation Model

To evaluate the overall performance of automated systems one must have available a suitable economic evaluation model. A good economic evaluation model must have the charateristic features described below.
Exhaustiveness of feasible alternatives : The evaluation model must have the ability to include all feasible alternatives in terms of price, performance and timing of investment. The alternatives must be formulated and structured such that to be mutually exclusive. The effects of postponing the aquisition of the selected automation must also be included in the analysis.

Discounted cash flow : A good criterion for evaluating the preceding alternatives (which are mutually exclusive) must be adopted. Some criteria available in the technical literature are :

- *accounting methods* (e.g. average rate of return) which generally do not take into account the opportunity cost (i.e. the rate of return which might be gained on alternative investment oppurtunities).

- *payback period method* (the payback period is the number of periods that are needed for cumulative gains to equal cumulative costs).

- *discounted cash flow method* (it includes both the time value of money and all cash flows over the project life).

Comprehensiveness : The model must be comprehensive, i.e. it must incorporate all the economic consequences of the evaluated system over the life cycle of the system (i.e. the model must include the initial investment cost, the development cost, the operating cost, the maintenance cost, and the disposal cost).

Credibility : The model must give confidence to the decision maker(s) in the sense that the results are in fact a reasonably accurate representation of reality.

Verifiability : The results from the numerical testing analysis must be verifiable or replicative.

Learning over the planning horizon : A good model must take into account the fact that people with machines learn to improve the product (output) when the volume of production is increased. This is formally expressed by the so called learning (or improvement, or experience or manufacturing-time forecasting) curve which is defined as follows (by T.P. Wright) : "every time the quantity produced doubled, the cumulative average cost at the doubled quantity bore a fixed and lower relationship to the cumulative average cost of the previously undoubled quantity".

Dynamic pattern for economic consequences : The model must show the time-dependent (dynamic) pattern for each economic consequence. Obviously not all economic consequences are kept constant with time.

Technological effect on economic consequences : The cost and capabilities of "high-tech" automation equipment change rapidly over time. Future systems are expected to be substantially different from the ones presently available. The model should take this fact into consideration (as much as possible)and also predict the operational and replacement costs.

Inflation and Uncertainties : The model must include in the analytical part the relative price changes (inflation) that are anticipated over time, as well as the economic uncertainties that are expected from other sources. Clearly, all consequences of interest lie in the future which cannot be predicted with certainty.

For more detailed discussions on these and other economic issues of automated manufacturing and robotics the reader is referred to [121-124].

8 Conclusions

During the last two decades many areas of information technology have been profitably applied to automated manufacturing and associated *intra*-and-*inter*-organizational/business areas. Due to the use of incompatible hardware and software systems, the original result was the so-called "islands of automation", i.e. isolated areas of automation which are of substantial value only for each individual company operating them. In Europe, manufacturing contributes to more than 60% of the gross national product of the highly industrialized countries. Therefore, the advances in computing, control, mechanization and automation as applied to the manufacturing sector can seriously increase the competitiveness of the European automotive industry in the international market. However, unlike the Japanese automotive industry, where vehicle manufacturers and their primary suppliers belong to the "same family" and are in close geographical proximity, the European automotive industry suffers from the fact that it involves a variety of distinct and unrelated companies with supply and distribution chains that have their own corporate objectives. Thus a technical, managerial, information, and communication integration will increase further the productivity and competitiveness of the European automotive industry.

In this chapter an attempt was made to highlight many of the IT, managerial and communication issues that must be addressed for achieving such integration. This was made through the review of the models developed in two major European projects, namely the CIM-OSA and CMSO projects [35,46]. In addition, several other aspects have been considered and reviewed such as the basic CIM functions, the contribution of IT to the enhancement of competitive advantage, the role of artificial intelligence in manufacturing, and the economic evaluation/justification of automated manufacturing.

In choosing the right control strategy for meeting its goals and needs, a manufacturing company must amog others address the following questions :

- What is the type and structure of the company ?
- What kind of managerial style do the company's managers use ?
- How accurate and reliable are the major performance measures used ?

The answers to these questions must be systematically evaluated in order to determine which strategy is the most appropriate. Usually a convenient trade-off between the different strategies is selected. The above approach is known as the "Informed Choice" among management and control strategies. In addition, the company needs to evaluate alternative approaches to designing and using control systems. The estimation of the potential effects of control systems and the consideration of the great importance of the *process of control* (as opposed to the technology of control) play a central role in making a manufacturing company and its people more productive and competent. A full discussion of issues of this kind can be found in [129].

References

[1] U. Rembold and R. Dillmann , Computer-Aided Design and Manufacturing : Methods and Tools, *Springer*, Berlin, 1986.

[2] Y. Ito , Human-Intelligence-Based Manufacturing, *Springer*, Berlin, 1993.

[3] R. Bernhardt, R. Dillmann, K. Hörmann and K. Tierney , Integration of Robots into CIM, *Chapman & Hall* , London/New York, 1992.

[4] Y. Koren , Robotics for Engineers, *McGraw-Hill*, New York, 1985.

[5] G. Doumeingts, M. C. Maisonneuve, V. Brand and C. Berard , Design Methodology of Computer Integrated Manufacturing and Control of Manufacturing Units, In : *U. Rembold and R. Dillmann, Computer-Aided Design and Manufacturing : Methods and Tools, Springer* , Berlin, pp. 137-182, 1986.

[6] G. Boothroyd and P. Dewhurst , Design for Assembly Handbook, *University of Massachussets*, Amherst, 1983.

[7] M. M. Andreasen and T. Ahm , The Relation Between Product Design, Production of Layout and Flexibility, *Proc. 7th ICAA* , Zurich, 1986.

[8] H. W. Stoll , Design for Manufacturing - An Overview, *Appl. Mech. Rev.* , Vol. 39, No. 9, pp. 1356-1364, 1986.

[9] A. Gairola , Design for Automatic Assembly : In : H. W. Warnecke and H. J. Burlinger, eds., Factory of the Future, *Springer* , Berlin/New York, 1985.

[10] T. C. Chang and R. A. Wysk , An Introduction to Automated Process Planning Systems, *Prentice-Hall* , Englewood Cliffs, N. J., 1985.

[11] G. Spur and F. L. Krause , Technological Planning for Manufacturing-Methodology of Process Planning, In : *Computer Aided Design and Manufacturing : Methods and Tools (U. Rembold and R. Dillmann, Springer-Verlag* , New York, Chapter 3, 1986.

[12] W. Eversheim, H. Fuchs and K. H. Zons , Anwendung des Systems AUTAP zur *Arbeitsplanerstellung*, Ind. Anz. H. 55 : 29-33 , 1980.

[13] C. H. Link , CAM-I : Automated Process Planning System (CAPP), *Tech. Paper* , Dearborn, Mich., 1976.

[14] G. Spur and E. Hein , Ergenbnisse zur recherunterstrutzten Prufplannung, *Endbericht P6.* 4/28 ;B-PRi/2 , KfK-BMFT, 1981.

[15] M.S. Dunn and S. Mann , Computerised Production Process Planning , *Proc. 15th Numerical Control Society Annual Meeting* , Chicago, 1978.

[16] G. Schaffer , GT. via Automated Process Planning, *Amer. Machinist* , pp 119-122, May 1980.

[17] J. Tulkoff , Lockheed's GENPLAN , *Proc. 18th Numerical Control Society Annual Meeting and Tech. Conf.* , Dallas, Texas, 1981.

[18] A. Rolstadås , Scheduling Batch Production by Means of an Online Microcomputer ,*SINTEF Report STF17 A77064* , November, 1977.

[19] S. M. Alexander and V. Jagannathan , Computer-Aided Process Planning Systems : Current and Future Directions, *Proc. IEEE Int. Conf. on Systems, Man and Cybernetics* ,New York , 1983.

[20] R. Dillmann , Computing Aids to Plan and Control Manufacturing, In : *Computer-Aided Design and Manufacturing : Methods and Tools (U. Rembold and R. Dillmann eds.), Springer-Verlag*, New York, Chapter 6, 1986.

[21] T. J. Williams , Developments in Hierarchical Computer Control Systems, *Proc. CAPE' 83* Amsterdam, 1983.

[22] R. Conterno , Hierarchical and Decentralized Control for Batch and Repetitive Manufacturing, *Proc. 1987 IEEE Int. Conf. Robotics and Automation* , Raleigh, N. C. March 30-April 3 , 1987.

[23] P. Ranky , The Design and Operation of FMS , *JFS (Pubications) Ltd. and North-Holland*, Amsterdam, 1983.

[24] P. Ranky , Dymamic Simulation of Flexible Manufacturing Systems, *Appl. Mech. Rev. ,Vol. 39(9) : pp. 1339-1344*, 1986a.

[25] P. Ranky , Computer-Integrated Manufacturing, Prentice-Hall , *Englewood Cliffs* , N. J., 1986b.

[26] S. S. Heragu and A. Kusiak, Analysis of Expert Systems in Manufacturing Design, *IEEE Trans. Syst. Man Cybern.* , Vol. SMC - 17, pp. 898-912, 1987.

[27] D. M. Miller and R. P. Davis , The Machine Requirements Problem, *Int. J. Prod. Res. Vol. 15, pp. 219-231* , 1977.

[28] A. Kusiak (Ed.) , Artificial Intelligence : Computer Integrated Manufacture, *IFS*, Kempston , Bedford, U. K. 1987a.

[29] A. Kusiak , The Production of Equipment Requirements Problem, *Eur. J. Oper. Res. Vol. 29, pp. 229-251* , 1987b.

[30] A. Kusiak and S. S. Heragu , The Facility Layout Problem, *Eur. J. Oper. Res. Vol. 29, pp. 229-251* , 1987.

[31] M. E. Porter , Competitive Strategy, *Free Press* , New York, 1980.

[32] M. E. Porter and V. E. Millar , How Information Gives You Competitive Advantage, *Harvard Business Review* , pp 149-159 , July-August, 1985.

[33] M. E. Porter , Competitive Advantage, *Free Press* , New York, 1985.

[34] F. Warren Mc Farlan , Information Technology Changes the Way You Compete, *HBR* , May-June, 1984.

[35] CIM-OSA : Reference Architecture Specification, *ESPRIT Project No. 668 (AMICE Consortium)* , Brussels, 1988.

[36] T. Klevers , The European Approach to an Open System Architecture for CIM , *Proc. 5th CIM Europe Conference* , pp 109-120, 1989.

[37] W. Eversheim and T. Klevers , CIM-OSA Peilt Integrierende Infrastruktur, *Computerwoche* , No. 7, pp 18-21, Mumch, 1989.

[38] M. Klittich , CIM-OSA : The Implementation Viewpoint, *Proc. 4th CIM Europe Conference* , IFS Publications/Springer, 1988.

[39] M. Klittich , CIM-OSA and its Relationship to MAP, *Proc. 5th CIM Europe Conference* , pp 131-142, 1989.

[40] MAP 3.0 : Manufacturing Automation Protocol, *North Amer. MAP Users Group , Ann Arbor, 1988, Also : European MAP Users Group*, Cransfield, 1989.

[41] ISO IS 9506 : Manufacturing Message Specification (MMS).

[42] ISO/IEC JTC1/SC21/WG7 No 47 :Basic Reference Model of Open Disrtibuted Processing. Configuration Programming for Distributed Systems.

[43] H. P. Godbersen, M. Matthiesen and W. Schaber , Modelling of Interorganisational Operations, *Proc. Telematics '90*, Bremen, pp 207-221, Dec. 1990.

[44] H. J. Schneider and S. G. Tzafestas , Integrated Approach to Computer-Aided Multi-Supplier/Multi-Distributor Operations in the Autonomotive Industry,*Proc. 23rd Intl. Symp. on Automotive Technology and Automation*, Vol III, pp. 264-273 , Vienna, Dec. , 1990.

[45] H. J. Schneider, M. Lock, M. Matthiesen and H. Rentschler , CMSO : CIM for Multi Supplier Operations, *Proc. APMS '90 : IFIP Intl. Conf. on Advances in Production Management Systems* , Espoo, Finland, Aug., 1990.

[46] CMSO-CIM for Multi-Supplier Operations, *ESPRIT II Project*, No. 2277, Final Report, March, 1992.

[47] W. Schader and M. Matthiesen , The CMSO EDI Architecture for Interorganisational Operations, In : *S. G. Tzafestas, ed., Engineering Systems with Intelligence*, pp. 505-513, Kluwer, Dordrecht/Boston, 1991.

[48] W. Schaber and M. Matthiesen , Interorganisational Manufacturing/Supplier Operations in the Automotive Industry, *Proc. ELEDIS '91 : Intl. Conf. on Electronic Data Interchange Systems*, Milano, May, 1991.

[49] S. G. Tzafestas and N. Konstantinidis, ENGEXP : An Integrated Environment for the Development and Application of Expert Systems in Equipment and Engine Fault Diagnosis and Repair, *Advances in Engrg. Software* , Vol. 14, No. 1, pp. 3-14, 1992.

[50] J. Honig and A. Vonk , Natural Language and Technical Information Systems-Problems Oriented Information Systems, *T. U. Delft CMSO Working Paper*, 1989.

[51] E. Charniac and D. McDermott , Introduction to Artificial Intelligence, *Addison-Wesley Reading*, MA., 1985.

[52] S. G. Tzafestas , AI Techniques in Computer-Aided Manufacturing Systems, In : *H. Adeli, ed. , Knowledge Engineering, Vol. II, McGraw-Hill* , New York, pp. 161-212, 1990.

[53] S. Albayrak and H. Krallmann , Distributed Artificial Intelligence in Manufacturing Control, In : *S. G Tzafestas and H. Verbruggen, Artificial Intelligence in Industrial Decision Making, Control and Automation, Kluwer* , Dordrecht/Boston, pp. 247-294, 1995.

[54] E. Post , Formal Reductions of the General Combinational Problem, *American J. Mathematics* , 65, pp. 197-268, 1943.

[55] M. Minsky , Semantics Information, *MIT Press* , Cambridge, MA, 1986.

[56] F. Puppe , Systematic Introduction to Expert Systems : Knowledge Representations and Problem Solving Methods, *Springer-Verlag* , Berlin, 1993.

[57] B. Roberts and I. Goldstein , The FRL Primer, AI Memo 408 , *MIT*, Cambridge, MA 1977.

[58] L. Daniel , Artificial Intelligence : Tools, Techniques and Applications, *Harper & Row*, New York, 1984.

[59] R. E. Fikes and N. J. Nillson , STRIPS : A New Approach to the Application of Theorem Proving to Problem Solving, AI 2, pp. 189-208, 1971.

[60] G. W. Ernst and A. Newell , GPS : A Case Study in Generality and Problem Solving, *Academic Press* , New York, 1969.

[61] D. H. D. Warren , WARPLAN : A System for Generating Plans, *DCL Memo 76, Dept. of AI* , Edinburgh Univ., 1974.

[62] I. Bratko , Knowledge-Based Problem Solving in AL3. In : Machine Intelligence (J. Hayes, D. Michie and J. H. Pao, Eds), *Harwood* , Chichester, U. K., pp. 73-100, 1982.

[63] D. E. Wilkins , Recovering from Execution Error in SIPE, *Comput. Intell. J. 1* , Berlin, 1986.

[64] R. Fikes, P. Hart and N. Nillson , Learning and Executing Generalized Robot Plans, *AI 3, pp.* 251-288 , 1973.

[65] R. S. Michalski, J. G. Carbonell and T. Mitchell , *Machine Learning*, Tioga, Palo Alto, Calif, 1983.

[66] A. Bundy , What Has Learning Got To Do with Expert Systems ?, *Paper No. 214*, Dept. of Artificial Intelligence , Univ. of Edinburgh, 1984.

[67] J. R. Quinlan , Discovering Rules by Induction from Collections of Examples. In : Expert Systems in Microelectronic Age (D. Michie, Ed.), *Edinburgh Univ. Press* , Edinburgh pp. 168-202, 1981.

[68] B. A. Shepherd , An Appraisal of a Decision Tree Approach to Image Classification, *Proc. 8th Int. Joint Conf. on Artificial Intelligence (IJCAI '83*), Vol. 1, pp. 473-475, 1983.

[69] R. Dechter and D. Michie , Structured Induction on Plans and Programs, *IBM*, Los Angeles , Calif, 1984.

[70] B. Dufay and J. C. Latcombe , An Approach to Automatic Robot Programming Based on Inductive Learning, *Int. J. Rob. Res. Vol. 3,* 1987.

[71] I. Bratko , AI Tools and Techniques for Manufacturing Systems, *Robotics Comput. Integrated Manuf. Vol. 4(1/2) : pp. 27-31* , 1988.

[72] I. Bratko, I. Mozetik and N. Lavrac , Automatic Synthesis and Compression of Cardiological Knowledge, Machine Intelligence , Vol. 11, *Oxford Universsity Press,* Oxford, England, 1988.

[73] G. Burle , The Role of Qualitative Reasoning on Modelling , *Proc. IMACS Symp. on AI, Expert Systems and Languages in Modelling and Simulation* , Barcelona, pp. 117-122, 1987.

[74] R. O 'Keefe , Simulation and Expert Systems. A Taxonomy and Examples, *Simulation, Vol. 46(1), pp. 10-16*, 1986.

[75] S. G. Tzafestas , AI Techniques in Control : An Overview. In : *AI, Expert Systems and Languages in Modelling and Simulation (IMACS Proc. 1987) (C. Kulikowski and G. Ferrate, Eds.)*, North Holland , Amsterdam, 1988a.

[76] C. B. Mouleeswaran and H. G. Fisher , A Knowledge Based Environment for Process Planning. *In : Applications of Artificial Intelligence in Engineering Problems, Vol. 2 (D. Shiram and R. Adey, Eds.), Springer-Verlag,* New York, pp. 1013-1027, 1986.

[77] C. B. Mouleeswaran , PROPLAN : A Knowledge-Based Expert System for Process Planning, *MS Thesis* , Univ. of Illinois, Chicago, III, 1984.

[78] P. M. Ferreira, B. Kochar and V. Chandru , AIFIX : An Expert System Approach for Fixture Design. In : *Computer-Aided/Intelligent Process Planning (C. R. Liu, T. C. Chang and R. Komanduri, Eds.) , ASME* , N. Y., pp. 73-82, 1985.

[79] J. R. Dixon, A. Howe, P. R. Cohen and M. K. Simmons , DOMINIC I : Progress Towards Domain Independence in Design by Iretative Redesign, *Eng. Comput., Vol. 2, pp. 137-145* , 1987.

[80] S. Mittal, C. L. Dym and M. Morjaria , PRIDE : An Expert Systemfor the Design of Paper Handling Systems. In : *Applications of Knowledge-Based Systems to Engineering Analysis and Design(C. L. Dym, Ed.), ASME AD-10, ASME* , New York, 1985.

[81] J. R. Dixon and M. R. Simmons , Expert Systems for Design : A Program of Research, *ASME Conf. on Design Engineering* , Cincinnati, Ohio, Paper No. 85-DET-78, 1985.

[82] J. R. Dixon, E. C. Libardi, E. C. Luby, M. V. Vaghul and M. K. Simmons , Expert Systems for Mechanical Design : Examples of Sympolic Representations of Design Geometries. In : *Applications of Knowledge-Based Systems to Engineering Analysis and Design (C. L. Dym, Ed.), ASME AD-10* , N. Y. 1985.

[83] J. J. Shah , Development of a Knowledge Base for an Expert System for Design of Structural Parts, *Proc. 1985 ASME Int. Computers in Engineering Conf. and Exhibition*, Boston, Mass, 1985.

[84] D. C. Brown and B. Chandrasekaran , An Approach to Expert Systems for Mechanical Design, *Proc. AI Trends and Applications* , Gaithersburg, MD., pp. 173-180, 1983.

[85] D. C. Brown and B. Chandrasekaran, Knowledge and Control for a Mechanical Design Expert System, *Computer, Vol. 19, pp. 92-100*, 1986.

[86] C. Fellenstein, C. O. Green ,L. M. Palmer and D. J. Wyler, A Prototype Manufacturing Knowledge Base in Syllog, *IBM Res. Develop., Vol. 29(4), pp. 413-421* , 1985.

[87] A. Walker , Syllog : An Approach to Prolog for Nonprogrammers. In : *Logic Programming and Its Applications (M. van Caneghem and D. H. D. Warren, Eds.).Ablex*, Norwood , N. J., 1985.

[88] Y. Descotte and J. C. Latombe , GARI : An Expert System for Process Planning. Solid Modelling for Computers : From Theory to Applications, New York, 1984.

[89] K. Matsushima, N. Okada and T. Sata , The Integration of CAD and CIM by Application of Artificial Intelligence Techniques. In : *Manufacturing Technology*. Techn. Rundschan,Berne , Switzerland, 1982.

[90] H. R. Berenji and B. Khoshnevis , Use of Artificial Intelligence in Automated Process Planning, *Comput. Mech. Eng.*, pp. 47-55, 1986.

[91] I. Darbyshire and E. J. Davies , EXCAP-An Expert System Approach to Recursive Process Planning, *Proc. 16th CIRP Int. Seminar of Manufacturing Systems,* Tokyo, 1984.

[92] A. Sluga, P. Butala, N. Lavrac and M. Gams , An Attempt to Implement Expert Systems Techniques in CAPP, *Robotics Integrated Manuf., Vol.* 4(1/2), pp. 77-82 , 1988.

[93] D. S. Nau and T. C. Chang , A Knowledge Based Approach to Generative Process Planning, In: *C. R. Liu, T. C. Chang and R. Comanduri, eds., Computer-Aided Intelligent Process Planning, ASME Publ.* , New York, pp. 65-71, 1985.

[94] V. R. Milacic and M. Urosevic , SAPT-Knowledge-Based CAPP System, *Robotics Comput. Integrated Manuf.*, Vol. 4(1/2), pp. 69-76 , 1988.

[95] K. Preiss and E. Kaplanski , Solving CAD/CAM Problems by Heuristic Programming, *Comput. Mech. Eng., Vol.* 2(2), pp. 56-60, 1983.

[96] D. A. Bourne and M. S. Fox , Autonomous Manufacturing : *Automating Job-Shop , Computer, Vol.* 17(9), pp. 79-86 , 1984.

[97] J. Vancza , Organizing Classificatory Knowledge by Induction : A Case Study in Manufacturing Process Planning, *Proc. 12th IMACS World Congress* , Vol. 4, pp. 258-260, 1988.

[98] F. Gliviak, J. Kubis, A. Milovsky and E. Karabinosova , A Manufacturing Cell Management System : CEMAS. In : *Artificial Intelligence and Information : Control Systems of Robots (I. Plander, Ed.)*, North Holland, Amsterdam , 1984.

[99] K. Svenes , An Application of a Parallel Systems Planning Language in Decision Support-Production Scheduling, In : *Advances in Production Management :Production Management Systems in the Eighties, Proc. IFIP WG 5.7 Working Conf.*, Bordeaux , France, pp. 241-249, 1982.

[100] M. S. Fox ans S. F. Smith , ISIS : A Knowledge-Based System for factory Scheduling, *Expert Syst., Vol.* 1, pp. 25-49 , 1984.

[101] G. Bruno and G. Marchetto , Process-Translatable Petri Nets for the Rapid Prototyping of Process control Systems, *IEEE Trans. Software Eng , Vol. SE-12(2)*, 1986.

[102] F. G. Mill and S. Spraggett , An Artificial Intelligence Approach to Process Planning and Scheduling for Flexible Manufacturing Systems , *Proc. Int. Conf. Computer-Aided Engineering , IEE*, London, 1984.

[103] M. Litt, J. C. H. Chung, D. C. Bond and G. G. Keininger , A Scheduling and Planning Expert System for Multiple Furnaces, *Eng. Appl. AI 1* (March), pp. 16-21, 1988.

[104] G. Boothroyd and P. Dewhurst , Computer Aided Design for Assembly, *Assembly Eng.*, Vol. 26(2), pp. 18-22, 1983.

[105] G. Boothroyd and P. Dewhurst , Design for Assembly-A Designers Handbook, University of Massachusetts, *Amherst*, Mass, 1984.

[106] D. De Winter and H. Van Brussel , An Expert System for Flexible Assembly System Design , *Proc. 8th Annual British Robot Association Conf.*, Birmingham, U.K., pp. 133-142, 1985.

[107] W. Jozewicz and A. Urbanski , Expert Systems and Automatic Component Assembling in CAD, *Proc. Int. Conf. on Computer-Aided Engineering , IEE,* London, U.K., 1984.

[108] K. Lee and D. C. Gossard , A Hierarchical Data Structure for Representing Assemplies : Part I, *Comput. Aided Design*, Vol. 17(1), pp. 15-24, 1985.

[109] A. P. Ambler, H. G. Barrow, C. M. Brown, R. M. Burstall and R. J. Popplestone ,A Versatile System for Computer-Controlled Assembly, *AI 6*, pp. 129-156, 1975.

[110] A. C. Kack, K. L. Boyer, C. H. Chen, R. J. Safranec and H. S. Yang , A Knowledge-Based Robotic Assembly Cell, *IEEE Expert, pp.* 64-83, 1986.

[111] K. H. Chang and W. G. Wee , A Knowledge-Based Planning System for a Mechanical Assembly Using Robots, *IEEE Expert, pp.* 18-30, *Spring*, 1988.

[112] H. B. Farber and E. L. Fisher , MATHES : Material Handling Equipment Selection Expert System, NCSU-IE, *Tech. Report 85-16* , North Carolina State Univ., Raleigh, NC, 1985.

[113] C. J. Malmborg, M. H. Agee, G. R. Simons and V. J. Choudhry , Selection of Material Handling Equipment Alternatives for CIM Systems Using AI, *Ind. Eng., Vol.* 19 (May), pp. 58-64, 1987.

[114] J. M. McGlennon, G. Cassidy and J. Browne , ROBOSPEC : A Prototype Expert System for Robot Selection, In : *Artificial Intelligence : Computer Integrated Manufacturing (A. Kusiak, Ed.), IFS* , Kempston, Bedford, U.K., 1987.

[115] S. G. Tzafestas and G. Tsihrintzis , ROBBAS : An Expert System for Choise of Robots, In : *Managerial Decision Support Systems and Knowledge-Based Systems (M. Singh and D. Salassa, Ed.), Elsevier/North-Holland* , Amsterdam, 1988.

[116] E. L. Fisher and O.Z. Maimon , Integer and Rule Programming Models for Specification and Selection of Robots, In : *Artificial Intelligence : Computer Integrated Manufacturing (A. Kusiak, Ed.), IFS*, Kempston, Bedford, U.K., 1987.

[117] T. J. Doll , An Expert System for Selecting Sensors and Grippers for Robot Applications, *Proc. 12th IMACS World Congress* , Paris, pp. 412-414, 1988.

[118] G. A. Fleisher , Economic Justification of Automation, *Proc. 24th IEEE Conf. on Decision and Control* , Ft. Lauderdale, USA, pp. 1978-1983, 1985.

[119] Industrial Robots : A Summary and Forecast, Naperville, *IL : Tech. Tran. Corporation,* 1983.

[120] G. A. Fleisher , A Generalized Methodology for Asserting the Economic Consequence of Acquiring Robots for Repetitive Operations, *Proc. AIIE Conference*, pp. 130-139, 1982.

[121] F. Leimkuhler , Economic Analysis of Computer Integrated Manufacturing Systems, In : *U. Rembold and R. Dillmann, Computer-Aided Design and Manufacturing : Methods and Tools , Springer* , Berlin, pp. 401-444, 1986.

[122] R. E. Gustavson , Engineering Economics Applied to Investments in Automation, *Proc. 2nd Intl. Conf. on Assembly Autom.*, Brighton, U.K., 1981.

[123] Z. J. Chzajkiewicz , Justification of the Robots Applications, Flexible Manufacturing Systems : Methods and Studies, *North-Holland/Elsevier* , Amsterdam, 1986.

[124] K. J. Meyer , A Cookbook Approach to Robotics and Automation Justification, *Proc. Robot 4th Conf.*, pp. 21-50, 1982.

[125] T. Jordanides and B. Torby , Expert Systems and Robotics, Springer , Berlin/New York, 1991.

[126] S. G. Tzafestas , Engineering Systems with Intelligence : Concepts, Tools and Applications, *Kluwer* , Dordrecht/Boston, 1991.

[127] S. G. Tzafestas , Expert Systems in Engineering Applications, *Springer* ,Berlin\N.Y. , 1993.

[128] S. G. Tzafestas and H. B. Verbruggen , Artificial Intelligence in Industrial Decision Making, Control and Automation , *Kluwer* , Dordrecht/Boston, 1995.

[129] C. Gamman and D. A. Nadler, Fit Control Systems to your Managerial Style, *Harvard Business Review*, pp. 65-72, Jan.-Feb., 1976.

[130] S. G. Tzafestas and C. Athanassiou, A New Class of Petri Nets for Fast Robot Cell Prototyping, *CC-AI : Communication and Cognition-Artificial Intelligence*, Vol. 12 (3), pp. 225-252, 1995.

2 Decision Support Systems in Manufacturing Systems Management

O. Moravčik and M. Mišút

1 Introduction

Competition in the international market place has meant that industrial production must become more structured and more efficient. In particular, the organisation of the production process has conflicting goals: production costs have to be kept as low as possible while specific customer demands should be satisfied to a maximum degree. Consequently, modern production management aims to satisfy high standards of product quality as well as meeting delivery dates under economic conditions requiring progressively smaller lot sizes. The continually changing objectives place increasing emphasis on the clear formulation and use of decision techniques supported by computers.

Computers have been used for design, manufacturing, control, diagnosis and monitoring in manufacturing industries. This allowed us to implement more advanced manufacturing ideas. The technological advances in manufacturing have also addressed the research interests of intelligence systems for planning, operation, control and diagnosis of manufacturing.

In the world of manufacturing there is a number of processes in which a decision has to be made. First, operation research (OR) techniques have been widely used to support decision making in industrial production. However, difficulties have been encountered with the formulation of models, management of data, and interpretation of results. Since many problems of manufacturing systems are usually unstructured or ill-structured problems that deal with non-numerical or non-algorithmic information, new methods had to be searched for. For solving this type of problem, artificial intelligence (AI) methods seem to be applicable. Expert systems (ES), which are one of the most attractive branches of AI, can provide a programming methodology for solving ill-structured problems.

Benefits from advances in computer science along with developments in new decision methods and techniques have been extensively involved in decision support systems (DSS) whose usage is limited not only to purely algorithmic solutions but also to unstructured or ill-structured problems.

The dramatic development of information technology, which has been progressing for the last 15 years, has changed the nature and cost of the decision making process. The spread of computer-tools across the entire office landscape and the use of computers in production has created good conditions for extensive exploitation of management decision tools. Decisions are now based on the models incorporated in DSS. These models can be built not only upon OR but also on heuristics or on expertise knowledge. Decisions can then be made by the selection of the best alternative from the alternatives provided by DSS.

The paper is divided into 3 main parts. First general basic information about DSS along with an explanation of the knowledge structure and the knowledge representation for intelligent DSS in section 2. The second part consists of sections 3 and 4. Section 3 gives a brief overview of the problem areas, focusing on the manufacturing, in which DSS can be implemented. In the next section the problem solving strategies and techniques suitable for implementation in DSS devoted for use in manufacturing are discussed. In the third part two examples are presented in the form of a discussion of two DSSs that have been developed to illustrate the topics under discussion. The first example was chosen from a very popular but hard the problem area - operation scheduling, the second one presents DSS for process planning.

2 DSS in General

2.1 DSS Background

The field of Decision Support Systems has already been intensively discussed for over twenty years. The first decade concentrated on the basic, "life important" issues of DSS. In 1976, Decision Support represented a radical new idea in the use of information systems and analytic tools. Now, twenty years later, it is not possible to imagine management and production without DSSs.

The basic idea introduced in 1976 was based on a combination of human judgement and the power of computer technology so it could improve the effectiveness of decision makers, without intruding on their autonomy. This was a new way of applying the analytical methods of management science instead of through the normative approach of optimisation science. Optimisation techniques ignore he needs, habits, experiences and preferences of the decision maker and try to replace judgement not to respect it.

Since the start of the DSS movement, there has been no established definition of DSS. In the literature a number of different definitions are given. Some of them emphasise the Decision Support aspect others stress System characteristics. The definition introduced by Keen [1] best fits the purpose of this work and will be used as a reference. Moreover, it is supposed that DSS involves the formulation of alternatives, the analysis of their impacts, and interpretation and selection of appropriate options for implementation.

Next Keen's definition of DSS is given:

DSS is the application of available and suitable computer-based technology to help improve the effectiveness of managerial decision making in semi-structured tasks.

As Stabell [2] said:

> ...DSSs are thus not a particular technology in a restricted sense, but primarily a perspective on management and the role of computer-based systems as a managerial tool...

Stabell [2] also introduced and analysed four relatively unique and distinct DSS "schools" whose draft description is as follows:

- *decision analysis* - the oldest and the best established DSS-school. Decision analysis was initially presented as a general methodology, without any reference to the use of computer-based systems. Decision analysis attacks complex problems by reducing them in to smaller, manageable components and focuses on the choice phase of the decision process. The computer-based system is referred to most often as a decision aid.
- *decision calculus* is a model-based set of procedures for processing both data and judgement. The model is the organising element and is designed to support the manager's use of judgement and experience in the decision process. Decision calculus focuses on the decision situation. The goal is better decisions through better models of the decision situation.
- *decision research* - considers primarily how to develop a DSS. Decision research views DSS development as an effort to improve the manager's decision making process- increase the effectiveness of how decision are made. Decision research focuses on the decision process and the decision maker. Computer-based support is seen as only one of several possible means to improve decision making.
- *implementation process* - focuses on the implementation phase of the DSS development process. The implementation process is development process orientated which is viewed from a builder perspective. This approach is particularly concerned with the builder's role in the development process.

In accordance with McLean [3] the development of DSS can be viewed as a process with the following stages:

In the early 1970s, DSS were characterised simply as computer-based systems to aid decision making.

In the middle to late 1970s, the growing DSS movement emphasised interactive computer-based systems which help decision makers utilise data bases and models to solve ill-structured problems. The emphasis was concentrated on the support for personal computing rather than on the decision process.

In the late 1970s to early 1980s, a strong emphasis was put on utilising operations research methods to improve the effectiveness of the decision process.

From the mid 1980s artificial intelligence methods in the form of expert systems have been accepted as a suitable technology to improve creativity in decisions.

Although there are some technical blockages, such as the lack of data base management software and hardware that can handle high-volumes of transactions cost-effectively, and problems of incompatibility, standards and integration technology there is no longer a bottleneck in development [1].

The expert systems are an emerging technology that DSS developers recognise as a key challenge in building DSS. There are several practical ways in which the DSS field can exploit the opportunity the expert system technology offers. The first is to use the new power tools to build standard DSS, the next is to utilise the experience of knowledge engineering methods in DSS design.

2.2 Decision Support System Structure

DSS can support humans in a variety of forms, from the simple provision of information, to more sophisticated systems that analyse data and actively help decision-making. For example optimisation DSS uses linear programming and another OR techniques with historical data. On the other hand, model-based DSS can contain a model of a part of business such as a manufacturing process, inventory etc. In more advanced concepts of DSS expert system technology is merged with conventional techniques (this type is called intelligent DSS). In these systems a set of rules is combined with the more usual alghoritmic- or model based systems. Another growing area among others is the application of simulation in DSSs. The main function of simulation interfaces is to allow interactive modelling and decision-making.

It is obvious that a particular DSS has special characteristics but there are some common components the DSS needs and these can be found in all kinds of DSS.

An architecture of typical DSS is composed of three main components:

- the dialogue system;
- the model and knowledge base;
- the data component.

A fundamental work and a useful framework for other researchers in this area was done by Sprague. Sprague [4] identified the main components of a DSS as a user-interface software, a data base, and a model base. However, this idea could not be easily adapted onto intelligent DSSs, thus Bonczek *et al.* [5] introduced the concepts of a language system, a knowledge system, and a problem processing system. This framework enables relatively easy utilisation of artificial intelligence methods in DSS.

DSSs used in manufacturing may have a different structure because of the particular purpose of this DSS and conditions in which this DSS is to be used but in all cases it is possible to identify the main parts as they were described in [4] or [5]. The speciality of manufacturing is, on the one hand, its dynamic nature and occasional unpredictable behaviour with frequent disruptions, which require approximate and heuristics control and decision procedures, and on the other hand, the possibility of using automatic devices and/or automated complexes, which allow the use mathematical models and optimisation procedures. These conditions impose greater requirements onto DSS speed. Models incorporated into DSS have to be able to cope with all the kinds of problems arising from the conditions in manufacturing. Since knowledge-based methods seem to be suitable for a number of ill-structured problems, intelligent DSS also become popular in the area of manufacturing. Expert systems can be useful in solving decision-making problems, where the decision-maker is treated as an expert on his own problem. The software tools should be interactive and easy to use, so that experts can express and acquire knowledge relevant to the decision with minimum supervision. However, the quality of the solutions they propose is highly dependent on the knowledge they posses. Knowledge based approaches to decision support have concentrated on capturing knowledge, knowing what to represent in a knowledge-base, and how to access it. The reduction of a large body of knowledge to a precise set of facts and rules, has already become a major bottleneck impeding the application of expert systems into new domains.

2.3 Knowledge Acquisition, Structures, and Representation for Decision Support

Intelligent DSSs, which enable us to implant knowledge into DSS, have one or several of the following characteristics [6]:

- coupling "deep knowledge" provided by casual models with "shallow knowledge" corresponding to the experience of experts;
- integration of data bases with models and AI tools to provide normative assistance to the decision maker;
- developing an intelligent user interface;
- providing methodological assistance such as decision analysis.

Knowledge acquisition is a bottleneck in the development of an intelligent DSS as it is time consuming and difficult. That is not only the problem of DSS design but a general problem which has to be tackled when knowledge-based systems are intended for use. The art and in the same time science of gaining knowledge is called knowledge engineering. The review by Hart [7] and Burton's work [8] give an overview of the knowledge engineering field. As follows from the literature it is hard to say what type of knowledge elicitation technique is the most suitable for a given problem domain. All the techniques which follow may be useful. Which technique can be used depends on the particular situation. The following methods are commonly used:

- interviewing of domain experts (structured interviews);
- Delphi technique;
- literature search;
- expert conversations;
- protocol analysis;
- scenarios;
- category sorting;
- simulation.

After this knowledge has been acquired it has to be represented in a knowledge-base. Knowledge representation research is concerned with identification, representation, and utilisation of knowledge in problem-solving. Knowledge representation schemes can be classified, according the Mylopoulos and Levesque (cited in [9]), into four categories: logic, network, procedural, and frame-based schemes. Each representation scheme has its advantages and drawbacks, therefore some authors proposed hybrid schemes, e.g. Kusiak and Wang in [9].

Knowledge representation can be also classified as having declarative or procedural characteristic. On declarative knowledge representations most of the knowledge is represented as a static collection of facts. General procedures are applied to domain specific knowledge to make deductions or interfaces. Procedural knowledge representation enable us to easily encode heuristic knowledge. A piece of knowledge in this type of system specifies how, and under what conditions, it is to be used. Much of the knowledge in expert systems is heuristic in nature - it consists of rules of thumb that are often, but not always true.

Several representation schemes and data structures have been proposed for heuristic knowledge e.g.:

- production rules and predicate calculus;
- frames;
- logic representation;
- semantic networks.

Production rules and frames are by far the most frequently used. Production rules have the form of the condition-action pair (IF - THEN). It is important to note that rules may form part of the knowledge base and part of the inference control mechanism, as well. Often rules for control test a particular context which may arise during operation. The action part then invokes a particular strategy for reasoning. Rules for general control of reasoning are often referred to as heuristics.

A frame is a part of knowledge which consists of a collection of slots to describe an object (Minsky's definition). Frames can be organised in hierarchies, using pointers to point to the next or previous level. Thus frames are suitable for representing categories of knowledge.

Procedural representation techniques are obviously already used when formalisation and specialisation of work processes become important.

To achieve the balance, the information can be abstracted into more than one level of detail. Thus hierarchical knowledge representation schemes arise. In those schemes the abstract information is used for ease of manipulation and the necessary details are used for the prevention of an unacceptable level of accuracy loss in representing problem.

3 Problem Domains

DSS originated to support managers in their decisions concerning more strategic and tactical than operational decision problems. As the information technology and other relevant methods (e.g. AI methods) were being developed, the chance of the use of DSS also in operation decision and industrial decision problems grew. The character of the DSS is determined by the type of decision for which DSS gives support.

According to Anthony [10] decisions can be classified into four types of decision:

- Strategic Planning Decisions : decisions related to choosing the highest-level policies and objectives, and associated resource allocations.
- Management Control Decisions: decisions made for the purpose of assuring effectiveness in the acquisition and use of resources.
- Operational Control Decisions: decisions made for the purpose of assuring effectiveness in the performance of operations.
- Operational Performance Decisions: day-to-day decisions made while performing operations.

Operational performance decisions in which expert skills, rules (heuristics), or formal reasoning are used for judgement, are more likely than strategic planning decisions to be prestructured. Thus, expert systems can usually be expected to be more appropriate for operational performance and operational control. Anthony's classification can be naturally extended to include problems of industrial and process control. Industrial problems can be defined as those where decisions are made and control actions are taken regularly for continued operation of a plant. A large number of operations control and industrial control problems are in the nature of decisions problems, which can be structured into specific procedural instructions, performed routinely or repetitively. However, not all operation and industrial control problems are those, e.g. operation scheduling requires deep analysis, judgmental input as well as a considerable amount of mental abstraction and reasoning skill.

There are many application areas in which the use of a DSS is promising. These include management and planning, command and control, system design, operations management, and in other areas in which a decision has to be made and the problem is ill-structured. It is obvious that decisions in manufacturing system management belong to operational performance decisions and operational control decisions. DSS in manufacturing system management is used in those domains, in which human judgement is still important, e.g. scheduling, system failure and diagnosis, allocation problems, and quality control. The higher the degree of automation the higher the degree of causality and the more the control is deterministic. Therefore DSS can be thought of as an intermediary stage in manufacturing control, which will exist until, the human will be needed to make a decision. In full automatic control DSS will be replaced by an automatic control system.

4 Problem Solving Strategy

Because of the different nature of the problems at each level as defined by Anthony, different types of solution may be required. Decision analysis is well applied to the strategic and tactical planning problems [11]. Mathematical programming is most successful in its application to tactical planning problems [12]. It can also be applied to a lesser degree to operation control and industrial control problems. Expert systems are claimed to have applications across a whole range of problems and the most successful applications are reported in operations control and industrial control [13].

Madey *et al.* [14] developed a classification model for intelligent hybrid systems. Problem-solving taxonomy specifies four types of problem-solving processing:

- for conventional problem-solving technologies:
  - information processing
  - numeric processing
- for intelligent problem-solving technologies:
  - symbolic processing
  - sub-symbolic processing

Most real world manufacturing problems are complex and ill-structured. So they cannot be easily and effectively solved by the majority of single-technique-tools. One of the possibilities to deal with real manufacturing problems is to integrate the use of two or more techniques.. Such an approach is based on the assumption that it is possible to merge the strengths of different techniques to produce a system that effectively solves manufacturing problems. The most popular approach is the combination of AI methods with convential methods in intelligent DSS. This provides a good possibility to cope with decision situations in which expertise as well as normative methods are needed. Expertise can be used for decision model selection and final decision making, while conventional methods can be used in decision models for preparing decision alternatives.

The techniques currently used to support decision making in manufacturing as were developed by the industrial engineering, operational research, and engineering management disciplines are based primarily on numeric and information processing technologies:

- mathematical programming;
- decision analysis;
- optimisation;
- stochastic modelling;
- queuing analysis;
- statistical analysis;
- heuristics;
- simulation.

These techniques cover the whole spectrum of techniques from sophisticated and mathematics based to heuristic and experience based ones. Recently also simulation and simulation like techniques have became popular for supporting decisions. Advances in information technology enable us to implement not only mathematical but also heuristic models and to simulate them in reasonable time so results can be used as a base for decisions. The design of simulation models can be speeded up by using of specific knowledge often stored in knowledge base of intelligent DSS.

5 DSS for Operation Scheduling and Dispatching

5.1 Setting up the Problem

As was mentioned in section three, operation scheduling for general job-shop is one of the areas, in which DSS can be used. This is due to the nature of operation scheduling. Operation scheduling as a traditional problem of mechanical engineering production can be viewed as a form of planning where the considerations of constraints, resources and time assume a particular significance. The main problem in scheduling is to timetable the processing of jobs by machines so that all the constraints are satisfied and some measure of performance achieves its optimal value. The processing of a job on a machine is called an operation. Determining a suitable solution procedure for the general scheduling problem is a tremendous task because of NP-complexity, thus the application of various pure algorithmic solutions seems to be improper, particularly in job-shop type operation scheduling. Therefore many researchers strive to find approximations or another method to reach feasible solutions, e.g. heuristics and heuristic-based algorithms, simulation techniques, knowledge-based methods, etc.

Operation scheduling involves operational performance decisions made on a day-to-day basis as was defined in the previous section. In the process of scheduling, the scheduler has to decide which strategy will be applied into scheduling to ensure fluent and effective production. In the scheduling strategy the domain-specific knowledge is contained in scheduling rules or heuristics. Because different strategies may have contradictory impacts on the production and often several different objectives are followed, decision making is highly dependent on the scheduler's skills. Heuristic knowledge can also be used by the scheduling system, if it exists.

Originally the operation research (OR) techniques were intended for use in supporting decision making in production scheduling. However, the difficulty is expressing all scheduling constraints, and the inherent complexity and the uncertainties present in most scheduling applications prevent the use of these models, which are seldom applied in industrial practice.

Simulation and simulation-like techniques, as another usable technique in scheduling, can represent quite realistic systems at modest computational costs and have the advantage of providing a more natural approach for interfacing with human expertise. However, the disadvantage is that it is not clear how good the obtained solutions are.

It has already been shown that knowledge-based technology, mainly expert systems (ES), would be appropriate for complex scheduling problems. The ES methodology looks attractive because it may help general purpose knowledge to co-operate with specific information provided by shop-floor managers. Dorn [15] lists three reasons why knowledge-based systems are preferred:

- First, in KB systems it is easier to apply heuristics to reduce the inherent complexity in scheduling.
- Second, KB systems offer the possibility to reason with incomplete, uncertain or inexact knowledge.
- Finally, in KB systems the existing knowledge is described explicitly and should therefore be more easily maintainable.

Simple heuristics and/or artificial intelligence (AI)-based approaches are perceived to be more capable of dealing with the precedence and capacity conflicts that result from dynamic changes in the system because it can incorporate pragmatic knowledge into the scheduling process. Artificial intelligence (AI) approaches involve heuristic programming and can handle dynamic situations. However, a completely knowledge-based solution tends to be ineffective when problems become large and complex and that is true for most real environments. Therefore a hybrid approach that combines the advantages of both methods, the generation of conventional schedule (OR and simulation techniques) with knowledge-based control, seems to be the most promising basis for providing a complete, effective and practical solution.

5.2 Solution: a DSS

There exist a number of attempts, described in the literature, to design an autonomous scheduling system, independent of humans which is able to schedule operations in automatic mode. However, operation scheduling for job-shops is a NP-hard problem, thus only the exploitation of heuristics and knowledge-based methods have resulted in usable systems for job-shop scheduling. Because of the dynamic nature of production, full automatic systems have problems to cope with it, thus they need human co-operation. Both facts, the human aids as well as the usage of appropriate scheduling methods are important for designing a proper job-shop type scheduling system.

With respect to these facts the Integrated Scheduling System (ISS), as a decision support system for operation scheduling that exploits artificial intelligence as an expert system, has been proposed.

5.3 Parts of ISS

The ISS was proposed as an interactive problem-solving and decision-support tool for managers of job-shops and can be used as a decision support system (DSS) for operation scheduling. The ISS is a tandem system that consists of two main subsystems: RAPO and ESPRA and combines the simulation approach (subsystem RAPO) with the knowledge-based system approach (subsystem ESPRA). The RAPO subsystem is responsible for schedule generation and is based on deterministic discrete-event simulation. A new algorithm, which speeds up schedule preparation, was developed and built into RAPO. The ESPRA subsystem is an expert system that supports selecting the most promising scheduling strategies for further use in RAPO. The inference engine of ESPRA is based on the method that optimises the solution's process for any rule-based backward-chaining diagnostic system. To facilitate knowledge acquisition, the GEPRO system was developed.

AI can be exploited in the scheduling in several ways. If the published results of KB-scheduling systems are generalised, then it can be shown that in almost all systems for scheduling AI is used in two manners as a *generative* or as an *evaluative* principle.

Artificial intelligence is used in the generative systems for creating a schedule instead of OR algorithms. This is possible to accomplish in two ways: first searching in the state-space or secondly constraint-based searching. In the literature there is a very small number of references to successful implementation of these type systems in the operation scheduling. But their application in higher-levels (e.g. master scheduling) provide the most promising results. Most of the generative systems are based on planning ES (searching for goal), which use forward-chaining reasoning.

Evaluative systems can exploit the principles of AI for:

- evaluation of schedule-variants and selection of the best one
- selection of scheduling model, method or rules
- rescheduling, maintaining or improving prepared schedules.

In evaluative systems the conventional algorithms are used for preparing schedules, e.g. OR methods, heuristics or simulation, etc. On the contrary, artificial intelligence is used in the evaluative systems to support classical methods to improve the solutions they are able to derive. It means, that evaluative systems are systems coupling both methods - AI and conventional methods. Most evaluative systems are based on the use of diagnostic ES (searching for a matching pattern), which uses mostly backward-chaining reasoning. This approach seems to be more promising since the results that are reported in the literature support this contention.

5.4 Procedure of ISS Scheduling

The need for a job to be rescheduled emerges when the availability of workers, tools and machines changes unpredictably over time. As a result of this a deviation between schedule and reality can come into existence. If the deviation between reality and the schedule is greater than the permitted difference, jobs must be rescheduled.

The procedure of scheduling starts with the choice of scheduling strategy(ies), which will be used further in RAPO's scheduling rounds. A scheduling strategy (as understood in ISS) consists of a selected scheduling rule and a number of parameters, which have influence on the scheduling process, (for example: usage of alternative machines, the time window for simulation, etc.) The choice of scheduling strategy depends on the objectives, which the production process control follows, as well as on actual state the controlled production system has. There is no definite dependency between scheduling strategies and their impacts on scheduling measures. The impact of the same strategy can be different for various production systems and even different for the same production system depending on the state of the production system. Therefore specific knowledge is needed about which scheduling strategy can be used for that particular production system and when. This kind of knowledge constitutes a part of the ESPRA knowledge-base.

The scheduler starts his/her work with ISS by consulting ESPRA about the most appropriate scheduling strategies. Depending on his/her answers ESPRA recommends a sorted list of scheduling strategies. These scheduling strategies are ranged according to their convenience of application in a given situation. A sequence of scheduling strategies is made on the basis of general as well as specific and domain-specific knowledge application. Once the scheduling strategies are selected, RAPO's schedule generation for each selected strategy can start. Then each schedule variant is appreciated and the scheduler can decide if one of them is applicable or the schedule generation with changed parameters will be repeated. This procedure can be repeated until a satisfactory schedule is reached, which can then be applied to production. These tactics come into force, when some disruptions in the production process causing disproportion between schedule and reality occur or at the beginning of each scheduling period.

5.5 Features of RAPO

The RAPO system, which allows the scheduling of job-shop type production systems, is based on deterministic discrete-event simulation and uses a new effective algorithm described, for example in [16]. The basis of the explored algorithm comes from the resource-based decomposition principle. Hence operations are selected from queues either according to scheduling rules or according to human decisions. By this method a real production process is simulated in advance with a recording of all the events in the system. The timetable is then one selected variant of the recorded events′ list.

To obtain a "good" solution a multi-pass scheduling method [17] was chosen as a basic principle of schedule improvement in ISS. As a production process is constantly disturbed by a number of exogenous and endogenous influences, consequently its time-trajectory starts to differ from the planned (required) trajectory. Multi-pass scheduling methods strive to minimise problems, which result from the dynamic and stochastic nature of the production process. On the basis of actual information, schedules are generated. Simulation yields the possibility of short-time prediction of future states and is used to select one schedule from a range of alternatives. The accomplishment of all the schedule alternatives allows us to perform an analysis of future scenarios, so defects in the schedule can be detected before schedule execution. If the time-period for prediction is short enough, it could be assumed that the production process is deterministic and static. From this supposition, a discrete-event simulation can be used for prediction.

The RAPO system operates interactively as a human support tool. A human has several degrees of autonomy of decision when operating RAPO. First, when all the decisions are human-made and RAPO serves only as a simulator on which the human can emulate his/her decisions on the production process with an evaluation of scheduling measures at the end, the human takes full responsibility of the scheduling. On the other hand, the human can grant some kinds of decision to RAPO. Then particular decisions concerning operation dispatching are made by RAPO which operates in automatic mode and finishes by preparing all the scheduling variants for human appreciation. All the decisions RAPO makes are made according to the selected scheduling strategy. The human operator is in both cases responsible for scheduling strategy selection and for choosing the best variant of schedule by apprcciating the results at the end of the RAPO run.

The steps in which the RAPO system performs the scheduling rounds can be understood from figure 1.

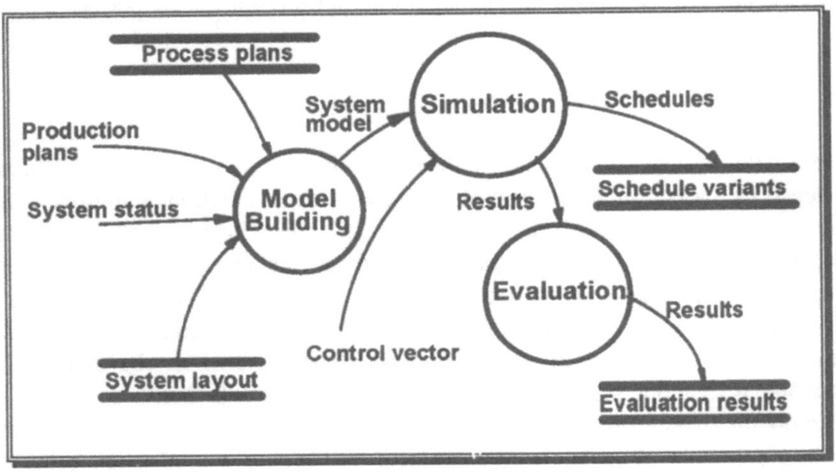

Fig. 1. The RAPO system data flows

In one simulation round it is possible to prepare a number of schedule variants according to chosen scheduling strategies. At the end of each simulation run some parameters of the generated schedule are evaluated. These parameters are related mainly to flow time criteria, due-date criteria, resource utilisation criteria and inventory criteria. From the value of these criteria, humans can propose either one schedule variant to realise or to repeat the simulation round with other scheduling strategies. Once suitable results have been obtained, the best scheduling strategy can be employed in manufacturing control. Generated schedules serve as a basis for the planning of set-up activities, transport, material and people' management.

The RAPO system employs such features as priority management (in which three types of priority are individually set-able for each order), crisis management (only important rush orders are scheduled, the rest can be manufactured during idle times), maintenance management and peoples' exceptions. For each machine, and for each time-interval it is possible to individually set the productivity parameter depending on the actual operator, to set transport times and other properties.

The advantage of the RAPO system is its ability to simulate a great number of process types because of its generative model-building capability. This means that the model of production system is build dynamically and it depends only on the data in the data-base it has. If any disruption of the process occurs, it is incorporated into the model without any difficulties.

RAPO has tools for updating the data-base, creating scheduling strategies and priority management. Input and output interfaces are also involved to communicate with the MRP II system and with an overall control system if automated control is used.

5.6 Features of ESPRA

ESPRA is a rule-based backward-chaining diagnostic expert system (ES) for the proposition of the most promising scheduling strategies for production systems, depending on particular situations. Subject to the state of the production system and to current goals, the ESPRA system suggests a set of scheduling strategies. The proposal is derived from the essence of the stored knowledge in the knowledge-base.

There are several tasks the operator of the RAPO system has to perform. He is essentially responsible for:

- the recognition of the needs of performing a scheduling round;
- the setting of external job priorities;
- the appropriate scheduling strategy selection;
- the selection of the most appropriate schedule and scheduling strategy for implementation in production;
- the determining of the point from which the new scheduling strategy will be implemented.

Because of the complexity of these duties, it is not very easy to replace a human with ES. Thus it has been decided to substitute human decisions with ES in several steps. In each step ES will cover more and more of these functions as listed above. If the ES shoulders only part of the operator's duties and acts as a decision support tool for the scheduler it is referred to as ES-Aided Scheduling. In contrast, full replacement of RAPO's operator can be referred to as ES-Guided Scheduling.

In ES-aided scheduling the ES serves as an aid for the human, who does not need to be as skilled as before. The first task that ES (ESPRA) is responsible for is the selection of the most promising scheduling strategies. As a result of ES-aided scheduling, less experienced operators can schedule with results as good as those of skilled ones.

In the ES-guided Scheduling the main tasks of ES will be:

- recognition of when it is desirable to generate a new schedule;

- set the scheduling strategy on the basis of the current-state-of the system, and goals of the system;
- evaluate and select the most suitable schedule.

ES-guided scheduling can be seen as a further goal of ISS development.

The first problem in ESPRA design, that was solved, was that of the knowledge-base structure. ESPRA employs a production rule scheme. To obtain efficiency and speed, production rules have been grouped into classes and organised in a hierarchical structure [16]. Rather than scanning a large number of unstructured production rules, depending upon the context, the inference engine refers to the required class of rules.

The intelligence of the experts′ decision-making knowledge of the domain is analysed and represented as rules of the following type: - "IF situation THEN lower level rules" and - "IF particular situation THEN scheduling strategy group". This means that the rules are divided into two levels. On the higher level are rules, which (on the basis of some parameters (e.g. set-up costs' attributes, process-plans features, etc.)) refer to a particular group of lower level rules. These rules, depending on system goals and production process parameters, select an appropriate group of scheduling strategies.

It is very easy to adapt a set of particular scheduling strategies to knowledge development, because each scheduling strategy has two parameters, which express its convenience of application in a particular situation:

- the first parameter determine a specific scheduling strategy group for the scheduling strategy in question;
- the second parameter states the place and sequence order of scheduling strategy in the chosen scheduling strategy group.

These two parameters represent the meta-knowledge in the model of expertise.

Up to the present time forty-nine rules have been proposed, implemented and approved [18].

Inference Mechanism. The problem of scheduling strategy proposition is a diagnostic problem. Therefore a backward-reasoning method was considered appropriate for this task. For the implementation of ESPRA two possibilities came into consideration: first - to use a general ES-shell or as a second possibility to develop your own specific ES using programming language. Since one type of difficulties with ES usage in manufacturing result from their slow speed of reasoning, the main emphasis was put on the development of an effective inference mechanism that would ensure sufficient speed of ES response. Therefore instead of using a general ES-shell at the end a second possibility was taken. After research a method was developed, which optimises the number of steps needed to reach the reasoning goal [19,20]. It was proved [19] that in this "optimal" knowledge base, the number of steps to goal assertion is about 50-70% of the steps needed in the knowledge base that is created without respect to this principle and it is valid for backward-chaining inference in the diagnostic systems generally. This method was implemented into ESPRA using the Prolog language.

5.7 Knowledge Management in ISS

Types of Knowledge Employed by ISS. In recent years several results have been published (e.g. [9,15,21]) that refer to the model of expertise. These try to formalise a structure of scheduling knowledge or new knowledge representation schemes.

For example Dorn [15] proposed a model of expertise with four levels that cover strategic knowledge, task knowledge, inference knowledge and domain knowledge. In contrast Sauer [21] divided knowledge into the following areas:

- The domain knowledge contains static information about the application environment;
- The situation knowledge represents the current state of scheduling, e.g. the existing schedule, the remaining capacities of resources, etc.;
- The meta knowledge contains the information necessary to determine the "best" algorithm for the current scheduling problem, i.e. which algorithms are applicable for which tasks (goals) and which ones are appropriate for which events;
- The scheduling knowledge is divided into static knowledge and dynamic knowledge.

A natural distinction exists between static (data) and dynamic (methods) scheduling knowledge. The static knowledge gives all the information about the workshop and the production objectives. The dynamic knowledge describes the expertise available for deriving of feasible or interesting schedules. It may involve three components:

- Theoretical expertise, which is very formalised and is in the hands of OR specialists, deals with the management of time and resources;
- Empirical expertise obtained by simulation of heuristic scheduling rules;
- Practical dedicated expertise provided by shop-floor managers who supervise the production methods, and are aware of the technological or human-originated constraints to be respected by the scheduler. In the case of new and/or very complex manufacturing systems, this type of knowledge can be very poor, and only express constraints on the manufacturing process

To perform scheduling tasks, the proposed system ESPRA needs to have incorporated into the knowledge-base the following kinds of knowledge:

- domain and situation knowledge about structure and the parameters of the production system;
- meta knowledge;

– empirical and practical expertise such as general heuristics, specific heuristics and domain specific knowledge (i.e. actual system load, bottlenecks, resource limits, allowed deviations between planned schedule and reality, etc.)

The heuristics that are generally valid were acquired from an exhaustive review of the published research results. They are related to usage ability, applicability and relative effectiveness of the scheduling rules.

The specific heuristics can be acquired by the simulation using the case-based learning method [22,23]. A tool for this method called GEPRO is described in the next section.

In the process of knowledge-base implementation the first general valid heuristics were implemented as a basis. This base can be extended and specialised for a particular production system with specific and domain specific knowledge, which will be acquired following ISS or GEPRO application.

The dynamic nature of the information process when a scheduling task is performed can be expressed by data and knowledge flows. Main flows can be seen in the figure 2.

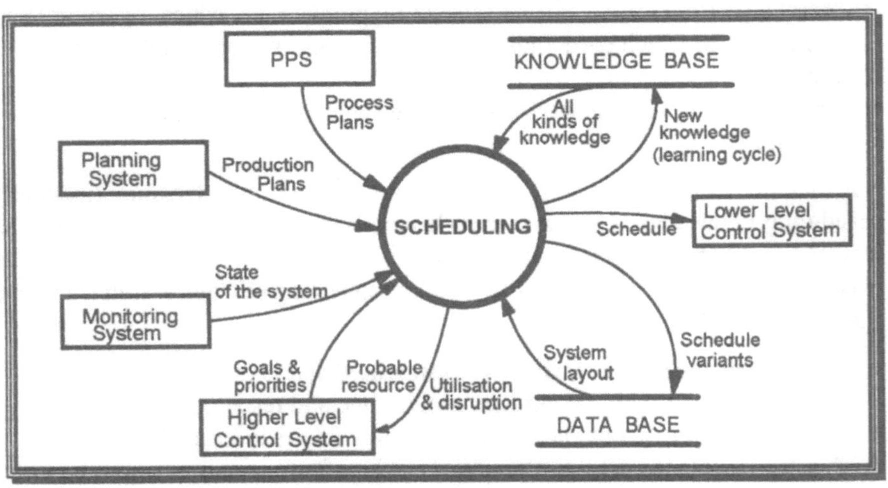

Fig. 2. Data and knowledge flows in scheduling

Knowledge Acquisition. 'Knowledge is power' and this is particularly true when applied to manufacturing. Manufacturing of any kind is a highly complex sequence of interactions between a number of requirements, resources and controls. Therefore one of the most difficult problems in the AI-based scheduling systems is knowledge acquisition, which hinders the use of AI in scheduling. This can cause (knowledge acquisition) the bottleneck of ES development [24].

Thus having proposed the ESPRA system it was necessary to build all kinds of necessary knowledge into the ESPRA knowledge-base. The quality of knowledge built in to ESPRA strongly influences schedule-quality. Since general heuristics were taken from results of literature review, it was necessary also to acquire specific heuristics that have to contribute to the improvement of schedule quality. The more specific knowledge about scheduling strategies of a given production system is implemented into the knowledge-base, the more precise and superior schedules are generated by ISS. To give the user a possibility of gaining specific heuristics, the idea of a case-based learning system was proposed.

One of the advantages of such a case-based system is the possibility to investigate scheduling strategies not only for the existing system but also for the systems being projected [23]. In advance it is possible to prepare guidelines on how to solve specific situations which can occur. The case based learning system can be used in two manners:

- for specific production systems to develop specific heuristics and to acquire domain related knowledge;
- as an experimental tool for investigating behaviour of general heuristics.

Since RAPO is able to simulate any problems that can occur in production it is possible to use it as a simulation tool. The only problem was to build an appropriate and consistent data-base for each case being solved. This was due to the industrial purpose of RAPO usage. The data-base structure has been designed regarding this industrial environment therefore it was necessary to develop a system capable of case-data preparation. This system has to replace input to the RAPO system from an MRP II system (planned task) and feed-back from the process (state of the system resources, tasks and commitments).

The GEPRO system was developed to fulfil the above stated requirements [22]. It works as data generator preparing case-data according to chosen distribution functions and parameters. These parameters serve as input to the GEPRO system and describe the desirable characteristic of the case (type of production system, utilisation level, kind of products, production rates, process plans, orders' parameters etc.). The level of work-in-process, maintenance calendar, operator exceptions are considered also. Having inputted these parameters, using statistical distributions the physical layout of the system can be generated (this can be inputted directly as an exception to describe the concrete system under consideration), along with product descriptions and the process plan for each product. Then the production tasks are generated which are balanced to the desirable level entered earlier. As a result of the GEPRO run, data is generated, which in full describe the system layout, current system status and tasks to be processed in the system together with tasks in-progress. Having prepared the case data-base, the simulation scheduling rounds can be performed. The principle of the knowledge acquisition cycle is illustrated on figure 3.

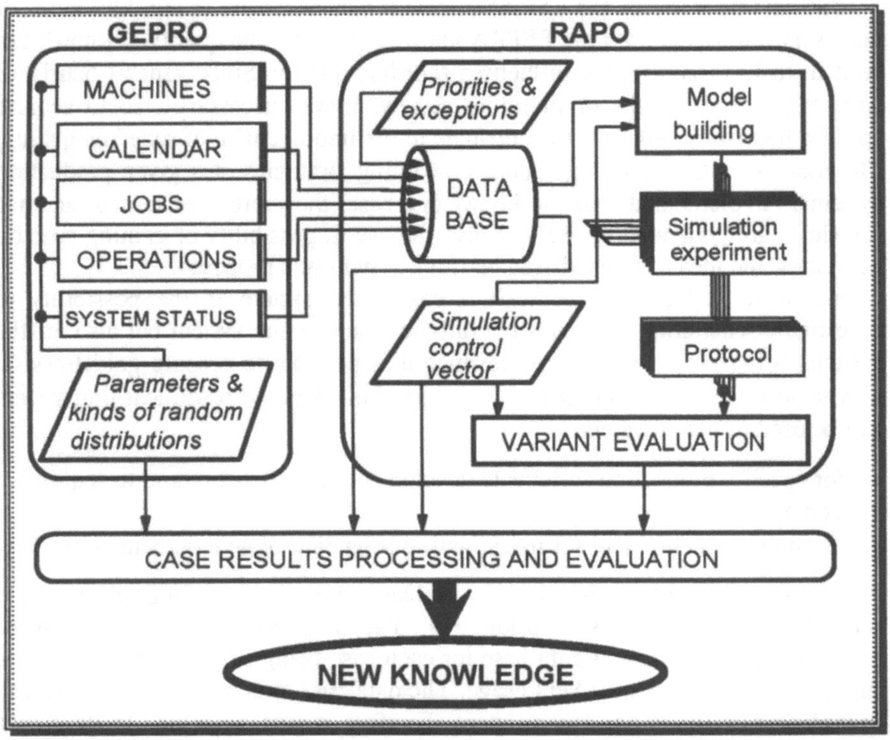

Fig. 3. Knowledge acquisition

After the initial phase (knowledge-base structure and consistency, inference engine and elementary set of production rules) work is now progressing on the development of new scheduling strategies and the extension of the knowledge-base.

5.8 Experimental Results and Implementation

Development of ISS was conceived due to the ZŤS Petržalka plant. ISS forms only one part of the complex information and control system for production management of ZŤS Petržalka. All the ISS was implemented on a PC platform as the first step necessary for verifying the principles used. After realisation, on the validation of the ISS for a real environment a number of schedulers and managers of job-shops from ZŤS Petržalka and ZŤS Dubnica have taken part. Theirs suggestions have been taken into account for further development on the

UNIX platform. People who have tested ISS reported that a primary advantage of ISS is its speed and improved schedule quality

6 DSS for Process Planning

6.1 What is Process Planning?

Process planning is the act of preparing detailed operation instructions to transform an engineering design to a final part [25]. The detailed plan contains the route, processes, process parameters, machines, and tools required for production. Traditionally the process planning activity has been experience based and performed manually. Manual process planning has many problems though it is well suited for small firms with a few parts. In order to implement an optimal process plan, the process planner must consider many factors which are involved in manufacturing. Thus most industries have several different process plans for the same part, which leads to inconsistent plans and additional paperwork. In recent decades, there has been a trend to automate process planning, since it increases production efficiency and parts can be produced more economically. To avoid problems of manual process planning, a computer-aided approach is employed. In spite of the fact that tremendous efforts have been made in the development of systems for process planning, the process planning activity has still not been completely automated. Instead of it, the computers can serve as a tool, that help process planners in decisions. DSS helps planners to gain quick and flexible access to data and models, which are potentially relevant to a particular decision. Planners have to only decide which data and models are to be used, when and in which order. Also the final decision has to be made by them, who take into account partial solutions obtained by the selected activation of several models.

6.2 Process Planning Procedure

The task of a process plan design involves several steps. The first consideration is the interpretation of the design data. The second step of process planning is to design stock. Although the properties of the raw materials are specified by the design engineer, the geometrical shapes of raw materials are designated by process planners. In the next step the selection of machining processes has to be made. Once the processes are selected, the machining equipment (machine tools, workstations, machining centres, etc.) which can perform one or more machining process, have to be selected. After the machine tools are selected, the cuttings tools, including clamping and measuring devices and auxiliary tools, must be chosen. If some specific tools, fixtures, or auxiliary devices need to be made, the design project must be proposed by the process planner. When all the tools have been selected, the operation sequence can be determined along with determination of operational dimensions and tolerances. Then the appropriate cutting conditions, such as the depth of cut, feed, and speed rates, have to be determined and total machining and non-machining times have to be calculated including batch set-up time, loading and unloading, tool changing, and inspection time. In the end, process plans are prepared and process sheets can be generated.

6.3 Approaches to Computer Process Planning

In general, two approaches to computer process planning are recognised traditionally:- the variant approach and the generative approach [26]. The variant approach to process planning is comparable with the traditional manual approach where a process plan for a new part is created by recalling, identifying, and retrieving an existing plan for a similar part (sometimes called a master part), and making the necessary modifications for the new part. In some variant systems, parts are grouped into a number of part families, characterised by similarities in manufacturing methods and thus related to group technology. For each part family, a standard process plan, which includes all possible operations for that family, is stored in the system. Through classification and coding, a code is built up by answering a number of predefined questions. These codes are often used to identify the part family and associated standard plan. The standard plan is retrieved and edited for the new part. On the other hand, in the generative approach, process plans are generated by means of decision logic, formulae, technology algorithms, and geometry based data, to perform uniquely the many processing decisions for converting a part from raw material to a finished state. The rules of manufacturing and the equipment capabilities are stored in a computer system. When using the system, a specific process plan for a specific part can be generated without the involvement of a process planner.

6.4 The Proposed DSS

To facilitate the process planning a software system have been proposed [27,28] and implemented [29], which can serve as a decision support tool for the process planner. This system is build upon the variant approach and is suitable for designing process plans of rotational parts. Decision models, which are built in to this system, arose from the subsequent approach: From the mathematical point of view, the problem of process planning can be interpreted as searching for appropriate models of elements and the parameters of machining processes and their relations. Those operators, which optimally transfer input quality into required output quality, have to be determined. Under the parameters of machining it is understood:

- elements of machine systems (i.e. machine tools, fixtures, cutting tools, raw material);
- parameters of these elements (i.e. performance, measurements, material type of tools);
- parameters of machining process (i.e. depth of cutting, cutting speed, and feed rates).

The final solution of the process planning task is represented as some combination of parameters, which are being searched for. In the decision situation a possibility of choice at least from two alternatives has to exist in order to apply decision procedures. The quality of alternatives can be appreciated regarding different goals. These express the usual global goals of production as:

- minimum tool costs;
- minimum machining costs;
- maximum profit;
- maximum output etc.

At the beginning of solution, the chosen goal is expressed as a function of the required parameters (as mentioned above). This function is called the "goal" function. A feasible solution with the extreme value of the goal function is an "optimal" solution. Determination of the optimal solution for global goals of production is not an easy problem. Instead of it the "optimal" conjunction of the required parameters is searched for. This results in the global goal functions being replaced by more simple ones. These can be obtained by the mathematical or logical analysis of impact of the required parameters on the goal function. The use of statistical models of machining parameters is a starting-point of such optimisation. Analytical methods, numerical methods, and their combination can be used to solve mathematical models of machining parameters optimisation.

The proposed system is dedicated especially for expert determination of machining conditions for each operation to produce a given rotational part and is capable, after the basic information from the user is inserted, to determine for one of the selected machining processes (turning, milling, drilling) or for their conjunction the optimal machining parameters regarding goal function and to calculate machining time needed for planned operations.

As was said, it is based on a variant approach. Thus all the product parts were grouped into subsets (part families) according to the similarities of the parts. For each part family the master part, which represents all similarities in the family, was designed. For each master part the original process plan, which is called the master plan, was made manually and stored in a database.

Whenever a process plan is required to be made for a new part, the process planner on the basis of the geometrical attributes will retrieve the appropriate master plan. Then the process planner can modify the master plan according to the specific attributes of the part. The proposed specific process plan is then explored and "optimised." The word "optimised" is actually misleading. The process planner is seeking a "good-enough-solution" which satisfies all the constraints and is as near as possible to optimal value for each parameter in question, i.e. number of tool changes. In the procedure of "optimisation" the process planner can change machining processes, processes sequence, machining and cutting tools, and their parameters. For each change made the appropriate machining conditions have to be determined. Following the given goal the "optimal" combination of the process and their parameters is selected by the process planner. It is not a trivial task, because of the vast possible alternatives, conflict in the objectives, and nature of a real manufacturing environment. The final modified plan will be printed out as a plan sheet.

The input data necessary for the system is entered into the system by means of a user-friendly and highly interactive session. The system allows for easy dialogue between the user and the system. The output data from the system can be printed out as an individual process plan or sheet or an individual operation plan sheet, which are well suited for shop floors. This DSS does not have an explanation ability as expert systems have, because in this the global solution strategy, which is supplied by the operator, is missing.

7 Conclusions

Computer assisted decision making falls into a very sensitive area where new developments in information technology, artificial intelligence and mathematical modelling play key roles in problem solving in industry. The potential of DSSs is still being investigated. Their benefits and usefulness seem very promising, so it is important that pure and applied research should continue as can be also seen from a vast number .of papers, research materials and books published on this topic in recent years.

DSSs have an enormous potential for improving industrial decision-making also for unstructured or ill-structured problems, whether it is in operation scheduling, production control, fault diagnosis or others. It is believed that until the point when the human has to take part in the decision making process in production management, DSSs will be useful to him.

References

1. Keen, P.G.W., "Decision Support Systems: The Next Decade," *Decision Support Systems: A Decade in Perspective*, McLean, E.R. and Sol, H.G. (eds.), Elsevier Science Publishers B. V. (North-Holland), IFIP, 1986, pp. 221-237.

2. Stabell, B.S., "Decision Support Systems: Alternative Perspectives and Schools," *Decision Support Systems: A Decade in Perspective*, McLean, E.R. and Sol, H.G. (eds.), Elsevier Science Publishers B. V. (North-Holland), IFIP, 1986, pp. 173-181.
3. McLean, R.E. and Sol, H.G., "Decision Support Systems: A Decade in Perspective," *Decision Support Systems: A Decade in Perspective*, McLean, E.R. and Sol, H.G. (eds.), Elsevier Science Publishers B. V. (North-Holland), IFIP, 1986, pp. v-xi.
4. Sprague, R.H., "A Framework for Research on Decision Support Systems," *Decision Support Systems : Issues and Challenges*, Fick, G., and Sprague, R.H. (eds.), Pergamon Press, Oxford, 1980.
5. Bonczek, R.H., Holsapple, C.W., and Whinston, A.B., *Foundations of Decision Support Systems,* Ademic Press, 1981.
6. Klein, M.R., "Embedding Knowledge in DSS," *Decision Support Systems: Experience and Expectations,* Jelassi, T., Klein, M.R. and Mayon-White, W.M. (eds.), Elsevier Science Publishers B.V. (North-Holland), IFIP, 1992, pp. 57-63.
7. Hart, A., *Knowledge Acquisition for Expert Systems*, Chapman and Hall, 1989.
8. Burton, M.A. et al., "A Formal Evaluation of Knowledge Elicitation Techniques for Expert Systems," *Research and Development in Expert Systems IV*, Moralee, S. (ed.), Cambridge University Press, 1987.
9. Kusiak, A. and Wang, J., "A Knowledge-Based System for Building Scheduling Models," *Artificial Intelligence in Optimal Design and Manufacturing,* Zuoming Dong (ed.), Englewood Clifs, 1993, pp. 173-193.
10. Anthony, R.N., *The Management Control Function,* Harvard University Press, 1988.
11. Phillips, L.D., "Decision Support for Managers," *Managerial Challenge of New Office Technology*, Otway, H.J. and Peltu, M. (eds.), Butterworths, 1984.
12. Bradley, S.P., Hax, A.C. and Magnanti, T.I., *Applied Mathematical Programming*, Addison-Wesley, 1977.
13. Efstathiou, J., and Mamdani, E.H., "Expert Systems and How they are Applied to Industrial Decision Making," Computer Assisted Decision-Making, Mitra, G. (ed.), Elsevier Science Publishers B.V. (North-Holland), IFIP, 1985.
14. Madey, G. R., and Weinroth, J., and Shah, V., "Hybrid Intelligent Systems: Tools for Decision Making in Intelligent Manufacturing," *Artificial Neural Networks for Intelligent Manufacturing*, Dagli, C.H. (ed.), Chapman and Hall, 1994, pp. 67-90.
15. Dorn, J., "Task-oriented Design for Scheduling Applications," *Scheduling of Production Process*, Dorn, J. and Froeschl, K.A. (eds.), Ellis Horwood Limited, 1993, pp. 163-179.
16. Mišút, M., *Operations Scheduling in Automated Production,* (in Slovak), Ph.D. dissertation, Trnava 1992.
17. Wu, S-Y.D., and Wysk, R.A., "Multi-Pass Expert Control System - A Control / Scheduling Structure for Flexible Manufacturing Cells,"*Journal of Manufacturing Systems,* Vol. 7, No. 2, 1988, pp. 107-120.
18. Schreiber, P., and Mišút, M., "Knowledge-Based System For Scheduling Strategy Selection," *MICRO-CAD SYSTEM 93, Proceedings of International Science Conference.* DT, Košice 1993.
19. Schreiber, P., *Classification Problems in the Diagnostic Expert Systems,* (in Slovak), Ph.D. dissertation, Trnava 1993.
20. Moravčík, O., et al., "Decision Support System for Scheduling*," Large Scale Systems: Theory and Applications,* Roberts, P.D., and Ellis, J.E. (eds), Preprints of the 7th IFAC/IFORS/IMACS Symposiom, Vol.1, Pergamon, 1995, pp. 295-300.

21. Sauer. J., "Meta-Scheduling Using Dynamic Scheduling Knowledge," *Scheduling of Production Process*, Dorn J. and Froeschl K.A. (eds.), Ellis Horwood Limited, 1993, pp. 151-163.
22. Mišút, M., "Preparation of simulation experiments for the RAPO-system," *Selected Problems of simulation models*, Proceedings of XIV. International Colocvium, DT Ostrava, 1992.
23. Mišút, M., "Results of the Experimental Verifying of Some Scheduling Strategies," *MICRO-CAD SYSTEM 93*, Proceedings of International Science Conference, DT Košice, 1993.
24. Kathawala, Y., and Allen, W.R., "Expert Systems and Job Shop Scheduling," *International Journal of Operations & Production Management*, Vol. 12, No. 2, pp. 23-35.
25. Chang, T.C., *Expert Process Planning for Manufacture*, Prentice Hall, Englewood Cliffs, 1989.
26. Zhang, H.-C., and Alting, L., *Computerized Manufacturing Process Planning Systems*, Chapman and Hall, 1994.
27. Doskočil, Z., Zongor, J., and Moravčík, O., "Support System for Cutting Conditions Determination," (in Slovak), *Proceedings of the 5. Conference on Machining*, Dom techniky ČSVTS, 1990.
28. Doskočil, Z., and Moravčík, O., *Support System for Experts' Cutting Conditions Determination*, Research paper HZ 77/89, StF SVŠT Trnava, 1989.
29. Doskočil, Z., and Moravčík, O., *Expert System for Cutting Conditions Determination*" Research paper HZ 77/89, StF SVŠT Trnava, 1990.

3 AI in Manufacturing : Application to FMS Simulation, Scheduling and Control

George L. Kovács

1 Introduction

If we had to define what should be meant under "AI in Manufacturing" we would say that all AI related means and tools which can be used during the life-cycle of a manufacturing system. A manufacturing system may mean everything from simple, single machine manual workshops to totally equipped CIM systems, however most commonly Flexible Manufacturing and Flexible Assembly Systems (FMS, FAS) are the typical examples. As FMSs are more spread out than FASs in the rest of this paper we shall speak mostly only about FMS, but almost everything concerning FMS can be understood concerning FAS, too. Sometimes CIM will be used with the meaning of FMS in the text.

The life-cycle of any system/product means the total life-span starting with an idea, requirements specification, design, implementation, test, operation, maintenance, ending with dismounting, reuse and recycling of the system or its parts. Each of these steps consists of several different components which may differ in the case of different products/systems. For instance design may be split into conceptual design, preliminary design, detailed design, etc. Simulation - what was not mentioned separately - may be important in the design and in the operation phase as well, however simulation of other steps may be useful, too.

AI in Manufacturing means the application of Expert Systems (ES), Knowledge Based Systems (KBS), Artificial Neural Networks (ANN), Fuzzy Systems (FS), Object Oriented Technologies (OO), and several other things for simulation, design, evaluation, production planning, quality control, diagnosis, failure detection and recovery, forecasting, control, etc. of FMS. As FMSs are responsible for production of products the life-cycles of products and of production systems (FMS) should be taken into account, too. This way the paradigm of Concurrent Engineering (CE) makes the picture even more interesting giving further application possibilities for AI applications.

A survey of some references on "AI in Manufacturing" was done by the author taking a look around in the CIM Research Laboratory (CIMLab) of the Computer and Automation Research Institute of the Hungarian Academy of Sciences (CARI

HAS). The following was found, which covers more or less the library of the institute, too:

- dedicated books of given authors [1-5]
- catalog and 1 dictionary on AI-tools and on AI [6-7]
- edited, multi-author books, including conference proceedings [8-28]
- dedicated journals [29-40]
- a list of about 1000 joint European R&D projects (BRITE, ESPRIT, EUREKA, COST, COPERNICUS, etc.) from which about 100 (10%) deals with aspects of AI in Manufacturing.

We believe that there are at least 10 more international and several national journals and periodicals publishing relevant papers, and there are about 15-20 international conferences and workshops every year (edited, organized or sponsored by IEEE, IFAC, IFIP, IFORS, ASME, etc.) dealing with topics as CIM, FMS, Control, Expert System's application, etc. All these forums deal with problems of AI in manufacturing.

Based on this huge amount of published information it seems to be irrealistic to give a survey on the given topics. Instead we decided to summarize which are the most common chapters of the books to see what are the most commonly tackled issues. This could be the skeleton of a new, "super-book" on "AI in Manufacturing", if someone decided to write it.

The next big part after this "survey"-like skeleton of this paper will give data on the efforts of CARI HAS in the fields of AI in Manufacturing with the goal of reaching intelligent control of FMS.

2 Important Issues in Application of AI in Manufacturing

One of the dedicated books [1] deals with process planning only, one with intelligence of manufacturing equipment [2], and the other two [3,4] with all aspects of AI in manufacturing. All of them were published in 1988 and in 1990. It is interesting to mention that most of the books in our list are dated between 1986 and 1991. We are proud to see that one of the earliest ones [8] was published in Budapest about a conference held in 1986, in Budapest. 1986 means the starting date when the problems were understood about 10 years ago, and then came a boom of relevant publications in 1988-89. We found only conference proceedings and books edited from selected papers after 1991. It may mean that a kind of saturation occurred, however the conferences are going on and journals are active, too.

The author has the feeling that by now all basic problems are understood, and the tedious and time consuming work of implementations and industrial applications are going on, and within an other 3-4 years the new wave of AI in manufacturing will show up. This will be strongly triggered by the world-wide co-operation project called Intelligent Manufacturing Systems (IMS), initiated by Prof. Yoshikawa of the Tokyo University. This project had some test cases to study the relevance of the project, and real results will come up in some years to rewrite the "AI in Manufacturing" scene.

In the following the main common topics will be given as the promised "skeleton" based on the book of A. Kusiak [4], however some titles will be omitted, and several topics will be added from other books and from all other resources, including the author's view.

Every chapter should have its own introduction, conclusions, problems and questions and solutions, references, examples, and comparisons with other systems in the literature and at the market, if possible. These will not be mentioned separately, unless we want to strongly emphasize them.

2.1 Introduction to manufacturing

The first part gives a short overview of manufacturing systems and system elements with the most common problems of manufacturing.

- Definitions of manufacturing means, as CNC, FMS, FAS, CIM, etc.
- Evolution of manual and automated manufacturing
- Control structures for manufacturing, holonic manufacturing
- Communication problems (proprietary, open, etc.)
- Integration and standardization problems
- Enterprise modeling, reengineering
  - Management issues, business system reengineering
- Intelligent Manufacturing (system components, system architecture, data flow, system operation)
- Flexible machining and assembly systems (FMS, FAS)
  - Design of parts/products for assembly
  - Process planning, scheduling,
- Application of robots, intelligent robots
- Tool management
- Modeling and simulation of FMS and FAS
  - Simulation and modeling methods and tools
- Agile and lean manufacturing

2.2 Introduction to intelligent systems

The second part is devoted to explain intelligent tools and means which are candidates to be used for manufacturing applications.

- Definitions of AI, expert systems, KB, etc.

- Knowledge representation (KR) and control
  - First order logic
  - Horn Clause subset of first order logic
  - Production rules
  - Structured production rules
  - Frames
  - Semantic networks
  - Combined knowledge representation schemes
- Object Oriented (OO) design methodologies, OO programming
- Blackboard architectures
- Application of CASE tools for design, etc.
- Comparison of KR schemes, comparison of CASE tools
- Inference engine
- Knowledge acquisition
- Knowledge explanation
- Optimization and KB systems
  - Stand alone KB systems
  - Tandem and hybrid KB systems
  - Artificial neural networks (ANN)
  - Fuzzy sets (FS)
  - Intelligent, co-operating agents
  - Hybrid ANN, FS and KB systems

2.3 Machine learning

Learning is always an important aspect of gaining intelligence, thus machine learning may be important in different applications in manufacturing, too.

- Conceptual learning
- Computational complexity of learning
- Learning and neural networks

2.4 Design of mechanical parts, mechanisms and machines

Now, as we have all basic knowledge on manufacturing and on AI application possibilities, the target fields in design should be defined first. Design of parts and of production systems has to be done together, and the human should be taken into account, too.

- Design of mechanical parts: a refinement approach
  - Overview of a given system (Dominic) to show the
    - flow of control,
    - design process,

performance

- Design of mechanical parts: a model-based approach
  - Overview of a given system (PROMPT) to show the
    - graph of models,
    - modification operators
- Design of mechanisms
  - Overview of the "configuration space method"
    - computing configuration spaces,
    - analysis of mechanisms,
    - design of kinematic pairs,
    - design constraints,
    - shape design from configuration space
- Feature based design
- KB design of automated assembly
  - Knowledge base
  - The explanation system
  - Entering new production rules
- Concurrent Engineering (CE) for product and facility design
- Building intelligent machines
  - Unifying intelligence for manufacturing tools
  - The machining brain
  - The manufacturing eye
  - The manufacturing hand
  - Application of voice processing
- The skills of the craftsman in machining
  - Extracting skills from the craftsman
  - The machine tool's setup plans
  - The machine tool's hand: jigs and fixtures
  - The machine tool's cutting sensors
  - The machine tool's team of experts
- Autonomous manufacturing

2.5 Process planning

Process planning (PP), group technology (GT) and Scheduling are the critical steps of the design of manufacturing processes for products and for the design of the manufacturing facility (workshop) as well. This design consists basically of the selection of machining and transportation and other equipment and of the planning of the layout of all equipment, including transportation.

- Material Resource Planning (MRP)
- Just in time (JIT) and Kanban
- Importance of process planning

- Feature recognition
- Phases of the process planning
- Volume decomposition
- Selection of alternative machines, tools and fixtures
- Machining optimization (single pass and multipass models)
- Decomposition of machinable volumes
- Selection of machinable volumes
- Generation of precedence constraints
- Sequencing of machinable volumes
- Selection of PPs in automated manufacturing systems
    - Problem background
    - Integer programming formulation
    - Construction algorithm

2.6 KBSS: A KB systems for equipment selection

The equipment selection process is a crucial part of manufacturing system design/planning. It will be demonstrated by an existing system

- Manufacturing system design
    - Equipment selection problem
    - Machine cell formation
    - Machine layout problem
    - Cell layout problem
- Modeling the manufacturing equipment selection problem
- Problem solving approach in KBSS
- Structure of the KB system for equipment selection
    - Data base
    - Knowledge base
    - Inference engine
- Concurrent Engineering (CE) for product design and equipment selection

2.7 Group technology (GT): models and algorithms

Group technology a and its application is very important, its application may result in great improvement in the design and in the performance of manufacturing systems. Most authors dealing with AI in manufacturing still do not deal with GT. In this chapter some general issues, in the next one an example will be given.

- Introduction
    - Visual method
    - Coding method

- Cluster analysis method
  - Matrix formulation
  - Mathematical programming formulation
  - Graph formulation
  - Nontraditional application of GT

A KB system for GT.

- GT in automated manufacturing
- Structure of the KB system (KGBT)
  - Input data
  - Grouping process
  - Output data
- Data base
- The KB subsystem (KBS)
  - Knowledge base
  - Inference engine
  - Request processor
- Clustering algorithm
- Application of KGBT
- Quality of solutions
- Industrial case study

2.8 Machine layout: models and algorithms

The next issue is to solve the machine layout and the workshop layout problem.

- Models for the single-machine layout problem
- Solving the single-row layout problem
  - Modified spanning tree algorithm
- Models for the multi-row layout problem
  - Quadratic assignment model
- Efficient models for the multi-row machine layout problem
  - Models with machines of equal area
  - Models with machines of non-equal area
- Algorithms for the multi-row layout problem
  - Optimal algorithms
  - Suboptimal algorithms
- The penalty function algorithm
- Concurrent Engineering (CE) for machine layout, equipment selection and product design

KBML: A KB system for Machine Layout. A system for the second step (after equipment selection) of facility design gives some ideas about application of a KB system.

- Data input for KBML
- Problem solving approach
- Structure of the KBML
- Systems from the literature (FADES, IFLAPS)

2.9 Aggregate scheduling of machining and assembly systems

Short term production planning (for one hour, one shift or one day) is called scheduling in most manufacturing literature. There are different one-machine, two machine, and more machines approaches, as well as solutions for one product, more products, and for one and more batches. Solution of the SP means the most important step towards solution of the control of manufacturing systems, as all timings of all equipment are available if the SP is solved.

- Typology of aggregate scheduling problem (SP)
- Mathematical scheduling
- Heuristic scheduling
- Scheduling by simulation
- Knowledge based scheduling
- The single-product SP
- The n-product SP
- The single-batch SP
- The n-batch SP

2.10 Scheduling models and algorithms

This chapter suggests some solutions to the above problems, partly through an example.

- Scheduling n operations on a single machine
- Scheduling a flexible forging machine
  - Features of the forging machine SP
  - Modeling the forging machine SP
- Two machine flow shop problem
  - Two machine job shop problem
  - Special case of the three machine job shop problem
- Modeling the SP of n operations on m machines
- Solving the SP in automated manufacturing systems
- Reactive scheduling (real-time rescheduling)

KBSS: KB system for scheduling in automated manufacturing. This chapter describes an example KB scheduling system for FMS.

- Structure of KBSS
  - Knowledge base
  - Algorithm base
  - Data base
  - Inference engine
- The heuristic algorithm
- Computational results

2.11 Quality control

In this chapter one of the most complex application areas of AI will be discussed. Both expert systems, ANN and fuzzy systems are used to reach safe, high quality production. International standards, as ISO 9000 are getting more and more important and inevitable.

- Total Quality Management (TQM)
- Taguchi's methodology
- Inspection, diagnosis, error detection
- Error forecasting
- Error recovery
- Maintenance

2.12 The Intelligent manufacturing systems (IMS) project

This world-wide co-operation will have great impacts on the future of AI in manufacturing.

- Robust Control
- Intelligent control
- Test cases of the IMS project
  - Clean manufacturing in the process industries
  - Global concurrent engineering
  - Global manufacturing in the 21st century
  - Holonic manufacturing systems
  - Rapid product development
  - Gnosis- Systemization of Functional knowledge
- Future plans of IMS

2.13 Conclusions, further development trends

- Vision of the factory of the future

3 Knowledge based simulation, evaluation, scheduling, quality control and real-time control of manufacturing systems

Flexible manufacturing systems (FMS) are used world-wide as the production parts of computer integrated manufacturing (CIM) systems. CIM means computer assisted management, marketing, design, manufacturing, testing, service, etc. integrated into one system.

FMSs are generally large-size, complex, dynamic and stochastic. There are no 'simple', methods (based on input/output relationships) and procedures to design and operate an FMS. This is why there is a world-wide research to apply modeling and simulation procedures and co-operative expert and traditional systems (hybrid systems) to assist FMS design and operation. The most promising systems are realized by means of co-operating expert systems on parallel computers [41].

Some rapid prototype programs for knowledge based evaluation, simulation, scheduling, control (supervision), networking and quality control of FMS are under test and the first practical example to which they are applied is the CIM pilot plant at the Technical University of Budapest (TUB). Fig. 1. shows the four, interconnected, holonic-like cells of this system: storage and AGV cell (7), manufacturing cell (2+3+4+5), assembly cell (1) and measurement cell (6).

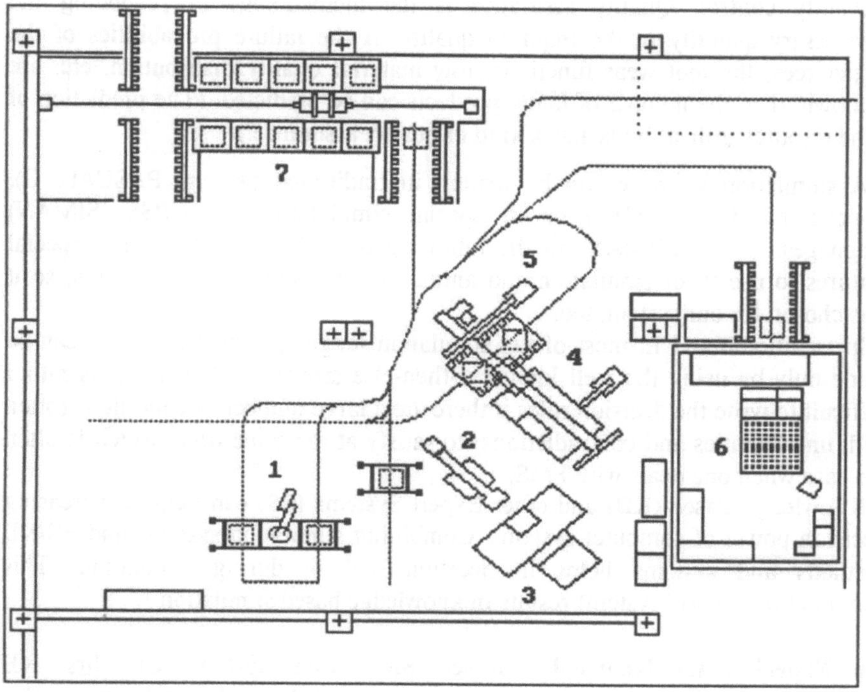

Fig. 1. The layout of the TUB CIM system

During the last years we have had some results in expert system base cell-configuration and reconfiguration [42], [43], in PROLOG based simulation [44], and in the application of the SIMAN/Cinema [45] system for simulation and animation [46].

3.1 Knowledge based simulation, scheduling and quality assurance

Computer based simulation can be used in different periods of the life cycle of an FMS, as:

- Capacity planning: the bottle-necks of the system can be detected. After evaluating the simulation results, the number of machine tools, transport facilities, buffer sizes, etc. can be increased or decreased.
- Performance evaluation of planned and of working FMSs.
- Checking given production schedules.
- Creation of manufacturing schedule. Scheduling based on simulation is generally fast and reliable.

- Quality control. Quality assurance is the methodology of producing the necessary quantity in the required quality. If the failure probabilities of the resources, the tool wear functions, row material quality distribution, etc. are known, then the number of faulty products can be predicted. (The prediction of the resource utilization is also a kind of quality assurance.)

A simulation software can be written in traditional (as e.g. PASCAL, C), object-oriented (e.g. C++) or in special simulation (e.g. GPSS, SIMAN) languages. Cinema/SIMAN (in the following SIMAN) [45] has several special features to use it for simulation and animation of manufacturing systems, so it was chosen for our system, too.

In traditional and in most of the simulation languages most decisions can be made only by using the well known if-then-else structure. However it is rather difficult to write the decision rules if there are a large number of conditions (often with uncertainties and contradictions) to satisfy at the same time, which is often the case when one deals with FMS.

Knowledge Based (KB) and other Expert Systems (ES) can help in increasing decision power of computer systems. Combining simulation systems and ES/KB methods and systems helps in decision making during simulation. This combination (hybrid system) results in knowledge based simulation.

An Experimental Knowledge Based Simulation System. Our first KB Simulation was implemented in a special rule based simulation language CS-PROLOG (Communicating Sequential Prolog) [47]. Then we started to develop a new system which reflects the idea of connecting a traditional simulation system to expert systems.

At the same time it was realized that some quality control and scheduling capabilities can relatively easily be incorporated into the system by applying separate advisor systems which are communicating with the simulation.

Architecture and knowledge bases of the system. The system consists of four main modules: Preparation Expert System (PES), Advisor Expert System (AES), Evaluation Expert System (EES) and Simulation-Animation System (SAS) (Fig. 2.). All the three expert systems are communicating with the traditional SAS system.

The Preparation Expert System (PES) will collect all input data and will create the simulation model. Its knowledge base will contain rules to chose the appropriate simulation model from the set of possible models (discrete or continuous, queuing model, Petri-nets, etc.) and to build up the model from the available model building tools.

The Simulation-Animation System (SAS) executes the simulation model generated by PES. It can be used for FMS evaluation, scheduling and quality control as well. Both functions need many decisions, which are partly made in the SAS itself and partly made by the AES. If the simulation needs help from the AES it stops, sends its question to the AES and waits for the reply. During the simulation a graphical animation helps to understand and follow the simulated manufacturing process.

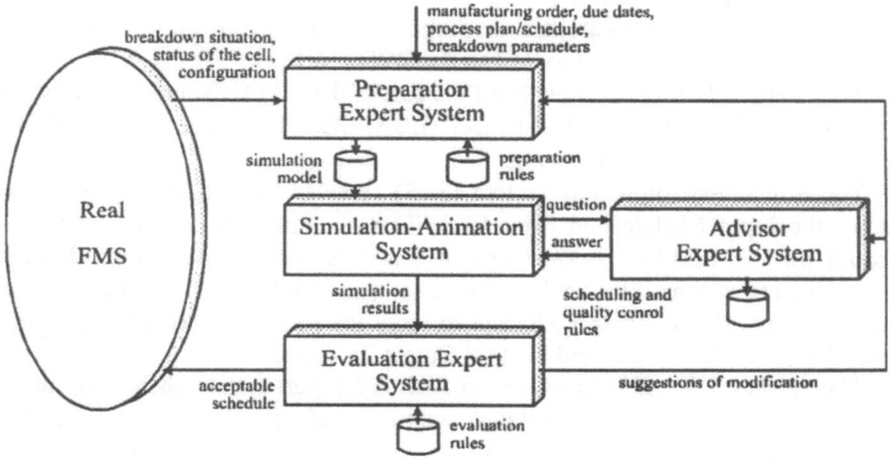

Fig. 2. Modules of the hybrid KB simulation system

The Advisor Expert System (AES) waits for the questions of the simulation on certain decision points. Receiving the question the AES starts its inference process and sends back the concluded answer. The knowledge base of the AES consists of separated scheduling and quality control rules. Workpiece and resource priority rules belong to the scheduling part of the knowledge base, and measurement evaluation rules to the quality control.

The Evaluation Expert System (EES) evaluates the results of the simulation which are utilization statistics and a manufacturing schedule. If there was an input schedule given to the system the resulting schedule will be a checked version of it. If the scheduling capability of the system is used an original schedule will be produced by the simulator and checked by the EES. The knowledge base of the EES contains all the evaluation rules (points of view, etc.) collected from human experts and from experience which help in deciding on acceptance or rejection. Depending on the knowledge base of the EES it can decide whether the schedule and the cell configuration are acceptable or not, and modifications can be suggested. These suggestions may be the application of new rules by AES or the modification of the original configuration of the manufacturing cell by PES. EES has an other way of operation, when it statistically evaluates several different simulation runs of a system.

The first prototype version of the system. The first applicable prototype version of the KB Simulation Scheduler System with Quality Assurance consists of five main modules:

The Input Module substitutes PES. It collects the schedule (if there exists) and the cell configuration and all other necessary data from the user. This menu driven program has been written in Turbo C using the Zortech C++ Windows library. It is running on IBM PC/AT under DOS.

The Graphical Layout Planner (GLP) is part of the SIMAN simulation software package (SAS) running on IBM PC/AT (EGA)/DOS. The input of the GLP can be an AutoCAD drawing, as in Fig. 1.

The Simulation-Animation System (SAS) receives the prepared information from the Input Module and from GLP. This simulation-animation module is based on the SIMAN package and it has five main parts: Model Frame, Experimental Frame, SIMAN Run Processor (extended with C routines to keep the contact with the Advisor ES, and to write the records of the manufacturing schedule), SIMAN Linker and SIMAN Output Processor.

The Advisor ES (AES) is written in ALL-EX expert system shell [48], under OS/2 operating system. It contains feature based expert knowledge on the production means and products and on quality control and scheduling. The quality control part of AES decides whether the workpiece is good, repairable or waste based on the simulated measurements. Scheduling rules in AES help in different decisions as e. g. which workpiece can leave the storage first, which resource is more advisable for a certain operation, etc. In the recent prototype version FIFO, LIFO and the well-known Round Robin strategies are implemented.

The Evaluation Expert System (EES) has the task to check basic inconsistencies of the outgoing schedule on one hand, and to give statistical evaluation, if necessary, on the other hand.

Some test runs with this prototype proved that knowledge based systems can effectively help the work of a traditional simulation system and that such a hybrid system can solve different tasks, depending on the applied expert knowledge. SIMAN proved to be a proper FMS simulation package. The application of the ALL-EX shell for the advisory systems was successful to have a rapid prototype, but because of memory, speed and communication problems we were forced to find another expert system shell.

Application of G2 to solve all advisory problems. With the working SIMAN/ALL-EX based hybrid prototype version of the Simulation Scheduler System we have proved that knowledge based systems can efficiently help the work of traditional simulation systems.

Then we decided to use other, more commonly used expert systems instead of ALL-EX. We had some experiences in this project with the production system CLIPS [49], but as the appropriate version could not be obtained we do not use it. On the other hand the application of the G2 real-time expert shell [50] was a success. At the beginning SIMAN/Cinema were running on a PC, while a SUN SPARCstation was used for G2. To run the two systems together we had to develop an interface between SIMAN and G2 [51]. This will be detailed next.

Connecting a simulation package with a real-time ES. As the remote goal of our research efforts is to reach real-time intelligent control and G2 is a real-time system which is able to produce control information to FMS we continued its application as KB system. On the other hand SIMAN is widely used to model (and simulate/animate) FMS. In spite of the fact that it is not too difficult to build an FMS simulation model in G2 - as we shall see later on - because of some special features of SIMAN we decided to keep using SIMAN and to connect it with G2.

- Timing problems of a combined system. The KB may run prior to, after, or during the simulation, or some combinations of the three possibilities can be applied. If the KB runs during simulation and may influence the simulation run on one hand, and the simulation may influence the parallel running KB on the other hand the connection of the two softwares is called deep-coupled and real-time. In our case, as it can be seen in Fig. 2. we plan to use a combination, where the simulation is prepared by a KB system (PES), and there is a KB system that gives advises during the simulation run (AES), and finally a KB system evaluates the results (EES).

- Physical interconnection possibilities. A communication between our two systems may be either a network connection, or direct data transfer if the two systems run on the same hardware, or files and data can be transferred using floppy discs, which is not a really good on-line solution. There are possibilities of applying KB routines in simulation, or applying simulation built into the KB reasoning system. These latter possibilities are used when the built in simulation of the G2 is applied.

- Interfacing and I/O possibilities of SIMAN and G2. There are three possibilities for SIMAN I/O: application of either the SIMAN Summary Report, or of C and FORTRAN language routines that can be linked to the SIMAN routine library, or of the SIMAN commands READ and WRITE. G2 is a system designed to have external connections. It has four interface possibilities: Foreign Function Interface, G2 File Interface (GFI), G2 Standard Interface (GSI), G2 Simulator Interface (GSPAN).

- G2-SIMAN real-time interfacing. For PES and EES purposes any of the previously given methods can be used for interfacing from both the SIMAN and G2 side, however the simplest implementation solution can be gained by using GFI together with the READ/WRITE facilities of SIMAN. This way all communications are performed via files, and there is no need to use the ICP networking of G2. For connecting G2 as Scheduler and Quality Advisor the only way in the SIMAN side is to write C (and FORTRAN) routines. From the G2 side GSPAN and GSI could equally be used, however, as SIMAN is not object-oriented, the advantages of GSPAN cannot be gained, so the application of GSI (via ICP) will be implemented. In our first experiment the solution was a simple message exchange via two files over a TCP/IP network using NFS. The logical queue is only one message deep and the handshaking conditions are the existences of the files.

SIMAN/Cinema and G2 on the same workstation. Now we are working on a version where the workstation version of SIMAN/Cinema runs on the same SUN, where the G2. This way there is no special interface problem, just the co-operation of the two different systems, as tasks has to be provided.

The structure of the G2 based system is in Fig. 3. Its principles are the same as in the case of the system in Fig. 1., however the two on-line, deep-coupled advisors for quality assurance and for scheduling are separated.

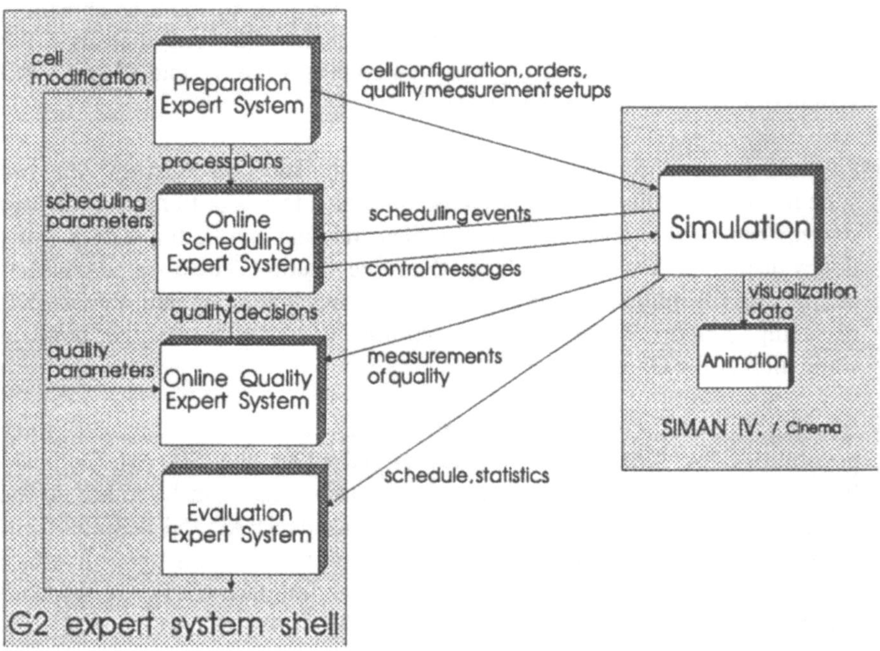

Fig. 3. The structure of the G2-SIMAN/Cinema system

All the knowledge of the advisors implemented in ALL-EX was transformed into the appropriate formats of G2 without major problems - automatically.

The knowledge base of the Advisor Expert System (AES) is now filled with some more scheduling knowledge as the knowledge which is now being built into the simulation model can hardly be modified and there is no inference mechanism to effectively apply scheduling rules. Quality control knowledge will be changed to more powerful feature based knowledge with several new rules in the near future.

The knowledge about the product and production system quality is described with objects. The system, after identifying the type of the actual part sends a message for the simulation about the quality of the part and what machining modifications should be done at the actual machine in order to keep (avoid waste) the part (batch) quality.

The information gained from the AES also can be used by other computerized systems like process planner, CAD.

Simulation and advisory systems in the same G2 shell. As a next prototype experiment we implemented a system where all programs were written in G2 to solve the tasks of knowledge based simulation, evaluation, quality control and scheduling (and control) of an FMS. The structure of the system is given in Fig. 4. The natural advantage of this system is that neither interfacing nor co-operation problems appeared in comparison to the previous prototypes, where G2 had to cooperate with SIMAN/Cinema.

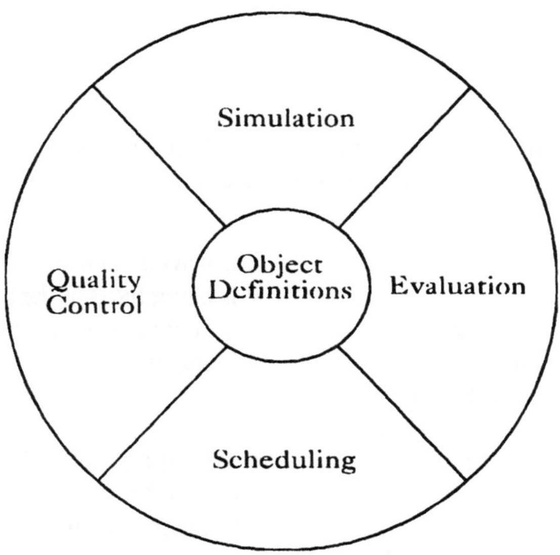

Fig. 4. The structure of the pure G2 system

The biggest difference was caused by the G2's object-oriented view of a system. All elements of the simulated system are now represented by G2 objects in object and class hierarchies. There exists a hierarchy between the objects which makes the attribute inheritance possible. In the following some of the classes we have defined are listed, together with their attributes (inherited attributes are not mentioned):

cell-element
- workpiece (state, present-station, next-station, current-station, type, color, process-plan, op-tion)
- process-plan(first-element, type, color, is-required, sum-of-is-in, sum-of-is-req, sum-of-is-prod)
- place (state, level, layer, machine-name, entry-x, entry-y, time-to-begin, time-to-finish)
  - workplace (state)
  - storeplace (state)
- cell-unit (state)
  - transport-device (state, dest-x, dest-y, speed, speed- level, radian)
  - cell-equipment (state, place-list)
    - store (wp-list, place-list)
    - machine (place-list)
      - meas-machine
      - lathe
      - manuf-center
      - portal-robot
      - asm-robot
      - fan-robot

Some explaining notes on the attributes of the classes mentioned above, for instance on the attributes of the workpiece class:

workpiece:

state	idle	it is staying on machine and it is idle
	fixed	it is being fixed in the store
	defixed	it is being defixed in the store
	in-store	it is staying in the store and it is idle
	transported	
	manufactured	
	waiting	it is waiting the AGV which has been reserved for it
present-station	the name of the cell-equipment which it is currently processed on	
next-station	the name of the cell-equipment which it is next processed on	
current-station	the name of the cell-unit which it is staying on	
type		
color		

process-plan
op-tion the name of the operation to be processed

In comparison with G2 we can say about SIMAN program that it is more likely a traditional program, consisting from a main program which contains commands, branchings, cycles and procedure/function calls and from further details of procedures and functions. On the contrary, when programming in G2 shell the main program is missing from the mentioned structure. The work of the system is directed by rules. There are rules which fire on their activating (initially ., unconditionally ... rules), others fire when a variable receives a value (whenever ..., if ..., when ... rules) and, finally, rules can be fired in an explicit way as well (invoke ... rules, focus on ... actions). The system can be told to execute an activity or to call a procedure by an user as well - for this purpose there are user-buttons and menu-choices. Procedures can be called also from within rules.

For better understanding how different processing may appear let us see the analysis of the next-station procedure which was designed to handle situations when a workpiece becomes idle or a transport-device becomes free:

This procedure may be started from the following rules (conditions), in the following situations:

Rule-6 a workpiece is in the store, its state is becoming in-store and the state of a transport-device is free

Rule-10 a workpiece is on a machine, its state is becoming idle and the state of a transport-device is free

Rule-8 the state of a transport-device is becoming free and there is a workpiece the state of which is idle

The next-station procedure calls the free-place procedure to find out if there is a place on the machine which is the next in the process-plan of the workpiece.

If the next-station procedure was not started by the Rule-6, that means the workpiece is not in the store, in the case of no free place on the next machine the next-procedure calls again the free-place procedure, this time to see if there is a free place for the workpiece in the store.

There are several possibilities on continuing the processing, depending on the result of the running of the free-place procedure:

- if a free place was found (on a machine or in the store), the agv-goes-to-wp procedure is called
- if no free place was found:
  - the state of the transport-device is set to free (which causes the Rule-11 and Rule-8 to fire)
  - the state of the workpiece is set to:
    - in-store, if the workpiece is in the store (which causes the Rule-6 to fire)
    - idle, if the workpiece is on a machine (which causes the Rule-10 to fire)

And this is the point where the next-station procedure ends.

3.2 Some experiments with the simulation schedulers

Application of expert systems allows to apply contradictory rules simultaneously, and probabilities and uncertainties can be applied, too. This way the power of the system can be increased and programming is easier.

For our experiments we applied simple rules of different priorities with a goal function of having the shortest throughput time (total production time). The results are evaluated based on total throughput times.

Scheduling was divided into two consecutive steps. We separately tried to optimized the sequence, as row parts arrive into the input buffers of the FMS (pre-schedule, or loading in some references), and then optimization is done when the parts are moving in the FMS (schedule).

Optimizing the input sequence of row parts. In this experiment a preliminary sequence of the different raw pieces is determined to the input storage of the FMS before of entering to the system to see the effects of these sequences to the final schedule.

The following rules/priorities were used:

- MQ, type of maximum quantity first (always that part type has the highest priority from which the biggest number is required, the basis of comparison is the number of required parts minus number of parts under /plus after/) processing of the given types)
- MC, multicolor input (as long as possible all types of parts should be in the input storage)
- UD, uniform distribution input (the distribution of different parts in the input storage corresponds to the distribution of part types to be produced)

Application of different scheduling algorithms. As it was mentioned earlier only a simple strategy was built into the SIMAN simulation system itself and a LIFO and a FIFO algorithm were implemented in the SAES first. Later on we changed these algorithms, and for the experiments the following ones were implemented (jobs are always associated to parts.):

- SP, job (part) with shortest process plan first
- LP, job (part) with longest process plan first
- SI, job (part) with the shortest next operation time first
- LI, job (part) with longest next operation time first
- ED, job (part) with the earliest due date first
- SR, job with the shortest remaining processing time first
- LR, job with the longest remaining processing time first

The time-spans of individual process-plans are measured by generating individual schedules for each part as they were the only single parts in the system.

Tests on a real 8 machine workshop, application of the hybrid system. For these experiments the hybrid program of Fig. 3. was used on the system given in Fig. 5.

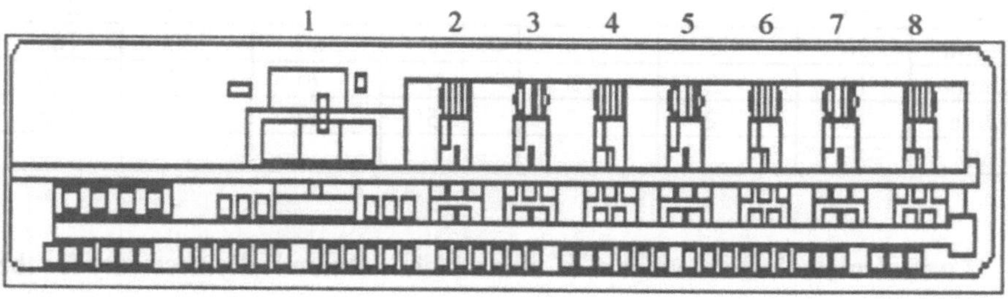

Fig. 5. Layout of the GoldStar FMS

The goal of the experiments was twofold: to test the hybrid simulation-scheduler and to make test runs with a real industrial example.

The system works in Korea in the workshop of the GoldStar Heavy Industries Cable factory near Seoul. It is in operation since 1992. There are 8 pieces of CNC machining centers with palette exchanger and tool magazines in the system situated in one long line, there are seven buffers for temporary palette storage, with 6 slots each. One transport carriage is moving on rails, with a speed of 60 m/min if loaded and with 80 m/min if empty and its accuracy is 0.2 mm. The palette size is 1m X 1m, the maximum weight of the workpiece is 3000 kg (3 tons), there is room only for one palette in the carriage. There is a load/unload station with four slots where parts are fixed in the palettes, and where they are unfixed to leave the system, fixing/defixing is done manually.

It is seen in Fig. 5. that the machine tool No. 1 is a little different from the others which are equally the same. However functionally the first one is the same, too.

To demonstrate the size of the workshop we give the distances of the machine tools from the load/unload station, as machine identifier (distance in meters):

1(12), 2(27), 3(26), 4(33), 5(40), 6(47), 7(54), 8(61)

The production task in a shift (one week) week will be given below. There are two operations to be done on every part. An operation is defined as all CNC programs on one machine tool with one palette position. The two operations can be done at any machine tools, and there is a need of a new fixing of the part between them.

We use the animation colors instead of the names of the parts to identify them.

color	batch size	package size	due date (day)	setup time (min)	operation times (min)		
					op1	op2	sum
red	100	6	6	5	105	77	182
yellow	100	5	6	5	65	50	115
blue	90	10	6	5	30	225	255
brown	30	3	6	5	60	120	180
green	10	2	3	5	70	65	135
white	10	2	3	5	80	65	145
Total	340						

If we calculate the total amount of time which is theoretically necessary for the production we have:

182x100+115x100+90x255+30x180+10x135+10+145=60850

If we add two setup times to each workpiece, we get:

TSUM=60850+340x10=64250 min
TSUM in days is 64250/60/24=44.6 days,

if we suppose 1 machine tool only, but we have 8 similar purpose machine tools:

TSUM8=44.6/8=5.57 days.

Now we can say that to produce all parts required 5.57 days were necessary if every machine tool worked full load all the time, i.e. theoretically the production can be done in a working week with 6 days to work with 24 hours work a day. The high amount (42) of buffer places makes such a vision realistic. As the speed of the AGV is high enough to say that the time periods which the palettes spend on the AGV are one-two order smaller than the machining times, transportation times are neglected. We can say that they are included into the setup times as averages.

Tests on the pilot plant of the TUB, application of the pure G2 program. For these experiments the program of Fig. 4. was used with the goal of checking as many test runs as possible with a complicated, multi-cell FMS, with simple, but realistic production tasks, to find out the differences between the possible combinations of pre-scheduling and scheduling rules.

Several different batch sizes of all parts were checked with several scheduling algorithms available (in the SAES system).

We made test runs with a real production task where 4 different parts should be produced and checked on the CIM pilot plant of the Technical University of Budapest (TUB). The layout of the system is in Fig. 1. This flexible manufacturing system consists of four different cells (manufacturing, storage and AGV, measurement and assembly) with similar cell-controllers, interconnected by a LAN.

All parts have different process plans which are the following:

part	operations (time) (min)	timespan (min)
red	fix(15)-turn(25)-mill(30)-dfix(5)	75
yellow	fix(15)-mill(18)-asm(30)-dfix(5)	68
blue	fix(15)-meas(25)-asm(30)-dfix(5)	75
green	fix(30)-turn(20)-meas(10)-mill(20)-meas(10)-mill(20)-asm(30)-dfix(5)	145

The meaning of the abbreviations are the following: fix-fixing, turn-turning, mill-milling, asm-assembly, meas-measurement, dfix-defixing.

We use the different colors to identify the different parts as it was implemented in the animation of the process. The respective due-date priorities of the part type were given instead of real due dates and they order was the same in all experiments:

yellow-green-blue-red

The yellow parts are the most important, the red ones are the least important, if due dates are concerned.

The batch sizes varied from 0 to 30 during the experiments. As the Input Storage of the system has 10 slots the experiments would have been meaningless if the sum of the four batch sizes had not been bigger than 10.

We made runs with several combinations of the priorities (3 times 4= 12 runs for each production task. All the three input sequence algorithms and the following scheduling algorithms were used: LP, SI, LI, ED) with the following batch sizes:

yellow:	3	4	5	6	7	8	9	10
green:	3	4	5	6	7	8	9	10
blue:	3	4	5	6	7	8	9	10
red:	3	4	5	6	7	8	9	10

and then with the same numbers, but one part was always missing. As example we give the table without yellow parts.

yellow:	0	0	0	0	0	0	0	00
green:	3	4	5	6	7	8	9	10
blue:	3	4	5	6	7	8	9	10
red:	3	4	5	6	7	8	9	10

This way we had 8 times 12=96 runs for each group, i. e. altogether we made 5 times 96 = 480 experimental runs, which took several hours . Comparisons are made only within the same groups, i. e. if the final evaluation will be done we expect results for five distinct types of production, what is naturally far from an exhaustive examination.

The results are rather interesting, the statistical evaluation will take some more time, as no direct relationships could yet be determined.

Some basic characteristic of the test-bed problems. There are some basic differences between the two problems we made experiments with:

- The layout of the TUB system is really two-dimensional, with complicated trace for the AGV to follow.
- The layout of the GoldStar system is one-dimensional, the machine tools are in one line, and there is a parallel line for the buffers, while the AGV moves on rails between the two parallel lines to easily serve both of them.
- The task of the TUB system is more complicated as not only machining, but measurement and assembly functions are done, too, and the system is cell-based.
- The GoldStar system does only machining with equivalent machine-tools
- There is no substitution possibility for any of the resources in the TUB system, thus there cannot be alternative routings for the workpieces.
- There is a very high substitution possibility in the GoldStar system as all machine tools (and all buffers) are the same, thus alternative routings can be essential.
- The TUB system has a 10 slot I/O buffer and each resource has a 2 slot buffer for palette exchange.
- The GoldStar system has a 4 slot I/O buffer and each machine tool has a 1 slot buffer, and the system has 42 more buffer storage places for temporary storage of palettes.

Based on the above differences it is clear that we had completely different goals with the two sets of experiments:

- We had the goal of trying to approach to the theoretical value of 5.57 days, which is a lower limit to the given production task in the Goldstar system.
- The goal of the several test-runs on the TUB system was to figure out whether there are relationships between the different production tasks and different pre-scheduling and scheduling algorithms on a given FMS.

Evaluation of the test-run results.

The GoldStar system. For this example we did not deal with the sequence of filling up the 4 slot I/O buffer, but a fixed sequence was used for all test-runs. This was the following: 2XWhite, 2XGreen, 6XBrown, 18XBlue, 20XYellow and 20XRed. This set of 68 workpieces was repeated 5 times.

We made runs where the following algorithms (priority rules) were checked: the original embedded scheduling algorithm of SIMAN, and the SI, LI, SR, LR, and LP.

It is interesting, that the SI and SR rules gave exactly the same results, as well as the application of the LI and LR rules.

It was surprising that all runs gave results within the range of 5.6 to 6.1 days, i. e. all were very closed to the theoretical minimum, to the 5.57 days. The differences are so small that no reason to specially evaluate them.

But, as we know that the FMS was designed to optimally produce the given parts in the given amounts we can tell that the surprising results are not really surprising. There are two main factors that influence these results: the high speed, and neglected times of the AGV and the high amount (42) of temporary buffers.

Now we are in the process of checking whether that FMS is really flexible or not, and we try to make test runs with different production tasks of the same magnitude, as the real one is. The evaluation of the results will take a longer time.

The TUB system. In spite of the very high amount of test-runs we still do not have enough results to have any kind of final conclusions concerning the FMS and the given production tasks on it.

Just to give an idea on the measured time values some results were the following:

Scheduling rules	workpieces 5,5,5,5 input sequence			workpieces 10,10,10,10 input sequence		
	MQ	UD	MC	MQ	UD	MC
ED	624	638	587	1119	1158	1142
SI	591	653	605	1143	1159	1175
LI	625	622	604	1162	1183	1207
LP	619	625	617	1170	1191	1194

Based on the above and several similar tables we can get only some rather general conclusions on the given FMS: generally the ED algorithms are better than the LI ones, there are no significant differences between MQ, UD and MC, etc.

3.3 Experiences with systems using and not using SIMAN

Since SIMAN is a language developed directly for FMS simulation it has some very pleasant properties which are missing from G2.

For instance, in SIMAN it is very convenient to define a transport-device to follow a zigzag route - it is only necessary to define the route in the animation module. Then the transport-device can follow the route only by telling it the start and final station. On the contrary, in G2 a route following algorithm has to be implemented.

Another feature that would be of great use in G2, too, were the presence of predefined statistics and distribution functions. Statistics functions are used in numerical evaluation of the simulation results and distribution functions are required when we wish to randomly generate numbers with a given distribution.

On the other hand program development is definitively easier in G2 than in SIMAN - even for simulation. The object oriented feature of G2 is one of its great advantages. And there is a big advantage if no interface is needed as everything is in G2.

To have a final opinion - and final decision - further experiments should be performed.

3.4 Knowledge Based Control and Open Communication of Advanced Manufacturing Systems

Problems of the real-time, intelligent FMS control. A rather simplified definition of Intelligent Control of FMS means a continuous or frequent observation and evaluation of the status and condition of the system performance, decision making based on the evaluation results and on pre-defined knowledge, and then the operation according to the decisions. In the case of so-called normal operation, which runs according to the given schedules there is no special need for intelligence, neither for interactions to modify the operation, but the above given procedure may help [52] if any kind of disturbance or irregularity happens, what is rather common in the case of highly sophisticated, complex systems.

Recent discrete manufacturing and/or assembly systems (FMS/FMA) are more and more often using MAP/MMS (Manufacturing Automation Protocol, Manufacturing Message Specification) [53], because this technology is widely available from many vendors and really gives a safe and open solution according to the demands of OSI (Open System Interconnection). Many users do not exactly know that they have such interconnections, they just enjoy the useful features of MAP. We must also admit that the MAP or related solutions are still rather expensive.

MMS is a network protocol that gives a tool to describe the manufacturing specific features of different industrial resources, as CNC, robot, PLC, etc. At the same time MMS provides a modeling method of the resources from the communication point of view, and it implements the communication messages, too. It is closed to the OO methodology as different objects have different type of services in MMS.

On the other hand the so called intelligent control is getting to be a general demand. There is a vivid discussion in literature and in private communications among control engineers about the existence and need of intelligent control [54].

Some experts claim that there is nothing that really could be called intelligent control. Most of them are speaking only about process control, and not about discrete manufacturing (FMS: Flexible Manufacturing System) or robot control, however the control tasks and problems of manufacturing systems are basically similar to those of batch-like process control. Now, without going into discussions in this issue we accept the need and necessity of intelligent control which we intend to solve by means of knowledge based (KB) systems, which we often call expert systems. We also say that some commercial expert systems provide good problem description and software development tools where the programming is closer to the problem to be solved and to the user to have to support with limited Real-time facilities [55].

To speak about Real-time control we need all scheduling data, as the starting and finishing date of all operations of all equipment (machine-tools, robots, transfer units, as, AGV, etc.) together with the possibilities of downloading (CNC) control programs to all equipment, and to accept and evaluate different signals from them, etc.

Real-time communication with expert systems. The practical problems of the communication of expert systems in CIM applications can be divided into two parts. One is the hardware-software connection (physical) and the other is the logical one between the controller(s) and controlled devices. This decomposition was very useful both in the design and implementation phase during the last projects of our CIM Laboratory. If this decomposition is not so sharp many problems may occur during the development and specially in maintenance later on..

There are relatively easy programming interfaces (etc. C/C++) in most available ES shells. These interfaces provide data transfer and communication possibilities with external tasks, stations, etc. They support clear and easy programming to reach objects, to call procedures, to set and get variables, etc.. The interfaces are dedicated to specific software tools of the ES and they are general towards the external world without being able to take into account the requirements of the given application. So nearly each CIM (FMS) implementation requires special software development to cover this gap between the external world and the ES.

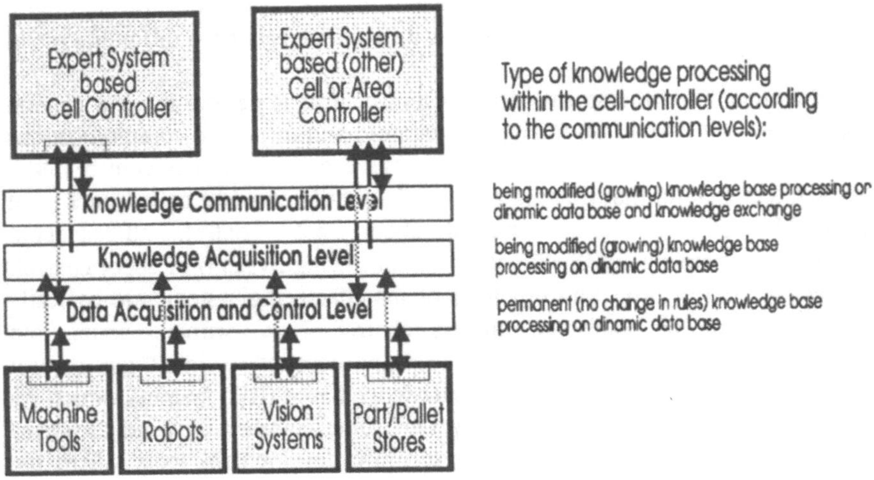

Fig. 6. Different logical communication levels of an intelligent cell controller

Clearly the communication functions depend on the capabilities of the expert system. The way of learning and knowledge handling determine the logical levels of the communication. Three different types of working mode and different levels of the communication of an intelligent cell-controller in a CIM environment are shown in Fig. 6. [56]. These levels are implemented - of course - within the same protocol. This picture just explains the different possible meanings of given messages. The lowest level has the basic control and data acquisition type messages. The other two levels have messages if and only if the 'intelligence of the cell controller is not hidden'. Hidden intelligence means in this term that the knowledge based technology is applied only inside the cell controller and it has no specific actions via the communication channel. A typical example of the hidden case if a KB system is built up on the top of a traditional control system using its original communication. The knowledge acquisition and the so-called knowledge communications were separated. The first one contains specific data for modifying or verifying the knowledge of the given controller. When a KB system shares its knowledge (new or modified) it belongs to the knowledge communication level [57]. The communication messages of the most real and pilot KB applications belong to the lowest, possibly to the middle logical level.

The object oriented view of MMS. Going back to the so-called physical connection there are many alternatives. Most controller and controlled device vendors offer good (proprietary) solutions to communicate and also vendor independent standards are available.

Also in the CIM area there are more accepted models or modeling tools to describe the objects of an FMS. In the communication point of view the most promising one is the object oriented view of MMS (Manufacturing Message Specification).

MMS gives a so-called VMD (Virtual Manufacturing Device) view about each resource of the FMS. Each resource in an MMS network is defined as a VMD. The VMD object itself has some general attributes, as physical and logical status, vendor or model name, etc. The VMD contains additional objects to define the resource specific elements as the domains, program invocations, variables and some more, but not widely used attributes as events, semaphores, journals, etc. In a VMD it is possible to create, read and write different types of variables, to up- and download domains which can be programs or machine data, to start and stop different tasks (with program invocations) and to handle events.

Fig. 7. shows a simple, but rather general type of VMD :a robot VMD. It was realized that this specification is good on the higher level of the FMS [58] to give a communication oriented view about the network elements and their resources.

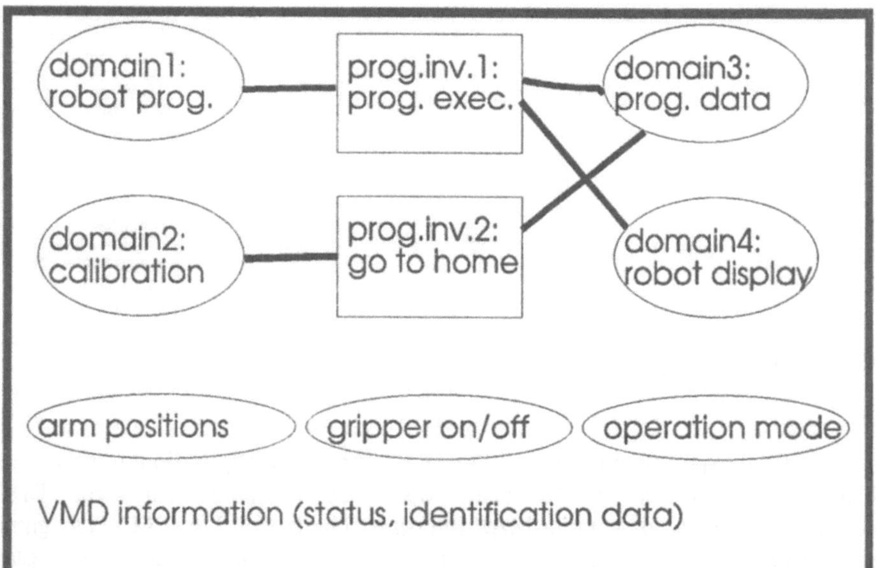

Fig. 7. Virtual Manufacturing Device model of a robot

The object-oriented view of MMS allows to design the VMD model of a certain device, and it is immediately possible to use this model as a specification of the communication where the services (what one can do with a given object of the VMD) are defined and they are working in the MAP networks.

A practical example to control system with OSI. As the goal of this work is to provide Real-time, intelligent control of CIM (FMS) systems the next step was to apply all development results given in the previous sections and to integrate the independently working and tested software-hardware modules.

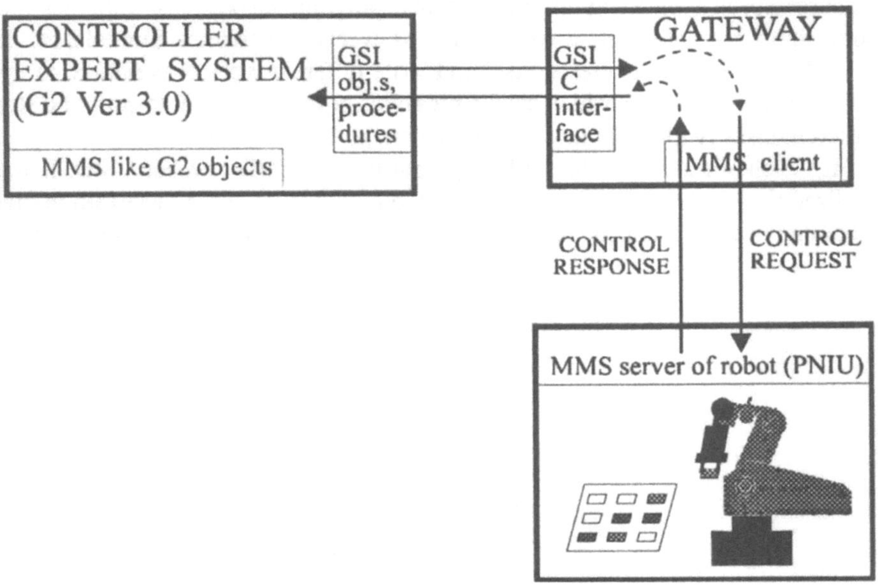

Fig. 8. KB control of a robot with OSI

As a first experimental set-up (Fig. 8.) a simple configuration was chosen. From earlier projects a Mitsubishi robot was available with serial DNC connection to a PC that had a MAP interface. The trivial TIC-TAC-TOE game was served by the robot according to the following manner: a user could play against the G2 system. Of course it is not a problem for that such a system needed. But this well-known game is able to demonstrate very clear the 'intelligent' features of the control. On the other hand the full capability of the MAP networking was used.

The structure of the setup shows that a special gateway was developed between the ES and the robot. It couples the G2 system interface (called GSI) to the MMS standard communication. In this case it was the 'gap software' mentioned in the second chapter. This special gateway of the G2 allows to develop any discrete manufacturing applications to develop in G2 using MMS standard interface without the need of modification of the interface. In the demo version the gateway supports the context management and some program invocation management services.

Two solutions were examined. In a work of a student [59] it was examined how the complete MMS object structure can be built up within the G2. It took a lot of efforts and gave a relatively complicated and not too useful structure. Because of the different services and their parameters many external procedure calls were defined. All implemented MMS services have their own GSI procedures and all MMS object types have their internal G2 object description. The complexity of the inner structure was the same as the MMS itself.

The second solution reduced the necessary objects and procedures to some general terms. It uses limited number of external GSI procedure calls also, but give a rather general (few limitations) interface towards MMS.

The implementation was the following: a special object hierarchy was built up within G2 to map the MMS objects in a more general way. We call it MLGO (MMS-like G2 Objects). MLGO is coupled to the MMS network by means of some special procedures via the GSI of G2. This way the user (who is the developer of the control system in G2) is able to handle all remote resources via the MMS network from his G2 application using MLGO.

4 Conclusions

A good, closed to optimal design and operation of complex, real-time, stochastic systems, as flexible manufacturing systems cannot be reached only by the application of traditional (procedural) programming, but the application of artificial intelligence means (expert systems) is required, too, together with sophisticated modeling and simulation.

Some examples and results were shown how we began to combine a traditional simulation/animation package (Cinema/SIMAN) with advisory expert systems written using expert system shells (ALL-EX, G2) and with C programming to get solutions to some aspects of simulation, quality control, scheduling and control of flexible manufacturing systems.

The implementation of our hybrid programs in SIMAN and G2 was fast enough, G2 was excellent to make the different experimental program runs. Data preparation is - however - a tedious and time-consuming activity.

There is a basic problem to be solved: if G2 and SIMAN are used together most data of the production plant, products and of the production have to be duplicated to be there in both systems. This should be avoided by using some kind of automatic data conversion between the object oriented G2 and the traditional SIMAN. This is one of the interesting challenges for our future research and development - if we decide to keep SIMAN/Cinema for simulation and animation.

It should be mentioned that the application of these up-to-date means (SUN SPARCstation, G2, SIMAN/Cinema, networks) will lead us to write industrially useful Knowledge Based Simulation-Scheduling-Quality Control Programs to support the evaluation and better performance and control of working FMS and to be used in the design (planning) of new FMS implementations.

Based on the test-runs we have to admit that experiments should be designed more carefully, and then more characteristic results can be obtained. In the recent set of tests we tested our programs, and we can conclude that they can be used for checking different possible production schedules of given FMS. On the other hand the test results are strongly FMS and production task related, so we could not yet find general ideas to suggest for scheduling new tasks. Further experiments should be done on different FMS with different production tasks to be able to define quality differences between the scheduling and dispatching algorithms (rules) discussed in this paper.

Real-time intelligent control of FMS is tackled in the last part of this paper as the most difficult, but most promising task to be solved within the frames of "AI in Manufacturing". As example the simulation of a real FMS and then a robotics cell (where the robot is defined as a virtual manufacturing device (VMD)) controlled by a Real-time computer system is presented. The control software is implemented using a Real-time expert system shell (G2).

We made some research and development in MAP/MMS/VMD problems, and the application possibilities of expert systems was studied in the OSI communication point of view. An interfacing problem was solved when the G2 system and the MAP/MMS network were connected to reach Real-time intelligent control.

These results cam be very useful and interesting in the future when they will be applied fore more complex and sophisticated large scale manufacturing and assembly systems which hardly can be designed and controlled without using simulation and expert systems.

As our recently developed system is working with feature based and object oriented data and knowledge for scheduling and quality control, the future integration of our planned FMS design system with CAD/CAM systems should not mean a major problem, so our systems will serve as good bases for Computer Aided concurrent engineering design of products and production systems.

5 Acknowledgment

This work was partly supported by the Hungarian Research Fund (OTKA), projects T 4122, T 14549.

References

[1] T.C. Chang: *Expert Process Planning for Manufacturing*, Addison-Wesley, 1990.

[2] P.K. Wright and D.A. Bourne: *Manufacturing Intelligence*, Addison-Wesley, 1988.

[3] R.K. Miller and T.C. Walker: *Artificial Intelligence Applications in Manufacturing*, SEAI Publications, Fairmont Press and Prentice Hall, Georgia, USA, 1988.

[4] A. Kusiak: *Intelligent Manufacturing Systems*, Prentice Hall, 1990.

[5] G. Halevi and R.D. Weill: *Principles of Process Planning*, Chapman & Hall, 1995.

[6] *Catalogue of Artificial Intelligence Tools*, ed. A. Bundy, Springer-Verlag, 1984.

[7] J.M. Rosenberg: *Dictionary of Artificial Intelligence and Robotics*, John Wiley & Sons, 1986.

[8] *Preprints of International Conference on Intelligent Manufacturing Systems*, ed. J. Hatvany, Budapest, June 1986. organized by the International Journal of Robotics and Computer Integrated Manufacturing

[9] *Artificial Intelligence in Manufacturing, Key to Integration? Proceedings of the Technology Assessment and Management Conference*, ed. T. Bernold, Zurich, Nov. 1985, North Holland, Amsterdam, 1987.

[10] *Advances in Artificial Intelligence, Proceedings of the 1987 AISB Conference*, ed. J. Hallam and C. Mellis, Edinburgh, April 1987, John Wiley & Sons, 1987.

[11] *Artificial Intelligence II, Methodology, Systems, Applications, Proceedings of the AIMSA'86 Conference*, ed. P. Jorrand and V. Sgurev, Varna, Bulgaria, Sept, 1986, North Holland, Amsterdam, 1987.

[12] *Intelligent CAD Systems I., Theoretical and Methodological Aspects, EUROGRAPHICS Workshop*, ed. P.J.W. ten Hagen and T. Tomiyama, Springer-Verlag, 1987.

[13] *Intelligent CAD Systems II., Implementation Issues, EUROGRAPHICS Workshop*, ed. V. Akman, P.J.W. ten Hagen and P.J. Veerkamp, Springer-Verlag, 1988.

[14] *Expert Systems for Engineering Design*, ed. M.D. Rychener, Academic Press, 1988.

[15] *AI Tools and Techniques*, ed. M.H. Richer, Ablex Publishing, 1988.

[16] *Artificial Intelligence, Manufacturing Theory and Practice*, ed. S.T. Kumara, A.S. Soyster and R.L. Kasyap, Industrial Engineering and Management Press, IIE, 1988.

[17] *Artificial Intelligence Implications for CIM*, ed. A. Kusiak, IFS (Publications) and Springer-Verlag, 1988.

[18] *Artificial Intelligence, Simulation and Modeling*, ed. L.E. Widman, K.A. Loparo and N.R. Nielsen, John Wiley & Sons, 1989.

[19] *Artificial Intelligence in Design*, ed. D.T. Pham, Springer-Verlag, 1991.

[20] *Complex Machining and AI Methods, Proceedings of the IFIP TC5/WG5.3 Conference on Process Planning for Complex Machining with AI Methods*, ed. D. Kochan, Gaussig, Germany, Nov. 1991, North Holland, Amsterdam, 1991.

[21] *Intelligent Computer Aided Design, Proceedings of the IFIP WG 5.2 Conference on IntCAD'91*, ed. D.C. Brown, M.B. Waldron and H. Yoshikawa, Columbus, OH, USA, Sept.-Oct. 1991, IFIP Transactions B-4, North Holland, Amsterdam, 1992.

[22] *Artificial Intelligence Applications in Manufacturing*, ed. A. Famili, D.S. Nau and S.H. Kim, AAAI Press/The MIT Press, 1992.

[23] *KNOWHSEM: Knowledge Based Hybrid Systems, Proceedings of IFIP/IFAC Conference 1993*, ed. I. Mezgár and P. Bertók, IFIP Transactions, North Holland, 1993.

[24] *IFAC Preprints of AIRTC'94: Symposium on Artificial Intelligence in Real-Time Control*, ed. A. Crespo i Lorente.
[25] *Preprints of the Second International Workshop on Learning in Intelligent Manufacturing Systems*, ed. L. Monostori, Budapest, April 1995, AKAPRINT, 1995.
[26] *Proceedings of the first World Congress on Intelligent Manufacturing, Processes and Systems*, University of Puerto Rico, Mayagüez, Puerto Rico, ed. V. Milacic, Febr. 1995.
[27] *Proceedings of AIENG: Applications of Artificial Intelligence in Engineering VIII*, , ed. G. Rzevsky, J. Pastor and R.A. Adey, Computational Mechanics Publications and Elsevier, 1995.
[28] *Proceedings of IFAC/IFIP/IFORS Workshops on Intelligent Manufacturing Systems*, Dearborn, 1993, Vienna, 1994, Bucharest, 1995.
[29] *Journal of Automated Reasoning*, Kluwer Academic Publishers, Netherlands.
[30] *Data and Knowledge Engineering*, North Holland (Elsevier), Netherlands.
[31] *Machine Learning*, Kluwer Academic Publishers, Netherlands.
[32] *Minds and Machines*, Kluwer Academic Publishers, Netherlands.
[33] *Artificial Intelligence*, Elsevier, Amsterdam.
[34] *Computers in Industry*, Elsevier, Amsterdam.
[35] *Engineering Applications of Artificial Intelligence*, Elsevier, Oxford, UK, Pergamon Press.
[36] *Artificial Intelligence in Engineering*, Elsevier, Oxford, UK
[37] *Applied Intelligence*, Kluwer Academic Publishers, Netherlands.
[38] *Journal of Intelligent Systems*, Freund Publishing House, UK.
[39] *International Journal of Intelligent Mechatronics, Design and Production*, Middle East Technical University, Ankara, Turkey.
[40] *Journal of Artificial Intelligence Research*, AI Access Foundation and Morgan Kaufmann.
[41] G. Riley: "Implementation of Expert System shell on a Parallel Computer", *Proceedings of the 3rd Annual AI and Advanced Computer Conference*, Long Beach, CA., USA, April 1987, pp. 45-49.
[42] G.L. Kovács, I. Mezgár and S. Kopácsi: "Concurrent Design of Automated Manufacturing Systems Using Knowledge Processing Technology", *Computers in Industry*, 17, pp. 257-267, Elsevier, 1991.
[43] G.L. Kovács and I. Mezgár: "Expert Systems for Manufacturing Cell Simulation and Design", in *Engineering Applications of Artificial Intelligence*, ed. M. G. Rodd, vol. 4, no. 6, pp. 417-421, Pergamon Press, 1991.
[44] G.L. Kovács, S. Kopácsi and I. Mezgár: "A PROLOG Based Manufacturing Cell Design System"; *Practical Application of Prolog Conference*, Abstracts of the Poster Session, April 1992, London, The Institution of Civil Engineers, pp. 3-4.
[45] C.D. Pegden et al.: *Introduction to Simulation Using SIMAN*, McGraw Hill, 1990.
[46] G.L. Kovács and S Kopácsi: "Cooperative Agents for the Simulation of FMC and FMS", *TIMS/ORSA Conference*, Orlando, USA, April 1992.
[47] I. Futó: "AI and Simulation on PROLOG Basis", *International Symposium on Artificial Intelligence, Expert Systems and Languages in Modeling and Simulation*, Barcelona, Spain, June 1987.
[48] *ALL-EX Ver. 3.1 Expert System Shell - Reference Manual*, Applied Logic Laboratory, Budapest, Hungary, 1990.

[49] J.C. Giarratano: "CLIPS User's Guide", Artificial Intelligence Section, Lyndon B. Johnson Space Center, 1989.

[50] *G2 Reference Manual, G2 Version 2.0*, Gensym Corporation, Cambridge, MA, USA, 1990.

[51] G.L. Kovács and S. Kopácsi: "Connecting G2 and SIMAN", *Proceedings of the Gensym Users Society Worldwide Meeting*, Santa Margherita Ligure, Italy, June 1992, pp. 9.1 - 9.10.

[52] P.E. Smith et al: "The Use of Expert Systems for Decision Support in Manufacturing", *Expert Systems with Applications*, 4, pp. 11-17, 1992.

[53] M. Brill and U. Gramm: "MMS: MAP Application Services for the Industry", *Computer Networks and ISDN Systems*, 21, pp. 357-380, 1991.

[54] G. Franklin, A. Meystel et al.: private discussion by e-mail in *Architectures for Intelligent Control Systems Discussion List*, <AICS-L@UBVM.CC.BUFFALO .EDU>, 1994-1995.

[55] T.J. Laffey et al: "Real-Time Knowledge Based Systems", *AI Magazine*, vol. 9, no. 1, pp. 27-45, 1988.

[56] J. Nacsa and G.L. Kovács: "Communication Problems of Expert Systems in Manufacturing Environment", in *Preprints of the IFAC Symposium on Artificial Intelligence in Real-Time Control, AIRTC'94*, ed. Crespo, Valencia, Spain, 1994, pp. 377-381.

[57] P. Buta and S. Springer: "Communicating the Knowledge in K-Based Systems", *Expert Systems with Applications*, 5, pp. 389-394, 1992.

[58] G. Nagy and G. Haidegger: "Object Oriented Approach for Analyzing, Designing and Controlling Manufacturing Cells", *Proceedings of AUTOFACT'94*, ed. Moody, Detroit, USA, 1994, pp. 12/1-10.

[59] Gy. Tömösy: "Implementation of MMS Services into an Expert System", unpublished *Diploma Work*, MTA SZTAKI, Budapest, 1994. in Hungarian

4 Modelling and Analyzing Processes in Production and Administration

Norbert Kratz and Thomas Rose

1 Motivation

The term *business process reengineering* has recently emerged as the most popular buzzword in the context of improving the competitiveness of enterprises in the increasingly competitive and complex environment of international markets [1]. Customer orientation of business processes has grown to be the primary focus for organizing operations—as opposed to the structures following the traditional tayloristic management theories. Departmental structures or individual job allocations require a revision in order to implement new structures and processes which assure maximum benefits to customers in terms of quality, cost or lead time.

Such a radical optimization of business processes carries a significant impact on the design of organisations and processes [2]. Moreover, it also poses tremendous requirements on the utilization of information and software technology within these processes and the design process. Quite frequently, new opportunities result from technological improvements, i.e. they are the key technology to elevate the performance of an organization [3]. Within this context, information systems are no longer limited to the support of isolated tasks, but information systems ought to entirely support processes by providing any appropriate and required information at the right time and place, to provide services for decision support and to improve the transfer of information and the quality of communications.

Another essential feature of an optimized business process is its flexibility, which embraces the capability of an organization to permanently adapt its processes according to changes of its environment. This agility is one of the key issues for competitiveness. In many cases, agility can only be accomplished by adequate utilization and employment of information and communication technologies.

In our opinion, a successful approach towards reengineering of business processes essentially requires a comprehensive set of tools for the formal representation and analysis of current business processes and those processes that are still on the drawing board [4]. This observation results from the fact that on the one hand organizations and processes tend to be reasonably complex. On the other hand, rather different but inter-dependent perspectives and optimization goals—such as time, cost, capacity, or quality—have to be considered in concert.

Yet, we do not claim a universal approach that intends to fit any bill. We rather propose a framework, that incorporates different views, while each view is founded in the specific requirements of the process objectives to be accomplished.

Within this engineering framework, it is important to have a closer look at the ways in which process modelling is applied (cf. figure 1.1.). Essentially, process modelling is the first step to come-up with a formalization of processes. This formalization serves as the basic platform to build upon.

Firstly, it provides the means to discuss the current status of processes and reachable improvements [5]. Based on this assessment, new software services might be designed to support certain process features. Process formalization furnishes the basis to engineer software components supporting certain tasks .

Secondly, process modelling will be utilized when analysing process performance; both in a static way, e.g., activity-based costing [6] or time-based analysis [6], and in a dynamic way, e.g., dynamic behaviour of processes.

A third application area is process implementation. In the course of implementing processes, process models are used in two ways. On the one hand, the process model supplies the control information as required by workflow engines [7]. On the other hand, process models provide the basic design of a process to be used in customizing standard software, such as SAP R3. In the latter, the designed process is matched against ideal reference models. Based on this matching and appropriate implementations of components of the reference model, operational systems are configured.

A single project typically shows these different aspects in concert. For example, one might think of a reengineering project. This project starts with a specification of the current process and designs improvements which involves an analysis of performance gains in order to justify investment in new technology such as workflow management, electronic document management or archives. Finally, parts of the process model translate into control information of the workflow engine, while other sections serve as requirement specification for software engineering activities in this reengineering project.

This short example suggests that an unconcerted set of process models—whilst each model dedicated to a specific task of process engineering—might not be appropriate. Such a bundle of process models (for a detailed overview on different approaches towards process modelling see [8]) will likely result in inconsistencies. Due to significant modelling efforts for each task, such a unconcerted approach will also imply high project costs because of the low degree of model reuse.

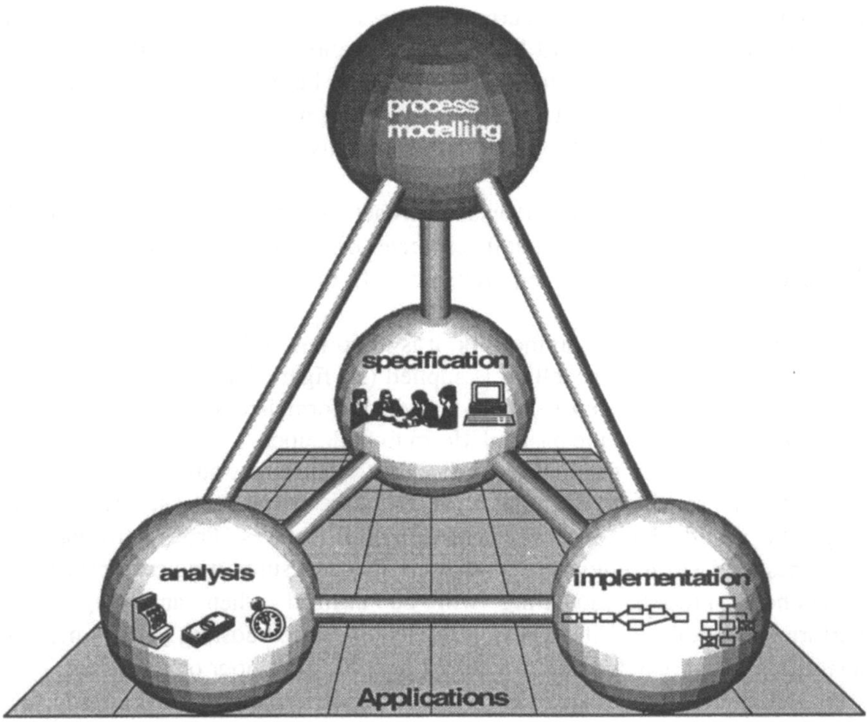

Fig. 1.1. Applications of process modelling

To reiterate, integration requirements are apparent for formal process representations, yet various ways of model utilization ought to be considered. In a nutshell, our approach considers process integration along two dimensions. *Vertical integration* refers to the migration of knowledge about processes from the early phases of the requirements specification down the reengineering project to operational process implementations, and vice versa with respect to system evaluation and requirements assessment. *Horizontal integration* refers to the communication of process assessments among various analysis approaches and tools, be it various ways of static and dynamic process analysis. Yet, horizontal integration also involves the implementation of processes in different environments.

This paper is organized as follows. Following the motivation of our process framework, section 2 elicits the process modelling requirements. These requirements have been drawn from several application-oriented projects in the fields of manufacturing, administration and engineering. They pose the design principles of our *enterprise design and execution environment*—named *ProMAX* (Process Modelling, Assessment and Execution Environment) [9]. The ProMAX approach to process modelling, analysis and implementation is presented in section 3—in accompany with its architecture and pertaining workbenches. Then, sections 4 through 6 are devoted to specific aspects of our process framework. Section 4 refers to the *impact on software engineering* with a specific emphasis on the *reuse of building blocks* in the operationalization of processes. These building blocks arise from software components that are already employed for certain tasks, or they are provided through means of commodity software use. Section 5 reports on process modelling experience and potentials in the realms of *administration*, be it public or private sector. Here the focus is on the *change in processes due to document management facilities*, *customer care* as an important objective for performance evaluation, and *electronic interfaces* to parties which are related, but outside the operational scope. Then, section 6 compiles experience on *processes in production environments* focusing primarily on process identification and software engineering. Finally, section 7 draws conclusions from our experience with ProMAX, points to future work for improving ProMAX and directs the attention to issues of process enginering that require further research and development efforts.

2 ProMAX Requirements

In this section, we will derive some functional and non-functional requirements for the design of ProMAX, i.e. what are the features of a process engineering environment in terms of expressiveness and functional scope. As already indicated, one way of structuring is the distinction of vertical and horizontal integration. Vertical integration addresses the inter-dependencies among phases of the engineering process, while horizontal integration rather copes with intra-dependencies of tools dedicated to a specific development phase. Vertical integration is one of our chief concerns, since we strive for process implementations on top of existing workflow management systems [10, 11]. As dictated by budget limitation—an essential, non-functional requirement—process implementations ought to reuse implementation platforms where available. Vertical integration should support the mapping of process models to control languages of different workflow engines, as far as possible.

The second question is horizontal integration, which in our opinion is the key issue of process modelling in order to penetrate a wider range of engineering applications. Horizontal integration provides the interoperability means for different forms of process analysis and assessment. Each of these applications—

called advisors in the remainder of the paper—are dedicated to the objective, which they are supposed to achieve based on their individual competence of the domain.

Common process modelling tools, such as ARIS [12], perform some kind of static analysis on cost and time. The results—although of course indispensable for the optimization of processes—tend to be of limited value for designing a good-quality process, because they do not help to answer the question of what actually constitues an effective process. Please note, that we do not argue on efficient processes, since efficient processes indeed can be determined by time and cost considerations. However, a good process or an effective one is also characterized by other factors.

The basic question is, whether a process designed is a good process or not. But what constitutes a good process? This question is definitely not just answered by time and cost attributes, but also by other criteria, such as customer orientation, flexibility to cope with changing markets and market changes, resources used, products and quality to be assured, and possible process adaptations to cope with varying workloads, e.g. number of customer requests. These criteria require much more sophisticated support coordinates than traditional calculations of process parameters.

However, there might also be some project contexts, where less powerful approaches—less powerful from a viewpoint of analytic reach—appear appropriate. Thus, one might want to keep the models as simple as possible. Hence, we argue that an appropriate architecture should provide a framework, which supports the calibration of concepts. Based on some basic enterprise and process model, a process designer is guided in order to add parameters, which she deems suitable in order to accomplish the reengineering goals, to her model.

In our opinion, an enterprise design environment should meet certain requirements, including a thorough representation of enterprises and their structures, user support for using features of the design environment and the ability to change the enterprise description easily. The functional requirements follow first, while the functional requirements imply representational requirements.

Modelling of Goals and Objectives. Process goals and objectives vary among different enterprises. Therefore enterprises have to specify their goals and objectives explicitly, in order to have appropirate measures for evaluating their compliance. Methods similar to quality function deployment are instrumental to derive goals, objectives and their relationships. Factors, which place goals into the right proportion, should be derived and used to find the optimal process in case of multi-criteria optimization.

Predefined and User Defined Evaluation Criteria. In addition to predefined process performance measurements (evaluation criteria), user defined evaluation criteria are conducive to increase flexibility of process and enterprise assessments. Varying criteria are important for process assessment in order to adapt to specific contexts. Moreover, a flexible specification of data to be captured during simulation runs is necessary.

Simulation Based on Enterprise Models. Evaluation of process and enterprise performance employs simulation. Simulation runs should be based on different sets of data, called "scenarios", which allow the representation of optimistic or pessimistic prognosis. Tools to instantiate simulation models with initial data are needed. Simulation model execution and data recording according to current specifications of evaluation criteria have to be implemented. Graphical animation of processes communicates the current status and performance of the processes to the users.

Performance Assessment Based on Goals and Simulation Results. Assessment of processes relies on the comparison of to-be values (resulting from goal /objective definition) and as-is values (gathered during simulation) by using charts and graphics. Strengths and weaknesses of the processes under consideration should first be identified by an analysor, which encodes knowledge on performance indicators and means of bottlenecks. Then, the analysor visualizes the strengths and the weaknesses to focus the designer's attention.

Identification of Possibilities to Improve Structures and Processes . If values of performance indicators do not comply with business goals, then it is necessary to identify the culprits of the mal-performance [13]. Users need tools that help them to identify places of process performance improvements. Such tools require rules, be it expertise in the domain or rules of thumb, that are derived from expert knowledge (e.g., business process reengineering cases). Once these rules are formally incorporated into the enterprise model, then analysors can determine the effects of process strutures on indicator values.

ProMAX is also intended to characterize and assess an enterprise's behaviour. At first glance, there appear two approaches to engineer the processes of an enterprise. The first one starts with *ideal processes*, i.e. processes are designed without any particular relationship to the enterprise. For instance, one might design processes for distribution and delivery services that meet goals that are commonly accepted. Here, the process, its structure and its quality is the focus of attention. The second approach focuses on the competence, the capablities and the facilities as they are currently available to the enterprise. Here, the primary focus is *current competence.* Then, processes are reengineered upon the competence. If an enterprise has a strong record on particular services or certain knowledge of the market, then this competence will become a key part of the redesigned process.

When following the second approach, one needs to model the processes of an enterprise and where its competence specifically contributes to the performance and objectives of the process, i.e. how the capabilities of an enterprise are used for producing products or services for customers. This approach recommends that the process models clearly separate the representation of processes and the capabilities, e.g., resources, products, services and organization structures, which they employ to meet their goals.

Levels of Abstraction and Concern. In order to understand an enterprise's structure and behaviour, it is necessary to represent any concept of the company and to describe their abilities and restrictions. Yet, many concepts are common throughout all types of companies. These common concepts will be captured in terms of *generic models*. Entities needed for all kinds of enterprises can be grouped together into generic models, e.g., all entities concerned with resources or all entities concerned with organizational structures belong to one generic model. Generic models should be structured according to concepts commonly used within enterprises.

Since enterprises vary with respect to their sector of industry, products, naming conventions, etc., it is not possible to prepare appropriate building blocks to represent every company. Instead of providing building blocks, which are easy to use, but inappropriate for detailed modelling, a layered approach is recommended, which structures concepts on different levels of granularity and detail. Branch specific models include concepts specific to certain sectors of industry or branches. Starting with each type of generic model, branch specific models are developed by refining existing concepts (predefined in generic models) and adding new concepts. It should be possible to create as many layers of branch specific models as needed.

Based on branch specific models, users should be able to define models tailored to their company. Users should make use of branch specific models, refine them and include additions to represent their companies structure and behaviour as accurate as possible. Once enterprise specific models are implemented, they can be reused. Enterprise specific models should be updated when the (real) enterprise changes.

The concept of generic models resembles that of reference models, such as the ones coming with SAP's R3. Reference models provide a "guided tour through ideal processes" for specific tasks. In this regard, both concepts are similar. But they distinguish with respect to content. Generic models also include a set of concepts to describe capabilities and competence available to an enterprise, as pointed out earlier in this section.

An overall classification of models is maintained across all layers. It classifies models with respect to certain topics to guarantee clarity. *Resource models* allow the description of concepts related to resources, e.g., availability of resources, resources that are usable many times, resources that are sharable and resources that are consumed. *Product models* describe concepts to represent products, e.g., parts, release dates, versions etc. *Organizational models* describe company structures in terms of departments, jobs, roles, employees and their relationships. *Process models* describe activities, their preconditions, results and relationships. *Service models* classify activities based on their type (e.g., writing services, machining services) and establish the connection between resources or employees and tasks (activities) they can perform.

Interface Between Design and Production Environment. Tools are required for the analysis of the performance of implemented processes in order to find out whether an implemented process mismatches the objectives or to detect impediments in a process's performance. Therefore interfaces are required to feed back real world data from the operational production environment into the evaluation and assessment tools of our environment.

In order to minimize implementation efforts it should be possible to translate concrete process models into the production environment without considerable effort. ProMAX provides methods to translate formally represented business processes into a workflow system representation for a set of workflow management systems.

3 ProMAX Architecture and Environment

This section describes the overall architecture of ProMAX and presents the accompanying enterprise engineering workbenches that come as part of ProMAX.

3.1 Overview of the ProMAX Components

ProMAX embodies four properties that seem decisive for the success of modelling business processes and corresponding process implementations.

- Firstly, ProMAX encompasses a concept taxonomy to represent enterprises, their objectives, capabilities and business processes. Equipped with accompanying simulation, performance assessment and optimization tools ProMAX allows the engineering of business processes. ProMAX can be used to engineer all kinds of processes, for example, administrative or manufacturing processes. Processes are represented within a process description framework.

- Secondly, ProMAX deploys a repository that manages standardized models and entities available for enterprise engineering. In the process of implementing business processes, these entities have to be considered for reuse. In order to implement processes as specified, predefined entities can be plugged into ProMAX´s process description framework.

- Thirdly, ProMAX will assist designers to evaluate process performance and provide heuristic guidelines to improve processes with respect to a set of performance criteria.

- Fourthly, ProMAX avoids gaps between process design and execution by connecting the enterprise design environment with operational environments, e.g., workflow management systems.

ProMAX can be used as a stand alone tool for business process engineering. It provides all evaluation, assessment and optimization facilities based on its own data representation, since process descriptions ought to be independent of any specific workflow management, office automation or production planning system. To connect ProMAX with operational production environments, assistant-guided mappings are intended.

3.2 Architecture

The overall ProMAX architecture is layered around an integration kernel, which comprises generic models for the representation of enterprises and concepts for the representation of the reengineering process, e.g., process design, evaluation, assessment and optimization. The architecture is designed to integrate process execution and software development methodologies and tools into the overall architecture.

To ease changes to models, ProMAX facilitates distinct representations of processes and related information. ProMAX comprises models to describe an enterprise, its processes, products, resources, services, organizational structure and business objectives. The process model includes references to information embodied in other models of the same enterprise. Since distinct but related models are used, interactions between processes and resources, organization and products become explicit.

Processes have to be described at different levels of abstraction according to the knowledge available at a certain point in time. Therefore different constructs to represent relationships are available.

ProMAX enables users to model an enterprise and its processes, to measure process performance and it offers guidance to improve business processes. In order to improve an enterprise's structure and to enhance process performance, we implement tools to support and advise users. Advisors point to opportunities of change based on model performance assessment and assessment of interactions between processes and modelled products, resources, objectives and organizational structures.

Representation of enterprises, processes and related information employs distinct models instead of one overall model in order to enable users to check different process designs and update models. ProMAX is based on a set of generic models. Each comprises entities of a certain domain and basic methods to handle these entities and to compute attribute values. These entities were constructed using our modelling environment and offered for future reuse by ProMAX.

Users will interact with ProMAX using five workbenches. Each workbench comprises different tools, while each tool offers a special service, e.g., calculation of quantitative information, visualization of models or editing special views on models.

The basic steps of enterprise design are mirrored by the ProMAX workbenches. The workbenches should help users to accomplish all tasks necessary to achieve the goal of each major process step. The use of workbenches in the engineering cycle and their position in the design and production environment are depicted in figure 3.1.

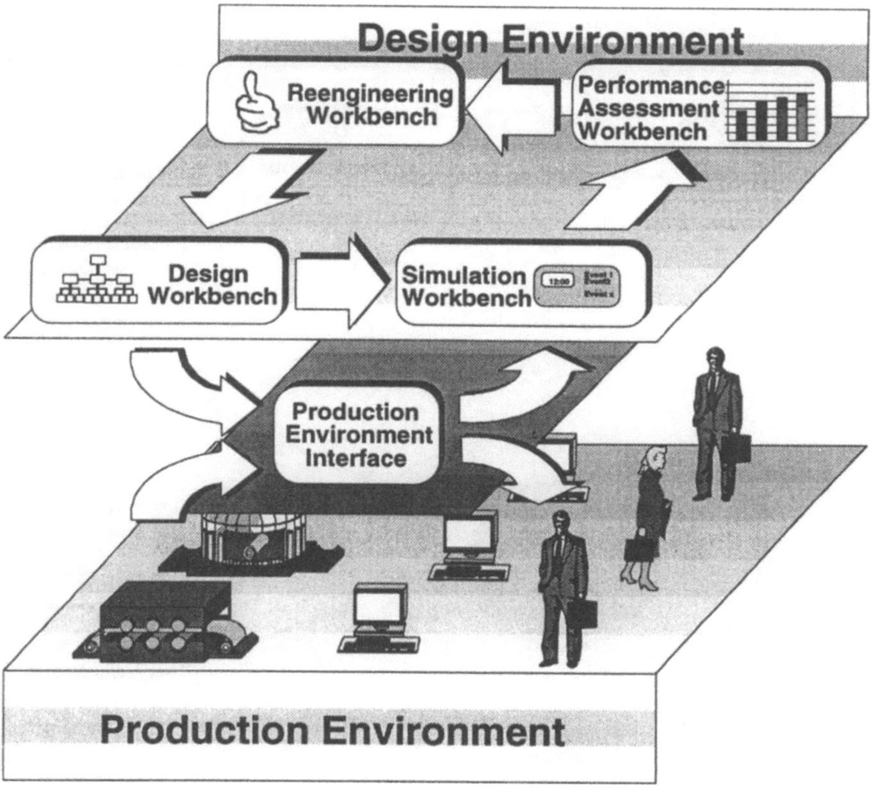

Fig. 3.1. Enterprise design environment and engineering cycle

If not merely information but advise is needed (for example to increase an enterprise's performance) tools comprised in the workbenches are not sufficient. Sophisticated reasoning functionality is offered by advisors which users can call to solve problems. Advisors are dedicated programs that incorporate knowledge to solve sophisticated problems. The knowledge needed is represented in terms of micro theory, method or set of rules. The advisors access enterprise models, gather data and draw conclusions based on retrieved information and their own knowledge. Figure 3.2. depicts the relations between workbenches and advisors and the enterprise model.

The thick drawn arrows represent enhancements on underlying models. The design workbench supports the creation of enterprise models and sub-models. Advisors can trigger the expansion of the knowledge base about the enterprise, if they require additional information to accomplish their task. Therefore, advisors can arrange for additions by requesting the Enterprise Design Workbench to inquire information every time an advisor is invoked with a model, which does not include appropriate information.

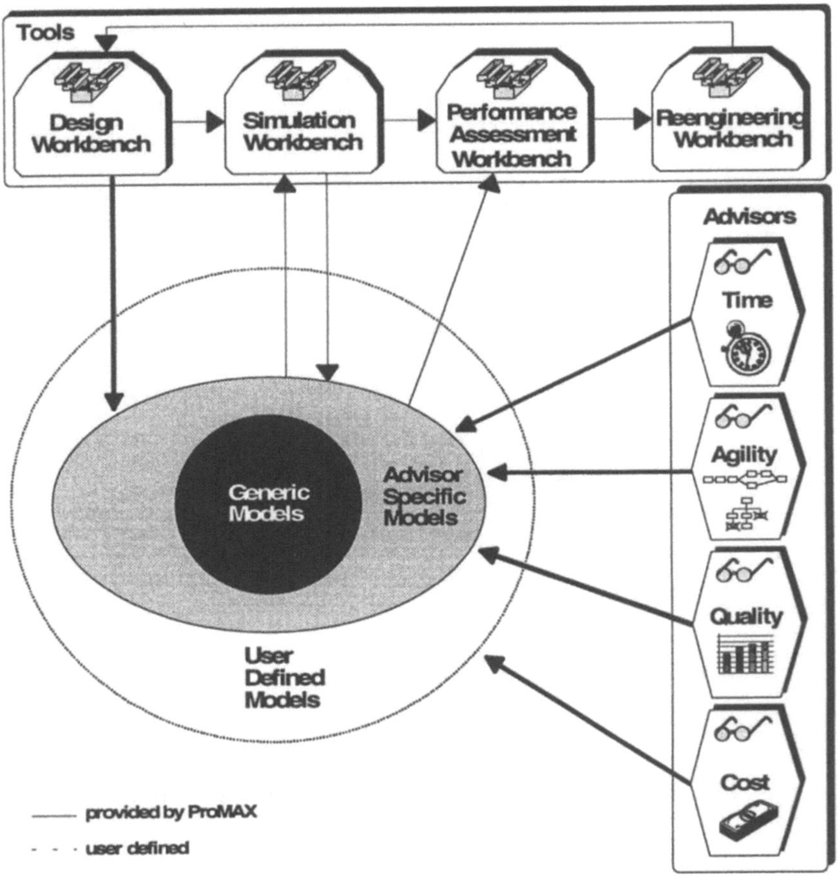

Fig. 3.2. Interactions among workbench, advisor and enterprise model

The thin arrows symbolize the use or change of data incorporated in the enterprise model. For example, the simulation workbench accesses and updates enterprise model data whereas the analysis and diagnosis workbenches use but do not modify model data. In figure 3.2. the use of advisors by workbenches is depicted by thin lines.

The main advantage of our architecture is the user friendly interface to enterprise models and advisors offered by ProMAX's workbenches. The functionality of workbenches supports users during all steps of the enterprise design process. According to the current knowledge about the enterprise and its processes models can be enhanced dynamically. New advisors can be added to the system without considerable effort.

3.3 Workbenches

Workbench	Tools
Design	Library Constructor Library Manager Organization Editor Resource / Product Editor Goal Editor Service Editor Constraint Editor Process Editor Design Rationale Checker Model Integrity Checker
Simulation	Evaluation Editor Scenario Editor Simulation Tool
Performance Assessment	Performance Evaluator Process Comparator Static Problem Identificator Dynamic Problem Identificator Causal Explanator
Reengineering	Rule Selector Model Optimizer
Production Environment Interface	Model Translator Real Process Monitor

Fig. 3.3. Decomposition of workbenches into task-oriented tools

3.3.1 Enterprise Models

The workbenches provide support for model building and modification (design and redesign). In particular, the tools allow interactive manipulation and editing of the enterprise model. The workbenches in the system address software and visualization requirements and reasoning tasks as major issues. The reasoning tasks of the various workbenches include the evaluation and analysis of properties of the enterprise model. The workbenches are intended to be used as an "enterprise model interpreter" - the analysis is performed incrementally as the model is being constructed. Each workbench comprises many complex tasks. Workbenches are implemented as a collection of tools to improve clarity and ease maintenance (cf. figure 3.3.).

Goal / Objectives Editor. The Goal / Objectives Editor defines the enterprise's goals and objectives in terms of objects, their attributes and required values. It allows for the assignment of values to predefined goals / objectives (e.g. Profit > 500.000 $), as well as the definition of new goals by selecting objects and their attributes. In addition, there are facilities supporting the definition of goals in terms of lower level objectives (e.g. throughput > 5000 parts per day) and the assignment of weights to goals to allow for multi-criteria optimization. The Goals / Objectives Editor provides basic facilities for integrity checking of goals with respect to the enterprise model. This ranges from ensuring that values are defined for all goals to the more complex tasks of the identification of user requirements to derive goals and their interdependencies. The Goals / Objectives Description Editor is primarily an acquisition tool; as such, there should be an appropriate interface for defining and editing goals. This includes providing a table of goals and their intended values.

Organization Editor. The Organization Editor offers distinct views for organizational modelling, depicting hierarchy and grouping of jobs. Inserting and deleting jobs using the graphical representation of organizational models in the Organization Editor automatically creates objects and relations.

Process Editor. The Process Editor allows users to model and visualize processes and their components. It therefore offers textual and graphical methods to describe processes using predefined primitives. Understanding of complex processes is supported by different views on a process (e.g., a temporal view to clarify temporal relations of activities, state view to represent preconditions, triggers and results of activities). The state view represents the relationship of processes and the resources, products and employees. The visualization of processes requires a symbology for objects, as well as a graphical representation of temporal constraints. In order to establish relations between activities or states and all other types of objects (like resources or jobs) different types of lines and arrows have to be offered in order establish connections by drawing lines / arrows between objects.

Resources / Product Editor. The Resources / Product Editor allows to edit existing resource / product descriptions, to create subtypes of predefined resource classes and to define attribute values for resources / products. Attributes and relationships can be depicted graphically or using a table field. Additionally it should be possible to visualize special relations graphically (e.g. the "contains" relation in order to generate a bill of material). According to user needs different types of relations can be presented graphically.

Service Editor . The Service Editor is used to create services and to construct hierarchies of services. Services are constructed by establishing relationships between agents (resources / employees) and activities. The Service Editor lists all available agents and activities and allows to create new activities. A service object is created (deleted) if a connecting line between the agent and the activity is drawn (removed). The Service Editor presents all services of a resource / employee as a list or diagram.

3.3.2 Analysis and Optimization

The tools in the Simulation Workbench assist in the specification of information required for simulation and in the execution of process models. The *Evaluation Editor* enables users to describe all parameters that shall be gathered during a simulation run.

The *Scenario Editor* is used to define the scenario which describes the current state of the enterprise under consideration. A scenario comprises the initial state of the enterprise that has to be simulated and the set of events that will change the initial state during simulation.

The *Simulation Tool* must schedule and execute events (perform start and end actions), as well as represent and advance simulation time.

3.3.3 Performance Assessment

The Analysis or Performance Assessment Workbench provides tools for the description of presentation preferences depending on the type of data and user requirements. In addition, it saves relevant performance metrics in order to compare alternative process definitions. This workbench consists of tools to evaluate and analyze processes with respect to the evaluation criteria. The Performance Assessment Workbench presents results of simulation runs in different ways (tables, charts, Gantt-Diagrams).

The primary task of the *Constraint Evaluator* is to evaluate the constraints and performance metrics of the enterprise at different points in time.

The *Process Comparator* compares metrics of different process.

Given an enterprise model, the *Static Problem Identificator* will identify potential problems that are independent of any given scenario.

The *Dynamic Problem Identificator* tool analyzes how a process is integrated into the enterprise and co-ordinated with other (concurrent) processes.

The primary task of the *Causal Explanator* is to provide explanations for conflicts between the results of simulation and the goals / objectives and constraints within the enterprise model.

3.3.4 Reengineering

Within the Reengineering Workbench rules based on business process reengineering strategies shall be used to derive hints to improve process and enterprise models. Basic principles of business process reengineering (c.f. figure 3.4.) will be incorporated into ProMAX's advisor modules. Therefore the business process reengineering knowledge has to be made available for computer support by translating these high level ideas in computer processable methods or rules. These rules will be used to improve a processes structure.

(1) Organize around outcomes, not tasks
(2) The process should be performed by the one who uses the output
(3) Information processing work should be included into the process that produces the information.
(4) Geographically dispersed resources should be treated as if they were centralized.
(5) Parallel activities should be combined instead of integrating their results.
(6) Decisions should be made where work is performed.
(7) Control should be build into the process.
(8) Information should be captured only once and at its source.

Fig. 3.4. Basic business process reengineering rules

The primary task of the tools within the Reengineering Workbench is to assist in the redesign of an enterprise model based on problems encountered during simulation. The tools in this workbench give hints to improve the process or enterprise with respect to encountered problems.

The generated suggestions are based on modelled heuristics and rules of thumb. Since there are many different heuristics or rules of thumb that could be applied in a specific situation a possibility to select one method or a combination of methods / rules should be available. The *Rule Selector* can be used to select the appropriate methods / heuristics to optimize the process and other parts of the enterprise model.

Beginning with the problems identified by the Causal Explanator, the *Model Optimizer* will augment the enterprise model by either giving hints to prevent the problems or to take counter measures to repair the inconvenient performance.

Figure 3.5. gives an overview about problems that can be identified and tips ProMAX generates to remedy this problems.

Problem	ProMAX's Suggestions
exceeding failure costs	insert test-activities before activities which produce a high value added.
high failure rate	insert test activities at points where the cumulated failure rate of activities would exceed
bottlenecks	provide additional or alternative resources to reduce bottlenecks.
long lead and waiting times, many queues	reduce the number of involved departments / employees e.g. by reorganizing departments according to processes
long lead and waiting times, high costs	avoid back and forth change of persons dealing with one object (procedure / form / part / product) e.g. by enlarging competencies
insecure processes, unpredictable lead times	enhance the security of the process by providing backup for resources with high failure rate.
priorized jobs are not processed in time	allow only a certain percentage of express jobs
an important job was delayed	insert objectives for this special job
only some goals could be achieved	change weighting of goals (decisions during process execution are adapted automatically)
throughput to low	change order release strategy
proportion of time needed for activities with net product and time needed for activities without net product (value added) is to low	identifies reasons for long durations of activities without value added, tries to find ways to avoid causes (according to other rules)
co-ordination of concurrent processes not possible without offending constraints	increase flexibility (and security) by providing alternative possibilities to perform a process.

Figure 3.5. Problems and resulting suggestions to optimize processes

3.3.5 Interface to Production Environments

The Production Environment Interface is used to transfer process models into the real production environment (e.g. using a workflow management system) and to use the reasoning and visualization facilities that come with the design and evaluation tools for real process data. Therefore the *Model Translator* translates the developed processes into a description usable by work flow management systems. The translated representation of the process can then directly be executed within the productive environment.

To allow for monitoring and assessment of implemented processes the *Real Process Monitor* is needed. It has to receive data gathered during execution of the real life process and to provide the Performance Assessment Workbench with information. If process performance is found to be not sufficient the Production Environment Interface has to initiate the reengineering process by invoking the Reengineering Workbench.

4 Impacts on Software Engineering

As already indicated, a universal approach for the *implementation of processes* in production and administration environments is indispensable for the success of business process reengineering. The implementation process ought to be both efficient and effective. *Efficiency* refers to the implementation of re-designed processes within reasonable short time frames, i.e. eventually within six months, while *effectiveness* alludes to the transition of re-designed processes into full-scale operation with a level of accuracy as high as possible.

The implementation of business processes basically follows two approaches. One centers around the utilization of workflow tools. Going down this avenue, re-designed processes are implemented upon facilities such as the ones supported by contemporary workflow management systems. This approach is in particular attractive for processes that mainly revolve around document processing and distribution—perhaps extended by some form of intelligent document routing and forwarding. A fine-tuned combination of workflow and regular standard tools—e.g., for archiving, word processing, etc.—has proven to be adequate for the implementation of business processes. Yet, appropriate interfaces among tools utilized in the course of the implementation process are necessary.

Despite the reuse of commodity software, individual software implementations prevail when the functional complexity of business processes increases or the set of intended services is rather specific to the business of the individual enterprise. The latter in particular embraces services and facilities deemed instrumental for the objectives and competitiveness of an enterprise.

In our opinion, the distinction of both approaches is blurred. Many situations recommend a combination of both approaches. This combination calls for an appropriate architecture for the reuse of ready-made software components from the shelf and the individual development of software components tailored to specific purposes.

Some software components will be designed to perform dedicated activities within a process. First of all, the relationship between process activity and software component ought to be managed. On the one hand, management of this relationship serves documentation purposes. On the other hand, management of this relationship is also instrumental for the reuse of already available software components in the course of implementing another business process. When a similar activity occurs in some other process, one can derive that there already exists a software component providing related services.

A repository is in the center of this approach as depicted in figure 4.1. The repository manages information about process activities, software components and their relationships. It basically documents the development process and encompasses a library of software components and building blocks for the implementation of business processes.

As part of its user interface, the repository is equipped with a browsing facility that aims at the identification of matching software components, i.e. building blocks to be (partly) reused. It supports a developer in browsing through available components and experience about how certain process features have been accomplished in earlier projects. The identification of matching items employs information accumulated during both the documentation of the analysis and the design of the business process and the corresponding software engineering process. This distinction of concern mirrors the assumption that reengineering of business processes should consider the services that are already in place in an enterprises, i.e. what is the specific competence of an enterprise, rather than optimizing processes independent of the specific competence of an enterprise.

Besides the reuse of implementation experience, the repository is also instrumental for configuration and change management in order to adapt to changing environments. In particular, it manages services and facilities installed and their corresponding parameters. Hence, it serves as enabling tool to utilize commodity software for implementing customized processes. This complies with an apparent development in the software engineering camp. Commodity software penetrates the market in significant numbers. Rather than developing individual software, one uses standard software packages and customizes regular services in order to attain individual solutions. This requires a common architecture of software modules and business processes to ease such adoptions of regular services. Some of the more complex standard software products, such as R3 from SAP, even come with an embedded workflow system. Thus, the scene resembles the one described at the beginning of this section.

There is ample evidence at customer sites, that heterogeneous approaches are necessary to combine the utilization and customization of commodity software together with the development of software components aiming at individual business process features. This requirement is even intensified by customer demands to reengineer business processes—from analysis to implementation—within time frames from six to twelve months. Contemporary software engineering practice will likely fail in meeting this requirement.

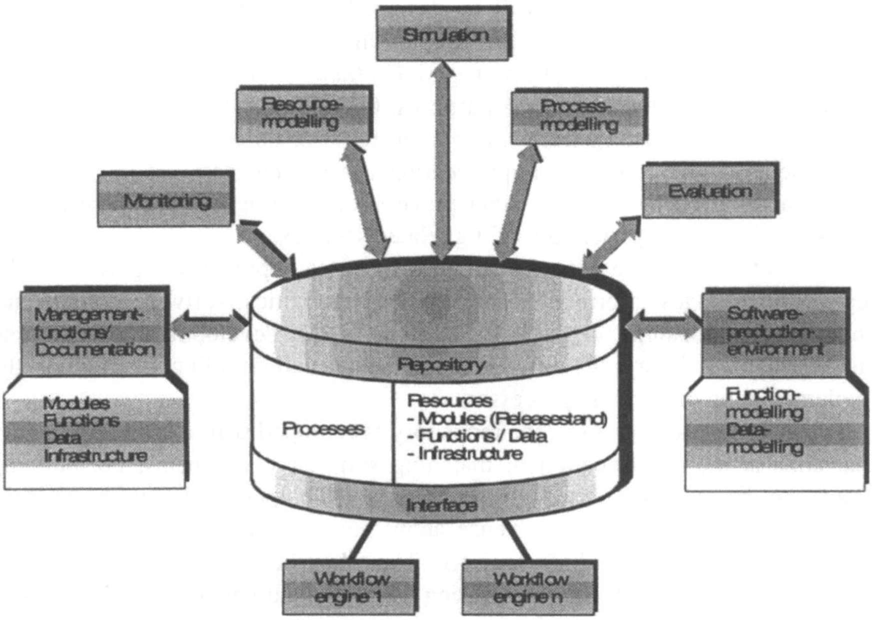

Figure 4.1. An architecture combining software and business process engineering

5 Applications in Administration

Business process reengineering in administration—be it public or private sector—is dominated by the utilization of tools for *communication* and *information management*. Empirical studies provide ample evidence, that lack of *document management* and *processing* facilities is the prime culprit for low levels of performance. Performance improvements—be they measured in terms of turn-around times, lead times, response time to customer requests, or customer satisfaction—are contingent upon an appropriate orchestration of communication and information management technology. Here it is, where technology is the prime contributor to reengineering. This is not to say that office automation did not carry an impact.

The current leap is rather contingent upon the technological improvements with regard to multimedia document representation—including tables, graphics, pictures—and sole capacity. In particular shared document processing, which basically leads to an increased freedom of organizing the operational flow, is indispensable for the reengineering of processes. The *virtual office* is at the brink.

Electronic document management [14] is the key factor to reengineering business processes in administration. Reengineering revolves around the *concept of documents*. Business processes in administration generate, access and update documents. These documents might include structured data, e.g., database entities, and unstructured data, e.g., plain letter received or generated by a word processor. Reduction of idle times and interleaving or tasks is an essential design objective. Electronic document management facilities provide the enabling means to support and partially automate business processes in administration.

Figure 5.1. depicts the basic architecture to support business processes in administration. In its core, it utilizes several document management facilities, including a filing facility for documents that frequently change, a document archive for long-term recording, a process engine and document router, and a knowledge base of the organizational structure.

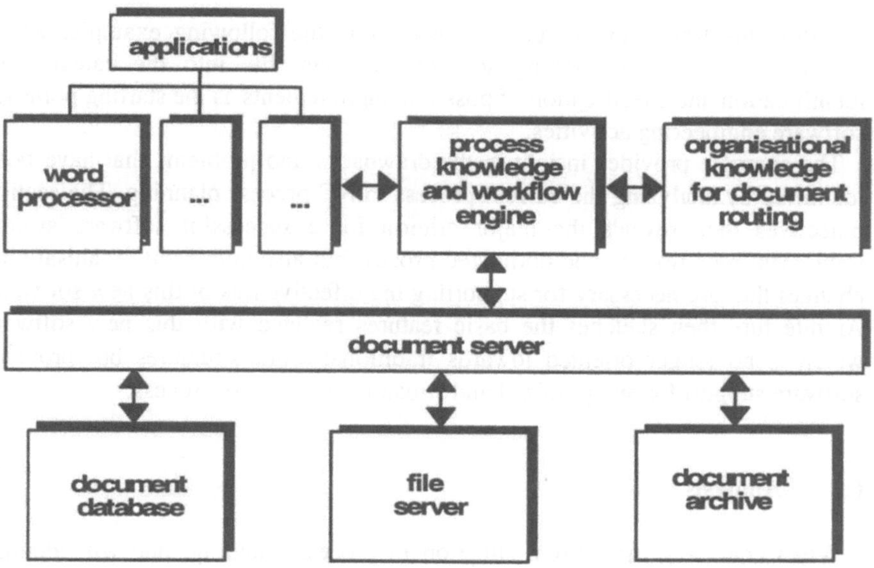

Figure 5.1. Electronic document management

Reengineering starts with a formal representation of processes and documents involved. Current process structures are highly affected by the flow of documents and the conceptual coherence of documents through means of folders, i.e. parallel processing is restricted to the accessibility of folders and documents. This picture naturally changes, once documents become accessible for every site involved—of course requiring an appropriate authorization. Then, process structures are defined by conceptual relationships and inter-dependencies of activities. Yet, additional inter-dependencies might be incorporated to synchronize the flow of documents. Moreover, re-design of processes has to obey process structures that are defined by regulatory and standardization bodies, which aim at a standardization of processes in administration, e.g., at government agencies.

Once electronic document management is in place, then new opportunities emerge in the context of interfacing. Interoperability with other departments and organizations, e.g., by using EDIFACT, is one opportunity. The other opportunity centers around the dissemination of electronic forms to citizens and customers, and electronic call for tenders, where within the legal boundaries.

6 Applications in Production

Within the framework of process modelling, the following example, which describes one of our ongoing software projects, fits into the category of identification and specification of possible improvements as the starting point for software engineering activities.

The example provides insight in the drawbacks and problems that have been identified by analyzing the current process in NC-process planning. The section objectives then reveals the major criteria for a successful software system implementing a revised and optimized process but also points out organisational changes that are necessary for supporting the effectiveness of this new software. Architecture then sketches the basic features realized with this new software which is no longer oriented towards traditional work structures but provides software support for an optimized and streamlined business process.

6.1 Analysis

When analyzing the current situation in process planning, one will identify some weaknesses and drawbacks in the way this process is supported. The current situation may be characterized by the following items:

• lead times in process planning are to long, especially due to necessary iterations,

• lot of rather repetitive work, and

• the need to manually input information already covered by some software system into another one.

The quality of resulting NC-data primarily results from the individual experience of the NC-progammer involved in this process which implies that it is not possible to exploit what could be called a companies' know-how on production and make it available to all engineers. Due to gaps in the process chain consistency between design data and NC-data often gets lost especially if there are changes in design or in the way a product is going to be manufactured.

With CNC-tooling machines becoming more frequently installed in industry together with reduced lot sizes and the demand for shorter lead times in the proces planning departments more and better NC-progams are needed for production. However this cannot be achieved with current organizational structures and todays informations systems.

6.2 Objectives

At least some of the drawbacks -especially the problem of inconsitencies- might be cured by organizational provisions but these again would be quite time- and cost-consuming. Thus the correct consequence should be to develop a new kind of process-oriented treatment of these tasks which however has to come along with a completely new software environment that is opimized towards supporting this new structure of the planning process. For this reason together with the project partners Daimler Benz, Mercedes Benz, MAHO and CAP debis the FAW has developed an information system for supporting the entire business process of NC-process planning starting from engineering design over the definition of setup to the final generation of NC-code.

As consequence of the project goals the basic objectives of the software developed are twofold. One basic idea is to provide a common databases that serves the communication between the individual modules. The other objective is to provide efficient support for the actual task of generating NC-data fron the CAD-model of the part. As this process eventually incorporates a companies know-how on manufacturing and is therefore essential for the competitiveness of a business the system has to provide possibilities for customizing strategies of manufacturing and for the selection of tools. Another basic requirement for system design is functionality for visualisation and modification of almost all intermediate results generated by the system.

6.3 Architecture

Architecture is based on feature based design. Features as e.g. hole are realised as fully described objects thus providing the basis for the automated processing of subsequent planning. In a first step following design the necessary information on setup and the the fetures to be processed with this setup is specified. Next step is the automated generation of manufacturing operations and the definition of precedence relations between these operations. For this purpose a decision table system is used thus providing not only effective support but also the capability to modify the manufacturing strategies according to the preferences of an individual company. In a subsequent process the possible ways between the different operation points are computed also considering possible collisions with the workpiece and generating intermediate tool positions to avoid these collisions. Then an optimized process plan with a minimal amount of non-productive times is generated which subsequently is translated into NC-data that can be used on the concrete machine.From a system point of view integration is achieved using a relational data base and a common data-model for the individal moduls. As far as possible standards are used within this system, including a STEP compliant data-model and SQL. Programming language is ANSI-C thus providing portability over a variety of UNIX-based computer systems.

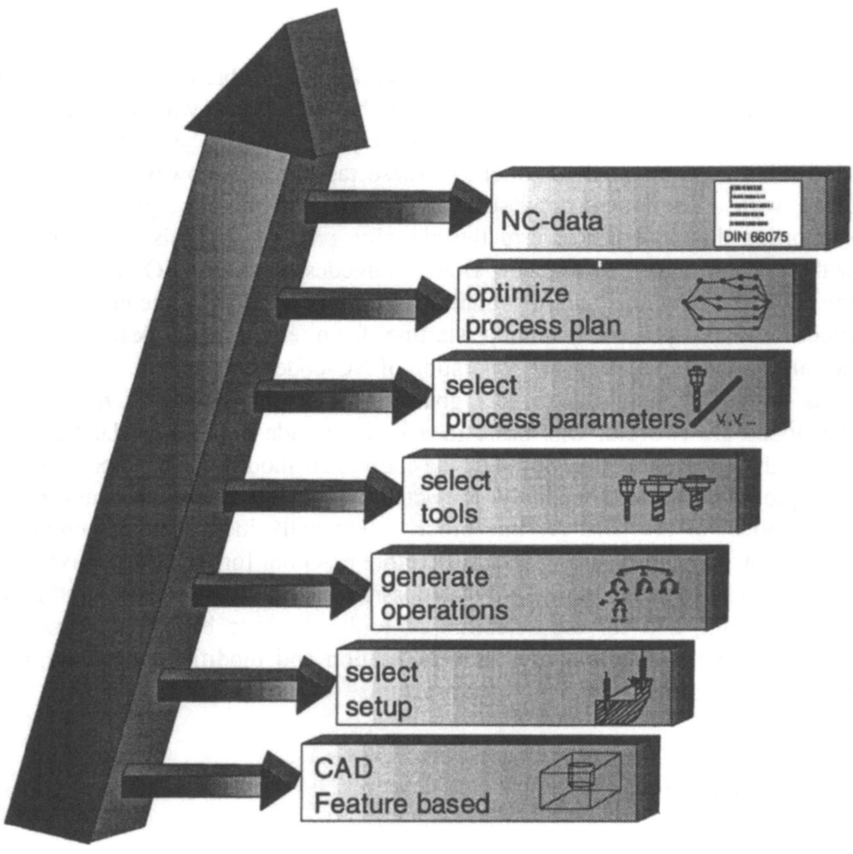

Figure 6.2. Structure of NC process planning activities

With this software the concept of business process oriented software has been realized which cures the current departemental oriented breaks between design, process planning and NC-programming.

7 Conclusions and Future Work

In this paper, we argued in favour of a holistic view on process engineering. Process engineering—if poised to be effective—should not be limited to mere process modelling. Process modelling is the first step. Process engineering should also include process analysis, simulation and optimisation as well as implementation and monitoring. Only an environment supporting this entire set of process engineering activities will turn process engineering into a successful business case. Going down this avenue, one has to consider two dimensions of integration as pointed out in this paper—vertical and horizontal integration.

With regard to future work, the most important challenge will be the enhancement of our advisor concept to improve analytic functionality and to provide a conceptual framework for the integration of process modelling and implementation. This implementation aspect may be at least partially covered with respect to integrating process modelling and workflow. Yet, the integration of business process and software engineering—including strategies and tools—still requires significant accomplishments to be achieved.

Acknowledgements.

The authors gratefully acknowledge the contributions of the members of the ProMAX team: Thomas Ernst, Ines Kittel and Joachim Mayer. Thanks are also due to Mark S. Fox and Michael Gruninger who helped us by offering many insights in enterprise engineering and formal approaches towards enterprise design and analysis.

References

[1] M. Hammer (1990). Reengineering work — don't automate, obliterate. *Harvard Business Review*, July-August 1990, 104-112.

[2] M. Hammer, J. Champy (1993). *Reengineering the corporation — a manifesto for business revolution*. Harper Collins Publishers, New York.

[3] T.H. Davenport (1993). *Process innovation — re-engineering work through information technology*. Harvard Business School Press.

[4] W.S. Humphrey (1989). *Managing the software process*. Addison-Wesley, Reading, Mass.

[5] M.S. Fox, J. Chionglo, F.G. Fadel (1993). A common sense model of the enterprise. *Proc. 2nd Industrial Engineering Research Conf. (IERC)*, Los Angeles, California, 425-429.

[6] M. Grunninger, M.S. Fox (1994). The role of competency questions in enterprise engineering. *Proc. IFIP WG5.7 Workshop on Benchmarking — Theory and Practice, Trondheim*, Norwegen.

[7] S. Jablonski (1994). Process Modeling and Execution in Workflow Management Systems. *Proc. Int. Working Conference on Dynamic Modelling and Information Systems,* Noordwijkerhout.

[8] B. Curtis, M.I. Kellner, J.Over (1992). Process Modeling. *Communications of the ACM,* September 1992, Vol. 35, No. 9.

[9] I. Kittel, T. Rose, M. Grüninger (1994). *ProMAX Basic Architecture,* FAW Ulm, FAW-TR-95003.

[10] T.A. May (1994). Know your work-flow tools. *Byte July 1994*, 103-108.

[11] Workflow Management Coalition (1994). *Workflow — glossary, reference model and application programming interfaces*. Brüssel, Belgien.

[12] A.-W. Scheer (1992). *Architecture of integrated information systems: foundations of enterprise modelling*. Springer, Berlin, Heidelberg.

[13] E. Yu, J. Mylopoulos (1994). Using goals, rules, and methods to support reasoning in business processes. *Proc. 27th Hawaii Intl. Conf. System Sciences*, Maui, Hawaii, Band IV, 234-243.

[14] A. Reinhardt (1994). Managing the new document. *BYTE Magazine*, August 94, 90-104.

5 Quality Management in CIM

E. Bajic and J. Richard

1 Introduction

Quality improvement represents a fundamental objective for industry which must continuously increase the performance of manufacturing plant in terms of costs, flexibility and quality products and services.

Purpose of quality management and requirement is not limited to the only product quality but includes the process quality by means of reliability, safety and perenniality as also economical and sociological criteria linked with production. During the all product life, quality management is a cyclical activity which is composed of four recurrent steps according to the Deming philosophy : Plan - Do - Check - Action, applied to all functions and services of the enterprise to reach the goal of Total Quality Management.

In a CIM context, this quality management approach must be integrated in the material and information processing system, federating all the whole range of the enterprise activities. Quality management activities must be conducted and based on both functional and informational modelisation approaches, according to pertinent and standardized modelisation framework as CIM-OSA, and also data representation model as STEP. Mastery over three topics is necessary to achieve efficient quality management in CIM :

- Quality Modelisation in Information System
- Quality Management in Production Control
- Communication support for Quality Assurance

2 Quality Management in Information System

2.1 Information System for Quality Assurance in a CIM Architecture

The information system concept has initially been defined for administrative systems [1] but its purposes can be translated to manufacturing systems. Its first purpose is to handle the communications inside the cell. Models have to define information flows according to both management rules and decisional activities. Thus, these models gather information which are enough representative of the considered system to ensure that the decisional activities are well informed.

Its second purpose is to define the means needed :

- to be representative of the organization, on how to get information,
- to provide information access facilities to the decisional activities.

In the case of a manufacturing system, any information must be available to any function. In other words, the information system integrates the manufacturing system from an informational point of view. In the case of control loop architecture of a flexible manufacturing cell, however, the information system must deal with the quality information. Therefore, information classes can be classified as follows :

- cell and workstations configurations,
- acceptable objects (products, pallets, tools, gauges, ...),
- physical and manufacturing capabilities,
- products and manufacturing processes models,
- feasible fixturing and handling tasks,
- works-in-progress and available resources,
- performed tasks,
- obtained qualities,
- performed corrective actions.

Finally, the information system must provide a uniform access to these information classes. The cell functions must not be developed taking into account whether some information is distributed or not, whether more than one network has to be used, ... In this way, the cell decisional system is completely independent of the operative system, that means the flexibility is further increased. That contributes to achieving a CME : Computerized Manufacturing Enterprise [2]. In order for the distributed control to take into account the control loops and the manufacturing exceptions, it must have access to any information relative to qualities, to manufacturing reports and to the cell configuration. This entails the following kinds of resources :

- communications,
- information storage,
- data processing.

These are the three facets of an information system study. The major purpose of an information system is to structure the data exchanges and the data storage in order to meet the users requirements.

2.1.1 Information Exchanges within the Control Loop Architecture

Applied to the control loop architecture, a conceptual communications model is necessary to depict (Figure 1.1) :

- the environment domains,
- the cell and workstations control systems which handle the control loops.

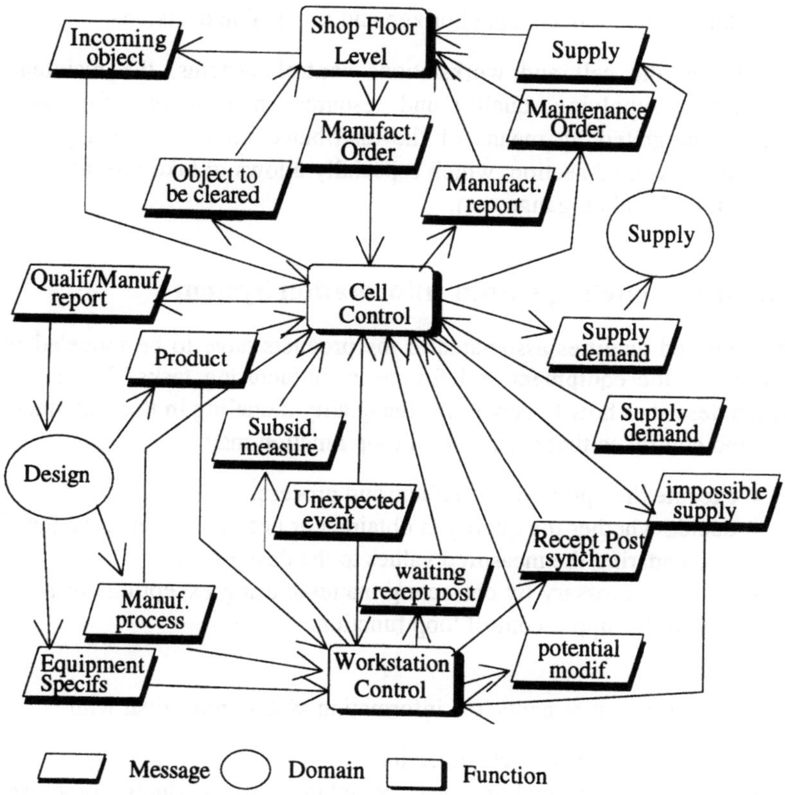

Fig. 1.1. Conceptual communications model of the manufacturing system

This diagram does not only define the information flows. It structures the interactions between the following functions (or domains) : design, manufacturing system control, supplying, without constraining neither design non manufacturing system control.

In this way, it points out the information flows of both kinds of integration :

Horizontal Integration :

- products, processes and equipment design,
- manufacturing and quality report,
- supplies orders, supplies integrated in the manufacturing control,

Vertical Integration :

1. the interactions between the shop floor and the cell control systems perform the maintenance, manufacturing, supplying, resources management functions; the shop floor level gives an autonomy of a decisional framework to the cell relative to both decisional horizon and acceptable decisions,

2. between the cell and workstations control systems, the maintenance, manufacturing, supplying, quality and resources management functions are completely integrated by means of the distributed control system previously depicted in the second section, which especially allows to take into account any object (part, tool, pallet, gauge, ...).

2.2 Quality Modeling within Information System

The required qualities associated to the products have to be modeled in the same way for the equipment and for the manufacturing tasks. The following paragraph describes how to generalize the quality modeling in the manufacturing process and control so that each control loop function may :

- execute the optimum manufacturing process,
- decide whether the quality is obtained or not for a product or for a task by comparing the measured values to the desire ones,
- react if necessary by correcting manufacturing exceptions or signaling them to the upper control loop function.

To meet these requirements, the information system must deal with :

- the controllable objects diversity,
- the required qualities variety relative to products, processes or equipment.

2.2.1 Required Qualities Diagrams

a) Default List to Model

Using topologies of both objects and manufacturing tasks limits the conceptual E-R diagram size. As regards the required qualities diversity, this size may be reduced by means of considering the usual controls relative to elementary data which are used in computerized systems. Thus, a required quality may be considered as :

- an order constraint between an attribute and a value, that means an imperative value, a threshold, ...,
- an interval of acceptable values, that means tolerances,

- a finite set of authorized values, which is a rather less frequent case,
- an order constraint between two attributes (>, <, ≥,≤, =, ≠),
- a compute rule between several attributes (operands and computed attributes).

Using only these constraints, a modeling principle can be established, so resolving the exhaustivity problem. This is described in the next section.

b) Modeling Principle

This analysis makes use of the Entity Relationship representation formalism [3], most represented in the international community.

Every attribute value may be considered by one of the listed controls. Entity relationship methodological environment allows to connect rules to entities, relationships and attributes but in an independent manner of their occurrences. Here, quality constraints must be expressed for particular occurrences of products features, processes parameters, ...

By applying this rule concept for expressing required qualities (see figure 1.2), a RULE entity can be the support of a QUALITY CONSTRAINT entity. This is associated to the object or to the task concerned by the expressed quality : only one occurrence is constrained. This principle is used for any of the five kinds of constraints. The figure 1.4. models a threshold constraint.

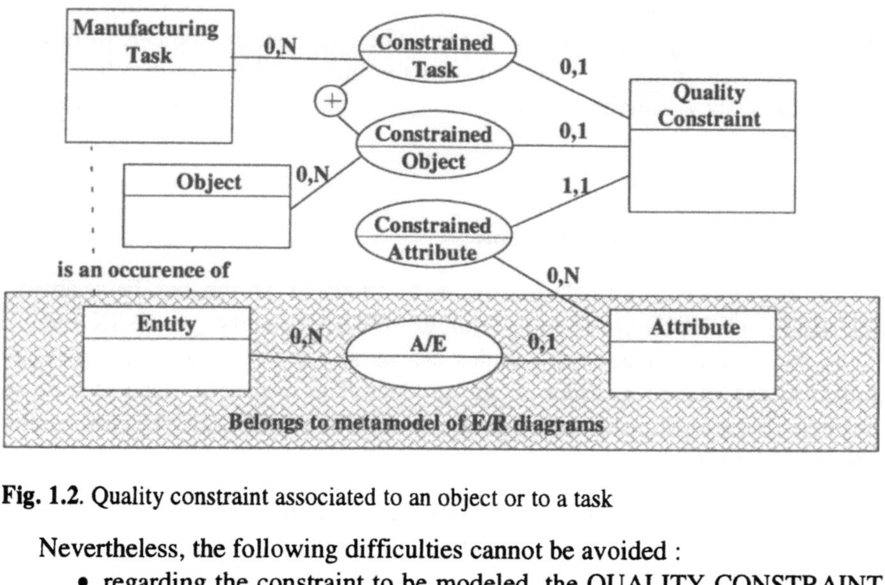

Fig. 1.2. Quality constraint associated to an object or to a task

Nevertheless, the following difficulties cannot be avoided :

- regarding the constraint to be modeled, the QUALITY CONSTRAINT entity may be linked to more than one attribute or may be more complicated (order relation, computing rule),

- the attributes type may differ : integer, real, ...; that entails more than one relationship although the real attribute is mostly used.

The E-R diagrams are further simplified by means of using a generalization/specialization link (see figure 1.3.).

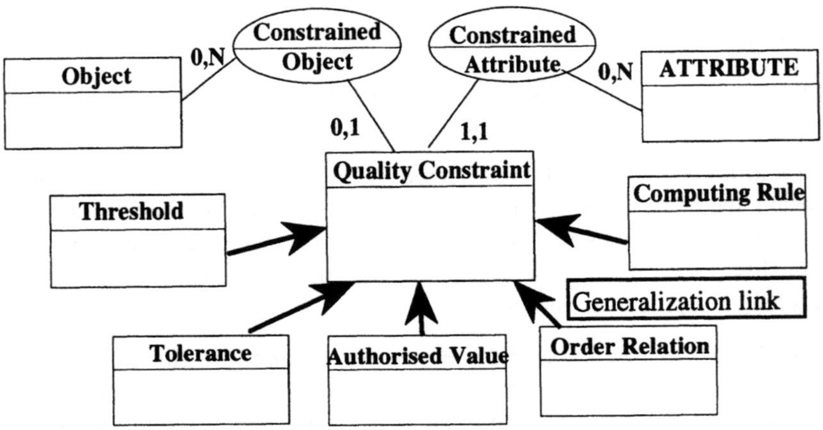

Fig. 1.3. Generalization of the kinds of constraints

In these diagrams, the following constraint cannot be modeled : an occurrence of the QUALITY CONSTRAINT entity can be linked to both an object and an attribute only if the object is an occurrence of the entity to which the attribute belongs. This constraint cannot be expressed by the E-R formalism extensions because these do not consider the links of the "is-a-member" kind, that means they cannot simultaneously handle an E-R diagram and its metamodel.

2.2.2 Measured Qualities Diagrams

To express a measured quality, a relationship is created between the QUALITY CONSTRAINT entity and the OBJECT entity (or TASK entity). Thus, for instance, measured qualities can be stored for any part and they are linked to the desired qualities of the product. The measured value is then an attribute of the relationship which connects the QUALITY CONSTRAINT entity and the OBJECT entity.

Another example may be proposed for a tool (see figure 1.4.).

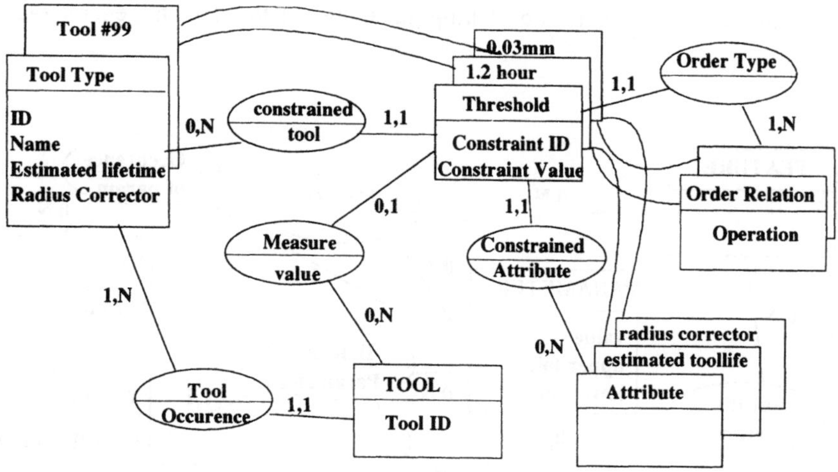

Fig. 1.4. Examples of constraints associated to machining tools

Its total time-in-use cannot be greater than an estimated life which is modeled by an attribute of the QUALITY CONSTRAINT entity. The attribute of the MEASURED QUALITY relationship is the current total time in use. The tool corrector value may be similarly constrained. In this case, the MEASURED QUALITY relationship is the current tool corrector value. The measured qualities are modeled in the same way for the other kinds of constraints.

2.3 Quality Assurance Modelization in a Flexible Manufacturing Cell

A feature-based product model includes a collection of valued and toleranced parameters [4]. A simple E-R diagram can be used to receive the design information from CAD/CAM [5].

These parameters values are here occurrences of the QUALITY CONSTRAINT entity; each of them is linked to both the kind of parameter and the feature to be controlled. The measured value is associated with both the desired quality and the performed quality control task (see figure 1.5).

2.3.1. Diagnosis Model

Unexpected events or drifts occur during the manufacturing stage and lead to non-quality output. Diagnostic functions may be implemented in the control loop architecture. They have to determine the sequences, the causes defects and to predict the process evolution [6]. This estimation is transmitted to the manufacturing control functions while the defects and their causes are stored by the information system to perform the quality report. A single relationship models the defects sequences by taking advantage of the defects generalization links.

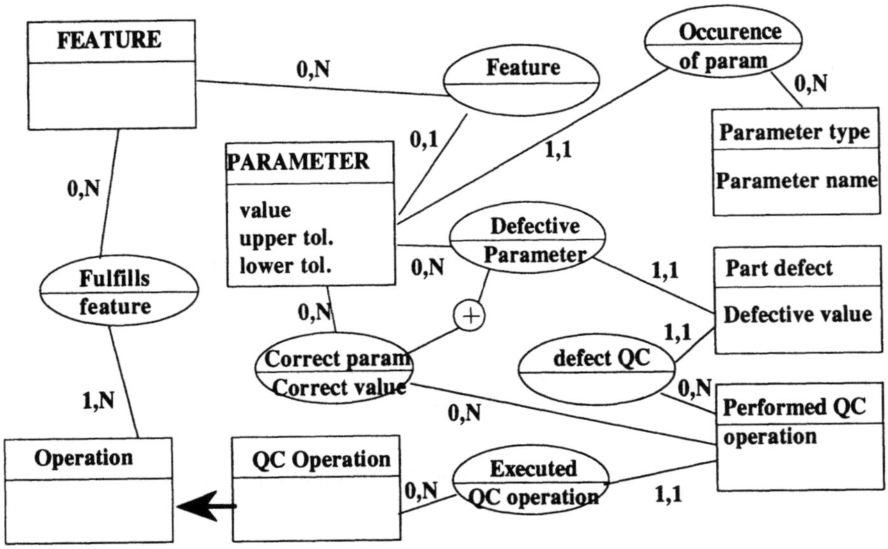

Fig. 1.5. Measured quality report, independent of both gauging process and controlled object or task

The kind of relationship : one-to-one, one-to-many, ... allows one or more estimated causes of a defect. The one that enables us to estimate many causes of defects at a time permits to link an attribute to the relationship : this attribute is a reliability factor which characterizes each envisioned cause (see figure 1.6).

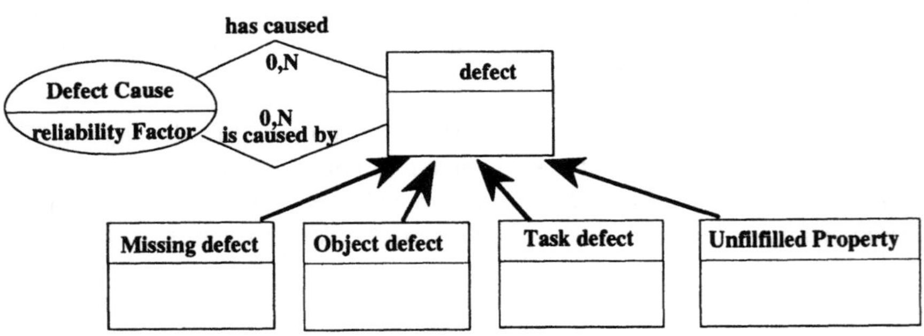

Fig. 1.6. Defects relationships E-R diagram

The following complementary remarks can be pointed out :

A- The defects may be modeled apart to facilitate their access. Figure 1.5. includes the way to separate obtained qualities from the defects. A second relationship is introduced : it handles the value which leads to the defect whereas the first one records the correct parameter values. That induces a redundancy. The first relationship would be sufficient to determine whether the desired quality is achieved or not. But this comparison should be performed each time the parameter is required.

B- The missing parts or equipment during a task may also be modeled as defects.

C- The defects may be qualitative. The diagnostic functions may indeed be supported by knowledge-based systems or may be based on fuzzy logic properties. So they may utilize non-valued causes. At least, the information system has to handle a collection of qualitative properties which are stored as text.

D- Corrective actions. According to the PDCA cycle, once the required qualities have been defined, and after the quality controls have been done, the cell reacts in a predictive or corrective way. The control loop functions do not have to determine the corrective actions if the information system can provide them [7]. That implies that the design function has previously established corrective actions and has communicated them to the cell information system. When this is done, the reaction response time can be significantly improved. Thus, the automatic exchange of equipment, especially tools, can be envisioned. This kind of decision may even be made by the information system : it becomes an active information system.

3 Quality Management in Production Control

3.1. Modelization Framework Application of CIM-OSA

Many firms are being confronted with increasing market pressures leaded by competitiveness, reduced prices, better qualities, minimal response times and a rise in product diversity.

In the last decades have emerged an increasing development of Flexible Manufacturing Systems (F.M.S.) and Flexible Manufacturing Cells (F.M.C.) to fulfill the requirements of efficiency, quality and flexibility simultaneously. F.M.S. are large and complex production systems constituted of a variety of Numerical Controlled machine tool, Coordinate Measuring Machine (C.M.M.), machining center, storage/loading /unloading / clamping / unclamping areas and pallet transport system. F.M.C. are reduced scale F.M.S. controlled as automation islands dedicated to the manufacturing of small sets of products class, and acting as computer controlled stand-alone entities of an overall F.M.S. Flexible Manufacturing Systems and Cells are considered as the shop floor start-up of a Computer Integrated Manufacturing (C.I.M.) architecture involving contribution and effort of all departments of the enterprise.

3.1.1 C.I.M. Architecture Analysis Methods

The last decades have seen evolution of technology, engineering and philosophy from Manufacturing Automation to Integrated Manufacturing with the C.I.M. concept as an integrated material - and information - processing system federating all the enterprise-wide activities.

Different approaches to C.I.M. components modeling exist as the CIM-OSA (Open System Architecture) Architectural framework [8], the CAM-I DPPM (Discrete Part Manufacturing Model), the NBS AMRF (American Military Research Facility) and others [9] covering for some of them the whole life-cycle of C.I.M. where flexibility and modularity are key elements (e.g. CIM-OSA). Development of those C.I.M. Reference Architectures needs a fully integrated Information System (I.S.) to dynamically and automatically links all the enterprise functions from design and manufacturing to marketing.

Flexible integration is expected to result from the use of a global information infrastructure that can lead to the definition of an enterprise-wide information system.

3.1.2 Quality Assurance in Shop Floor Management

Designs of both physical and informational architectures of the F.M.C. have been realized on the basis of flexibility and autonomy requirements of the manufacturing system with respect to the products, the shop floor level and the control methodology.

Cell Autonomy

In order to increase the cell autonomy, the production management of the shop floor and the cell must be coupled as closely as possible. Depending on the production management method, the manufacturing orders are created differently be either :

- received from a next cell (Kanban),
- determined by the shop floor (MRP),
- decided by the cell (OPT).

These differences have repercussions on the data flows. Once the cell has received manufacturing orders, it can sequence them but only in a very short-dated way. The integration of defects and manufacturing exceptions in the planning regarding the computing time is quite impossible.

This does not consider the due-dates which are computed by MRP (Material Requirement Planning) as priorities and thus use the FIFO criterion. As regards Kanban and OPT, the estimated processing times of the manufacturing operations may serve to compute due-dates. The cell may also employ the Kanban and OPT criteria or any other criterion in keeping with both its autonomy and the decisional framework which is defined by the whole manufacturing orders. Especially, if the cell is not overloaded or if it is not a bottleneck, it may decide to perform more quality control tasks. The cell also fulfills the reporting activity which concerns both executed tasks and obtained qualities. Nevertheless, the reports must be suited to the production management method because these methods do not consider the same indicators.

Flexibility According to the Product

The manufacturing abilities and the cell control must be able to manage varying manufactured articles (for example, using group technology configuration), or different products conceived depending on the customer request.

The equipment flexibility in the cell is managed by the product system design, and the control flexibility by the chosen control loop architecture. The first one is more relative to the integrated management of the quality, and the second needs a modular software and hardware.

At long-dated, the control reconfiguration depends on the facility to substitute the software in the computers, the numerical controllers, the programmable logic controllers ...

In manufacturing process, the new incoming product involves :

- to be able to explain the manufacturing specification using the design stage information. CAD/CAM product data exchange standards enables the consistency of data format from design to manufacturing and inspection;
- to perform the report of the adapted manufacturing, at least to apply Statistical Process Control methodology (SPC) .

Flexibility According to the Control Methodology

The manufacturing system fits on the control variation following the quantity and the due-date, in JIT (Just In Time) and OPT ways. The abilities of manufacturing means must change following the control estimates. That is not in the cell power, but the manufacturing management method applied in the factory or the shop floor. On the other hand, the control criterion taken relative to this order can progress following the methods and the manufacturing systems. So the cell control must be adaptive : its programming is parameterized by the criterion.

3.2 Quality Assurance in CIM

As one of the major goals in C.I.M., quality assurance concerns the whole life-cycle of both product and process, thus covering all quality management activities including quality planning, control and monitoring with appropriate feedback actions. The quality objective stands on horizontal and vertical integration flows in a C.I.M structure. For the horizontal view, in-process quality assurance and process certification methods needs a large use of in-process quality sensors and deterministic metrology methods supported by a reactive architecture. The QIA project has made big efforts on this subject [10]. Vertical integration of quality assurance operates from the cell automation to Computer Aided Design stage by mean of an information system with quality topics management features. Our works presents quality management in F.M.C. corresponding to this double approach guided by a reactive manufacturing architecture and a reactive information system in accordance with C.I.M.-O.S.A.

Quality control has to be integrated into the whole manufacturing process, the closer to the production operations to induce as soon as possible feedback actions on operational and/or informational stage of the system

An optimal management of manufacturing operations requires to set up feedback control-loops within the system architecture. To achieve this goal, the cell must be equipped with loop-controlled functions. Problem is to survey cell control loop solutions using existing and heterogeneous equipment, to specify necessary modifications for their integration and to propose adaptable and reconfigurable solutions for various types of equipments.

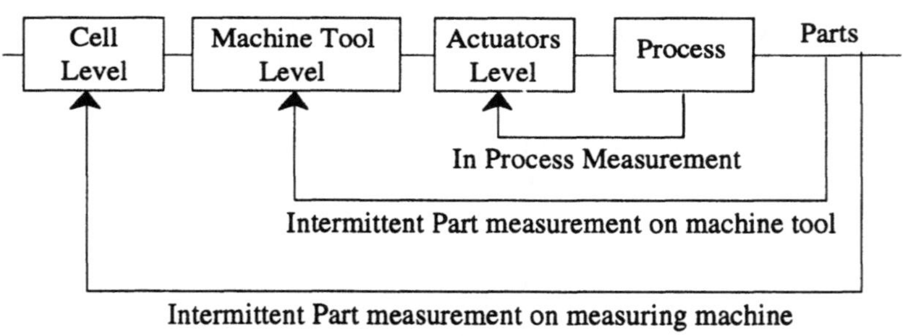

Fig. 1.7. Quality loop-controlled structure

Thus a solution based on a centralized production monitoring organization, coordinating N.C. machines without initiative would be easily implemented but too much sensitive to perturbations : the smallest equipment failure can lead the cell towards a degraded and some times incoherent functioning mode. For these reasons we have favored a distributed monitoring solution which attempts to grant more intelligence and also autonomy to the machines. Many works have demonstrated the profitability of such an approach [11] [12] [13].

3.2.1 Quality Control Feedback Methodology

Feedback of quality control architecture joins the PDCA cycle (Plan-Do-Check-Action) [14]. According to this cycle, the quality control function ensures that every manufacturing task is well performed and that the required quality of every work in progress is obtained.

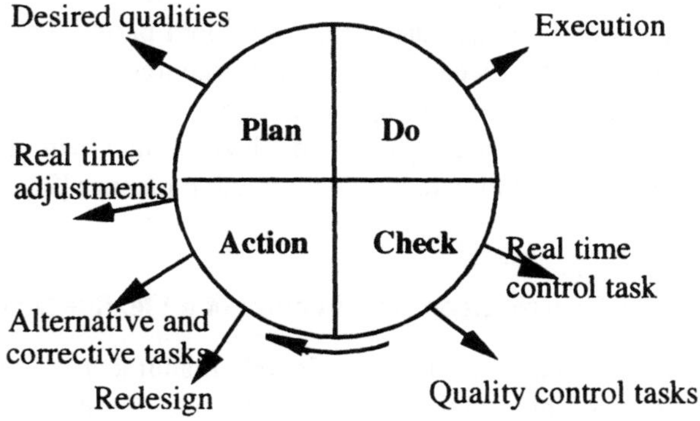

Fig 1.8. Plan-Do-Check-Action cycle applied to the control loop cell architecture

The relationships with both CAD/CAM and manufacturing management include the feedback loops which reconsider the design specifications and the planning of works in progress. The corrective actions of the cell depend on the CAD/CAM and manufacturing management capabilities.

Inside the cell, the control loop architecture has effects on the manufacturing monitoring. This must be able to take into account the manufacturing exceptions and to verify the effects of its orders. Then, the non-quality relative to parts or to the equipment induce reactions. The main problem is to organize these reactions in a coherent architecture which includes informational and decisional abilities.

3.2.1.1 Quality Control Loops Relative to Product and Process Design

The cell receives the design specifications which must deal with the quality control tasks. It executes them regarding the shop floor manufacturing orders. The specifications transmission may use the CAD/CAM data exchange standards

: IGES [15], CAD*I [16], SET [17], PDES [18], STEP [19], ... These standards generally include the quality data in the products and processes models, but do not always consider corrective actions.

In keeping with the Action stage of the PDCA cycle, it should indeed be possible to envision corrective actions depending on manufacturing exceptions and drifts. In other words, the manufacturing process must contain alternative solutions, especially those generated by the process planning but not recorded because they are not optimal.

The control loop from the cell to the design function consists of redesigning the manufacturing process, as well as the product. In this regard, the cell reports any quality data for statistical analysis or for an exhaustive analysis. This feedback enables TQM (Total Quality Management activities) [20]. If the report activity deals with the whole data relative to both parts (dimensions, roughness, ...) and equipment (tool wear, machine axis accuracy, ...), the first versions of the Ishikawa diagrams may be improved. Reports contain qualitative parameters that only knowledge-based systems or fuzzy logic can now process [21]. Redesign can now be automated in an interactive manner [22].

3.2.1.2 Quality Control Loop Architecture of a Flexible Manufacturing Cell

The control loop architecture of a flexible manufacturing cell firstly proposes a set of functions which process work-in-progress, equipment, actions and their sequencing. The quality control function is therefore distributed throughout the whole manufacturing system and is as close as possible to the equipment level.

The control loop of parts and equipment reviews as often as necessary their design and process plans. The control loop of actions (handling, machining for instance) adjusts their progress in real time if possible. The control loop of the actions monitoring modifies the real time control of the handling inside the cell depending on the problems.

Secondly, the cell tries to correct the recognized defects itself. To avoid the operator's intervention, it must be able to interpret any risk, to know which operation has to be corrected and to determine the corrective action.

4 Communication support for Quality Assurance

Distribution of CIM control loop functions realizes a vertical integration of manufacturing and quality functions. This possibility is offered, at the program level by allowing the machines a maximum of choice in the exploitation of feedback data, and in the planning function by allowing the coordination of the machines with respect to the cell specifications.

Thus, we propose a structure of the cell control architecture which can be detailed on two levels :

- the cell control level manages the input and output parts. It describes the sub-phases to realize on a part by a station and the job due-date to execute this work.
- the station level selects the work to realize on the parts depending on its abilities.

So, the control loop of the cell level allows to carry on with horizontal integration. From quality report task thus realized, the products and processes design can be taken back into consideration, particularly part-fixture positioning and handling tasks.

The cell level and the station level do not communicate directly together for the information relative to the manufacturing monitoring. The reception post concept is used to synchronize these two levels. The reception post corresponds to a physical location (handling equipment) which is accessible by both cell level and station level. It can be represented with state graph which allows to synchronize the cell and station level. The information conceptual model of figure 1.9 describes a reception post.

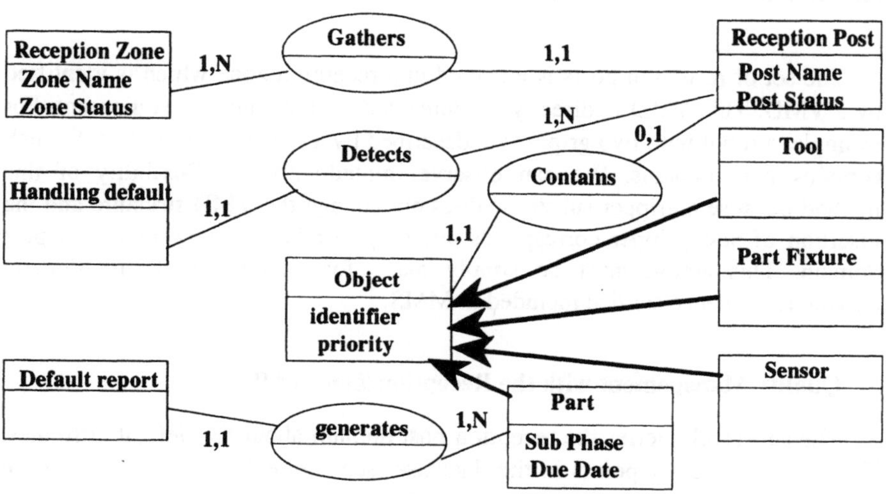

Fig. 1.9. Information conceptual model of reception post.

4.3. Distributed Management Using MMS

In order to present the capabilities and applications of MMS [23] (Manufacturing Message Specification) , it is necessary at first to explain some special terms and their corresponding definitions, which will be repeatedly used after. Some of them denote, and this often holds for the development of real devices, data and functions. In MMS, such models include, for instance, the Virtual Manufacturing Device (VMD), the Client and the Server, as well as the MMS objects.

With the MMS approach, a real system maps the concepts described in the model, to the real device. An object is then represented by a data structure.

The Virtual Manufacturing Device Object serves as a model that represents the behavior of a real device. A VMD is an abstract representation of the common characteristics of all real manufacturing devices (generalization of manufacturing devices) and represents the externally (from the network) visible behavior relevant to MMS (see figure 1.10). All standardized services refer to this virtual device, which by the implementation refers to real functions. MMS only describes the effects of the services on the VMD and does not prescribe a specific implementation or transformation to real functions.

Semaphore	Program Invocation	Variable and Type	Event	Domain	Journal

Fig. 1.10. Real and Virtual Manufacturing Device

The set of reception posts is gathered in a reception zone which is identified by a VMD. This VMD is directly implemented on the handling equipment. For example, a robot with two grippers is described by a reception zone VMD which contains two domains. We can observe immediately the flexibility of this method because a gripper substitute does not change the VMD structure and the insertion of new gripper corresponds uniquely to adding a new reception post domain. This action must be simply made for example with the domain download sequence service included in MMS.

Quality Management with the Reception Zone VMD

The cell level queries the reception post domain about the default attributes. These ones are entry points in the database (see figure 1.11) for retrieving the kind of defaults and their corresponding values.

In each reception post domain, a Default Report corresponds to an occurrence of the OBJECT DEFECT entity which is depicted in figure 7. A Handling Default corresponds to an occurrence of the DEFECT entity and may be an occurrence of the following entities : MISSING OBJECT, TASK DEFECT, UNFULFILLED PROPERTY.

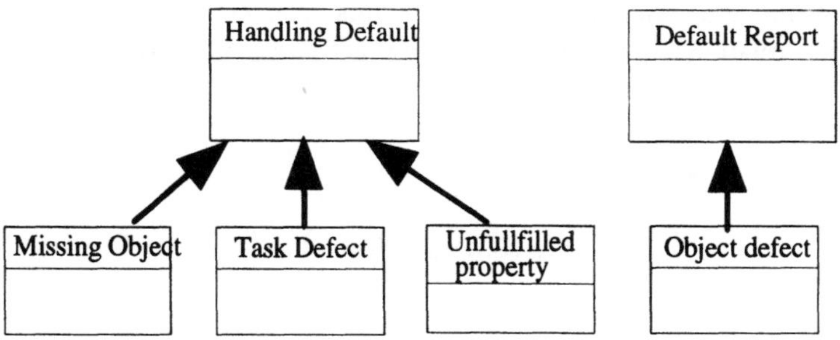

Fig. 1.11. Default identification

5 Product Based Quality Management and Manufacturing Control

Some research in the field of CIM concerning architecture and distributed system are putting the emphasis on the product, seen as an intelligent entity. New paradigms as heterarchical and multiagents control architecture [24] initiate the approach of intelligent product, acting as a manager of its own transformation process in which machines perform manufacturing tasks under the leadership of the product, as a mobile agent in the environment.

Such oriented research works are being carried out in the IMS Esprit project "Holonic manufacturing system" [25].

This novative approach has a major role to play with quality management functionnalities by means of product information and quality objectives distribution, production-oriented negotiation between product and machine for best quality fitting, as also production tracability.

5.1 Nomadic object concept

Part product can be considered as an actor in a surrounding environment and operating process, by supplying functionnalities locally and autonomously supported in order to reach its final objective with respect to quality specifications, and system capabilities. These functionnalities can be listed as follows :

- acting in decision-making
- cooperating with other entities and actors
- adapting capability according to its environment
- making acquisition of process knowledge, it is exploring.

This characteristics are supported by the nomadic object concept shown in figure

1.12. Adapted from the client/Server model, this concept reinforce the Customer/Supplier relationship in manufacturing activities in between part product and production machines.

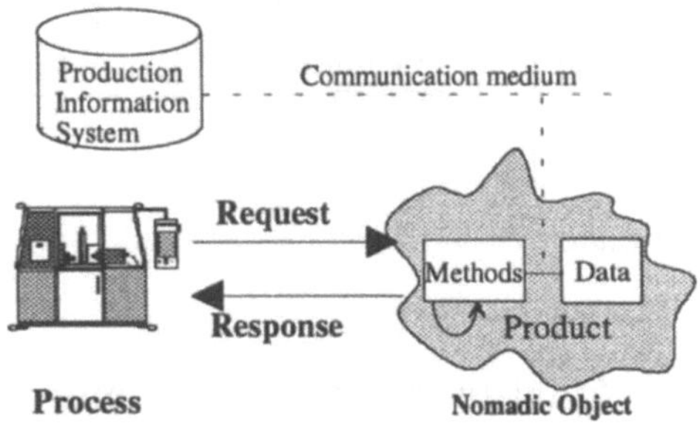

Fig 1.12. Nomadic Object Concept for Manufacturing Management

Part product as any other mobile object carries its own information and plan on a programmable device fixed on it, name tag or escort memory. Communication controller provides access to data by means of non contact technics : radio frequency, inductive waves, infra-red light.

5.2 Product Models for Manufacturing and Quality Management

One of the most significant approach today in product data management is the development of ISO 10303 standards called STEP (STandard for Exchange of Product Model) which defines models, database access and neutral files format for representation and exchange of product data. The goal is to define complete models for product life cycle in a CIM context, as well as the means for exchange of these data between enterprise functions along a product life cycle definition from design to production ,maintenance, ... EXPRESS is the object oriented modeling language associated to the STEP philosophy, dedicated to information representation.

Methodology can be built upon the STEP approach and Nomadic Object concept for product based manufacturing quality and control management [26]. Figure 1.13 shows a STEP-based methodology to perform translation of product models directly to product information retrievable anywhere in the architecture, provided by the product with the nomadic object concept, to perform manufacturing operations control, manufacturing tracability, quality check, product tracking, ...

Product models, oriented to manufacturing applications, can be described in the EXPRESS formalism according the STEP approach, to generate :

- Functions library dedicated to product information access, supporting direct requesting of target application on the Express schema data, named FApI, Functional Application Interface;

- The database structure definition, managed on a remote location in the CIM architecture.

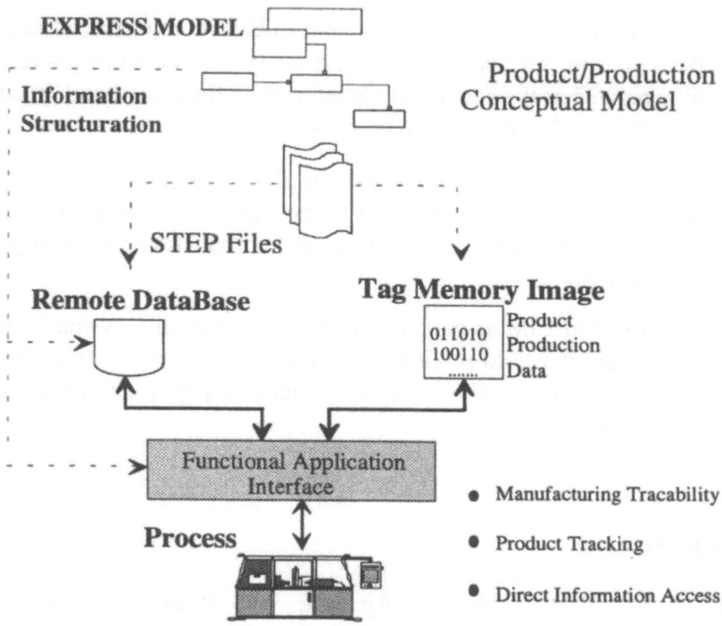

Fig 1.13. Product information

The product definition within the STEP neutral file is derivated to :

- create the tag memory image for the first state of the product, considering its current phase in its life cycle;
- populate the remote database.

This approach allows high level quality information management and manufacturing control in a distributive and with full respect of product model specification according to as many different life cycle phase described in the product data model. The nomadic object concept bring the customer/supplier relationship directly in the CIM architecture as a step to quality assurance objective.

6 Conclusion

Quality management activity largely depends upon quality-based product data management, over all product life cycle. For manufacturing activity, C.I.M. approach provides a logistic aupport for technical data management, and thus permits quality-oriented data exploitation as goals for production control.

Major axes of research for quality management in C.I.M. concerns on one hand quality modelisation in information system, directly derived from product data specification generated at the conception level, and on the other hand, integrated quality control loops within production process.

The nomadic object concept magnified by the emergence of automatic identification technologies, is an efficient support product quality management within CIM architecture, offering extension to after production phases of product life cycle management.

Quality management in CIM as part of enterprise-wide quality system, provides large access and importance to product data, and provides reactivity capabilities in manufacturing process. So offering quality asurance functionnalities it allows to reach Total Quality Control goal.

References

[1] G.B. Davis, M.H. Olson, J. Ajenstat, J.L. Peaucelle "Management information systems", Ed. Mc Graw Hill Inc., 2nd edition, 1985

[2] C. Hsu, L. Rattner, "Information modeling for computerized manufacturing", IEEE Transactions on Systems, Man and Cybernetics, vol. 20, n°4, p. 758-776

[3] P.P. Chen, The Entity-Relationship Model : toward a unified view of data, ACM/TODS vol.1 n°1, 1976, pp. 9-36

[4] T. Laakko, M. Mäntylä, R. Mänty lä, J. Nieminen, R. Sulonen, J. Tuomi, "Feature models for design and manufacturing", 23rd Annual Hawaii International Conference on System Sciences, Kailua-Kona, january 1990, Ed. IEEE Computer Society Press, pp. 445-454

[5] J-F. Couturier, J. Richard, M. Véron, "Control loop architecture and information system of a flexible manufacturing system", 5th International Conference on the manufacturing science and technology of the future, Enschede, Netherlands, juin 1991

[6] C. Jacquin, D. Gien, R. Stepourjine, "An expert system for diagnosis in a flexible manufacturing cell" AFE PROCESS'91, IFAC/IMACS/IFORS International Symposium, Baden-Baden, septembre 1991

[7] J-F. Couturier, J. Richard, E. Bajic, C. Doiteaux "Information system and quality assurance of a flexible manufacturing cell", International Conference EURISCON'91, Corfu, Greece, june 1991

[8] ESPRIT Consortium AMICE Ed, "Open System Architecture for CIM", Springer-Verlag Berlin 1993.

[9] G.J. OLLING, "CIM Status and direction in the USA", CAPE' 91, pp 33-40, Elsevier Publisher, September 10-12 1991, Bordeaux, France.

[10] Progress report of the Quality In Automation Project FY88, NISTIR 89-4045, Gaithersburg, Maryland, April 1989.

[11] N.A. Duffie, R.S. Piper, "Nonhierarchical control of manufacturing systems, Journal of Manufacturing Systems", vol. 5, n° 2, p. 137-139

[12] J.G. Maley, "Managing the flow of intelligent parts" Intern. Conference Manufacturing Systems and Technology of the Future, MIT, Cambridge, MA, juin 1987

[13] C. Doiteaux, J.F. Couturier, J. Richard, E. Bajic "Integration and control loop of a flexible manufacturing cell", CAPE'91 4th IFIP Conference on Computer Applications in Production and Engineering, Bordeaux, France, september 1991

[14] W.E. Deming, "Out of crisis", MIT, Cambridge, MA, 1986

[15] Initial Graphics Exchange Specification v4.0, National Institute of Standards and Technology, U.S. Department of Commerce, Gaithersburg, MD, 1988

[16] I. Bey, U. Gengenbach, " The CAD*I interface for solid model exchange", Computer & Graphics, vol. 12, n° 2, 1988, p. 181-190

[17] G. Stil, "S.E.T. Standard for Exchange and data Transfert", 3rd International Conference CAD-CAM-CAE Integration Technologies in the Automotive Industry, Torino, Italy, november 1988

[18] Product Data Exchange Specification, First working draft, National Technical International Service,U.S. Dept of Commerce, Gaithersburg,

MD, december 1988

[19] Standard for the Exchange of Product Data Model, ISO 10303-11, TC 184/SC4, 1994

[20] K. Ishikawa, "Le TQC ou la qualité à la japonaise" Collection AFNOR Gestion, Ed Eyrolles, 1984

[21] V. Akman, P.J.W. ten Hagen, T. Tomiyama, "A fundamental and theoretical framework for an intelligent CAD systems", Computer-aided design, vol.22, n° 6, p.352-367

[22] J.R. Dixon, M.J. Guenette, R.K. Irani, E.H. Nielsen, M.F. Orelup, R.V. Welch, "Computer-based models of design processes : the evaluation of designs for redesign" NSF Engineering Design Research Conference, Amherst, MA, juin 1989, Preprints, pp. 491-506

[23] "Manufacturing Message Specification, Part 1 : Service definition" ISO International Standard, ISO/IEC 9506/1

[24] D.M. Upton, "A flexible structure for computer controlled manufacturing system", Manufacturing review, Vol 5, N°1, pp 58-74, 1992

[25] L. Bongaerts, P. Valckenaers, H. Van Brussel, and al, "Schedule execution for a holonic shop floor control system", preprints Advanced summer institute '95, ASI'95, Lisbon, Spain, 25-28 June 1995

[26] F. Chaxel, E. Bajic, J. Richard "A Step to Mobile Database Nodes Integration " Architecture and Design Methods for Balanced Automation Systems, BASYS'96, Lisbonne, Portugal, 17-19 Juin 1996

6 Best Practice in Shop Floor Scheduling

David Little

1. Introduction

This work is based on a major research project undertaken the author and Dr John Kenworthy with the able support of Dr Keith Porter and Peter Jarvis in collaboration with the British Production and Inventory Control Society (now the Institute of Operations Management), Cincom Systems (UK) Ltd and ICI Group Engineering and supported by funding from the CDP Group of the UK Engineering and Physical Sciences Research Council (EPSRC). The project extended over three years and reviewed planning and scheduling procedures in some 35 industrial companies to establish best practice in short term scheduling and to identify the key factors for scheduling success.

This chapter begins by examining the dominant position held by MRP and MRPII within manufacturing industry today and then goes on to outline the objectives of the research undertaken and the method developed to provide a framework for the analysis of scheduling performance both by industrial sector and by scheduling approach. It presents the main findings of the research and then attempts to account for the comparatively low penetration of finite scheduling approaches within the UK when compared to the evident current success of MRPII and MRPII hybrid models. It then goes on to examine a number of identified weaknesses of the MRPII paradigm in meeting the challenges of agile manufacture and provides some new reference models which show how different business sectors are adapting the paradigm to provide a more dynamic scheduling capability.

Important elements of the project were the review and comparison of scheduling performance in conventional MRPII environments, typically using manual expediting or shop floor control, with that of more recent finite capacity based tools and a classification of production control approaches.

2. The Dominance of MRP and MRPII

A survey conducted by the writer in 1990 with the support of BPICS [1] showed that material requirements planning (MRP) and Manufacturing resource planning (MRPII) were the most widely used manufacturing control approaches within the UK. In the survey, some 80% of the sample of 311 UK plants drawn from a wide range of industrial sectors were found to be using MRP or MRPII to plan and control their manufacturing operations and that over 90% were achieving "better" or "much better" performance as a result.

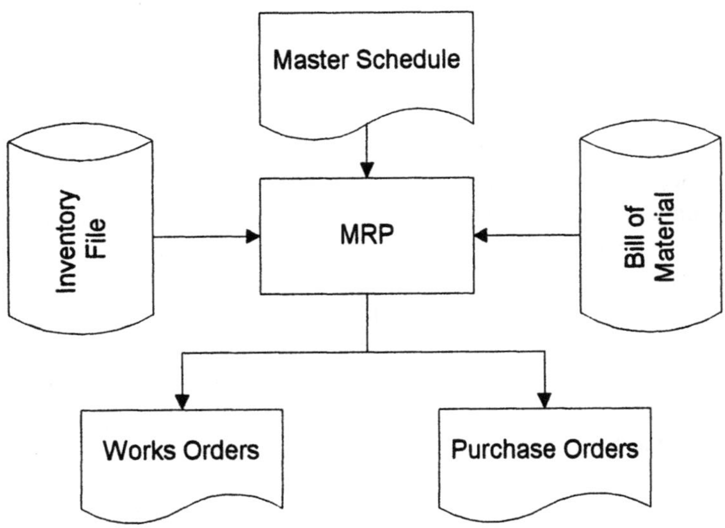

Figure 1: The MRP process

This should not be surprising since much UK manufacture is concerned with the production of relatively complex assembled products in batches and MRP evolved over a considerable number of years as a logical development of the established approaches to inventory control and materials management to address these requirements, particularly for assembled items. Since then much unnecessary confusion has arisen between MRP and MRPII and their respective roles. MRP is still used by many, often smaller, plants as a basic requirements planning tool which explodes the requirements for products or end items as expressed in a master schedule (or some other form of customer requirement) and "explodes" these on an item by item basis to generate gross component requirements. These are then netted against any available inventory and planned orders are generated on a time phased basis to meet the requirements of the master schedule. See figure 1.

In this way works orders are only released for items that need to be manufactured during a specific period and this helps to avoid tying up capacity unnecessarily and reduces work-in-process inventory. For many plants, this level of system is sufficient to meet their needs for manufacturing control.

MRP also forms the logical kernel of MRPII which has developed further to provide a more extensive information system which also takes into account factory resources (usually in the form of work centre capacity) to provide a capacity as well as a material plan. Capacity planning at factory level is undertaken at two main stages: rough-cut capacity planning, where the load imposed by the master schedule upon key work centres is checked to ensure that it is feasible and by capacity requirements planning, where the detailed load upon all work centres is checked against planned capacity for those resources. The load profiles thus generated give management a short term warning of excessive load and allow time to plan overtime or other contingency measures.

This capacity planning ability overcomes one of the basic weaknesses of the basic MRP model, an overstated master schedule which leads to bottlenecks and late deliveries. A typical modern MRPII package also provides interfaces to a range of business functions other than production (purchasing and manufacturing engineering, for example) and provides facilities for the effective management of these areas. In this way, MRPII packages have been extended over the years to provide comprehensive facilities which provide the majority of requirements for the management of multiple manufacturing sites. A simple view of some of the main MRPII facilities is given in figure 2 below. Some vendors have recently claimed that the model has been extended considerably further to provide full enterprise modelling capability and named this Enterprise Resource Planning. However, ongoing research which involves interviews with leading MRPII practitioners has yet to find one application which can justify this claim.

Although modern MRPII facilities are extensive, are applied to a wide range of manufacturing contexts (mechanical engineering, chemicals, pharmaceuticals, electronics) and appear to provide considerable benefits for their users, they tend not to be strong in the control of the execution of the plans which they generate. They may still be referred to as "push" rather than "pull" systems. There is some evidence from the research discussed later that manufacturing control approaches other than MRPII may be more appropriate for some industrial sectors.

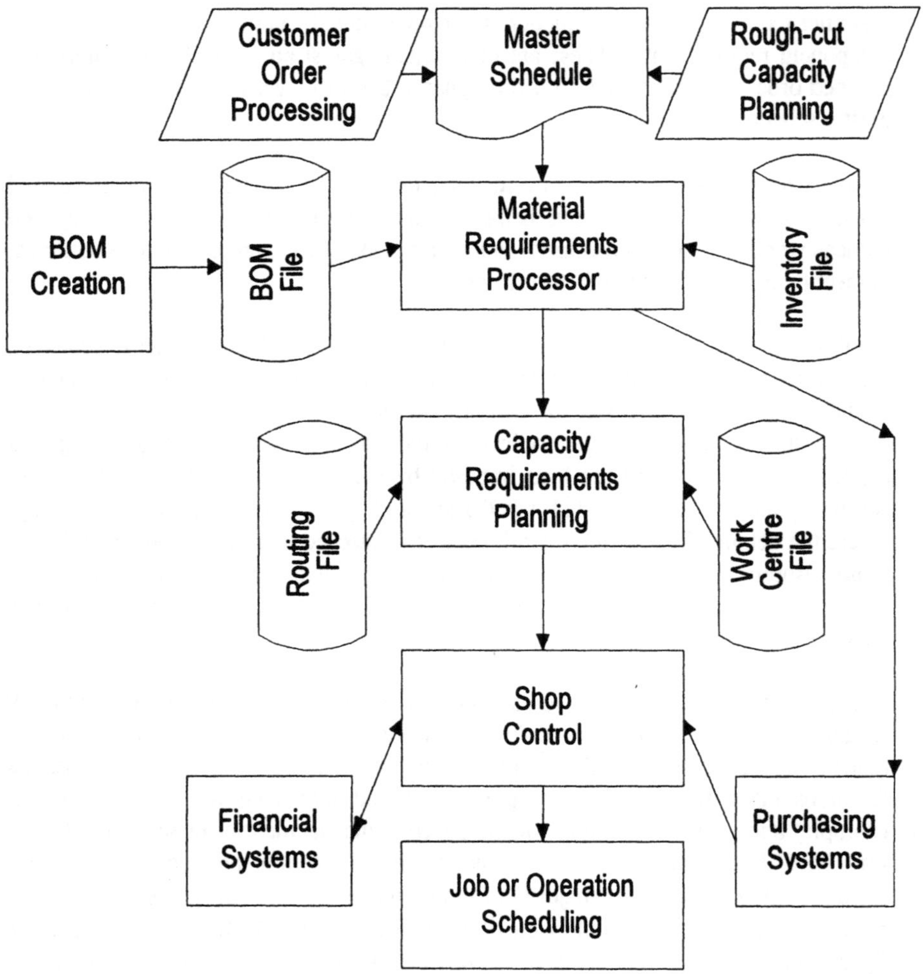

Figure 2: Elements of a typical MRPII Package

Whilst MRPII remains an excellent planning tool, the facilities it offers for shop floor control are limited and resource intensive. This continued dominance is surprising considering the more recent innovative approaches to planning and scheduling which have included Just-in-time (JIT), the use of kanbans between distributed cells, Optimised Production Technology (OPT), simulation, Leitstands [2] and the use of a variety of finite scheduling tools. This research project (completed in 1995) also supported the view of MRPII dominance, where eight out of the ten top performing companies were found to be MRPII users.

3. Modern Scheduling Approaches

The allocation of resources to jobs (scheduling) is an activity which is fundamental to manufacture. It is also an activity which takes place at all stages in the manufacture of products and components and at different points within the manufacturing planning time frame and at different levels in the organisational hierarchy, see figure 3. Despite regular (and frequently ill-informed) criticism by consultants and academics, the MRPII paradigm still appears to be effective in companies where markets are uncertain and products or processes complex. The majority of companies using MRPII still produce weekly or daily work-to-lists from their computerised systems and pass these on to supervisors and foremen who make local scheduling decisions often supported by no more than their experience and the back of an envelope.

Many of these companies are beginning to realise that the most carefully prepared plan has little or no value if it is not put into practice and that this execution phase, although critical to planned performance, usually lies outside the span of control of the planner. This has led some companies to focus on the selection and implementation of additional execution systems [3] which can be "bolted-on" to an MRPII package.

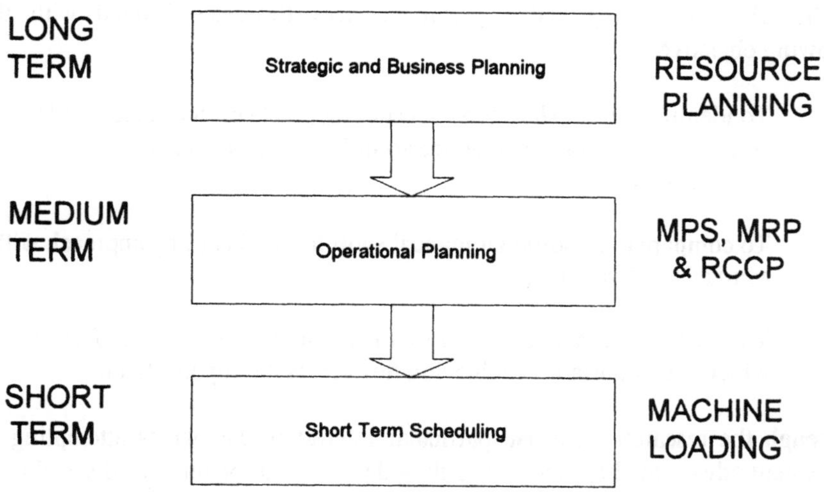

Figure 3: Main Planning and Scheduling Levels

Much previous work has concentrated upon the mathematical and algorithmic aspects of the scheduling process because of the limitations of processing power available to develop simulation based heuristic methods. However, in the last ten years processors and networks of sufficient power have become available to make feasible the development of effective shop floor scheduling tools. However, despite the volume of literature on the subject, this research shows that these have, as yet, achieved limited applications penetration within manufacturing industry.

The adoption of a scheduling tool was usually found to be an attempt to improve shop floor control within the factory although some novel applications were found by the project: including the use of a scheduling tool as an automated master scheduler. A previous survey of shop floor scheduling practice [4] undertaken by the author examined most aspects of shop floor control. However, there were few best practice examples of the application of finite and integrated scheduling tools to be found. Indeed, the lack of experience of the implementation and operation of such innovative methods led to many unsuccessful attempts to apply these. Unfortunately the literature gives little help to firms seeking guidance on the most effective approach to scheduling.

Further investigation based on a series of in-plant discussions and output from several well attended courses covering Advanced Shop Floor Control delivered by the writer, showed that short term scheduling practice was, for the majority of firms, to be something of a black hole. In an attempt to shed some light into this particular hole, a 60 man-month research project was proposed to the CDP Group of the UK Engineering and Physical Sciences Research Council with the following objectives:

(a) To identify current short term scheduling practices in a sample of UK companies and to grade practice according to the degree of success achieved.

(b) To compare the traditional manual short term scheduling approach with available software tools.

(c) To distil "best practice" in short term scheduling and identify factors which are important in achieving this across the supply chain.

Although the research team had particular interest in companies attempting to implement advanced shop floor scheduling decision-aids or integrated scheduling systems and to contrast their performance with other plants using the more conventional methods supported by MRPII, it soon became apparent that planning and scheduling had to be viewed within the context of the overall supply chain, since the difficulty of the scheduling task was heavily dependent upon both customer demands and upon supplier performance. It was also intended that the research answer a number of supplementary questions related to MRPII performance which were seen as important for the development of manufacturing planning and control approaches within industry:

* Is MRPII still a useful plant management tool in today's business climate?

* Have finite scheduling tools superseded MRPII either completely or in part for certain applications ?

* To what business sectors can finite schedulers be effectively applied?

* Are finite scheduling tools cost-effective?

* What are the key factors when choosing and implementing a scheduling system ?

4. Research Method

In the assessment of scheduling performance within a range of widely different companies [5], there are many difficulties to be faced. Companies may be of different sizes, may serve different markets and operate different types of plant. A particular problem identified was the difficulty of separating planning and scheduling activities. These are frequently closely related and may be difficult to distinguish. Another issue is the comparison of the effectiveness of different scheduling approaches used in what were found to be very diverse business contexts.

To address these and other problems and to reduce the level of subjectivity arising from any company visit programme, a research method which minimises the subjective viewpoint of the interviewers is essential. The method adopted relied on a series of structured interviews with at least four people responsible for different functions within the planning and scheduling hierarchy including production supervision. Interviews with different members of staff with different responsibilities and perspectives ensured that a range of views were expressed. and these views were later reconciled by the method to provide a composite picture of company operations.

These views were then encapsulated within a structured case study report and analysed using a performance assessment method based upon four key areas of analysis:

- The operating environment of the business
- The high level planning system used
- The short term scheduling process
- Measurement of overall performance of production control system

Informal face-to-face interviews with the different company representatives, each of whom could give an expert account of one such area and also informed opinion about other areas was found to be the most effective form of data capture. The informality of the interviews was considered necessary because of the diversity of different production control approaches in use. A rigid questionnaire was not sufficiently flexible and informality allowed the interviewers free rein. However this increased the difficulty of interview management to ensure that the base data is gathered but compensates by ensuring that areas of interest are raised and then probed in sufficient detail. It was found to be preferable to interview in pairs to compensate for the lack of structure within the interview. This "pairing" was also vital to the validation of performance scores generated by the scoring method which was developed.

Interviews were normally conducted as two visits of one days duration (i.e. four man-days in company per study) with the days typically spaced about ten days apart. The first day emphasised the business context and the high level planning systems, whilst the second day focused upon the scheduling approach and tools in use and their effectiveness in practice. Factory tours were requested as a means of quickly coming to an understanding of the processes and problems of the company.

4.1 Viewpoint Structuring

Another important factor in the information capture process is the balancing of opinions. Throughout an organisation people with different viewpoints often have disparate ideas as to how production control, and other functions, operate. If practices were to be evaluated the evaluated the first task was therefore to identify the true current operational activities. This is done using a technique called viewpoint structuring, which involves taking a cross-sectional slice of interviews through the organisation, encompassing people at each level of the production control hierarchy. In attempting to establish the true processes each persons viewpoint on any particular topic is weighted according to their role and how closely related they were to the area of question

4.2 Scoring System

A performance measurement framework consisting of 130 measurement parameters was developed which are scored on an integer range from 0 to 10 inclusive. With the full listing of questions comments and corresponding scores are given as benchmarks to assist consistency in the scoring process both over time and by different people. The two interviewers generate scores individually, these pairs of scores are then compared and where discrepancies occur they are reconciled by discussion to an agreed value. Performance is assessed against the seven broad categories described below:

Complexity (C) Complexity questions are intended to assess factors which are static within the short to medium term and define the basic boundaries within which production control operates. Complexity is a reflection of the underlying business conditions determined by parameters which cannot normally be controlled or manipulated to any great extent in this period. The basic parameters considered include the product range and physical composition, production processes, level of customer expectations or requirements, product; range and processing complexity and change of product base.

Uncertainty (U) Uncertainty relates to unpredictable short term changes inflicted on production control, the effect of which is directly related to both the frequency and level of change. The result of uncertainty is a more dynamic situation which necessitates reactive management. Uncertainty acts on top of complexity to increase the degree of difficulty of the manufacturing environment. The key areas of uncertainty identified are assessed for both frequency and level of impact on the schedule, they include customer demand fluctuations, internal component supply for dependent operations and external material supply issues. Also included is reliability of labour, machinery and plant capability to meet quality levels required.

Buffering (B) Buffering or moderating factors are designed to insulate production control and operations from the effects of complexity and uncertainty. Buffering questions relate to strategic business decisions which the company may choose to take, in order to insulate itself from any short term mismatch between output capacity and the market demand. These typically entail additional investment in one or more resource, so that the organisation can be responsive, or is buffered by a time delay giving it more opportunity to react. Buffers can take two basic forms; either a physical excess such as finished goods stock, labour or machinery, or by utilising flexibility which creates additional effective usable capacity when required. The questions cover the basic resource types, materials, production equipment, labour and relative lead-times of production and market expectation.

Disciplines & Structure (DS) Discipline and structure questions are concerned with the framework within which effective day to day operations are performed. Issues include business planning, the provision of all the necessary resources to produce goods efficiently, good inter and intra departmental communication, accurate and accessible information. Also definitions of job boundaries and objectives, known levels of authority within the scheduling operations and the existence of performance measures which steer all parties in a focused united company direction. In many cases discipline questions relate to the extent to which the organisation's activities are under formal control.

High Level Planning (H) High level planning scores are designed to examine the effective operation the of planning systems converting the sales orders into manufacturing plans. The aim is to assess the general ability of the function to give due consideration to all the parameters which may be of importance in the process of planning in line with the business focus and in a structured manner. Also the ability to effectively communicate and meet the needs of all other functions as an internal "supplier". Topics covered include forecasting, data management, ongoing improvement activity, system capability in terms of the range of resources and factors considered, ability to accommodate changes and react in a responsive manner, use of inputs from and communicate plans to other related functions. This is an important section as the quality of the planning process may determine the base information with which scheduling will operate. In conventional MRP logic the more effective the planning process, the easier subsequent scheduling task becomes.

Due to the complexity of defining the division between planning and scheduling the questions in this section are divided into two parts, questions which are considered to refer only to **H**igh level planning called "Pure Planning", they are labelled with an **H** prefix. Others, which refer to activities that may be executed in the context either of a planning or a scheduling activity, or in some cases both, are referred to as "Variable Production Control Parameters", numbered with the **PC** prefix for Production Control and a **h** suffix for high level planning. Directly equivalent questions to these can be found in the scheduling section with the same **PC** number but an **s** suffix. Questions are scored in either set according to the context in which they are performed and will then be scored zero in the other set unless the activity is specifically performed again in parallel and from the other perspective.

Scheduling Techniques (ST) Scheduling questions are concerned with the actual Short Term Scheduling procedures and the suitability of these techniques for the task in hand. Coverage includes the maintenance of accurate detailed data, consideration of all relevant factors, ability to react to change, usability of output, evaluation of alternatives, change control for issue of schedules, and time slots and horizons used in relation to the operations being controlled. Also evaluated is the thoroughness of Scheduling, how the function is perceived by others in related functions and whether it learns and develops with experience in a process of continuous improvement.

As with the Planning scores, there are two sections to these questions the pure scheduling questions which will apply to scheduling only, labelled **S** numbers, and Production Control Parameter questions applied to scheduling labelled **PC** numbers with the **s** suffix.

Performance (P) This is the area of performance measures which are directly affected by the quality of the production control activities following the application of both planning and scheduling activities. It is important to point out that this is not concerned with the overall commercial performance of the company but in this case specifically performance of the production control activity. Hence classical financial measures are not relevant. The emphasis is placed strongly upon the ability to satisfy the customer by understanding and controlling the supply chain effectively and also interfacing well with production facilities. The prime areas of consideration are, on time delivery performance, order backlog, use of resources, material control and flow, schedule adherence, quality standards etc.

Degree of Difficulty (D) Degree of difficulty (or Difficulty) is not a scoring category in its own right but a compound measure formed by grouping complexity, uncertainty and buffering sections. (For more detail refer to "Score Groupings" in the Analysis section of this document below). It is a measure of the difficulty of the production control task created by the production operations and market environment within which the company operates. This cannot be altered by production control but is influenced primarily by strategic operating policies. Difficulty is created for use in analysis because it provides a single measure of difficulty of the production control task, allowing comparison of companies operating in dissimilar, on a common scale.

4.3 Selection of Case Study Companies

It was deemed desirable to minimise effort spent investigating unsuitable companies. Companies were chosen to exemplify local scheduling practices across the range of scheduling approaches identified, from manual in the foreman's head, to simple decision aids, to fully automated finite scheduling systems, and also spanning a variety of business sectors and manufacturing environments.

Several methods were used to create a database of target companies, with various criteria then being applied to rank them in order of preferred study. These methods comprised:-

(a) Personal contacts.
(b) Analysis of "FORCAST" Eureka Project questionnaire.
(c) Magazine articles, eg Management Today - Best Factories

(d) Previous work by the research assistants.
(e) Companies accessed via software houses.
(f) Contacts within BPICS member companies.
(g) Companies who became aware of the research and contacted us.

Expansion of the database initially created was an on-going process and, as the project progressed, the focus became towards specific companies for reasons of gap filling to provide a broader selection of tools or industrial sectors. These companies were located primarily through the network of contacts established during the project with both industrialists and academics.

In contradiction with the experience of many other research groups, little difficulty was encountered in gaining access to companies. In some cases this has been due to the nature of the in company contact (i.e. friends and colleagues), but also, more importantly, the research approach adopted. Following an initial informal contact, companies were sent a one page summary of the research to introduce them to the work. This was then followed with a brief meeting with a senior employee to outline in a little more detail the terms of reference for the case study and to establish credibility. A range of other factors were also made clear to each potential case study company in order to encourage participation:

5. Assessment of Planning and Scheduling Performance

The generation of an effective measure of overall planning and scheduling performance was fundamental to the work. Initially planning performance had been considered independently to that of scheduling. However it quickly became apparent that this was not feasible for many companies. Planning and scheduling could not always be seen as a two stage process. Examples emerged of case study companies successfully using scheduling packages without the support of any planning system. This brought into question one of our initial assumptions: that good planning is required in order to enable effective scheduling. This assumption lies at the heart of the MRP/MRPII generic model which assumes a top down approach to material and capacity planning [6] and scheduling, with increasing refinement and detail as time horizons approach the point of production. Clearly the **Plan > Schedule > Execute** model does not apply in all situations.

It was surprising to see that any company could occupy the position shown by Company X on the planning/scheduling continuum (see figure 4 below). This led to a modification of the analysis method with planning and scheduling being assessed using two separate but related question sets and a significant change in the way in which the researchers viewed the whole planning and scheduling process.

Having decided to assess planning and scheduling activities independently, overall production control performance still needed to be measured and a series of suitable questions and associated scores were devised. This Production control performance was expected to depend on the quality and quantity of effort deployed by the company (as evaluated by planning and scheduling score sets), in relation to the difficulty of the planning and scheduling task. Therefore a way of measuring the difficulty of the planning and scheduling environment, which was heavily dependent upon supply chain aspects, was also necessary. When a full assessment system had been devised based upon the product and process complexity and demand and process uncertainty, three initial pilot studies were conducted to test the validity of the model. This led to a significant addition to the method: the consideration of the buffering which a company may employ to reduce the difficulty of the environment in which it operated.

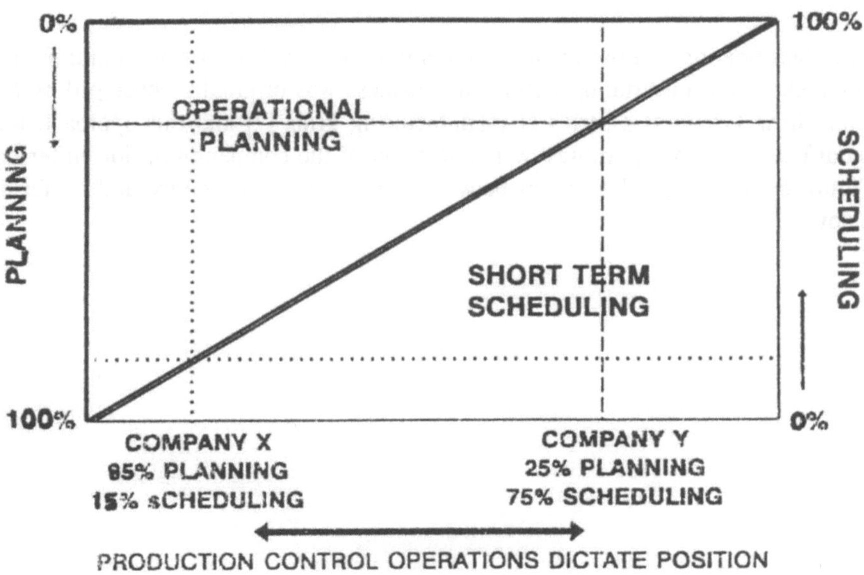

Figure 4: Planning and Scheduling Continuum

The assessment approach was based on a series of graded criteria to establish the company performance in five main parameters. These are given below along with their identification code:

- Degree of Difficulty (DD)
- Disciplines and Structures (DS)
- High level Planning ability (H)
- Scheduling Techniques (S)
- Overall production control Performance (P).

Degree of difficulty

The Degree of Difficulty parameter is a function of the complexity and uncertainty of the environment in which the company is operating less the degree of 'buffering' (spare resource, e.g. capacity or stocks) employed to reduce the impact of these two factors. It is much harder to achieve successful scheduling performance in a complex and uncertain environment with only minimum resource availability than in a firm making relatively simple products with a very low level of unpredictable change, spare resources and large finished goods buffer stocks.

The classification of a particular business into industrial sectors by examining the complexity and uncertainty within the business was originally developed by PA Consulting Group as a series of manufacturing grids [7] and this approach was modified to suit our purposes by the addition of the compensation for buffering. Figure 5 gives the relationship between complexity, uncertainty and buffering below:

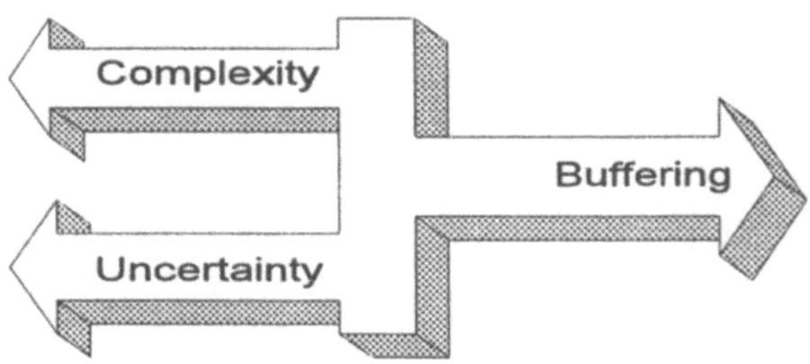

Figure 5: Components of degree of difficulty

The three components of Degree of difficulty give the final scoring system matrix which consists of seven sub-sections in total and can be related back to the original four prime areas of investigation identified earlier. See figure 6, below:

5.1 Performance Scoring

Following the series of interviews, each company was written up as a case study report which included a high level data flow diagram for production control information. Each case study report was then sent to the company for validation and, on return and after any agreed amendments, each individual company was "scored" independently by each interviewer against the various performance parameters. The scoring process involved analysis of 130 factors across the five categories (seven sub-sections) indicated above. All questions were scored against agreed benchmarks to aid consistency and to maximise objectivity. Each interviewer independently scored each aspect of the company and any differences between the two sets of scores were then reconciled by discussion.

Statistical analysis was undertaken to validate the original scoring method adopted and good correlation was found considering the relatively small sample of companies.

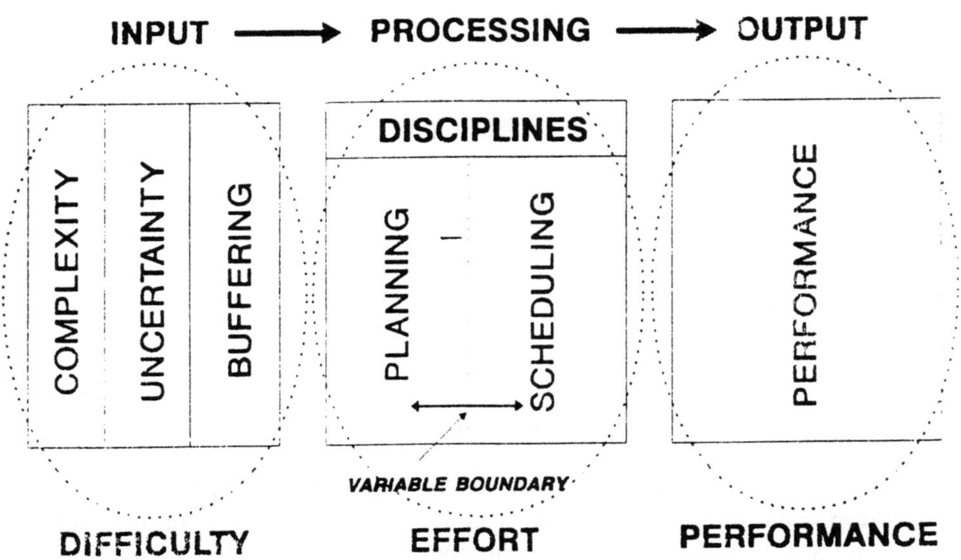

Figure 6: The seven sub-sections of the assessment method

The resultant scores were then mapped onto the five axis 'footprint' diagram as shown in figure 7 below. Overlapping the diagrams of different companies provides a quick comparison and is particularly useful for comparing different companies within similar business sectors. Unexpected differences were quickly highlighted and encouraged an analysis of the basic data set to establish their cause.

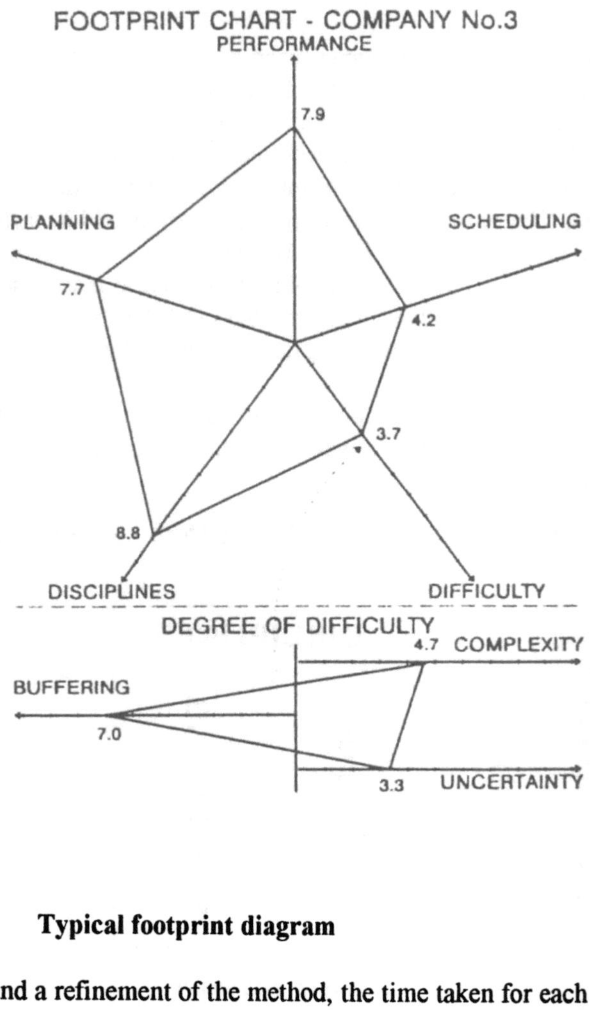

Figure 7: Typical footprint diagram

After experience and a refinement of the method, the time taken for each series of interviews could be reduced to three man-days although the time required for each case study, including write-up and scoring was considerably more. This limited the number of case studies which could be undertaken during the period of the project. Practical considerations limited the study to 30 companies which is much better than the original planned target of 20 case studies, but is recognised as a relatively small sample. Consequently statistical methods using SPSS and SAS packages have been used to examine the validity of the results and to determine the significance of the factors measured.

6. Classification of Planning and Scheduling

An examination of the case study companies identified a key difference in approach to planning and scheduling between two specific groupings of the various industrial sectors. This depended upon whether the main problem faced by the company was the control of material supply (typical of engineering and pharmaceutical companies) where MRPII would typically be used, or the effective use of capacity, often the issue in the FMCG companies, when MRPII did not prove effective in the scheduling of resources. This led to a differentiation between "materials led" and "capacity led" approaches. From this starting point the classification given in figure 8 was developed.

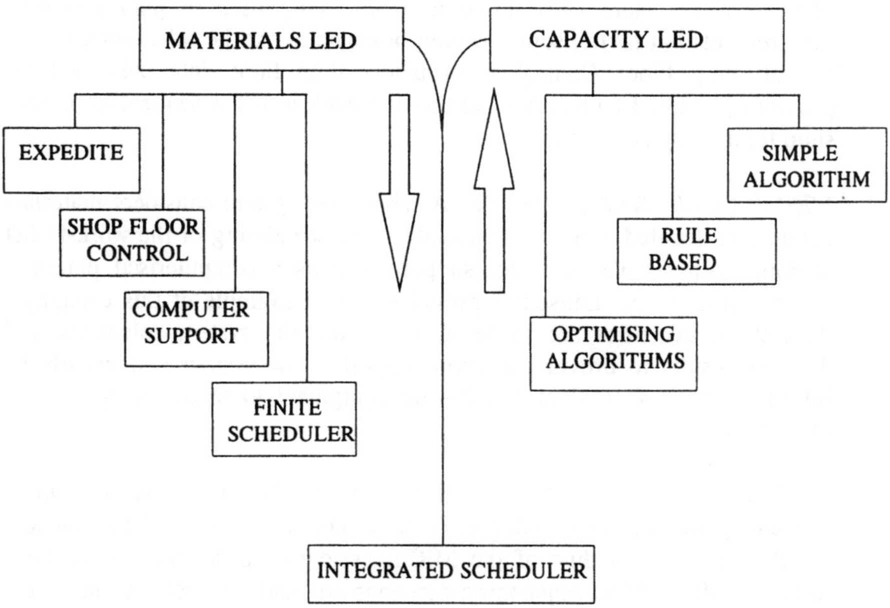

Figure 8: Classification of Production Control Systems

The classification has two main branches. The materials led side reflects the conventional top down MRP approach where the planning of material supplies is critical, for example in the manufacture of complex assembled engineering products such as tractors. In such companies the management of work centre capacity, whilst important, tends to be handled as a secondary issue. On the other hand, capacity led companies tend not to have major material supply problems and focus on the management of capacity for optimised output. In these industrial sectors finite capacity scheduling tools are particularly appropriate. While efforts have been made to identify distinct categories it is recognised that the boundaries are not always concrete and a software tool may fall into different classifications according to use. The most closely related categories have been drawn adjacent on the diagram and a brief explanation of each category is given below:

MRP & EXPEDITE This is the least sophisticated of our categories, basic MRP logic is being used to plan material and component requirements and to release manufacturing and purchase orders. Beyond the basic plan there is a work-to-list but only manual expediting of orders.

MRP & SFC Here MRP logic is again being used to plan material requirements but taking some account of capacity. At the scheduling level, formal Shop Floor Control is in place, shop floor data collection is providing a closed loop system so that the work-to-list is sequenced on the shop floor using defined rules.

MRP & COMPUTER SUPPORT As above the system considers materials before any detailed account of capacity. The scheduling of the work-to-list is done with simple computer support such as a computerised planning board software. A Leitstands provides a good example of this category. Almost all decisions are made manually but the system calculates and displays results to provide decision support. The system will usually be interfaced with shop floor monitoring equipment to maintain up to date information.

MRP & FINITE SCHEDULER Here MRP handles materials and broad capacity planning but scheduling is performed by a finite scheduling tool which may be a module of the MRPII system or a stand-alone package linked to MRP. Most sequencing decisions are made initially by automated decision logic driven by rules within the system with retrospective manual refinement.

OPTIMISING ALGORITHM BASED In functionality this category of system is similar to the finite capacity scheduler, but the system draws customer order information directly from the business system database and schedules this using pre-determined rules.

RULE BASED Here the core of the production control system is a scheduler which operates primarily from a rule or knowledge base as opposed to the mathematically based optimising algorithm systems above. This category includes expert systems [8] and other artificial intelligence technique based software.

INTEGRATED SCHEDULER A development of the rule based scheduler but with much greater functionality which considers other resources including material in addition to the primary capacity constraint implicit within its logic. This theoretically negates the need for a supporting MRPII system based on the scheduled output although, typically, the scheduler will be linked to an existing MRPII database. OPT is the leading product in this category.

In many ways the most interesting approach is the integrated scheduler. In this case the systems attempt to schedule material and capacity simultaneously by building a plant model which is specific for a particular work mix generated by the orders and forecasts for the period under consideration. Optimised Production Technology and DBR are examples of products which act in this way.

7. Analysis of results

Our original impression that shop floor scheduling is indeed a black art has been fully confirmed by our results. A small ray of light is provided by the relatively few companies found who exhibit best practice in the area. The main problem encountered is the lack of effective FST applications coupled to the different ways in which companies have implemented and applied them to make direct comparison difficult. The most interesting illustration of this is the use of finite schedulers for automating the master scheduling activity.

Another major difficulty found by the study is that, in practice, the majority of the tools used are operated in conjunction with MRP/MRPII packages. It was disappointing to find that too few FSTs were in use on their own to give us a direct comparison between them. This applies to all finite schedulers including OPT, which is usually run as an MRPII-hybrid system with the OPT software drawing on the MRPII database. An overview of the case study companies classified both by industrial sector and the particular planning and scheduling methods employed is given in figure 9.

Several different analytical approaches are being used because of the complexity of the problem. These range from simple data modelling to statistical analysis using correlations, stepwise regressions and cluster analysis. All methods use the basic 130 question scores either in raw form or various groups as described earlier.

Industrial sector / Systems classification	MRP and expedite	MRP and SFC	MRP and computer support	MRP and finite scheduler	Algorithm-based	Rule-based	Integrated scheduler
Mechanical engineering	1	1	1	2	1		2
Electronics	1	1					1
Pharmaceuticals	1	1		2			
Chemicals	1		1		1	1	
FMCG and related	3				3		
Other retail goods		1			1	1	2

Figure 9: Case Study Companies by Industrial Sector and Production Control Approach employed

Footprint diagrams have been used as a simple diagrammatic tool to show major imbalances in the various performance parameters and also to highlight areas of difference between the operations of companies within the same industrial sector. This has proved especially useful in comparing companies in the same sector employing different scheduling approaches or companies in different sectors employing the same scheduling tool. An important output from the work is the development of a novel set of reference models.

Early analysis was based on testing of hypotheses developed intuitively from on an understanding developed over the 30 case studies. As well as the footprints, use of simple Cartesian graphs to investigate the relationships between the various score groups was made in an attempt to confirm our insights.

This first level analysis has produced results such as Figure 10, which displays a significant relationship between performance and disciplines (0.76 correlation). In the context of the research, companies which exhibit discipline are those with clearly defined planning and scheduling objectives which are published and communicated throughout the company. Such companies hold regular meetings (often informal) to monitor schedule performance and have strong communications links between sales and production which ensures strong demand management. Companies of this type tend to achieve strong schedule adherence assisted by regular review meetings between the planners and the shop floor. They also have highly developed systems of forecasting demand.

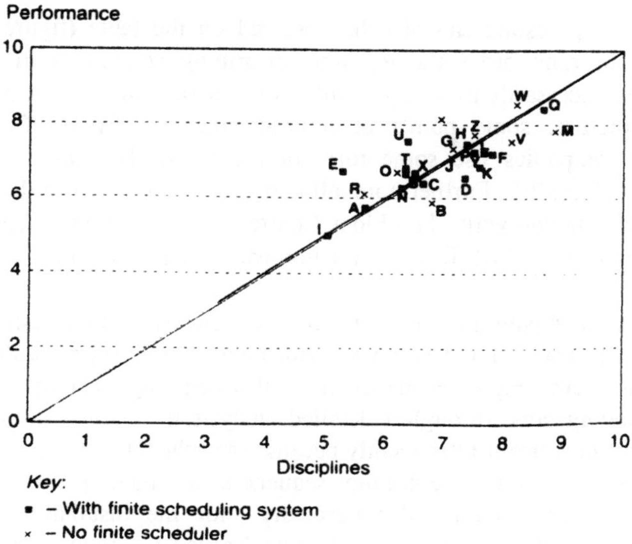

Figure 10: Disciplines versus Performance for all companies

The analysis suggests that finite scheduling tools can give improved performance for a given level of effort when applied to a task of equivalent difficulty. This statement can imply that equivalent performance can be achieved for less effort which, in practice, means fewer production control staff. This is suggested by figure 11 where the plot shows a clear bias in favour of scheduling tools. Twelve out of fourteen companies with finite scheduling tools were noted to be using a below average number of production control employees.

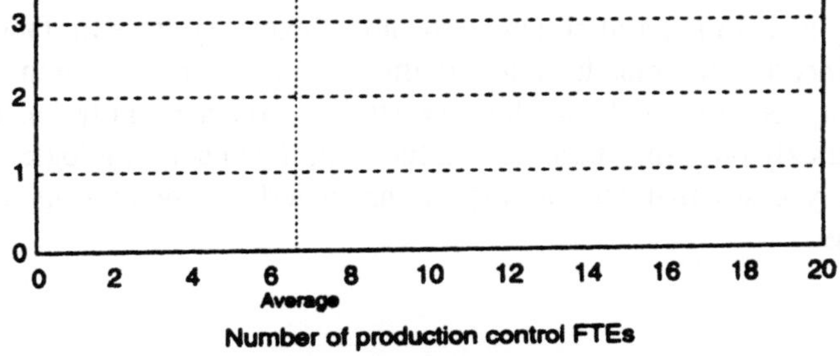

Figure 11: Difficulty versus Number of full time equivalent employees (FTEs) working in Production Control

Some interesting results can also be observed on the table (figure 12, below) which lists in rank order the top ten performing companies in the study. Although the necessarily limited amount of information on the table gives an over simplified picture, some points of interest emerge. Eight of the top ten performing companies use some form of MRPII whilst only two use FSTs independent of MRPII. Furthermore, all four of the companies in the study using MRPII in conjunction with Shop Floor Control feature in the top ten whilst only one company is using MRPII in conjunction with a scheduling tool.

The "best" two companies on our performance scale are also interesting. The top one developed, installed and has run a scheduler in conjunction with MRP for a period before reverting to primarily manual scheduling systems. The second company, having gone through a detailed analysis in order to install their rule based system, now use it infrequently because the schedulers have now learnt the logic necessary to create optimum sequences on their plant. Both have established the sequencing rules necessary for the installation of a finite scheduling tool and discovered that, having learned the sequencing rules, they now no longer find the tool essential.

RANK	CODE	SECTOR	CLASSIFICATION
1	W	FMCG	MRP & Expedite
2	Q	Chemicals	Rule Based Scheduler
3	#	Other Retail Goods	MRP & Expedite
4	M	Pharmaceutical	MRP & SFC
5	Z	Other Retail Goods	MRP & SFC
6	V	Electronics	MRP & SFC
7	U	FMCG	Optimising Scheduler
8	H	Mechanical Engineering	MRP & Finite Scheduler
9	P	Chemical	MRP & Expedite
10	G	Mechanical Engineering	MRP & SFC

Figure 12: Top ten performing companies in Rank Order

There is an interesting parallel here to the use of expert systems for fault diagnosis. Lucas Engineering Systems have found after the installation of a number of such systems that operators and maintenance staff use them initially but that usage drops off quickly once they become familiar with the underlying logic. It appears that in many cases that the capture of the underlying processing rules is the critical issue.

8. Conclusions

In spite of the smallness of the sample and the difficulties associated with the comparison of companies operating in very different industrial sectors, the work has produced some important results. Firstly the answers to the questions posed in the introduction can be given. Secondly several clear conclusions about the application of FSTs can be made. Thirdly, the ability of MRPII to meet the future needs of a wide range of industrial sectors can be questioned and, most importantly, a warning can be given to companies considering the adoption of FSTs.

The answers to the questions set out in the introduction are given below:

* **Is MRPII still a useful plant management tool in today's' business climate?**

  MRP/MRPII when properly installed and managed continues to produce good results, especially when used with a full implementation of the shop floor control which includes shop order progress monitoring. Eight out of the top ten sites in our survey were employing MRP/MRPII. See figure 13: Enhanced MRPII below:

* **Have finite scheduling tools superseded MRPII either completely or in part for certain applications ?**

  Finite scheduling tools were found to be used in conjunction with MRPII ("bolted-on") in the majority of case studies. However several studies suggested that for some industrial sectors where material supply is not a problem, they become a logical alternative likely to provide significant benefits. See Figure 14: Post MRP Scheduling.

* **Are finite schedulers cost effective ?**

  There is clear evidence that in some sectors users of finite scheduling tools achieve very effective results in circumstances of equivalent difficulty with lower staffing levels and less overall "effort" than in a conventional MRP/MRPII installation.

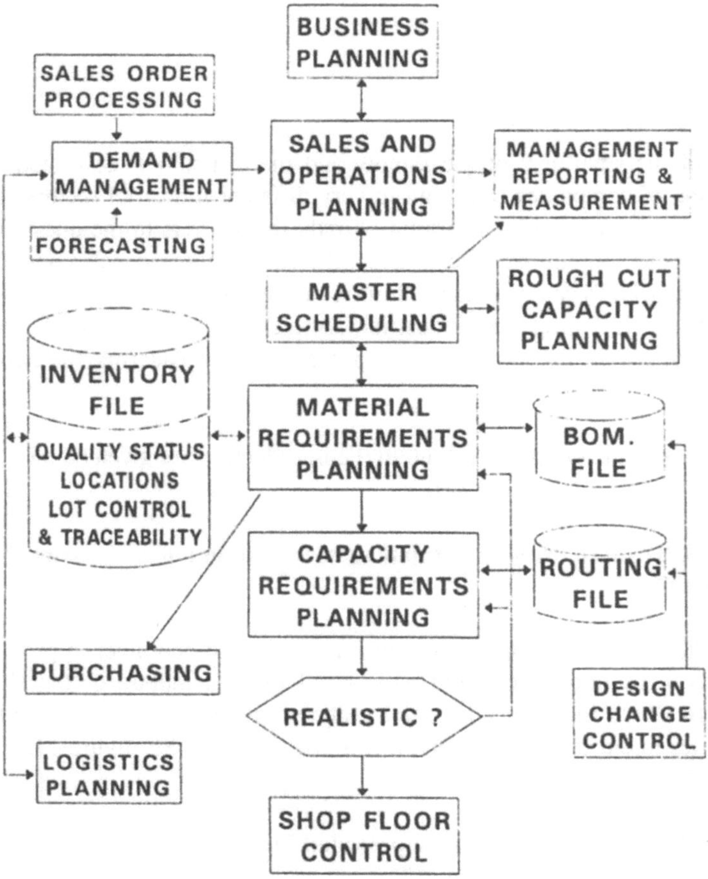

Figure 13: Enhanced MRPII

* **In what types of company are finite schedulers working effectively** ?

Finite scheduling tools were seen to work effectively in Fast Moving Consumer Goods (FMCG) sectors and in the scheduling of capitally intensive plant. They also were used in batch engineering companies with some degree of success. The use of an integrated scheduler to replace the core logic of MRPII as a hybrid scheduler were particularly appropriate for the manufacture of complex engineered products. See figure 15.

* **What are the key factors when choosing and implementing a scheduling system?**

The use of finite scheduling tools in production control is only likely to produce real benefits if applied in an appropriate industrial context and operated in a structured and disciplined manner. One reason for this is the reliance of these systems on a significant volume of detailed and accurate data which has to be collected and monitored in a dynamic environment. Another factor is the need to clearly understand the processes which are operated and the necessary sequencing rules with which they are associated and to select a scheduling tool which matches these requirements; a rate based as opposed to a works order based scheduler (or vice-versa), for example.

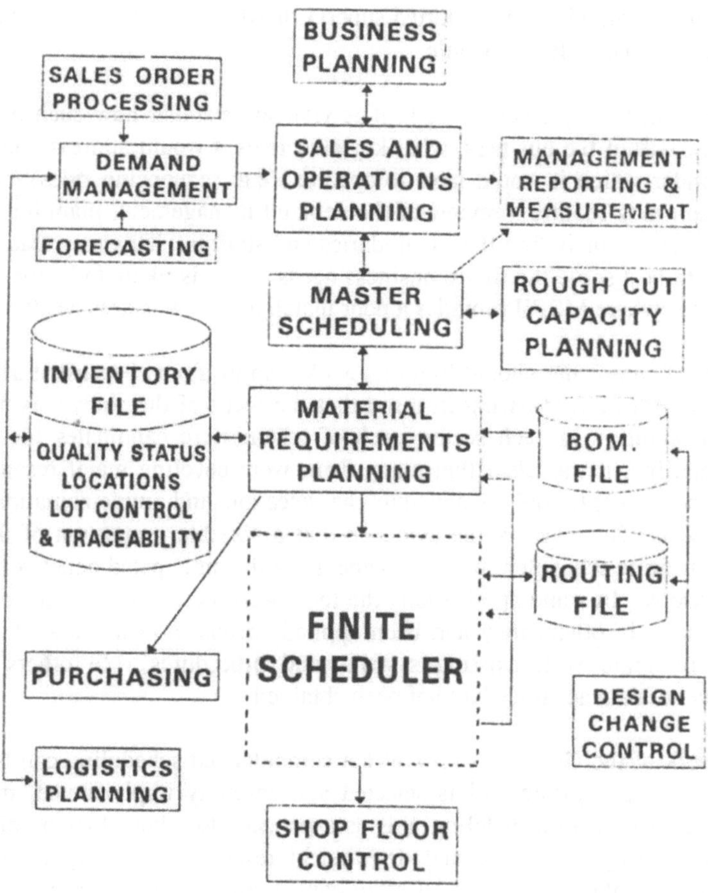

Figure 14: Post MRP Scheduling

There is strong evidence from this research that the use of finite scheduling tools and integrated schedulers has important implications for the staffing of the production control function, particularly in a dynamic environment operating with high data volumes. This is typical of the FMCG sector. Figure 11 shows that the use of such tools requires a lower level of staffing for an equivalent level of difficulty. This reduction, however, emphasises another finding, the need for high calibre staff in planning and scheduling who must therefore be suitably recognised, rewarded and supported. Given the increasing levels of complexity and uncertainty observed at all levels in the supply chain, the increasing demands on the function will necessitate better qualified and more enlightened people. Staff quality should be seen as a function of both ability and experience, one without the other would not suffice for such an important role. This has important implications for many businesses where the scheduling function is often left to an untrained an unsupported supervisor who may have no clear idea of the criteria for an effective schedule.

Before attempting to implement a FST, it is vital to establish how such a tool would be used within the business, what key objectives it would address and the constraints under which it would have to operate (e.g. sequencing rules). Any implementation may require a review of the demand management, planning and scheduling processes including their boundaries and staffing. This may lead to a re-alignment to best serve the future business needs. This is likely to be the case for MRPII users where MRPII provides a poor match to their business needs.

A wide range of companies should find this work helpful if only because it delays their purchase of a FST. Bearing in mind that the focus of the study was upon leading manufacturers in each sector, each of whom were candidates for best practice scheduling, it was clear that many firms were devoting major resources in terms of both people and money into the selection and implementation of scheduling tools. There was some evidence that a sizeable proportion of these investments in finite schedulers were not generating the anticipated benefits for a variety of reasons. In some applications the tools were not well matched to the applications area, in others they had been applied without a clear view of how they would integrate with existing systems and procedures. In others the necessary level of data accuracy had not been obtained.

Whilst the work of the Group has shown that computerised scheduling can be of benefit when an appropriate tool is selected and carefully implemented, many companies are looking to scheduling aids as a panacea for shop floor problems and are not devoting the intellectual or physical resources necessary for their successful implementation and use. Before considering the adoption of such a tool, a manufacturer should consider carefully the objectives of the implementation, and whether suitable data collection capability is in place providing timely data of sufficient integrity. They should also examine the constraints upon the scheduling process and ask themselves what realistic benefits can be achieved and at what cost. If they are existing MRPII users, they should also review the operation of their system and consider a more full and rigorous application of the shop floor control module before considering the implementation of a finite scheduling tool.

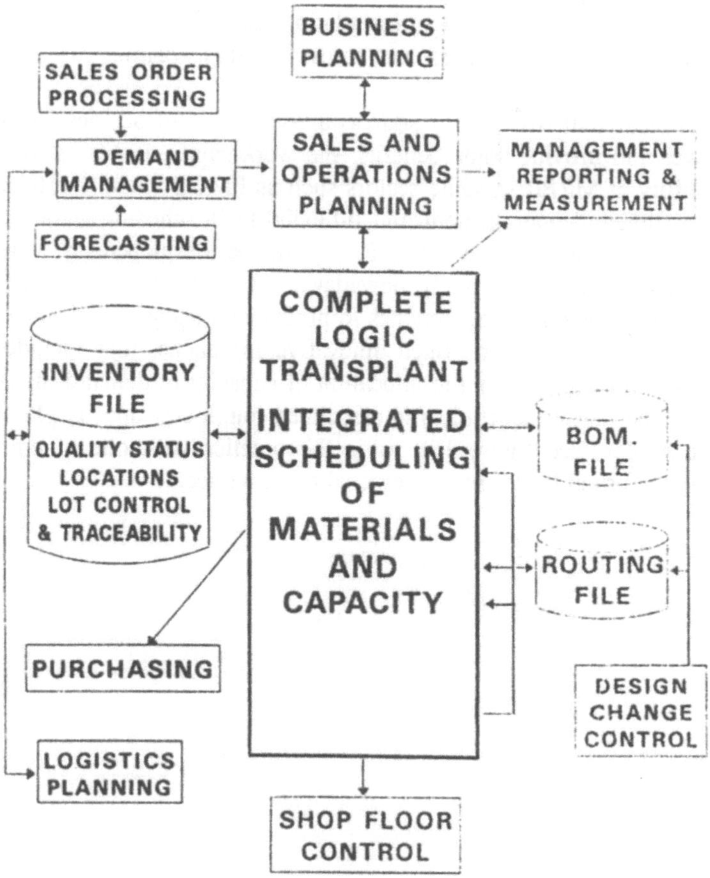

Figure 15: Hybrid Scheduler with Replacement of MRPII Core Logic

However, it is important to sound a cautionary note. Whilst the research found a disappointingly low level of FST implementation, some of the applications were most interesting. It must be borne in mind that in general the finite scheduling systems investigated in this study had been implemented much more recently than the MRPII systems with which they are compared and the tools themselves were typically subject to frequent revision. This tallies with a software market which in many ways is still in an adolescent phase as opposed to the MRPII market which may now be considered to be comparatively mature.

It is reasonable to expect the benefits of FST implementation to increase as tools and applications mature and some further work is already underway to investigate this hypothesis. Although MRP systems have been shown to deliver results in a range of businesses, this does not mean they are the best solution for the future in all environments. In certain situations scheduling tools are much more in line with the real business pressures and installation may offer a much more effective way to operate planning and scheduling across the supply chain.

Whilst it is clear that MRPII can address a wide range of business sectors at a generic level and provide some benefits, the work raises doubts as to the long term suitability of MRPII in some sectors such as FMCG where the fit is not so strong poor, demand management and inflexibility in scheduling for example. Either more specific variants of MRPII will be developed by vendors or companies themselves (less likely nowadays) or companies will look towards more innovative scheduling approaches to produce an adequate schedule quickly rather than seek the optimum. Such innovative approaches may include the use of a finite scheduling tool for the allocation of capacity in demand management by replacing the master schedule and associated rough-cut-capacity planner with a finite scheduler (see figure 16), as is the practice with one of our current collaborators, a brand leader in the personal toiletries sector.

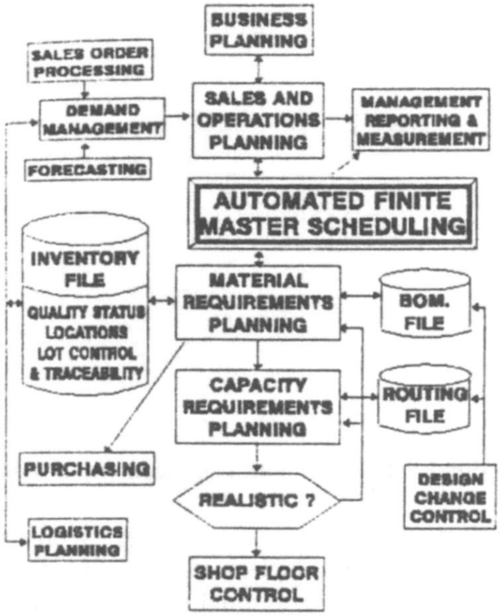

Figure 16: **Replacement of Master Scheduler and Rough-cut Capacity Planner with Finite Scheduler**

9. References

[1] Little D and Johnson S
"Survey of Manufacturing Control Practice: a brief report."
Control, BPICS, Vol. 1, no 2, 1990, 31-33.

[2] Adelsberger H H and Kanet J J
The Leitstand - A New Tool for Computer Integrated Manufacturing"
Production & Inventory Management Journal, 1st Qtr, 1991, 43-47.

[3] Freeman T
"Finite Capacity Scheduling"
Proceedings of the 27th Annual Conference, BPICS, 1992, 469-483.

[4] Little D and Jarvis P
"Survey of current UK shop floor scheduling practice."
Control, BPICS, Vol 19, no 1, 1993, 24-28.

[5] Wiendahl, Hans-Peter
"Logistics Performance Measurement of Shop Floor Activities"
Annals of the CIRP, Vol. 42, no.1, 1993, 509-512

[6] Blackstone J H Jnr
"Capacity Management"
South Western Publishing Co, APICS, 1989

[7] PA Consulting Group
"Manufacturing into the late 1990s",
HMSO, 1989.

[8] McIntosh T
"MERLIN - An application using expert systems for real-time scheduling within British Steel."
Proceedings of 28th Annual Conference, BPICS, 1993, 175-181.

[9] Little D
"Short Term Scheduling in the Process Industries."
Proceedings of the IChemE Symposium on Process Scheduling, UMIST, Manchester, 8 december, 1995.

7 A Stable, Distributed Routing Policy for Flexible Manufacturing Systems

J. Fenchel and Y. H. Chen

1. Introduction

Consider a multi-machine Flexible Manufaturing System (FMS) FMS, where a set of n input part types (PTs) is processed subject to processing and system constraints. The m servers are arranged in a parallel fashion, with identical operating conditions for a particular part type (PT), but nonidentical processing capabilities. In other words, there is no preference between servers for processing of a PT, whenever the particular PT can be processed on the server. This gives rise to processing constraints, namely routing decisions during system operation. All policies developed share the following set of assumptions: (1) process times, buffer bounds, and routing conditions are known in advance and deterministic, (2) no job preemption, (3) the machines allow for model processing breakdowns, (4) no precedence relations between PTs of one class exist, and (5) there are implicit setup times between PTs. For the considered type of model an activated breakdown incurs processing discontinuation of a particular PT model, resulting in unprocessed parts of a batch being transfered back into the input inventory buffers. This allows other models of a PT to be processed even though a certain PT model currently cannot be manufactured.

At this point, earlier research on real-time, stable scheduling of FMS [11, 12, 2, 13] has not directly addressed the case of routing multiple PTs to different servers in a multi-model environment. Chuah [4], and Cruz and Chuah [5] considered a system of two servers and two arriving PTs, studying the worst case delay performance under dynamic routing. From an applicational standpoint routing PTs to servers is of relevant importance, and significant research has centered on this problem [9, 3, 1]. Current results on dynamic routing consider an optimality condition, which is subjected to certain topological and distribution assumptions [8, 14, 15, 16]. The development of a stable, dynamic routing policy for a class of Discrete Event Dynamic Systems (DEDS) is currently not known to have been researched.

Distributed scheduling enables to apply local and global policies, while maintaining overall system properties. A local policy is able to strictly schedule a single machine, whereas a global policy further models the interaction between several servers. Consequently, it is of particular importance to develop a strategy to enable consistent interaction between local and global policies. Characteristic properties of the developed routing policy are (1) its ability to maintain or return to prescribed buffer bounds, and (2) the ability for controlled interaction between local, i.e. single server scheduling policies [6].

The formulation of the system model belongs to the class of DEDS, for which the concept of stability in the sense of Lyapunov was introduced by Passino et al. (1994) [10], and is reviewed in Section 2. In Section 3 the foundations for the new routing policy, the formal policy description, and corresponding proof of stability in the sense of Lyapunov are stated. A strategy to dynamically select the virtual queue bounds used within the concept of the policy is presented in Section 4. The introduced scheduling policy is applied to a 3 server 2 PT example in Section 5. Conclusions and future research conclude this work.

2. Lyapunov Stability of DEDS

Passino et al. [10] respresent a DEDS by a set G with

$$G = G(\mathcal{X},\mathcal{E},f_e,g,E_v) \tag{1}$$

where $\mathcal{X}$ denotes the set of states , $f_e : \mathcal{X} \to \mathcal{X}$, $\forall e \in \mathcal{E}$,
$\mathcal{E}$ denotes the set of events , $g : \mathcal{X} \to \mathcal{P}(\mathcal{E}) - \{0\}$,
f_e and g are operators, with a set of valid event trajectories $E_v \subset \mathcal{E}^{\mathcal{N}}$,
$\mathcal{P}(\mathcal{Z})$ for an arbitrary set $\mathcal{Z}$ states the power set of $\mathcal{Z}$.
The operator $f_e(x)$ is only defined for an event $e \in g(x)$. At a certain time $k \in \mathcal{N}$, the state is represented by $x_k \in \mathcal{X}$ and an enabled event $e_k \in g(x_k)$. The subsequent state x_{k+1} is determined by $x_{k+1} = f_{e_k}(x_k)$ if for the state x_k the event e_k lies on a valid event trajectory.

In general, an event sequence f_e uniquely defines the observed state trajectory, which is limited by physical constraints. Thus it is of particular interest to delimit the state trajectory to an invariant set, such that the physical constraints are not violated. A metric on $\mathcal{X}$ is defined by $\rho : \mathcal{X} \times \mathcal{X} \to \mathcal{R}$ and a metric space by $\{\mathcal{X},\rho\}$. The distance between a point x and a set $\mathcal{X}_z$, $\mathcal{X}_z \subset \mathcal{X}$, is expressed by $\rho(x,\mathcal{X}_z) = \inf\{\rho(x,x') \mid x' \in \mathcal{X}_z\}$. A mapping from an arbitrary set to $\mathcal{R}$ is called a functional. The set $S(\mathcal{X}_z,r) = \{x \in X \mid 0 < \rho(x,\mathcal{X}_z) < r, r > 0\}$ expresses the r-neighborhood of an arbitrary set $\mathcal{X}_z \subset \mathcal{X}$.

Theorem 1 [10]:
For stability in the sense of Lyapunov in E_a for a closed invariant set $\mathcal{X}_m \subset \mathcal{X}$ in G, in a small neighborhood of $S(\mathcal{X}_m,r)$, it is necessary and sufficient that a functional V satisfies 3 conditions:

(i) positive boundedness of V: for an arbitrary small $c_1 > 0$ *, a* $c_2 > 0$ *can be determined in such a way that* $V(x) > c_2, \forall x \in S(\mathcal{X}_m, r)$ *,* $\rho(x, \mathcal{X}_m) > c_1$*.*
(ii) boundedness : for an arbitrary small $c_4 > 0$ *, a* $c_3 > 0$ *can be specified , such that* $\rho(x, \mathcal{X}_m) < c_3$ *, and* $\forall x \in S(\mathcal{X}_m, r)$ *it follows:* $V(x) \leq c_4$*.*
(iii) negative definiteness of $\dot{V}$ *: for* $k \in \mathcal{N}, x_0 \in S(\mathcal{X}_m, r), \forall k \in \mathcal{N}$*,* $X(x_0, E_k, k) \in S(\mathcal{X}_m, r)$ *,* $\forall E_k, E_k E \in E_a(x_0)$ *, then* $V(X(x_0, E_k, k))$ *is a nonincreasing function.*

In Theorem 1, the functional $V(x)$ can be interpreted as a system energy measure, which, if it satisfies the necessary and sufficient conditions in Theorem 1, formally proves system stability. The real-time routing policy is developed and shown to be stable based on the result stated in Theorem 1.

3. Part Type Routing Policy

The fundamental characteristic of routing is the inherent interaction between global and local scheduling policies. Global scheduling affects through the choice of routing a PT to a server the local strategy, applied at a single server. An approach to reduce the level of interaction, thus decoupling global and local policies introduces a set of separate PT buffers between the decision point and every server, in conjunction with a link event. Conceptually, this can be interpreted as a reservation of parts for processing at a particular server, without physically removing units. The sets of supplementary buffers are called virtual buffers. Every server, regardless of processing/routing constraints, has a set of virtual buffers equivalent to the number of PT in class C_0^0.

Definition 1: Part Type Class
A set of PTs. With respect to the entity flow either the earlier or subsequent server or the dynamic link element (DLE) is shared by all PTs.

Definition 2: Dynamic Link Element - DLE
A DLE represents an *interaction logic* to enable distributed scheduling for multi - machine systems with local scheduling policies. The structure of a DLE in terms of the material flow is represented by a set of directly preceding and succeeding PT classes, i.e. buffers, which are either of a *physical or virtual* characteristic. A *link event* replaces the set of local, but interacting events, and thereby enables distributed scheduling using local policies.

With the introduction of the DLE it is then possible to enable distributed scheduling, applying local and global policies, while maintaining overall system properties. The modified system to model routing of PTs to various servers which includes a DLE is depicted Figure 1.

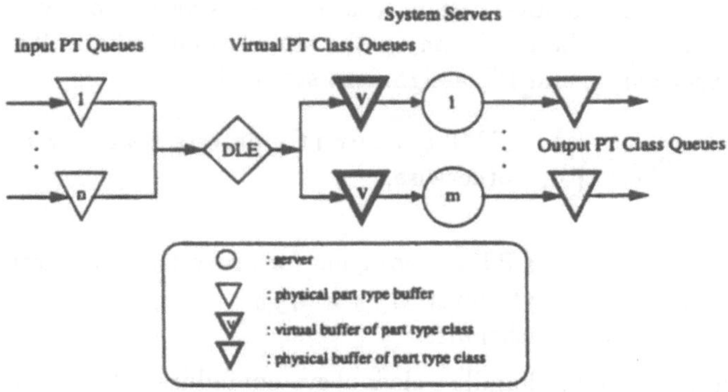

Figure 1: Virtual Part Type Buffer Concept

The proposed system modification offers the following benefits, which are further elaborated in the description of the scheduling policy: (1) increased flexibility if routing conditions change, (2) distributed application of existing policies (e.g. Single PT Branching (SPTB), and Multiple PT Branching (MPTB) [6]), (3) avoidance of system blocking for a class of system failure, and (4) distinction between two characteristic types of system servers. Based on the introduced virtual queue concept, a stable, balanced routing policy is developed next.

3.1 Part Type Routing Algorithm

The overall system objective is to maintain lower and upper buffer bounds for the state trajectory of the PT and PT-model inventory levels. The inventory level of PT i, at server j, and model k is represented by $x_{i,j,k}$, where lower bounds are represented by $a_{i,j,k}$, and upper bounds by $b_{i,j,k}$, where $i \in [1,n]$, $j \in [0,m]$, $k \in [0,L_{i,j}]$. Furthermore, $L_{i,j}$ denotes the number of models of PT i, at server j that are to be processed. The buffer bounds of the set of virtual buffers are determined through the minimum lotsize above which a PT is eligible for processing, and the maximum batchsize, such that $0 \leq a_{i,j,0} \leq b_{i,j,0}$ holds. Alternatively, the bounds on $a_{i,j,0}$, and $b_{i,j,0}$ can be determined by applying an economic order quantity (EOQ) argument, based on known values of inventory holding cost, demand rate, and order cost. A strategy to select buffer bounds for the virtual queues is presented in Section 3.2 of this work, which is further motivated to determine economic processing batch sizes for the individual servers. A violation on the bounds of the buffer levels is represented by the indicator variable $\zeta^r_{i,j,k}$, $r \in \{(l)ower,(u)pper\}$. Then

$$\zeta^r_{i,j,k} = \begin{cases} 1, & \text{if no bound violation;} \\ 0, & \text{if bound violation.} \end{cases} \tag{2}$$

Notice that $\zeta^r_{i,j,0}$ denotes the batchsize processing eligibility of the virtual buffers, and $\zeta^r_{i,0,0}$ the condition on the set of input buffers. To monitor the system processing, and PT assignment status, let

$$\sigma^P_{i,j} = \begin{cases} 1, & \text{if PT } i \text{ is currently processed at server } j \\ 0, & \text{otherwise,} \end{cases} \tag{3}$$

and

$$\sigma^A_{i,j} = \begin{cases} 1, & \text{if PT } i \text{ is assigned for processing to server } j, \\ & \text{whenever } x_{i,j,0} > a_{i,j,0} \\ 0, & \text{otherwise.} \end{cases} \tag{4}$$

The system enables to handle a class of system failure, where due to a machine failure further processing of an assigned model batch cannot be completed. Unprocessed parts are subsequently marked in order to be transfered back to the input buffer. The condition of a failure during processing of a PT model $(\sigma^P_{i,j} = 1)$ is symbolized by

$$\nu_{i,j,t} = \begin{cases} 1, & \text{after failure all unprocessed parts have been transfered} \\ 0, & \text{otherwise,} \end{cases} \tag{5}$$

where $t \in [1, L_{i,j}]$. The condition on *communication blocking* [17] of the system is expressed via the indicator variable:

$$\gamma_{i,j} = \begin{cases} 1, & \text{PT } i \text{ is eligible for placement or reservation at server } j; \\ 0, & \text{otherwise.} \end{cases} \tag{6}$$

Furthermore denote an indicator variable as to whether the reserved inventory level maintains the calculated virtual queue bound:

$$\epsilon_{i,j} = \begin{cases} 1, & \text{if } x_{i,j,0} \geq a_{i,j,0}; \\ 0, & \text{otherwise.} \end{cases} \tag{7}$$

Beside the above system state indicator variables, it is further necessary to model the PT routing constraints. The routing matrix Θ^{RM} contains routing eligibility of PTs to servers. Rows of Θ^{RM} represent input PTs, and columns represent the servers. If a PT i is admissible for processing on server j, then $\theta^{RM}_{i,j} = 1$, and 0 otherwise. The matrix Θ^{RM} comprises the conditions for the assigment of PT units to a particular server. A dynamic adaptation of PT routing conditions to the considered set of servers reflects the inherent FMS properties. A change of $\theta^{RM}_{i,j} = 1$ to $\theta^{RM}_{i,j} = 0$ is only possible if $\sigma^A_{i,j} = 0$. The lower bound of the virtual queue is to be set to zero, if no residual parts are to remain in the virtual queue. Assigning a PT i to a server j is dependent on processing, and buffer bound conditions which are combined in the virtual buffer matrix Θ^{VB}. Before developing the definition of $\theta^{VB}_{i,j}$ elements, two processing strategies are introduced which distinguish between *base* and *peak* operation of servers within a system.

The notion of base and peak capacity servers is used in systems whenever a set of machines is dedicated to perform specific tasks, and additional flexible servers are used for specified tasks or during peak demand periods. Consequently, both groups require considerably different scheduling approaches in terms of the length of the look-ahead planning period. *Base capacity servers (BCS)* in their local policies choose the PT to be processed from a defined set, whereas *peak capacity servers (PCS)* strictly have one PT assigned. When only one PT is assigned it is furthermore necessary to distinguish whether or not the PT can be processed concurrently on other PCS.

Strategy 1: Base Capacity Server (BCS) - Preload Scheduling
The maximum number of BCS is denoted by m_B. Rather than limiting the processing look-ahead decision to the strictly next PT, preloading a set of PTs may be advantageous. This situation is found whenever a server is used for base workload, and alternative machines only process during demand peaks. Another case is found when hourly operating cost between machines substantially differ, thus requiring a costly machine to have a higher fractional utilization in order to reduce cost per unit. The degree of concurrent processing of a PT i at various servers is controlled by the variable $PP_i^B \in \mathcal{N} \setminus \{0\}$.

Strategy 2: Peak Capacity Server (PCS) - Concurrent Scheduling
The maximum number of PCS is denoted by m_P.

Case 2a: no concurrent processing.
Inhibition of concurrent processing is desirable for PTs of comparatively high setup times, thereby reducing unproductive system idle time. A trade-off may occur, if e.g. due dates require parallel processing even though a large setup fraction of time is necessary, subject to technical feasibility.

Case 2b: concurrent processing.
A particular PT i can be assigned to several PCS. Introducing the variable $PP_i^P \in \mathcal{N} \setminus \{0\}$ denotes the bound on the maximum number of concurrent servers of PT j. The choice of PP_i^P is a system design variable which allows to control the degree of processing flexibility of a PT i in the system.

An element of the virtual buffer matrix $\theta^{VB}_{i,j}$ is then defined according to

$$\theta^{VB}_{i,j} = \begin{cases} 1, & \text{if } \theta^{RM}_{i,j} = 1 \text{ and } \zeta^{l}_{i,0,0} = 1 \text{ and} \\ & \underline{\text{Strategy 1:}}\ \sum_{\substack{k=1,\\ k\neq j}}^{m_B} \sigma^A_{i,k} < PP^B_i \le \sum_{k=1}^{m_B} \theta^{RM}_{i,k} \le m_B, \text{ and} \\ & \sigma^A_{i,j} = 1, \text{ and } \sum_{l=1}^{n} \sigma^P_{l,j} \le 1 \text{ where } j \in [1, m_B] \\ & \underline{\text{Strategy 2a:}}\ \sum_{\substack{l=1,\\ l\neq i}}^{n} \sigma^A_{l,j} = 0,\ \sum_{k=m_B+1, k\neq j}^{m} \sigma^A_{i,k} = 0, \\ & \sigma^A_{i,j} = 1 \text{ and } \sum_{\substack{l=1,\\ l\neq i}}^{n} \sigma^P_{l,j} = 0 \text{ where } j \in [m_B+1, m] \\ & \underline{\text{Strategy 2b:}} \\ & \sum_{\substack{k=m_B+1,\\ k\neq j}}^{m} \sigma^A_{i,k} < PP^P_i \le \sum_{k=m_B+1}^{m} \theta^{RM}_{i,k} \le m_P, \\ & \text{and } \sum_{\substack{l=1,\\ l\neq i}}^{n} \sigma^A_{l,j} = 0, \text{ and } \sum_{\substack{l=1,\\ l\neq i}}^{n} \sigma^P_{l,j} = 0 \\ & \text{where } j \in [m_B+1, m],\ \sigma^A_{i,j} = 1 \\ 0, & \text{if } \theta^{RM}_{i,j} = 0 \text{ or } \sigma^A_{i,j} = 0 \text{ or } \left(\theta^{RM}_{i,j} = 1\, and\, \zeta^{l}_{i,0,0} = 0\right) \text{ or} \\ & \underline{\text{Strategy 1:}}\ \sum_{\substack{k=1,\\ k\neq j}}^{m_B} \sigma^A_{i,k} = PP^B_i \\ & \underline{\text{Strategy 2a:}}\ \sum_{\substack{k=m_B+1\\ k\neq j}}^{m} \sigma^A_{i,k} > 0 \text{ or } \sum_{\substack{l=1,\\ l\neq i}}^{n} \sigma^A_{l,j} > 0 \text{ or} \\ & \sum_{\substack{l=1,\\ l\neq i}}^{n} \sigma^P_{l,j} > 0 \text{ where } j \in [m_B+1, m] \\ & \underline{\text{Strategy 2b:}}\ \sum_{\substack{k=m_B+1,\\ k\neq j}}^{m} \sigma^A_{i,k} = PP^P_i \text{ or } \sum_{\substack{l=1,\\ l\neq i}}^{n} \sigma^A_{l,j} > 0 \text{ or} \\ & \sum_{\substack{l=1,\\ l\neq i}}^{n} \sigma^P_{l,j} > 0 \text{ where } j \in [m_B+1, m] \end{cases} \tag{8}$$

From Θ^{RM} and Θ^{VB} generate the PT availability matrix Θ^{AM}, where

$$\theta^{AM}_{i,j} = \begin{cases} 1, & \text{if } \sigma^P_{i,j} = 0 \text{ and } \theta^{VB}_{i,j} = 1 \text{ and} \\ & \text{(1) PCS: } \sum_{k=1}^{n} \sigma^P_{k,j} = 0, \\ & \text{(2) BCS: } \sum_{k=1}^{n} \sigma^P_{k,j} \le 1 \\ 0, & \text{if } \sigma^P_{i,j} = 1 \text{ or } \theta^{VB}_{i,j} = 0 \text{ or PCS: for some } k \ni \sigma^P_{k,j} = 1 \end{cases} \tag{9}$$

The status information contained in Θ^{AM} is then used to decide PT access to a particular server. The following example for 4 PTs, and 4 servers (2 BCS, and 2 PCS) illustrates this procedure further.

<u>Example:</u>

Let $\Theta^{RM} = \begin{pmatrix} 1 & 0 & 1 & 1 \\ 1 & 1 & 1 & 1 \\ 1 & 1 & 0 & 0 \\ 0 & 1 & 0 & 1 \end{pmatrix}$. Where servers 3, 4 are PCS and 1, 2 are BCS. Delimit the processing of PT on servers by: $PP^B = [1\,1\,2\,1]$, and $PP^P = [2\,1\,0\,1]$. Assume at time t_1 the following processing conditions: $\sigma^P_{1,1} = 1, \sigma^P_{3,2} = 1, \sigma^P_{2,3} = 1, \sigma^P_{4,4} = 1$. Let $\sigma^A = \begin{pmatrix} 1 & 0 & 0 & 0 \\ 0 & 1 & 1 & 0 \\ 1 & 1 & 0 & 0 \\ 0 & 1 & 0 & 1 \end{pmatrix}$. Then

$\Theta^{VB} = \begin{pmatrix} 1 & 0 & 0 & 0 \\ 0 & 1 & 1 & 0 \\ 1 & 1 & 0 & 0 \\ 0 & 1 & 0 & 1 \end{pmatrix}$, and $\Theta^{AM} = \begin{pmatrix} 0 & 0 & 0 & 0 \\ 0 & 0 & 0 & 0 \\ 0 & 0 & 0 & 0 \\ 0 & 0 & 0 & 0 \end{pmatrix}$. At t_2 server 2 finishes processing of batch, thus $\sigma^P_{3,2} = \sigma^A_{3,2} = 0$. Then $\sigma^P = \begin{pmatrix} 1 & 0 & 0 & 0 \\ 0 & 0 & 1 & 0 \\ 0 & 0 & 0 & 0 \\ 0 & 0 & 0 & 1 \end{pmatrix}$, and $\sigma^A = \begin{pmatrix} 1 & 0 & 0 & 0 \\ 1 & 1 & 1 & 0 \\ 1 & 0 & 0 & 0 \\ 0 & 1 & 0 & 1 \end{pmatrix}$. Furthermore assume no arrival/reservation has occured. Therefore, $\Theta^{VB} = \begin{pmatrix} 1 & 0 & 0 & 0 \\ 0 & 1 & 1 & 0 \\ 1 & 0 & 0 & 0 \\ 0 & 1 & 0 & 1 \end{pmatrix}$, and $\Theta^{AM} = \begin{pmatrix} 0 & 0 & 0 & 0 \\ 0 & 1 & 0 & 0 \\ 0 & 0 & 0 & 0 \\ 0 & 1 & 0 & 0 \end{pmatrix}$.
Notice, the situation where processing of same PT as earlier is not allowed, i.e. forced setup.

When developing a set of events to enable routing to a set of servers it is imperative to first identify local and global system events. Local events have the property of not directly affecting the state trajectory of neighboring servers, whereas global events enable controlling the system state for shared resources. In Figure 2 local and global events are identified for the routing policy.

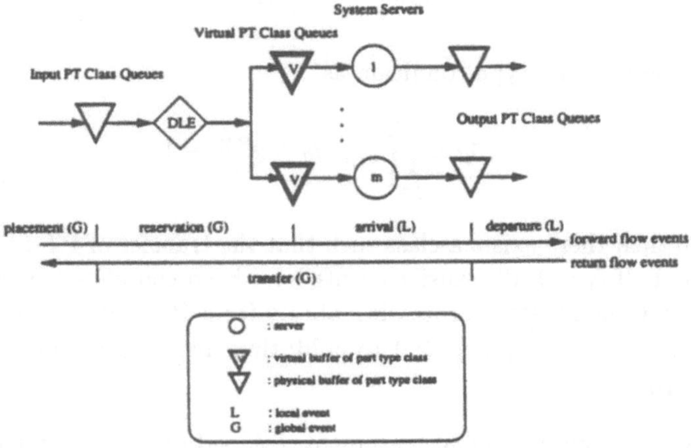

Figure 2: Local / Global Events - Routing Policy

Notice the placement, reservation, and arrival events identify both, a primary and secondary material flow direction. The primary flow is attributed to a system during non-failure operation, whereas the secondary flow results from a system failure which requires parts to be transfered back into the initial set of buffers. For the developed policy a transfer of parts can only occur between output and input PT buffers. Virtual buffers are not allowed for intermediate transfer of parts. The capital letter in parenthesis after an event symbol indicates whether an event is considered local (L) or global (G). A local event is to be defined by the applied distributed policies at a server (e.g. branching policies), whereas global events are to be developed in this chapter.

In order to avoid for several PTs attempting to access a server for processing a condition on mutual exclusion is required. Definition 3 extends the mutual exclusion system property introduced for the branching policies [6], to a multi-server policy.

Definition 3: Generic Parallel Mutual Exclusion (GPME)
At a particular server j, where $j \in [1,m]$ a condition on the parallel mutual exclusion is locally defined by either $\tilde{\psi}_S^j$ or $\tilde{\psi}_M^j$ as defined in the SPTB and MPTB policies [6]. A system is in a valid operating mode, whenever for all j the local conditions on parallel mutual exclusion are satisfied.
Similar to the GPME property, it is furthermore necessary to monitor the aggregate system condition with respect to the failure operating mode $\nu_{i,j,k}$.

Definition 4: Aggregate Transfer Condition - (ATC)
Let the part type - server condition be

$$\tilde{\nu}_{i,j} := \prod_{l=1}^{L_{i,j}} \nu_{i,j,l} \; , l \in [1, L_{i,j}] \; , j \ni \theta_{i,j}^{RM} = 1, \tag{10}$$

and denote the part type condition as

$$\tilde{\nu}_i := \prod_{j=1}^{m} \tilde{\nu}_{i,j} \; , \forall j \ni \theta_{i,j}^{RM} = 1. \tag{11}$$

If $\tilde{\nu}_i = 0$ then there exists a class such that the transfer of PT i has not been completed. If $\tilde{\nu}_i = 1$ all transfer events have been completed. In addition, to ensure continued system processing ability for a PT i, it is necessary at any time $\sum_{\forall j \ni \theta_{i,j}^{RM}=1} \sum_{l=1}^{L_{i,j}} \nu_{i,j,l} \geq 1$ to hold; thus avoiding system deadlock due to a failure.

As shown in Figure 2 global placement, reservation, and transfer events are to be developed for a multi-server routing policy. Therefore for the original local policies the placement, and transfer events are no longer valid in the global system. For the new placement event three different cases are to be distinguished: (1) PT failure for insufficient buffer capacity for a transfer, (2) PT failure for sufficient buffer capacity for a transfer, and (3) no PT failure, i.e. standard placement of a batch into the buffers. There is no distinction necessary between BCS and PCS for the placement event. A distinction becomes important for implementing the reservation event. The BCS provide an additional degree of freedom for adapting the system to a dyamic environment, because of the set of different preloaded PTs. Denote L_j as the maximum number of PTs to be preloaded into a BCS, where $j \in [1,m_B]$. Recall for $x_{i,j,0} > a_{i,j,0}$ then $\sigma^A_{i,j} = 1$. Consequently it is necessary to satisfy $\sum_{i=1}^{n} \sigma^A_{i,j} \leq L_j$. If for some PT_i no concurrent processing is admissible, then $\sum_{j=1}^{m_B} \sigma^A_{i,j} \leq 1$ or $\sum_{j=1}^{m_B} \sigma^A_{i,j} \leq PP^B_i$ otherwise.

Since for the considered system different processing capabilities are assumed, it becomes crucial to implement a strategy preventing servers with low capabilities to become underutilized compared to machines with high capabilities. An approach using *dynamic priorities* is proposed to overcome this issue.

Definition 5: Server Capability Index
Define a server *capability index* C_j as

$$C_j := \sum_{i=1}^{n} \theta^{RM}_{i,j} \ , \forall j \in [1,m] \, . \tag{12}$$

The choice of assigning a batch quantity to a server j is then based on the following strategy, which does not distinguish between *BCS* and *PCS*. Assign a PT i to a server subject to routing admissibility and strategy (PCS, BCS) conditions (for simplicity not explicitly mentioned here) according to:
(1) If $\exists j$, where $\theta^{RM}_{i,j} = 1 \ni \min_{\forall j \ni \theta^{AM}_{i,j} = 0} C_j$ then select server j.
(2) If $\forall k$ where $\theta^{RM}_{i,j} = 1 \ , \theta^{AM}_{i,k} = 1$ then choose smallest C_j for which there exists reservation capacity.

In conjunction with the capability concept, balancing the workload between servers is included into the reservation event. Workload balancing is considered for the set of virtual buffers, thus does not incur additional transportation time between servers. Enabling server balance assumes that the cumulative virtual inventories do not exceed the cumulative upper virtual buffer bounds, i.e.

$$\sum_{j=1}^{m} x_{i,j,0} \leq \sum_{j=1}^{m} b_{i,j,0} \, , \forall j \; where \; \theta^{RM}_{i,j} = 1. \tag{13}$$

The considered measure of imbalance between two servers is defined by the maximum of the difference between upper, or lower buffer bounds. In order to reflect queues of equal bounds a balance off-set of one unit is included. Thus

$$B_{i,j,h} := \max_{\forall h \, , \theta^{AM}_{i,j} = 1 \, , \theta^{AM}_{i,h} = 1} [\mid b_{i,j,0} - b_{i,h,0} \mid \, , \, \mid a_{i,j,0} - a_{i,h,0} \mid] + 1. \tag{14}$$

In order to enable a balancing of workload between two servers it is imperative to restrict balancing to servers not currently processing the particular PT. With this it is assured that a current processing condition of a PT batch size is not affected through the balancing of workload between servers. This implies to enable a balance event after an arrival event has occured. Further details for workload balancing are presented with the event description in the algorithm.

The transfer event allows to transfer unfinished model part quantities from the system output buffer into the input buffer, if a failure occurs during processing of the PT model. A transfer between the output buffer, and the set of virtual buffers is considered not admissible.

Both PP_i^B and L_j are design parameters which allow to adapt a system to various configurations. A change in PP_i^B or L_j to a lower value can only be enabled if $\sum_{j=1}^{m_B} \sigma_{i,j}^A < PP_i^B$, and $\sum_{i=1}^{n} \sigma_{i,j}^A < L_j$ hold. Deciding on L_j and PP_i^B is a multifactor relationship which involves a set of trade-offs, which can be investigated by simulation experiments on various system configurations.

In the case of PCS it is necessary to satisfy $\sum_{j=m_B+1}^{m} \sigma_{i,j}^A \leq PP_i^P$, where $PP_i^P = 1$ if no concurrent processing is allowed, and $PP_i^P > 1$ for admissible parallel processing.
A transfer event can only be enabled whenever a model k of PT i is currently processed at server j, and the input buffer of PT i has available capacity. Unprocessed parts form the batch being processed at server j are transfered back from the model output queue to the input buffer, which can subsequently be processed into an alternative model, or by an alternative server.

<u>Definition 6: State and Event Space</u>
$\mathcal{X} := \mathcal{N}^{n+mn+\sum_{j=1}^{m}\sum_{i=1}^{n} L_{i,j}}$ denotes the state space
$x_k := \left[x_{1,0,0},\ldots,x_{i,j,l},\ldots,x_{n,m,L_{n,m}}\right]^T$ denotes the system state at time k
$\mathcal{E} :=$ event space at time k. Where :
$e_{p_i} :=$ place a batch of size α_{p_i} in input buffer B_i ,$\forall i \in [1,n]$ - global event,
$e_{r_{i,j}} :=$ reserve a batch of size $\alpha_{r_{i,j}}$ of PT i at server j - global event,
$e_{b_{i,j,h}} :=$ balance the inventories between virtual PT queues - global event,
$e_{a_i} :=$ arrival of a batch of PT i - local event dependent upon the local server policy, e.g. SPTB, MPTB,
$e_{d_j} :=$ departure of a batch of PT model j - local event, dependent upon the local server policy, e.g. SPTB, MPTB,
$e_{t_j} :=$ transfer of a batch from the output buffer to the input buffer - global event.

<u>Definition 7: Part Type Routing Algorithm</u>
The following declaration of the routing algorithm formulates the identified set of global system events. Necessary extentions to the local policies are

stated in the context of the respective local events with reference to the local events and their respective cases.

<u>(i) Compound Placement:</u> e_{p_i}
<u>Case 1:</u> If $\tilde{\nu}_i = 0$ for some i, then $\nu_{i,j,k} = 0$ for some j,k where $x_{i,0,0} < b_{i,0,0}$, and $\sigma^P_{i,j,k} = 0$ then $e_{p_{i,j,k}} \in g(x_k)$ thus

$$f_{p_{i,j,k}}(x_k) := \left[x_{1,0,0}, \ldots, x_{i,0,0} + \alpha^t_{p_{i,j,k}}, \ldots\right]^T, \tag{15}$$

where $\alpha^t_{p_{i,j,k}} \in \mathcal{N} \setminus \{0\}$ such that

$$\alpha^t_{p_{i,j,k}} \leq x^t_{i,j,k} \, and \, 0 \leq \alpha^t_{p_{i,j,k}} \leq \zeta^u_{i,0,0} \mid x_{i,0,0} - b_{i,0,0} \mid \tag{16}$$

is satisfied. If $\alpha^t_{p_{i,j,k}} = x^t_{i,j,k}$ then $\nu_{i,j,k} = 1$ (reset).
<u>Case 2:</u> If $\tilde{\nu}_i = 0$ for some i, then $\exists\, j,k \ni \nu_{i,j,k} = 0$ where $x_{i,0,0} \geq b_{i,0,0}$. Futhermore, if $\sigma^P_{i,j} = 0$ for some $j \in [1,m]$, and $\exists \psi_{i,j} = 0$ then enable a reservation event.
<u>Case 3:</u> A batch of size α_{p_i} for PT i can be placed if $x_{i,0,0} \leq b_{i,0,0}, \gamma_{i,j} = 0$ for some $j \in [1,m]$,$\tilde{\nu}_i = 1$ then $e_{p_i} \in g(x_k)$. Thus

$$f_{e_{p_i}}(x_k) := [x_{1,0,0}, \ldots, x_{i,0,0} + \alpha_{p_i}, \ldots]^T, \tag{17}$$

where $\alpha_{p_i} \in \mathcal{N} \setminus \{0\}$ with

$$(1 - \zeta^l_{i,0,0,}) \mid x_{i,0,0} - a_{i,0,0} \mid \leq \alpha_{p_i} \leq \zeta^u_{i,0,0} \mid x_{i,0,0} - b_{i,0,0} \mid . \tag{18}$$

<u>(ii) Compound Reservation:</u> $e_{r_{i,j,k}}$
<u>Case 1:</u> If $\tilde{\nu}_i = 0$ for some i, then $\nu_{i,j,k} = 0$ for some j,k, and $x_{i,0,0} < b_{i,0,0}$ then $e^t_{r_{i,j,k}} \in g(x_k)$, thus

$$f_{e^t_{r_{i,j,k}}}(x_k) := \left[x_{1,0,0}, \ldots, x_{i,0,0} + \alpha^t_{p_{i,j,k}}, \ldots\right]^T, \tag{19}$$

where $\alpha^t_{p_{i,j,k}} \in \mathcal{N} \setminus \{0\} \ni$

$$\alpha^t_{p_{i,j,k}} \leq x^t_{i,j,k} \, and \, 0 \leq \alpha^t_{p_{i,j,k}} \leq \zeta^u_{i,0,0} \mid x_{i,0,0} - b_{i,0,0} \mid \tag{20}$$

holds. If $x^t_{i,j,k} = 0$, then $\nu_{i,j,k} = 1$ (reset).
<u>Case 2:</u> If there exists a PT $i \ni \tilde{\nu}_i = 0$ and $x_{i,0,0} \geq b_{i,0,0}$ and $\exists j \ni \nu_{i,j,0} = 1$, and $x_{i,j,0} < b_{i,j,0}$ then select PT i for reservation, else for some $i \exists k$, where $k \in [1,n]$,$k < i \leq n \ni x_{k,0,0} \leq a_{k,0,0}$, $x_{i,0,0} > a_{i,0,0}$. Thus, a batch of size $\alpha_{r_{i,j}}$ from PT i can be reserved by server j, where $j \in [1,m]$ if $\gamma_{i,j} = 1, x_{i,0,0} > a_{i,0,0}, \nu_{i,j,l} = 1, \forall l \in [1, L_{i,j}], \sigma^P_{i,j} = 0, \theta^{RM}_{i,j} = 1$, furthermore distinguish
(i) $\exists j$ for $\theta^{RM}_{i,j} = 1 \ni \theta^{AM}_{i,j} = 0$ and $\min_{\forall j} C_j$, if $\sigma^A_{i,j} = 0$ then temporarily set $\sigma^A_{i,j} = 1$. If $\theta^{AM}_{i,j} = 1, and\, \psi_{i,j} = 0$ then $e_{r_{i,j,0}} \in g(x_k)$, reset $\sigma^A_{i,j} = 0$

otherwise.
(ii) If $\forall k$ where $\theta_{i,k}^{RM} = 1, \theta_{i,k}^{AM} = 0$, or $\theta_{i,k}^{AM} = 1$ and $x_{i,k} < a_{i,k}$ then choose smallest C_j for which reservation capacity exists.

Thus

$$f_{e_{r_{i,j,0}}}(x_k) := \left[x_{1,0,0}, \ldots, x_{i,0,0} - \alpha_{r_{i,j,0}}, \ldots, x_{i,j,0} + \alpha_{r_{i,j,0}}, \ldots\right]^T, \tag{21}$$

where $\alpha_{r_{i,j,0}} \in \mathcal{N} \setminus \{0\} \ni$

$$\min\left[\zeta_{i,0,0}^l \left(1 - \zeta_{i,j,0}^l\right) c_{i,j}, \zeta_{i,0,0}^l \left(1 - \zeta_{i,j,0}^l\right) c_{i,0}\right] \leq \alpha_{r_{i,j,0}} \tag{22}$$

$$\alpha_{r_{i,j,0}} \leq \min\left[\zeta_{i,0,0}^l \zeta_{i,j,0}^u \mid x_{i,j,0} - b_{i,j,0} \mid, \zeta_{i,0,0}^l \zeta_{i,j,0}^u \mid x_{i,0,0} - a_{i,0,0} \mid\right] \tag{23}$$

where $c_{i,j} := \mid x_{i,j,0} - a_{i,j,0} \mid$, und $c_{i,0} := \mid x_{i,0,0} - a_{i,0,0} \mid$ holds.
Case 3: Consider the case where $\tilde{\nu}_i = 1 \, \forall i \in [1,n]$.
First, select a PT. If the condition on $\tilde{\nu}_i = 1$ holds, then select PT i for a reservation at server j, if $x_{i,0,0} > a_{i,0,0}$, and $\forall k$, where $i < k \leq n$, $x_{k,0,0} \leq a_{k,0,0}$ holds.
Next, select the server. If $\gamma_{i,j} = 1, \nu_{i,j,l} = 1, \forall l \in [1, L_{i,j}], \sigma_{i,j}^P = 0, \theta_{i,j}^{RM} = 1$, and $x_{i,j,0} < b_{i,j,0}$ then $\forall k$ where $\theta_{i,k}^{RM} = 1, \theta_{i,j}^{AM} = 0$, or $\theta_{i,j}^{AM} = 1$ and $x_{i,j} < a_{i,j}$ choose the smallest C_j for which reservation capacity exists. If for the selected PT i and server j, $\epsilon_{i,j} = 1$, then determine if there exists a k at $j \ni \epsilon_{k,j} = 0$ for $\theta_{k,j}^{RM} = 1$, and select k for subsequent reservation if so. Then $e_{r_{i,j,0}} \in g(x_k)$ and

$$f_{e_{r_{i,j,0}}}(x_k) := \left[x_{1,0,0}, \ldots, x_{i,0,0} - \alpha_{r_{i,j,0}}, \ldots, x_{i,j,0} + \alpha_{r_{i,j,0}}, \ldots\right]^T, \tag{24}$$

where $\alpha_{r_{i,j,0}} \in \mathcal{N} \setminus \{0\} \ni$

$$\min\left[\zeta_{i,0,0}^l \left(1 - \zeta_{i,j,0}^l\right) c_{i,j,0}, \zeta_{i,0,0}^l \left(1 - \zeta_{i,j,0}^l\right) c_{i,0,0}\right] \leq \alpha_{r_{i,j,0}} \tag{25}$$

$$\alpha_{r_{i,j,0}} \leq \min\left[\zeta_{i,0,0}^l \zeta_{i,j,0}^u \mid x_{i,j,0} - b_{i,j,0} \mid, \zeta_{i,0,0}^l \zeta_{i,j,0}^u c_{i,0,0}\right] \tag{26}$$

holds. For all cases ensure balanced loading at the servers.

(iii) Load Balancing : $e_{b_{i,l,k}}$
If $\sum_{j=1}^m x_{i,j,0} > \sum_{j=1}^m b_{i,j,0}$ for $\theta_{i,j}^{RM} = 1$ then enable local arrival events until $\sum_{j=1}^m x_{i,j,0} \leq \sum_{j=1}^m b_{i,j,0}$ holds. Define $\tilde{x}_{i,l,k} := x_{i,l,0} - x_{i,k,0}$. If $\exists$ a pair $l,k, l \neq k$, $\ni \mid x_{i,l,0} - x_{i,k,0} \mid > B_{i,l,k}$, $\sigma_{i,l}^P = \sigma_{i,k}^P = 0$, and $\sum_{j=1}^m x_{i,j,0} \leq \sum_{j=1}^m b_{i,j,0}$, $x_{i,l,0} > a_{i,l,l,0}$, $x_{i,k,0} > a_{i,k,0}$ then enable a balance event $e_{b_{i,l,k}} \in g(x_k)$ with

$$f_{e_{b_{i,l,k}}}(x_k) := \left[x_{1,0,0}, \ldots, x_{i,l,0} \oplus_1 \alpha_{b_{i,l,k}}, \ldots, x_{i,k,0} \oplus_2 \alpha_{b_{i,l,k}}, \ldots\right]^T, \tag{27}$$

where $\alpha_{b_{i,l,k}} \in \mathcal{N} \setminus \{0,1\}$. If there exist multiple pairs of servers for which the workload is imbalanced, select the pair l,k where $\max_{l,k \, l \neq k} |\tilde{x}_{i,l,k}|$ holds.

Distinguish two cases:
If $\tilde{x}_{i,l,k} > 0$ and $\tilde{x}_{i,l,k} > B_{i,l,k}$ then let $\oplus_1 := -$ and $\oplus_2 := +$, thus

$$\alpha_{b_{i,l,k}} := \min \left[\lfloor \frac{1}{2} \mid \tilde{x}_{i,l,k} \mid \rfloor , \min \left\{ \zeta^u_{i,k,0} \mid x_{i,k,0} - b_{i,k,0} \mid , \zeta^l_{i,l,0} c_{i,l} \right\} \right] \tag{28}$$

$c_{i,l} := \mid x_{i,l,0} - a_{i,l,0} \mid$.
If $\tilde{x}_{i,l,k} < 0$ and $\mid \tilde{x}_{i,l,k} \mid > B_{i,l,k}$ then $\oplus_1 := +$ and $\oplus_2 := -$, thus

$$\alpha_{b_{i,l,k}} := \min \left[\lfloor \frac{1}{2} \mid \tilde{x}_{i,l,k} \mid \rfloor , \min \left\{ \zeta^u_{i,l,0} \mid x_{i,l,0} - b_{i,l,0} \mid , \zeta^l_{i,k,0} c_{i,k} \right\} \right] \tag{29}$$

$c_{i,k} := \mid x_{i,k,0} - a_{i,k,0} \mid$.
Repeat $e_{b_{i,l,k}}$ until $\forall i \in [1,n], j,k \in [1,m]$, such that $l \neq k$, $\theta^{RM}_{i,l} = \theta^{RM}_{i,k} = 1$, $\tilde{x}_{i,l,k} \leq B_{i,l,k}$, where $\sigma^P_{i,j} = \sigma^P_{i,k} = 0$ holds.

<u>(iv) Local Compound Arrival:</u> $e_{a_{i,j,0}}$
Whenever a local arrival of a batch of size $\alpha_{a_{i,j,k}}$ has occured, initiate a balance event.
Based on this algorithm it is necessary to modify case 3 of the local arrival event. The modification concerns the lower bound of the arrival batch size $\alpha_{a_{i,j,l}}$ such that

$$\min \left[\zeta^l_{i,j,0} a_{i,j,0}, (1 - \zeta^l_{i,j,l}) \mid x_{i,j,l} - a_{i,j,l} \mid \right] \leq \alpha_{a_{i,j,0}} \tag{30}$$

and

$$\alpha_{a_{i,j,0}} \leq \zeta^l_{i,j,0} \min \left[x_{i,j,0}, \zeta^u_{i,j,l} \mid x_{i,j,l} - b_{i,j,l} \mid \right] \tag{31}$$

holds. This is possible whenever $\zeta^u_{i,j,l} \mid x_{i,j,l} - b_{i,j,l} \mid \geq a_{i,j,0}$ is true. Notice the local event cases 1 and 2 (see SPTB, MPTB policies for details) are not modified since the primary objective in the situation of system failure is to ensure inventory levels within their prescribed bounds.

<u>(v) Local Departure Event:</u> $e_{d_{i,j,k}}$
An alternative to the local departure event of the SPTB, and MPTB policies is proposed next.

A depature event of batch size $\alpha_{d_{i,j,k}}$ of model k, at server j, and PT i, can occur whenever $\sigma^p_{i,j} = 0$ or $\sigma^p_{i,j} = 1$ but the considered model k is not currently being processed at the server j. Then $e_{d_{i,j,k}} \in g(x_k)$ and

$$f_{e_{d_{i,j,k}}}(x_k) := \left[x_{1,0,0}, \ldots, x_{i,j,k} - \alpha_{d_{i,j,k}}, \ldots \right]^T, \tag{32}$$

where $\alpha_{d_{i,j,k}} \in \mathcal{N} \setminus \{0\}$

$$\alpha_{d_{i,j,k}} = \zeta^l_{i,j,k} \mid x_{i,j,k} - a_{i,j,k} \mid \tag{33}$$

holds.

(v) Global Transfer Event: $e_{t_{i,j,k}}$
A batch of size $\alpha^t_{i,j,k}$ of model k, from PT i, at server j can be transfered to the input buffer if based on the local parallel mutual exclusion conditions for PCS: ($\psi^i = 1$, and $\tilde{\psi}_S = 1$), and for BCS: ($\psi_i = 1$, and $\tilde{\psi}_M = 1$) then $e_{t_{i,j,k}} \in g(x_k)$ for the respective PCS or BCS. Then

$$f_{e_{t_{i,j,k}}}(x_k) := \left[x_{1,0,0}, \ldots, x_{i,0,0} + \alpha^t_{i,j,k}, \ldots, x_{i,j,k} - \alpha^t_{i,j,k}, \ldots\right]^T, \tag{34}$$

where $\alpha^t_{i,j,k} \in \mathcal{N} \setminus \{0\}$, and $x_{i,j,k} := x^r_{i,j,k} + x^t_{i,j,k}$ such that

$$0 \leq \alpha^t_{i,j,k} \leq \zeta^u_{i,0,0}\zeta^l_{i,j,k} \mid x_{i,j,k} - a_{i,j,k} \mid \leq x^t_{i,j,k} \tag{35}$$

and

$$\alpha_{t_{i,j,k}} \leq \zeta^u_{i,0,0}\zeta^l_{i,j,k} \mid x_{i,0,0} - b_{i,0,0} \mid . \tag{36}$$

If $x^t_{i,j,k} = 0, \forall k \in [1, L_{i,j}]$ then let $\nu_{i,j} = 1$ otherwise $\nu_{i,j} = 0$. Repeat $e_{t_{i,j,k}}$ until $\nu_{i,j} = 1$ or $x_{i,j,k} = a_{i,j,k}$, then $x_{i,j,k} = x^r_{i,j,k}$ and $x^t_{i,j,k} = a_{i,j,k} - x^r_{i,j,k}$. To develop a formal proof for stability of the proposed algorithm define next the admissible event trajectory and invariant set.

Definition 8: Trajectories
From the definition of g and f_e , where $e \in g(x_k)$, the set of all event trajectories is defined. Let

$$E_v = E. \tag{37}$$

Definition 9: Invariant Set
The invariant set includes the set of state constraints for the formulated scheduling policy: buffer bounds, cumulative inventory, buffer balance, PT failure, server-model failure, and parallel mutual exclusion. Let

$$\mathcal{X}_m := \mathcal{X}_{m_1} \cup \mathcal{X}_{m_2} \cup \mathcal{X}_{m_3} \cup \mathcal{X}_{m_4} \tag{38}$$

where $\forall i \in [1,n], j \in [0,m], l \in [1, L_{i,j}]$

$$\mathcal{X}_{m_1} := \left\{ x \in \mathcal{X} \middle| a_{i,j,l} \leq x_{i,j,l} \leq b_{i,j,l}, and\, x_{i,j,0} \leq b_{i,j,0} \right\}, \tag{39}$$

$$\mathcal{X}_{m_2} := \left\{ \sum_{i=1}^{n}\sum_{j=0}^{m}\sum_{l=1}^{L_{i,j}} x_{i,j,l} < M, \sum_{j=1}^{m} x_{i,j,0} \leq \sum_{j=1}^{m} b_{i,j,0}, for\, \theta^{RM}_{i,j} = 1, \right\}, \tag{40}$$

$$\mathcal{X}_{m_3} := \left\{ \tilde{\psi}^j_S \leq 1\, for\, PCS, \tilde{\psi}^j_M \leq 1\, for\, BCS, \sum_{i=1}^{n} \tilde{\nu}_i > 0, \sum_{k=1}^{L_{i,j}} \nu_{i,j,k} > 0 \right\}, \tag{41}$$

$$\mathcal{X}_{m_4} := \{ \mid x_{i,j,0} - x_{i,l,0} \mid \leq B_{i,j,l}, \forall\, pairs\, j,l,j \neq l \}. \tag{42}$$

The invariance of $\mathcal{X}_m$ is maintained by showing for any time $k, x_k \in \mathcal{X}_m$, then $x_{k+1} \in \mathcal{X}_m$. Stability in the sense of Lyapunov for a bounded discrete state space can be shown for some neighborhood of $\mathcal{X}_m$. In the sequel it is shown that an extension to all neighborhoods of $\mathcal{X}_m$ remains a stable system.

Theorem 2: Stability of Part Type Routing

Stability in the sense of Lyapunov for the scheduling policies formulated in Definitions 7, 8, and 9, in terms of the set of admissible event trajectories $\mathcal{E}_a$ is established.

Proof:

Define

$$\tilde{\rho}(x_k,\mathcal{X}_m) := \sum_{i=1}^{n}(x_{i,0,0} - \overline{x}_{i,0,0})^2 + \sum_{i=1}^{n}\sum_{j=1}^{m}\sum_{l=1}^{L_{i,j}}(x_{i,j,l} - \overline{x}_{i,j,l})^2 \quad (43)$$

$$+\sum_{i=1}^{n}\sum_{j=1}^{m}\sum_{\substack{h=1,\\ h\neq j,\\ h>j}}^{m} \mid (x_{i,j,0} - \overline{x}_{i,j,0})^2 - (x_{i,h,0} - \overline{x}_{i,h,0})^2 \mid \quad (44)$$

Define the metric

$$\rho(x_k) := \inf \tilde{\rho}(x_k). \quad (45)$$

Define the Lyapunov function

$$V(x_k) := \rho(x_k,\mathcal{X}_m). \quad (46)$$

Furthermore define the set

$$\mathcal{X}_{m_s} := \{x_{i,j,l} < a_{i,j,l}, i \in [1,n], j \in [0,m], l \in [1,L_{i,j}]\}. \quad (47)$$

Stability in the sense of Lyapunov in any neighborhood of $\mathcal{X}_m$ is established, if for the proposed policies $V(x_k) \geq V(x_{k+1})$ holds. The chosen definition of the Lyapunov function satisfies the necessary conditions on positive boundedness and boundedness.

For the compound placement and transfer events the proofs for stability are identical to the proofs given for the branching policies SPTB, and MPTB [6]. The only significant modification beside different index sets is associated with the choice of $\overline{x}_{i,h,0}$ in the Lyapunov function. Since it is required for the sets of virtual buffers that the cumulative inventory level is below the cumulative upper bounds, it is always possible to establish $\overline{x}_{i,h,0} \in \mathcal{X}_m$ by setting $\overline{x}_{i,h,0} = x_{i,h,0}$ for a balanced set of virtual buffers. The required link event, i.e. reservation introduces an extension to the structurally similar arrival event (see equation (17)), by including information to balance workload between admissible servers for a considered PT. It is therefore to be shown, that the definition of the balancing approach maintains the properties of the considered system. Further notice, equation (19) is identical to the transfer in the compound placement event.

Before the formal mathematical proof for the balancing in the context of the compound reservation is developed, a brief outline of the involved main conceptual steps introduces the approach taken.
STEP 1: The goal of the proposed balancing procedure attempts to reduce the cumulative level of assigned workload imbalance. Specify a bound on the level of imbalance of server j and server l.
STEP 2: Show for the balance batch quantity $\alpha_{b_{i,j,l}}$, a contracting sequence on the cumulative buffer imbalance is defined.
STEP 3: Show for $x_{i,j,0}\,(x_{i,l,0})$ either $> b_{i,j,0}\,(b_{i,l,0})$ or $< a_{i,j,0}\,(a_{i,l,0})$ it is possible to find $\overline{x}_{i,j,0}$,$\overline{x}_{i,l,0}$ which correspond to the invariant set $\mathcal{X}_m$.

The considered admissible level of imbalance between two servers for a PT i is defined in equation (14). To show admissibility of the balance event for all neighborhoods of $\mathcal{X}_m$ distinguish between (1) local, and (2) global virtual buffer bound violation. Local violation: for some j: $x_{i,j,0} > b_{i,j,0}$ and $\sum_{j=1}^{m} x_{i,j,0} \leq \sum_{j=1}^{m} b_{i,j,0}$. Global violation: $\sum_{j=1}^{m} x_{i,j,0} > \sum_{j=1}^{m} b_{i,j,0}$. The local bound violation is eliminated through a concatenation of balance events, until for all j, $x_{i,j,0} \leq b_{i,j,0}$ is achieved. Thus it is sufficient to show a reduction in imbalance by the balance event, thus leading to $\mathcal{X}_m$. For the global bound violation the balance event is, according to the conditions, never enabled. A sequence of local arrival events approaches $\mathcal{X}_m$. Whenever $\sum_{j=1}^{m} x_{i,j,0} \leq \sum_{j=1}^{m} b_{i,j,0}$ is achieved, the global bound violation condition holds, and the same argument as above can be applied, thus $\mathcal{X}_m$ is reached. To show a contracting sequence on the cumulative buffer imbalance it is necessary to also consider the resulting levels of imbalance for servers $w \neq j \neq l$. Define the following metrics

$$d_{i,j,w} := \left| \left(x_{i,j,0} - \overline{x}_{i,j,0}\right)^2 - \left(x_{i,w,0} - \overline{x}_{i,w,0}\right)^2 \right| ,w \neq j \tag{48}$$

, and

$$d_{i,l,w} := \left| \left(x_{i,l,0} - \overline{x}_{i,l,0}\right)^2 - \left(x_{i,w,0} - \overline{x}_{i,w,0}\right)^2 \right| ,w \neq l \tag{49}$$

where $\forall w \in [1,m]$. Recall that the choice for initiation of balancing considers the current maximum level of workload imbalance. Thus, it is then interesting to ensure simultaneous reduction for both (48) and (49). One can show for metric spaces [7]

$$|d_{i,j,w} - d_{i,l,w}| \leq d_{i,j,l}\ \forall w \in [1,m] \setminus \{j,l\}\,. \tag{50}$$

The left hand side of equation (50) can be interpreted as a measure of simultaneous imbalance between the entire set of servers. This essentially states that the absolute value of the difference between the levels of imbalance is bounded from above by the level of imbalance between the servers considered by the balance event. From the choice of the balance batch size $\alpha_{b_{i,j,l}}$ the reduction in the level of imbalance is ensured. Therefore the right hand side of equation (50) is bounded from above. For the entire system

$$\tilde{d}_{i,j,l} := \sum_{w=1, w \neq j}^{m} d_{i,j,w} \tag{51}$$

, and

$$\tilde{d}_{i,l,j} := \sum_{w=1, w \neq l}^{m} d_{i,l,w}. \tag{52}$$

From (51), and (52)

$$|\tilde{d}_{i,j,l} - \tilde{d}_{i,l,j}| = \sum_{w=1, w \neq j,l}^{m} |d_{i,j,w} - d_{i,l,w}|. \tag{53}$$

By (50)

$$\sum_{w=1, w \neq j,l}^{m} |d_{i,j,w} - d_{i,l,w}| \leq \sum_{w=1, w \neq j,l}^{m} d_{i,j,l}. \tag{54}$$

Thus

$$\sum_{w=1, w \neq j,l}^{m} d_{i,j,l} = (m-2) d_{i,j,l}. \tag{55}$$

Therefore, to show reduction in the overall level of imbalance it is sufficient to show $V(x_k) > V(x_{k+1})$ for the set of servers j,l. From equation (44) select $\overline{x}_{i,0,0} = \overline{x}'_{i,0,0}$, and $\overline{x}_{i,j,k} = \overline{x}'_{i,j,k}$, for $k > 0$, such that the infimum is achieved, and $\overline{x}_{i,0,0}, \overline{x}_{i,j,k} \in \mathcal{X}_m$ holds. Next recall two different cases for the direction of balance between servers j, and l (48, 49). Combining (46) and (55) it follows

$$|(x_{i,j,0} - \overline{x}_{i,j,0})^2 - (x_{i,l,0} - \overline{x}_{i,l,0})^2 > \tag{56}$$

$$|(x_{i,j,0} \oplus_1 \alpha_{b_{i,j,l}} - \overline{x}'_{i,j,0})^2 - (x_{i,l,0} \oplus_2 \alpha_{b_{i,j,l}} - \overline{x}'_{i,l,0})^2|. \tag{57}$$

Then

$$|(x_{i,j,0} - \overline{x}_{i,j,0}) + (x_{i,l,0} - \overline{x}_{i,l,0})|\,|(x_{i,j,0} - \overline{x}_{i,j,0}) - (x_{i,l,0} - \overline{x}_{i,l,0})| > \tag{58}$$

$$|(x_{i,j,0} - \overline{x}'_{i,j,0}) + (x_{i,l,0} - \overline{x}'_{i,l,0})|\,|(x_{i,j,0} - \overline{x}'_{i,j,0}) - (x_{i,l,0} - \overline{x}'_{i,l,0}) \oplus_1 2\alpha_{b_{i,j,l}}|. \tag{59}$$

Next distinguish two cases, (1) where $\oplus_1 = +$, and (2) where $\oplus_1 = -$.
<u>Case (1):</u> Select $\overline{x}_{i,j,0} = x_{i,j,0} + \alpha_{b_{i,j,0}}$, and $\overline{x}_{i,l,0} = x_{i,l,0} - \alpha_{b_{i,j,0}} + 1$, thus $\overline{x}_{i,j,0}, \overline{x}_{i,l,0} \in \mathcal{X}_{m_1}$. Then

$$|-1|\,|1 - 2\alpha_{b_{i,j,l}}| > \tag{60}$$

$$|(x_{i,j,0} - \overline{x}'_{i,j,0}) + (x_{i,l,0} - \overline{x}'_{i,l,0})|\,|(x_{i,j,0} - \overline{x}'_{i,j,0}) - (x_{i,l,0} - \overline{x}'_{i,l,0}) + 2\alpha_{b_{i,j,l}}|. \tag{61}$$

Selecting $\overline{x}'_{i,j,0} = \overline{x}_{i,j,0}$, and $\overline{x}'_{i,l,0} = \overline{x}_{i,l,0}$ maintains $\overline{x}'_{i,j,0}, \overline{x}'_{i,l,0} \in \mathcal{X}_{m_1}$. Thus

$$|1 - 2\alpha_{b_{i,j,l}}| > 1 \tag{62}$$

holds, since $\alpha_{b_{i,j,l}} \geq 2$.

Case (2): Select $\overline{x}_{i,j,0} = x_{i,j,0} - \alpha_{b_{i,j,0}} + 1$, and $\overline{x}_{i,l,0} = x_{i,l,0} + \alpha_{b_{i,j,0}}$, thus $\overline{x}_{i,j,0}, \overline{x}_{i,l,0} \in \mathcal{X}_{m_1}$. Then

$$|-1|\,|2\alpha_{b_{i,j,l}} - 1| > \tag{63}$$

$$|\,(x_{i,j,0} - \overline{x}'_{i,j,0}) + (x_{i,l,0} - \overline{x}'_{i,l,0})\,|\,|\,(x_{i,j,0} - \overline{x}'_{i,j,0}) - (x_{i,l,0} - \overline{x}'_{i,l,0}) - 2\alpha_{b_{i,j,l}}|. \tag{64}$$

Selecting $\overline{x}'_{i,j,0} = \overline{x}_{i,j,0}$, and $\overline{x}'_{i,l,0} = \overline{x}_{i,l,0}$ maintains $\overline{x}'_{i,j,0}, \overline{x}'_{i,l,0} \in \mathcal{X}_{m_1}$. Thus

$$|2\alpha_{b_{i,j,l}} - 1| > 1 \tag{65}$$

holds, since $\alpha_{b_{i,j,l}} \geq 2$. ◇

In the following Section 3.2 an economic order quantity argument is developed to select the bounds on the virtual queues.

3.2 Virtual Buffer Bound Selection

As indicated earlier in this work, the choice of the buffer bounds for the virtual queues can be approached by an economic order quantity (EOQ) argument. For the development of this strategy it is assumed for relevant cost components to be known. The considered decision variable is the lotsize quantity for an arrival event at a server. Based on admissible cost deviations a cost sensitivity measure is used to determine appropriate buffer bounds. Cost components included into the decision making process are: inventory holding, ordering, and processing cost. Denote the following variables: r_i inventory carrying charge [\$/\$/unit time], $d_{i,j,k}$ demand rate of PT i at server j, A_i fixed order cost, v_i variable order cost, and $q_{i,j,k}$ order quantity of model k of PT i at server j, and model k. The development of the subsequent approach is based on the assumptions of constant demand rate $d_{i,j,k}$, and unit processing time $\tau_{i,j,k}$.

The total cost equation for the included components:

$$TC_{i,j} := \sum_{k=1}^{L_{i,j}} \sigma_{i,j}^{A} \left[\frac{q_{i,j,k} v_i r_i}{2} + \frac{A_i d_{i,j,k}}{q_{i,j,k}} + c_{i,j,k} \right] \tag{66}$$

where $c_{i,j,k} := \tau_{i,j,k} q_{i,j,k} p_{i,j,k}$, and $p_{i,j,k} := \left(1 + \beta e^{\frac{-\sum_{i=1}^{n_j} \sigma_{i,j}^{A}}{L_j}}\right)^{-1} \tilde{p}_{i,j,k}$. Recall L_j denotes the maximum number of PTs that can be processed into a BCS, and $L_{i,j}$ corresponds to the number of models of PT i, at server j.

The parameter β is used to adjust the initial cost level, and $\tilde{p}_{i,j,k}$ represents the maximum admissible processing cost. The definition of $p_{i,j,k}$ includes the dependency of processing cost upon the overall level of congestion at a server, where processing of different PTs results in higher operating cost. For example, more PTs to be processed result in smaller lot sizes, and thus an increase in setup time proportions, and consequently to less system utilization unless an additional strategy is included to reduce the setup time proportion. Furthermore, the equation for $p_{i,j,k}$ allows to investigate scenarios where the number of PT temporarily exceeds L_j.

By standard optimality argumentation it follows for the optimal lotsize with respect to the considered cost components

$$q^*_{i,j,k} = \left[\frac{\sigma^A_{i,j}}{A_i d_{i,j,k}} \left[\frac{v_i r_i}{2} + \tau_{i,j,k} p_{i,j,k} \right] \right]^{\frac{-1}{2}}, \tag{67}$$

where $q^*_{i,j,k} \geq 1.0$ is to be maintained. Notice that $\sigma^A_{i,j}$ corresponds to a lower bound specification from earlier stages of the operating process. To initiate the procedure of calculating bounds on the virtual buffers set $p_{i,j,k} = 0$, and determine $q^*_{i,j,k}$ without including the processing cost aspect. To test whether at some point during operation the current bounds correctly reflect the overall system state temporarily assign $\sigma^A_{i,j} = 1$, calculate $q^*_{i,j,k}$, and corresponding buffer bounds according to the procedure outlined next. If the current inventory level is between the new calculated bounds, then let $\sigma^A_{i,j} = 1$, if below the calculated lower bound then reset $\sigma^A_{i,j} = 0$.

To develop respective upper and lower bounds on the virtual buffer, introduce the lot size perturbation

$$q'_{i,j,k} := (1 + \Delta)\, q^*_{i,j,k} \;, \Delta \in (0,1), \tag{68}$$

and the cost sensitivity measure

$$CS_{i,j} := \frac{TC_{i,j}(q'_{i,j,k}) - TC_{i,j}(q^*_{i,j,k})}{TC_{i,j}(q^*_{i,j,k})}. \tag{69}$$

With equations (68) and (69) it follows

$$CS_{i,j} = \sum_{k=1}^{L_{i,j}} \frac{\Delta \left[\frac{q^*_{i,j,k} v_i r_i}{2} + \tau_{i,j,k} q^*_{i,j,k} p_{i,j,k} - \frac{A_i d_{i,j,k}}{q^*_{i,j,k}(1+\Delta)} \right]}{\frac{q^*_{i,j,k} v_i r_i}{2} + \tau_{i,j,k} q^*_{i,j,k} p_{i,j,k} + \frac{A_i d_{i,j,k}}{q^*_{i,j,k}}}. \tag{70}$$

From an applicational standpoint the value of $CS_{i,j}$ can be defined by an alternative cost analysis, which is for operational purposes to be translated into a condition on the variability of the lotsize. Therefore solve equation (70) for Δ to determine the buffer bound variability. Some analysis shows

$$\Delta = \frac{1}{2\omega_1} \left[-\omega_3 + \left[\omega_3^2 + 4\omega_1\omega_4 \right]^{\frac{1}{2}} \right], \tag{71}$$

where

$$\omega_1 := \sum_{k=1}^{L_{i,j}} \left[\frac{q^*_{i,j,k} v_i r_i}{2} + \tau_{i,j,k} q^*_{i,j,k} p_{i,j,k} \right], \tag{72}$$

$$\omega_2 := \sum_{k=1}^{L_{i,j}} \frac{A_i d_{i,j,k}}{q^*_{i,j,k}} \tag{73}$$

$$\omega_3 := \omega_1 - \omega_2 - CS_{i,j}(\omega_1 + \omega_2), \omega_4 := CS_{i,j}(\omega_1 + \omega_2). \tag{74}$$

Thus, equation (71) in conjunction with (68) gives an approach to selecting bounds on the virtual buffer queues for a given admissible cost variation. The proposed virtual buffer bound values for $\Delta \neq 0$ are then

$$a_{i,j,0} := \lfloor q^*_{i,j,k} (1 - \Delta) \rfloor, \tag{75}$$

and

$$b_{i,j,0} := \lceil q^*_{i,j,k} (1 + \Delta) \rceil. \tag{76}$$

If $\Delta = 0$, then let $a_{i,j,0} := q^*_{i,j,k}$ and $b^*_{i,j,k} := q^*_{i,j,k} + 1$. Furthermore notice (75) and (76) are unique solutions only for single input - single output systems, whereas for multiple output models a secondary step trading off the set of solutions for the respective models is necessary.

4. Simulation Example

A simulation study of the balanced routing policy has been conducted. The following table summarizes the considered factors, factor levels, and used response measures.

Table 1: Routing Policy Experimental Factors, and Factor Levels

FACTOR	Level 1	Level 2
Initial Inventory	empty	within bounds
Arrival/Departure Process	deterministic	compound Poisson
Failure Process	no failure	compound Poisson
Input Batchsize Perturbation	10% bound	25% bound

The considered included: server utilization, part type waiting times, and queue length. Beside a change in the considered experimental factors no other adjustments have been implemented into the simulation model. The configuration of the simulated FMS is shown in Figure 3 below.

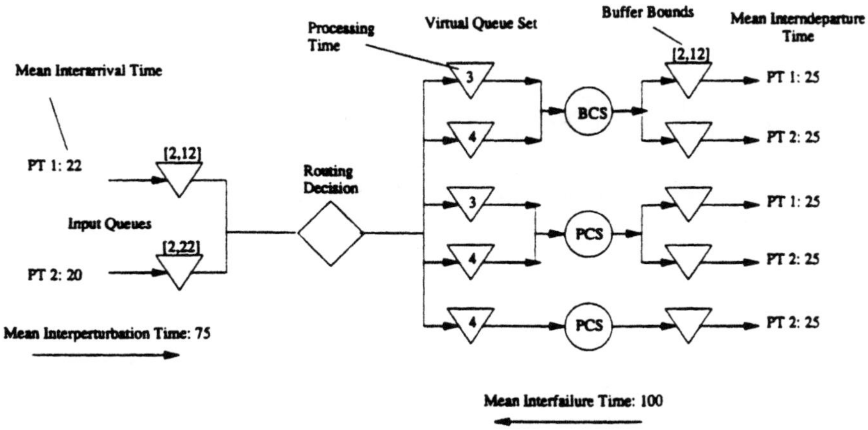

Figure 3: Balanced Routing simulation model and configuration

From analyzing the simulation output it was found that the equal variance assumption of balanced ANOVA does not hold, thus leading to apply the nonparametric Kruskal-Wallis test procedure for statistical inference.
The following Figures 4, 5, and 6 illustrate the obtained p-values for the considered set of simulation factors, and response measures.

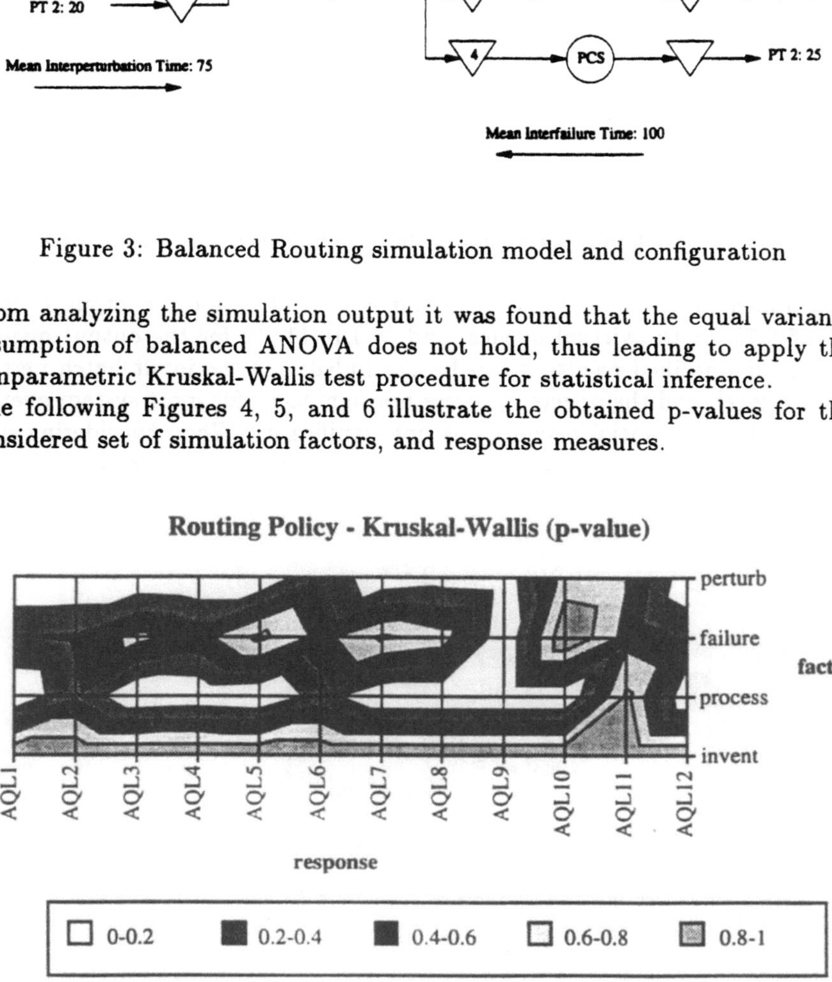

Figure 4: Balanced Routing Policy - Kruskal-Wallis Test Part 1

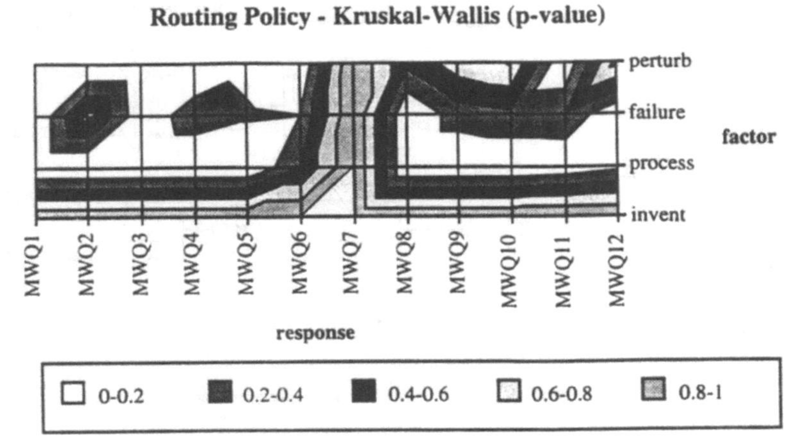

Figure 5: Balanced Routing Policy - Kruskal-Wallis Test Part 2

The index notation in Figures 4, and 5 uses (1,2) for the input queues, (3,4,5,6,7) for the virtual queues, and (8,9,10,11,12) for the output queues. From the obtained results it can be concluded: (1) the input batch perturbation level, and the selected process are significant for most system queues, (2) the level of initial inventory is not significant. When considering only the respective server utilization level in Figure 6 the observations are slightly different.

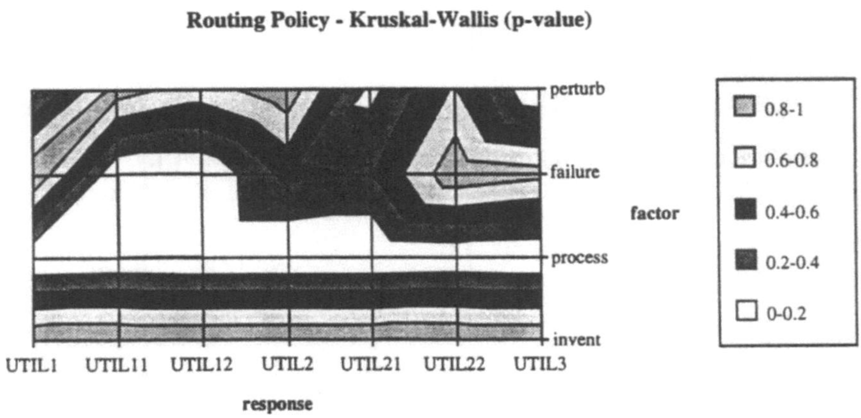

Figure 6: Balanced Routing Policy - Kruskal-Wallis Test Part 3

Notice that UTILj provides the aggregate utilization of server j, whereas UTILji expresses the utilization proportion of PT i at server j. Similar to the results obtained for the average queue length (AQL), and mean waiting time (MWQ) the utilization is statistically significant for the selected process, and insignificant with respect to the initial inventory level.

From the statistical analysis it follows for the initial inventory level to be insignificant for all considered response measures. The choice of the process interarrival / departure distribution is significant with respect to the utilization performance measure and most average queue lengths and waiting times. Perturbing the input buffer level is found to significantly affect both average input queue length and queue waiting time. At the aggregate server utilization level no significance of system failures can be identified.

Detailed observation of results for aggregate server and fractional PT at a server utilization also show compensating effects between PTs versus an aggregate view. This is a direct result of the PCS structure, and the balancing of workload between servers. The overall trend for the server responses is similar as was to be expected from the policy.

5. Conclusion

A routing policy for a generic Flexible Manufacturing System (FMS) has been developed. The policy is shown to maintain prescribed buffer bounds over time, and is applicable in a real-time environment. Balancing assigned workload between parallel servers is included to provide high utilization of the FMS. Production batch quantities are determined based on a modified EOQ argument. Further research is needed to adjoin stability of buffer state trajectories and specific system performance measures.

References

[1] Baker, K.R., 1974, *Introduction to Sequencing and Scheduling*, Wiley, New York.

[2] Chen, Y.H., 1992, Real-Time Scheduling of Flexible Manufacturing Systems, *Journal of Intelligent and Robotic Systems*, Vol. 6, pp. 51-63.

[3] Chow, Y.-C., and Kohler W.H., 1979, Models for Dynamic Load Balancing in a Heterogeneous Multiple Processor System, *IEEE Transactions on Computers*, Vol. c-28, pp. 354-361.

[4] Chuah, M.C., 1994, Analysis of Networks of Queues via Projection Technique, *IEEE Transactions on Automatic Control*, Vol. 39, pp. 1588-1599.

[5] Cruz, R.L., and Chuah, M.C., 1991, A Minimax Approach to a simple Routing Problem, *IEEE Transactions on Automatic Control*, Vol. 36, pp. 1424-1435.

[6] Fenchel, J., 1995, Stable, Distributed Real-Time Scheduling of Flexible Manufacturing Systems: An Energy Approach, *Ph.D. Thesis Georgia Insititute of Technology.*

[7] Kreyszig, E., 1978, *Introductory Functional Analysis with Applications*, Wiley, New York.

[8] Luo, Z.-Q., and Tseng, P., 1994, On the Rate of Convergence of a Distributed Asynchronous Routing Algorithm, *IEEE Transactions on Automatic Control*, Vol. 39, pp. 1123-1129.

[9] Morton, T. and Pentico, D., 1993, *Heuristic Scheduling Systems*, Wiley, New York.

[10] Passino, K.M., Michel, A.N., and Antsaklis, P.J., 1994, Lyapunov Stability of Discrete Event Systems, *IEEE Transactions on Automatic Control*, Vol. 39, pp. 269-279.

[11] Perkins, J.R., and Kumar, P.R., 1989, Stable, Distributed, Real-Time Scheduling of Flexible Manufacturing/ Assembly/ Disassembly Systems, *IEEE Transactions on Automatic Control*, Vol. 34, pp. 139-148.

[12] Perkins, J.R., and Kumar, P.R., 1990, Dynamic Instabilities and Stabilization Methods in Distributed Real-Time Scheduling of Manufacturing Systems, *IEEE Transactions on Automatic Control*, Vol. 35, pp. 289-298.

[13] Perkins, J.R., Humes, C., and Kumar P.R., 1994, Distributed Scheduling of Flexible Manufacturing Systems: Stability and Performance, *IEEE Transactions on Robotics and Automation*, Vol. 10, pp. 133-141.

[14] Sparaggis, P.D., Towsley, D., and Cassandras, C.G., 1994, Routing with Limited State Information in Queueing Systems with Blocking, *IEEE Transactions on Automatic Control*, Vol. 39, pp. 1492-1497.

[15] Towsley, D., Sparaggis, P.D., and Cassandras, C.G., 1994, Optimal Routing and Buffer Allocation for a Class of Finite Capacity Queueing Systems, *IEEE Transactions on Automatic Control*, Vol. 37, pp. 1446-1451.

[16] Tsitsiklis, J.N., and Bertsekas, D.P., 1986, Distributed Asynchronous Optimal Routing in Data Networks, *IEEE Transactions on Automatic Control*, Vol. AC-31, pp. 325-332.

[17] Walrand, J., 1989, *An Introduction to Queueing Networks*, Prentice Hall, Englewood Cliffs, New Jersey.

8 Shop Controllers-Managers for Intelligent Manufacturing

R. C. Michelini, G. M. Acaccia, M. Callegari, R. M. Molfino and R. P. Razzoli

1 Introduction

Intelligent manufacturing has been consolidating over the past ten years as a suitable means for sustaining competitiveness of industrialised Countries. The chapter presents contributions related to activity in the field done by the Industrial Robot Design Research Group of the University of Genova, aiming at the development of expert controllers-and-managers for shop-floor operations, to fully exploit flexibility. Starting points have been: • the recognition of innovating potentiality supported by the technological versatility of robotic fixtures [1]; and: • the need of re-stating the work-organisation through knowledge intensive set-ups [2] to comply with society-driven requests that in recent time have modified workers-market and trade-unions demands.

The spread over of computer-based solutions was offered as technology-driven option, whose appropriateness, however, required leanness check-ups for enabling economy of scope issues. Intelligent manufacturing, indeed, after initial forecasts too much related to redundant options supported by information technologies, is now more and more pricing the policy of offering economical solutions to actual cases. Few concepts are summarised in the second section, with focus on the goals of minimising resources and inventory and of avoiding functions and accomplishments not strictly tied to the principal process. The spin off of goals is split into four lines: piece wise continuous betterment; cooperative knowledge processing; diagnostics and monitoring maintenance; lean engineering check-up assessment. Then the enabling methodology aiming at efficiency through intelligent manufacturing is related to the decision cycle: intelligence/design, testing/feedback and choice/redesign, and to knowledge-based programming aids supporting the decisions of production engineers as: off-line consultation opportunities or on-line governing logic.

The attention is thereafter restricted, in the third section, to the knowledge-based software to be included *on-process* for enabling shopfloor control-and-management, that achieves higher efficiency by combining restoring flexibility at execution range, tactical flexibility at coordination range and strategic flexibility at organisation range. This software needs, typically, englobe AI modules for adapting the decisional mechanisms to the current situations. The validation of governing policies refers to assessed efficiency figures (net production or utilisation ratios) by iteration of the decision cycles.

The fourth section provides cases examples, as explanatory reference. The presentation singles out three applications that profit of integrated solutions, supported by knowledge intensive lay-outs to enable factory automation, namely:

- adaptive scheduling at emergencies, by integrated control-and-management;
- design and robotised (unattended) assembly of contractors' driven part batches;
- expert govern of the assembly section for customers' adapted product mixes.

Discussion of the quoted examples introduces to the logics used for properly setting the control structures and fitting the management policies, according to the application-driven requirements of each case development.

2 Efficiency Through Economy of Scope

Manufacturing competitiveness is measured by the continuous progression of market figures, through customers' loyalty, under fair trade conditions. A firm obtains profits from offering products having positive difference 'price *less* cost'; progression relays on customers' satisfaction, due to better 'value *less* price' figures as compared to competitors. In the past, the capability of offering goods and services was undersized with regard to potential customers' needs; market expansion was, then, related to the ability of lowering prices through mass production, to have more people that could afford purchasing. The affluent society faces market saturation; new buyers are won by creating new needs and by offering something typically personalised on individual whims.

Economy of scope replaces economy of scale. Fixed assets, namely, are amortised by a new work organisation aiming at flexible-manufacturing (as compared to earlier craft-manufacturing and mass-manufacturing, Fig. 1) based on complexity preservation, with series of objectives, such as: · to extend product mix variability to agree with larger amounts of consumers' wishes; · to avoid investments in specialised facilities and exploit robotisation to make diversified products; · to suppress redundancies and set-apart resources, to limit inventory and to apply bottom-up just-in-time schedules; · to enhance customers' driven responsiveness by minimal time-to-market; · to enable cooperative team work and distributed knowledge for suppressing duplication of not strictly necessary functions; · to exploit work creativity for continuously improving products-and-processes by shared interest; · to exclude sectorialisation of competencies and use decentralised responsibility to solve problems where they arise.

	Craft manufacturing	Mass manufacturing	Flexible manufacturing
Productivity base	Economy of skill	Economy of scale	Economy of scope
Work - organization	Evolving, learning and training	Scientific job - allotment	Intelligent task - assessment
Product definition	Design while producing	Specialized serial processing	Simultaneous engineering
Decision style	Craftmen committment	Hierarchical (off - process)	Distributed (on - process)
Knowledge context	Personal contribution	Addition of team work	Cooperative processing
Motivation system	Individual creativity	Division of competencies	Collaborative reward

Fig. 1. Characterising traits of manufacturing organisations

The economy of scope, summing up, enables procedures, such as the ones listed, to increase market share by:

- diversifying offers, widening delivered mixes (instead of mass-production) and satisfying the requests of broader spectra of buyers with certified-quality items;
- concentrating resources into the primary manufacturing process, avoiding redundancy and idle set-apart resources by exploiting recovery flexibility and getting rid of side-investments (procedural accomplishments, instrumental assets or operative functions) unable to transfer value to the offered product or service.

The policy is a combination of flexibility and cooperation; it takes account of the entire production chain, from customer's fancies, to ecological disposal; it includes process re-engineering and corporate strategies. It requires empowerment with team-work, and the transfer of decision-making as low as possible, with supportive management style. It is based on the possibility of resetting factory lay-out for modified production requirements (organisational flexibility), and on the capability of enabling recovery plans without set-apart redundancies at emergencies (executional flexibility).

2.1 Qualifying for economy of scope

Comparative studies on Japanese and Western Countries manufacturing industries ascribe differences in effectiveness to sets of factors, interlacing material and information flows, that entail a work-organisation based on knowledge intensive issues, namely: · piece wise continuous betterment; · cooperative knowledge processing; · diagnostics and monitoring maintenance; · lean engineering check-up assessment. Few topics are recalled, Fig. 2, with attention on the conditioning requirements of the software implementation.

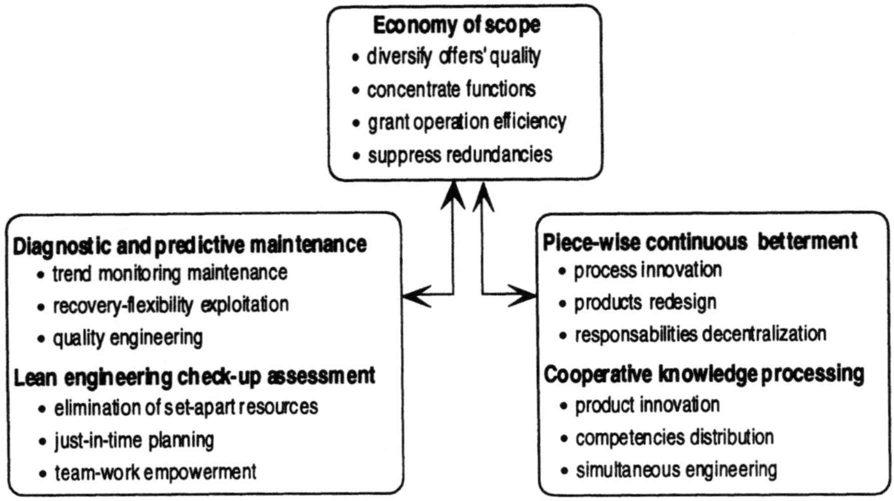

Fig. 2. Characterising features to obtain returns on investments in intelligent manufacturing

Piece wise continuous betterment: *to yield the successful effort of adapting products to consumers' wishes (increasing quality and lowering price).* Traditional optimisation of products tackles complexity by splitting down into sub-problems, solved by transferring constraints and interactions to reach coherent balancing of requirements. With customer-oriented manufacturing, unitariness of objective does not exist, complexity split cannot be done, and flexible specialisation [3] is tracked as issue granting the capability of adapting product properties to individual taste and fancy.

The effectiveness is enabled by knowledge distributed organisations, that exploit work praxis to improve product and process by shared interest and diffused responsibility. The method is put into practice permanently enabling decision cycles to redesign products and upgrade processes. It is exploited by Japanese manufacturers to adapt traditional goods (rather than to disclose innovation).

Within large enterprises, the option should carefully be supported by cooperation (to avoid functions overlapping) and balanced against creativity blunting (with stress on the autonomy of the concept design stage). Smaller enterprises have been using the approach as complement of flexible specialisation to improve niche qualification of manufacts by the skilful betterment of the process. In any case, the policy is reference hint for on-process combined exploitation of production control and management, aiming at rising plant effectiveness by interlacing strategic, tactical and recovery flexibility.

Cooperative knowledge processing: *to enable a reward system granting individual and team creativity, which aims at innovating products and processes.* Cooperative knowledge processing is investigated as opportunity for increasing the capabilities of intelligent manufacturing, according to simultaneous engineering rules. These offer conceptual aid for the upgrading tasks, providing consistent reference for: · design of product/process systems (approach and terminology); · representation and modelling codes (for unified virtual collocation); · knowledge frames and decision modes (repository, distribution and control of declarative / procedural data); · concurrent problem solving style, with multi-way computer communication; · information exchange and sharing, for cooperative team aid; · knowledge-based simulation/emulation (and related human-and-data interfaces); · decision steps logic (requirements on structural/behavioural contexts); · etc.

For efficiency, efforts of manufacturer's personnel need be coordinated; only pertinent information should be shared and collaboration ought to progress with involvement of (human and machine) expertise, within a cooperative framework that starts solving in parallel all sub-problems. The method tracks art and craft comprehensiveness preserving industrial effectiveness. Competency conflictuality and responsibility overlapping are common situations even within small companies; they are overridden, restoring efficiency, by cooperative knowledge processing [4]. Since design, market and production data are shared making possible the solution of problems where and when they arise.

Anticipation of product-engineering at design reduces time-to-market. Diversified manufacts are developed in parallel, to satisfy buyers tastes. Collaboration of experts with different functions and competencies is fostered, not bridled by splitting into departments. People needs yield, not hamper; sharing knowledge is aimed at adding up activities, managed by distributed computational resources. The common understanding grants visibility on the conditional knowledge, and helps making decision for the on-line govern of flexible manufacturing facilities.

Diagnostics and monitoring maintenance: *to aim at company-wide quality-control, and at predictive maintenance policies.* Knowledge intensive set-ups bring to exploit monitoring data: * for process, to enable recovery flexibility, to promote predictive maintenance, etc.; * for product, to grant intrinsic quality, selectively adapting technical specifications, etc. Diagnostics operates on-line by: · *situational analyses*, to give visibility on local situations by measuring quality indices on statistical consistent samples (data specify performance over tactical spans); · *relational analyses*, to transparently account for the conditioning effects on processes' behaviour through models that incorporate randomness [5]. Actions are taken on the processes; the products profit of results.

Recovery modes, of course, depend on facilities' technological versatility. Condition monitoring maintenance supports possibilities for managing flexibility so that restoring actions are performed if, and only if, necessary; facilities are exploited all along their efficient life; set-apart resources are avoided, since transparent access to reliability and availability data supplies reference information for enterprise strategic planning.

Quality assurance came out as a company-driven issue in mass-production; it is a fall out of monitoring every process-conditioning factor. The evolution toward flexible manufacturing leads to total quality, based on continuous betterment and cooperative processing; it is an achievement of 'robust' design, *ex post* assessed by testing. Enterprise effectiveness looks for zero-defects production; diagnostics and monitoring maintenance are necessary complements in front of continuously diversified products.

Lean engineering check-up assessment: *to remove material and information additions, that do not improve enterprise profitability*. The previous factors bear direct connections with knowledge diffusion; economy of scope further requires leanness, summarised by the two aphorisms: · set-apart physical resources are waste, the continuity of the principal process being based on technological versatility of manufacturing facilities; · redundant logical resources are nuisance and could create confusion by addition of unexploited functions or capabilities.

It is by now useful practice to perform audits on the enterprise leanness, verifying characterising objectives such as [6]:

a) *in terms of strategic perspectives at the organisational range:* - to integrate the overall business management processes: purchasing, design, production, sales, etc.; integration needs cover suppliers and dealers in order to avoid unnecessary duplication of functions; - to promote team activity for solving current problems by multi-disciplinary and multi-skilled members, instead of functional departments with fixed specialisation and hierarchical tasks allocation; - to build-up cooperative interpersonal relations, in order to accumulate insights on the enterprise organisation as team culture, and to make difficult, and not desirable, the switching of the worker to a new company; - etc.;

b) *in terms of tactical opportunities at the coordination range:* - to distribute staff and support functions into the primary processes; size of staff and support functions are continuously monitored on their ability of adding value, with careful attention on preservation of effectiveness; - to enable the quick visibility of problems, with solving responsibility decentralisation at the point of occurrence, to enhance piece wise continuous improvements, tied to the distribution of the various tasks and responsibilities; - to expand enterprise-information systems, supporting knowledge processing capabilities, with transparency of the decisional contexts and continuous monitoring of costs; - etc.;

c) *in terms of the recovery possibilities at the executional range:* - to remove over-protection and over-production by suppressing not necessary functions, by reducing idle resources, by avoiding piling-up of stock, etc., through the appropriate use of recovery flexibility; - to activate just-in-time logistic control with balancing of the distributed capacities, and to grant a stable production volume with variable products mix output, with on-line exploitation of tactical flexibility; - to acknowledge condition-monitoring maintenance standards, ensuring customer-oriented quality and zero-defects production, by means of strategic flexibility enterprise policy; - etc.

2.2 Knowledge cycles for intelligent manufacturing

Improvement of performance depends on suitable exploitation of flexibility. The goal takes principal part in widening product mix variability, and critical role in avoiding idle resources. Return on investments depends on the sets of well established paradigms, that directly follow from the above recalled objectives: · *functional integration* along the principal manufacturing process, to support synergetic cooperative behaviour of every factory resource; · *total quality*, for globally conditioning the enterprise organisation to be customers' driven; · *flexible specialisation*, to assure intensive exploitation of the facilities by expanding the products mix; · *lean engineering*, to minimise investment and personnel, in relation to the planned production requirements over the enterprise strategic horizon.

Intelligent manufacturing, therefore, characterises by knowledge intensive set-ups aiming at govern-for-flexibility achievements; it possesses, among others, the capability of building-up shared understanding, in order to support cooperative knowledge processing. The cognitive/operative lay-outs mutually condition each other, Fig. 3, with actions patterns for decisions embedded into the structures patterns used for knowledge. In such a frame, piece wise product/process quality upgrading is faced by closing concurrent decision cycles based on three steps: intelligence/design, testing/feedback and choice/redesign.

• The *intelligence* phase is the process of identifying consistent alternatives; the *design* phase is based on the evaluation of alternatives and confrontation of each one. The initial step, related to reference knowledge, requires meta-processing abilities and corresponds to the sentential conceptualisation layer.

• The *testing* is the assessment of performance; the *feedback* is the review process of the outcome of decision and is used to influence the acknowledgement of alternatives. The step needs understanding assessments and provides the monitoring reference measurements of efficiency, obtained through flexibility.

Cognitive lay-out	Problems	Instruments	Theories	*Knowledge manifolds*
	Communication	Sharing	Collaboration	*Knowledge layers*
	Structures	Results	Processes	*Knowledge instances*
Operative lay-out	Organization	Coordination	Context	*Decisional manifolds*
	Strategies	Tactics	Recovery	*Decisional layers*
	Actions	Setting	Fitting	*Decisional instances*

Fig. 3. Cognitive/operative lay-outs for cooperative knowledge-processing

• The *choice* is the selection and implementation of a decision alternative; the *redesign* is the piece wise process of improving the selected alternative. The closure step profits of cooperative knowledge decision style and exploits computer simulation as virtual reality reference, to enhance interactive consultation.

The application of the decision cycle models to intelligent manufacturing is basically concerned with the issues of governing flexibility, so that varying market-driven requests are satisfied whenever they emerge according to bottom-up planning, to continuously expand consumers' satisfaction and win new customers. Demanding aspect is that flexible plants are used not as they are, rather after setting and fitting:

a) The setting is concerned by *structural frames*, **CFC**: **c**omponents, **f**acility-configuration and **c**ontrol. Setting the **CFC** frames presents as everlasting activity aiming at exploiting versatility by enabling flexibility; the choices provide the reference for identifying current product-and-process arrangements all along the facilities operational life.

b) The fitting deals with the options of **m**onitoring, **d**ecision-manifold and **m**anagement, or: **MDM**, *behavioural frames*. Continuous reference to **MDM** frames and incremental acknowledgement of plant operational situations and functional trends offer data for manufacturing efficiency betterment, along with its economical exploitation.

The reset and the outfit decision logic needs be assessed and confirmed, with supports based on the measurement of plant performance and the comparison of current figures against the expected levels of efficiency. Indeed, it is not possible to, either, control and manage flexibility as current requirement, either, pre-set and implement flexibility as enterprise global policy, unless its effects are measured and the processes are tuned to exploit flexibility issues. To offer this kind of decision aid, functional modelling and computer simulation of manufacturing processes, providing virtual reality display of actual production plans, are the main analysis techniques.

Simulation software is based on two sets of modules. A set supplies the dynamical history of the material process (structural frames); the second provides the time evolution of the decisional logics (behavioural frames). The former requires the functional models of every relevant physical transformation; the latter, the generation of the consistent decisional modes. The first set of modules assures experimentation on several set-ups by computer simulation; the engineer exploits the option for developing the manufacturing facility. The second set provides the support for assessing flexibility effects on facility efficiency, when the plans are enabled combining decentralised control and supervisory management, as case arises, along the strategic, tactical or executional horizons. With simulation, any running condition of flexible plants can be tracked and performance assessed [7].

The characterising features of these computer aids are, by now, fully appreciated [8], and practical developments have been tested. The following section summarises typical requirements, making reference to the investigations carried on by the Research Group for different industrial applications.

3 Govern-for-Flexibility Knowledge Architecture

Manufacturing facilities are provided with information systems, (usually) in the form of distributed resources, with decentralised processing units, data-bases and interfacing modules. A functional reference scheme of a generic flexible manufacturing facility is shown by Fig. 4. Concurrence of the material and information flows is a characterising feature of intelligent manufacturing. Management of flexibility requires direct visibility of every relevant process quantity; programming is performed by referring to stored knowledge, and comparing the plant behaviour, on the selected strategic horizons, for competing tactical schedules. For flexible automation, the command flow operates on-process. Hereafter, the discussion is limited to the information aids, aiming at govern-for-flexibility issues by integrated control-and-management policies.

For sake of clarity, the software for factory automation mainly distinguishes into off-line and on-line codes. The off-line codes are typically represented by emulation·simulation packages necessary for the build-up of the reference knowledge on the effects that flexibility induces on plant productivity. These computer aids are used as decision support for selecting the integrated control and management policy assuring efficiency through flexibility; they require generation

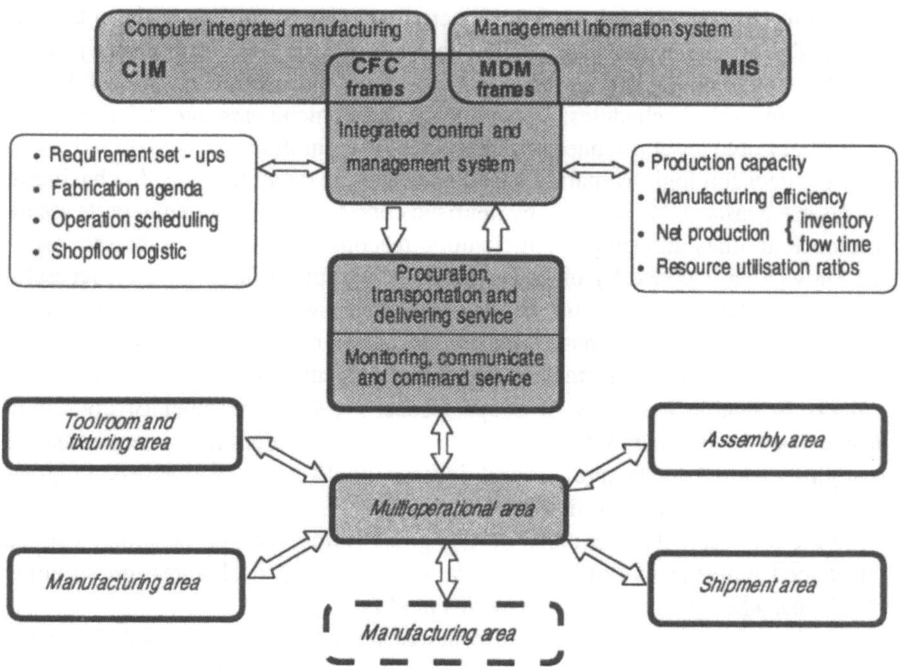

Fig. 4. Functional scheme of the generic flexible manufacturing plant

of causal (non-anticipatory) dependence to simulate the dynamics of material processes, and the addition of heuristic inference abilities to emulate anticipatory responses by the judgmental schemata of the decisional logics. The emulation modules, whether suitably structured for real-time operation, could be transferred to work on-line, directly interfaced with the manufacturing processes. At the moment, the factory automation directly ruled by govern-for-flexibility modules has only partially been implemented. Most of governing loops still close through human supervision, leaving unattended operations to be ruled by pre-set schedules, with management of the restoring flexibility limited to simple situations (quite similar to earlier fixed automation cases). The example cases, presented as explanatory references, on one side confirm the current state of the art; on the other side show the potentialities of existing technologies for further developments.

3.1 Decision aids for shop-floor govern

Computer simulation provides off-process consultation aids to production engineers to select 'best' govern-for-flexibility policies; the selection is based on productivity data assessed *a posteriori*, comparing flexibility effects on competing production programmes. The set of modules, performing emulation of the governing logic, when interfaced to the manufacturing facilities, generates a decision manifold granting *flexible* automation. This leads to the integrated control and management of production flows, with the real time scheduling and re-scheduling that represent the 'best' productivity choice, according to the enterprise policy. The efficiency needs an *on-line* decision-making style; to govern a *flexible* facility a combined logic is active: · to enable optimal decentralised *control*, for the steady state tactical schedules; · to switch on hierarchic *management*, for restoring or recovery policies at (planned or unexpected) discontinuities.

These decision logic is mainly based on empirical data; the related behavioural modes are consistently emulated by heuristic contexts. The judgmental parts assure inference abilities and supply 'anticipatory' responses according to the scheme: * the knowledge is coded by objects: "facts+beliefs+procedures"; * the successful response is the most plausible solution among answers consistent with system hypotheses. 'Intelligent' manufacturing is based on these perspectives, with software innovation, that emulates decision logic manifolds, re-introduced *on-process* to enable economy of scope issues. Now, the number of consistent answers is usually quite high, the attribute of plausibility does not generally lead to straightforward results and several indices could apply for assessing plant effectiveness. Conventional figures are:

- production capacity: estimate of the gross number of parts which can be produced with the enterprise organisation presetting; it changes, on strategic horizons, when production requirements modify;
- manufacturing efficiency: forecast from machines, labour, etc. availability, scrap ratios, set-up delays, etc. data; it is optimised by just-in-time policies;
- net production: ex post ratio 'current inventory/mean flow time'; the current

inventory is given by the shop information system as global figure, or as balance for given product items or work-sections; the mean flow time is computed, at constant inventory levels, for the reference tactical spans;

• utilisation ratios: fraction of activity-time of each manufacturing resource; statistical figures are computed: · during transients, as spot characteristics; · during steady schedules, as cumulative averages.

Cross-coupling effects are common occurrence for time-varying, low-inventory, lean-production set-ups. The effects of flexibility have an impact on net production in such a way that factory output cannot simply be inferred from the expected efficiency of really engaged resources. The continuous monitoring of plant effectiveness and the shared understanding of flexibility effects are reference data for choosing the **CFC** and **MDM** frames. The knowledge cycle illustrates the interactive nature of decision-making, and how the choice of an alternative influences which alternatives are identified for the next loop.

The experimentation of cross-coupled effects on real plants cannot be carried out with generality; limited example cases are, at must, used as validation bench-mark. Computer aids are thus general reference for the build-up of the basic knowledge and are exploited as consultation instrument for devising, selecting and assessing flexibility, by comparing alternatives singled out by the decision loops for fixing/resetting the **CFC** frame, and for fitting/restoring the **MDM** frame. The closure of the decisional cycles for control and management of plant operations makes extensive use of heuristics, and the effectiveness of a policy is only assessed by measuring really reached results. Computer-simulation, therefore, is the only economical way for gathering sufficient experimental evidence, of advantages and drawbacks that flexible automation, based on integrated control and management might offer by respect to fixed automation.

3.2 Integrated control and management decision loop

The distributed knowledge processing organisation is used as aid for moving back on-process the decisional manifold, consistent with flexible manufacturing. The expert simulation-emulation software, to support flexible manufacturing through the decision cycle approach, mainly exploits: model-based reasoning, during the testing/feedback phases; case-based reasoning, during the intelligence/design and the choice/redesign phases. At the first step, the preliminary setting of the manufacturing facility is fixed referring to past experiences. At the second step, the causal inference provides the assessment of plant performance by means of categorical features (patterns of actions); the heuristics is called for to implement govern-for-flexibility real-time procedures. At the third step, statistical inference can be used for the build up of framed databases to be back exploited, as consultation aid, to orient choices and to redesign the plant by patterns of structure. The reference knowledge continuously evolves, supplying the structures-and-actions patterns so that the manufacturing enterprise could be set/fit into position of tracking economy of scope conditions (Fig. 5). Few features are recalled [7].

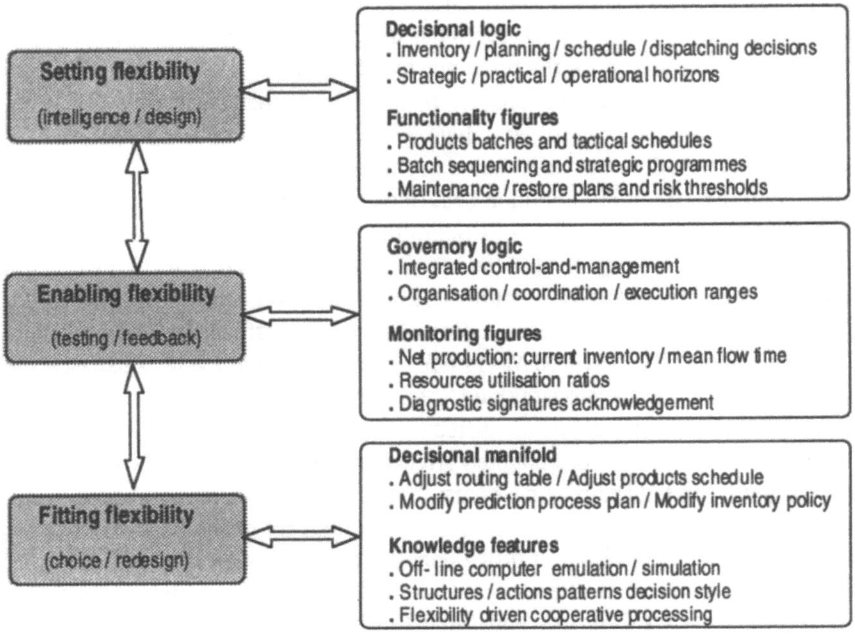

Fig. 5. Schematic of the knowledge trends supporting integrated control-and-management of intelligent manufacturing

The upshots of the coordinated control-and-management of shop-floor operations are acknowledged, for upgrading the manufacturing efficiency by govern-for-flexibility strategies, fully enabling the embedded technological versatility with due account of leanness. For capacity allocation, the control and management of flexibility is made up of decision options in connection with: · shop-floor logistic; · operation-cycles schedules; · production agendas planning; · capacity requirement setting. Each option offers several alternatives for how manufacturing capacity can be allocated to meet requests, according to the simultaneous engineering practice of mutually adapting products and processes with piece wise continuous betterment.

For upgrading, the decision loop begins with the issues of managing flexibility, over the selected horizon (strategic, tactical, operational), according to an enabling logic: · inventory decision, · planning decision, · schedule decision, · dispatching decision. Flexibility-dependent effects are related to previously measured performance. Results, structures and processes cannot be acknowledged separately. On these premises, suitable fabrication agendas are programmed, specifying: • products batches (part assortments, work cycles, shop logistics, delivery dates, etc.) and related tactical schedules; • batch sequencing policy, according to customers' requirements, with criteria for managing transients down to one-of-a-kind production; • maintenance and restoring plans with indication of monitored signatures and risk thresholds; • and the likes.

Feedback in capacity allocation decision cycle is built by monitoring flexibility effects; it provides reference data for the over-all procedure. This step measures the effectiveness of competing options by means of a manufacturing functional model, through the use of simulation work sessions. The testing with actual plants is expensive and unpractical; the use of pilot plants is restrictive, with , most of the times, unrealistic constraints. Results properly established by computer-simulation, on the contrary, provide a uniform basis to compare (for each **CFC** frame) the economic benefits enabled (and acknowledged) by means of the pertinent **MDM** frame.

The locally most efficient alternative is recognised at the last step in the decision loop. Each choice represents an optimised configuration of the flexible capacity allocation and yields the time-varying management policy, adapting production programmes through convenient selection criteria, such as: · adjust the routing table; · adjust the products schedule; · modify the production process plan; · modify the inventory policy.

Once a design-feedback-choice cycle is completed, the governor is ready to operate on-line. The software should include the ability of recursively running test and redesign cycles. Adaptive governors, interacting with the processes, Fig. 6, present multiple-layer build-ups [8], with vertical connections among relational, generative and information layers.

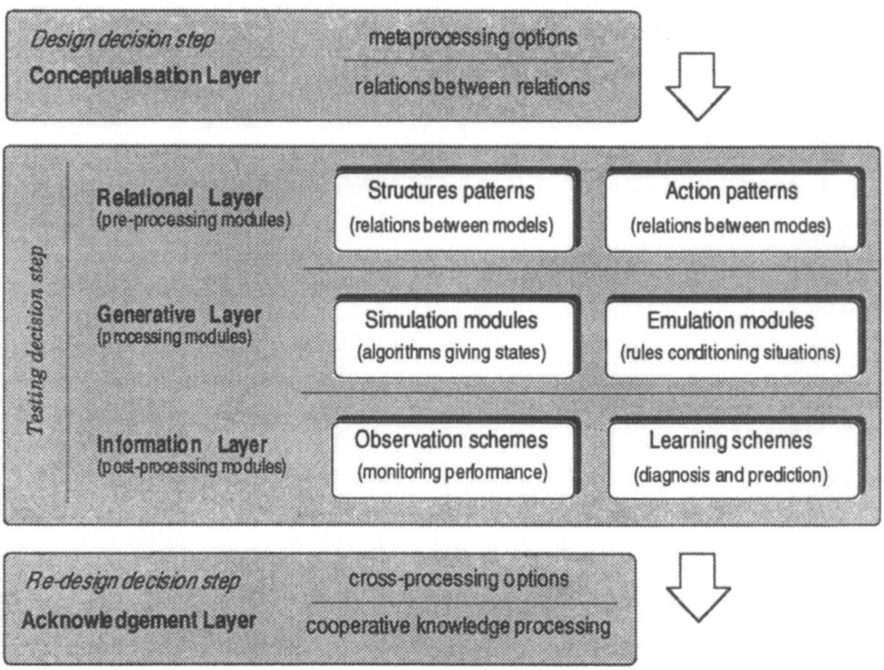

Fig. 6. Multilayer organisation of the adaptive governors for flexible automation

The *relational layer* exploits patterns of structures, and the expert governor expands on pre-processing modules for fixing the **CFC** and the **MDM** frames, with interfaces for structural model setting and for behavioural mode fitting. The *generative layer* exploits patterns of actions and propagates the causal responses with algorithmic blocks while acknowledging consistent suppositions through heuristic modules. The decision supports achieve best exploitation of flexibility by incorporating on-process data from the *information layer*; this monitors the performance evaluation figures and provides the diagnosis of facts for enabling trend-monitoring (predictive) maintenance.

Cooperative knowledge processing is, perhaps, still more related to research projects than to engineers' concerns. However, production engineering resulted these past years to be a quite stimulating application field for many artificial intelligence researchers, and a few implementations entered into practice, to enable economy of scope. The basic motivation is related to the ability of referring to information objects encoding 'knowledge', rather than simply 'data' [4]. 'Knowledge' is information with attached relational context, it is interpreted once the procedural conditions are duly instantiated by pertinent declarative data. With knowledge-based architectures, a twofold advantage arises: · the implementation of heuristic modules is straightforward, and the emulation of human-like decisional platforms provides the software instruments for enabling the factory automation with govern-for-flexibility features; · the goal: 'flexible automation' is coherently obtained by merging patterns of action and patterns of structure, provided that structures carry the signature of the originating changes, and actions grant the consistency of the generated configurations.

Knowledge structure-patterns	Data-bases	Models	Behaviours	*Reference structures*
	Results	Contexts	Reasonings	*Conditioning modes*
	Cases	Procedures	Methodologies	*Assessment tools*
	Communication	Sharing	Collaboration	*Information circuits*
Knowledge action-patterns	Activation	Scheduling	Planning	*Programmed actions*
	Execution	Synchronisation	Cooperation	*Functional situations*
	Operation	Coordination	Organisation	*Flexibility ranges*
	Recovery	Tactical	Strategic	*Time-span horizons*

Fig. 7. Knowledge structure /action patterns for intelligent manufacturing

The opportunity is generated by equivalent lore buildings, Fig. 7. On the knowledge actions-patterns side, we look after adaptive governors which transform control actions according to current process data, on condition that the process structure consistency is preserved; the resulting issues lead to the integrated control-and-management expert governors, based on real-time decisional manifolds, directly operating on manufacturing processes. On the knowledge structures-patterns side, we face reconfigurable build-ups combining patterns of the **CFC** structural frames with the **MDM** behavioural frames; the related issues provide knowledge support for closing the decision cycles, and re-setting the lay-outs (with improved performance) through off-process virtual reality simulation.

4 Applications and Example Developments

The inclusion of adaptive-governors operating on-line, within real manufacturing plants, is a relevant issue assuring consistency of flexibility and automation. The fitting of govern-for-flexibility options, in the practice, cannot be separated from the setting of flexible manufacturing facilities. With the approach summarised in the present chapter, fitting is done by knowledge-based software operating in real-time while production progresses; plant setting, moreover, exploits computer emulation·simulation results and uses such knowledge-based software as off-line consultation aid. Reset and outfit require shared understanding of process and results, grounded on broad-band experimentation of consistent fabrication plans.

The knowledge organisation requires recurrent closure of learning loops, to perform case-based reconfiguration of process structures and govern actions. The interlaced nature of on-process and off-process software packages is, perhaps, the most stimulating perspective of the factories of the future. The off-process aids cover a large variety of practical situations, and many issues of production engineers' concerns have already reached high levels of sophistication. Knowledge-based software, included on-process for governing unattended facilities, instead, is still restricted to face simpler problems.

Some typical developments will be shortly commented in this section, with hints on situations mainly supporting **MDM** fitting and other mainly aiming at **CFC** setting. The first case application deals with the integrated control-and-management of a production facility at emergencies so that manufacturing continuity is assured by recovery flexibility. It should be a paradigmatic reference, because still quite few facilities incorporate fully developed expert governors that automatically perform restoring attendance. The second case shows how simultaneous engineering paradigms support flexible specialisation by product-and-process innovation. The automatic assembly is programmed to continue during the night (unattended) shift, at least to face orders burden; the implementation of efficient robotic rigs required the previous careful analysis of the functionality given in charge to the products, to establish the correct balancing of technological investments and leanness requirements. As third case, the assembly of mechanical manufacts is, finally, considered under customers' driven requests, to be performed by multi-purpose units suitably balancing the parallel flows with on-process ruled adaptive scheduling. The examples are derived from industrial applications; the programming supports are different, since the individual case was developed with due account of user's habits.

4.1 Expert-govern to exploit recovery flexibility at emergencies

To increase operation autonomy and continuity margins, the manufacturing sections need be ruled by expert governors, enabling restoring procedures in front of exogenous occurrences requiring the modification of on-going schedules at unexpected (failures) or at strategic (market driven) interrupts. The functional characterisation of manufacturing processes requires two extensions of the model to englobe: · the generation of driving disturbances (as compared to optimal settings); · the availability of recovery procedures, to be automatically executed.

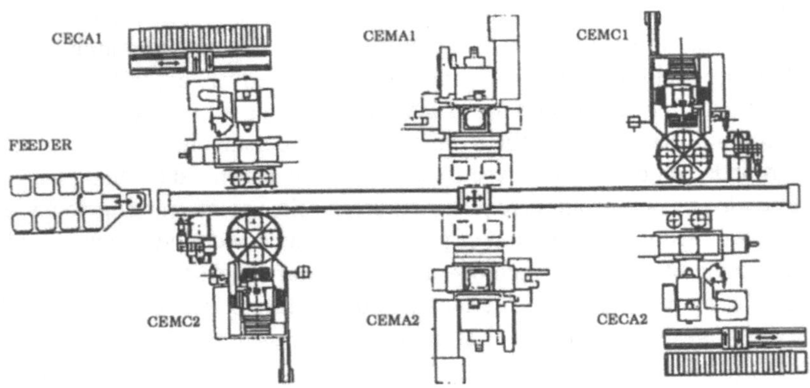

Fig. 8. Layout of the 6 work-stations with the travelling shuttle

The example section of Fig. 8 is composed by 6 work-stations (five axes machining centres with I/O units, whose main attributes are collected in Fig.9), a travelling shuttle and a local sorting/storing station working as a feeding unit; this manufacturing section is endowed with an expert governor, able to react in front of an extended set of external occurrences. The decisional logic was programmed with rules, written in OPS5, and tested through simulation, for singling out **CFC** setting and **MDM** fitting policies, that grant recovery and efficient predictive maintenance.

A typical fabrication agenda could be represented by a mix of 6 components A,

B, C, D, E and F, arranged into batches of 330 (40, 80, 40, 40, 80, 50) parts, whose work-cycles are collected in Fig.10. The shop-governor assures the ordered coordination of elemental jobs, namely: input unit loading (from shuttle), rotation of input unit, work-station loading, machining phase, work-station unloading, rotation of output unit, output unit unloading (to shuttle), travelling of shuttle, delivery of the (finished) part to the sorting-storing station; withdrawal of the (new) part from sorting-storing station.

MACHINING CENTER	INPUT & OUTPUT STATIONS
machining sequences codes machining sequences duration set-up codes set-up duration	kind of access duration of rotation loading time unloading time shift between I/O stations capacity

Fig. 9a. Attributes of the machines

MACHINING CENTER	MACHINING CODES
CECA1/CECA2	FP1 AF2 FI1 F4 SA2 M6 F10 FA12 F14 A15 M16 FA18 F20 F21 F22 M23 F25 M26 FP27
CEMA1/CEMA2	FP7 FS8 FS9 FS11 FS19 FS29 FS32 FS41 FS42 FP43 FS45
CEMC1/CEMC2	T13 T17 M28 FA30 T31 FP33 F34 F35 SA36 M37 T38 SA39 SA40 T44 M46 FP24

Fig. 9b. Machining codes for the work stations

UNIT		CAPACITY	LOAD TIME	UNLOAD TIME	ROTATE TIME
CECA1/CECA2	I	1	12	40	5
	O	1	8	8	5
CEMA1/CEMA2	I	1	12	40	5
	O	1	8	8	5
CEMC1/CEMC2	I	2	12	5	5
	O	1	5	10	5
FEEDER	I	40	8		5
	O	40	8	8	5

Fig. 9c. Parameters of the buffering units

Part "A" : THROTTLE

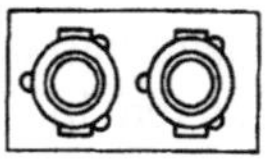

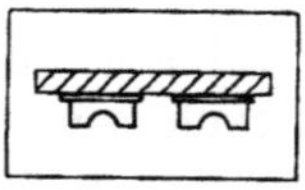

Machining	*code*	*tool*	*mach. time (s)*	*posit. time (s)*
planes milling	FP1	face mill	474	37
pre-drilling	AF2	drill	117	45
flange contour.	FI3	mill	625	30
drilling	F4	drill	335	190
rough boring	SA5	reamer	136	27
tapping	M6	screw tap	66	39
upp. plane mill.	FP7	face mill	354	30
undercutting	FS8	form cutter	570	30
seal seats rough.	FS9	mill	674	30
drilling	F4	drill	333	190
drilling	F10	drill	47	25
finishing	FS11	finish. mill	741	19
finish. boring	FA12	fin. reamer	240	28

Part "B" : CASE

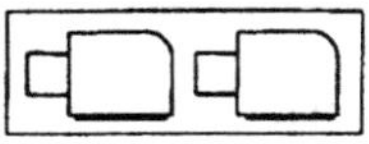

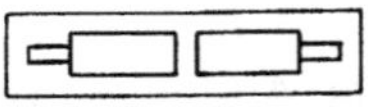

Machining	*code*	*tool*	*mach. time (s)*	*posit. time (s)*
turn. & underc.	T133	headstock	300	41
pre-drilling	F14	drill	104	96
pre-drilling	A15	insert drill	78	33
tapping	M16	screw tap	66	74
finish. turning	T17	reamer	56	17
finish. boring	FA18	reamer	74	33
ear removal	FS16	form cutter	450	32
pre-drilling	F20	drill	52	40
drilling.	F21	drill	13	14
pin pre-drilling	F22	drill	101	20
tapping	M23	screw tap	33	21

Part "C" : THROTTLE BODY DN 800

Machining	*code*	*tool*	*mach. time (s)*	*posit. time (s)*
roughing planes	FP24	plane mill	190	40
drilling	F25	drill	81	41
tapping	M26	screw tap	46	33
drilling	F25	drill	81	41
tapping	M26	screw tap	46	33
finish. planes	FP27	plane mill	129	34
roughing boring	SA5	reamer	136	27
drilling	F21	drill	13	14
tapping	M28	screw tap	13	26
seats milling	FS29	2 mill spin.	456	72
finishing	FA30	fin. reamer	237	72

Part "D" : THROTTLE BODY DN 300

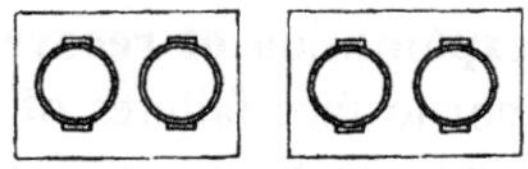

Machining	*code*	*tool*	*mach. time (s)*	*posit. time (s)*
rough. planes	T31	headstock	183	17
flange contour.	FS32	cyl. mill	550	36
planes milling	FP33	plane mill	116	33
pre-drilling	F34	drill	171	69
roughing boring	SA36	reamer	152	33
tapping	M37	screw tap	78	55
fac. pl., seal seat	T38	headstock	472	29
drilling	F35	insert drill	236	96
finishing boring	SA39	fin. reamer	162	33

Part "E" : CASE

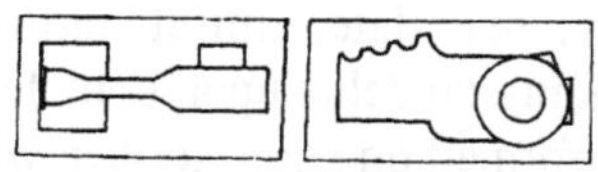

Machining	*code*	*tool*	*mach. time (s)*	*posit. time (s)*
rough. planes	FP24	plane mill	190	40
roughing	SA40	reamer	60	36
grooving	FS41	part. reamer	40	22
pre-drilling	AF2	drill	117	45
tapping	M6	screw tap	66	39
pre-drilling	F20	drill	52	40
tapping	M28	screw tap	13	26
contour tapping	F42	drill	23	39
tapping	M28	screw tap	13	26
drilling	F25	drill	81	41
tapping	M26	screw tap	46	33
finish. turning	T17	trn. reamer	56	17
finish planes	FP43	face mill	60	39

Part "F" : PRESSURE REDUCER

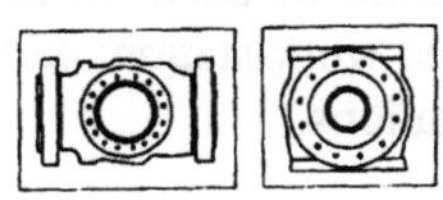

Machining	*code*	*tool*	*mach. time (s)*	*posit. time (s)*
planes milling	FP33	plane mill	116	33
turning	T44	trn. reamer	360	37
rough. boring	SA5	reamer	136	27
seats	FS29	2 mill spin.	456	72
grooving	FS45	mill	150	37
pre-drilling	F20	drill	52	40
tapping	M23	screw tap	33	21
finish. planes	FP27	face mill	129	34
rough. boring	SA36	reamer	152	33
drilling	F4	drill	333	190
tapping	M46	screw tap	363	93
finishing	FS11	fin. mill	741	19
finishing	FA30	fin. reamer	237	72

Fig. 10. Manufacturing cycles of the 6 parts

The exploitation of recovery flexibility. Let consider the reactions of the governor to face failures of work-stations, loading or unloading units, shuttle, buffering mechanisms. Decisions cover: · operation interrupt; · operation inhibition; · operation enabling. Interrupted operations are assumed inhibited, unless explicitly enabled. The enabling can force actions (e.g.: fixturing a work-station, switch part feeding among work-stations, modify sorting order from storages, etc.) or can require operator's help. Example comments on the decision logic are referred to storing and handling capabilities.

Management of the sorting/storing service. Part drawing is performed sequentially or can be driven: by the number of operations that still have to be performed; by such figure, plus a priority index. Priorities are set: statically, according to part family, due date, initial and/or transformation costs, etc.; or dynamically, depending on delay upon completion time or time margin of the operations still to be performed on parts, service wait time or stations' delay to finish the machining in progress, etc.; the second opportunity is used to tune the schedules according to delivery requirements or to reduce work-in-progress congestion. Example inhibited operations are, for instance: the use of the input unit of a station not fixtured for the particular work-cycle; the withdrawal from an output unit of parts that do not have assigned destination; etc. Priority is forced at a low value, when a work-station at least is being fitted out and no alternative exists, and even below if all work-stations are unavailable and the only storing station has vacancies. In this case, the station operates as central storage, shared by the whole section; the action should be avoided, whenever alternatives allow task completion on 'already in-charge' inventory. The 'downgraded' situation includes failure and unavailability. A part should not be forwarded to a location next to become 'downgraded' (e.g., in monitoring maintenance, when critical thresholds are approached). A symmetrical restriction of decision happens at the marking out of vacancies of a properly fixtured work-station; parts to be drawn and forwarded are looked for not to leave idle stations.

Management of the material handling service. Part handling aims at: · removing waits of shuttle in front of a busy station; · minimising periods of vacant/congested buffers; · avoiding excessive unbalance between work-stations loading; · preventing critical jobs with regard to diagnostic forecast; · etc. (minimisation of travelled distance is sometimes neglected, in the example application, due to the particular lay-out). The handling service splits into: · a fetch job, by getting to requested location and drawing the selected part; · a delivery job, by transporting the part and giving it out to a new location. The shuttle stops at failures (or expected failure). At service completion, the shuttle selects the subsequent task using the priority list: · to supply idle (rightly fixtured) work-stations; · to withdraw full output buffers; · to feed empty input buffers; · to prepare servicing work-stations next to end the machining; · to perform quick drawing from storage; · to perform quick delivering to storage; · to operate ordinary withdrawals. Subsequently, the handling supervision considers the filling up ratios of stations and the distance to be travelled; main goal is, always, to minimise the time that parts await to restore optimal schedules.

The controller is enabled once the scanning of expected occurrences is completed; the time margins are compared with the work-loads given in charge to the individual work-stations. The set of planned jobs omits occurrences allowing fixed ruling (since they do not generate activity perturbances) and occurrences depending on 'not yet defined' constraints. The control-and-management exploits distributed knowledge bases, to decentralise the checks; it may slow down the access of parts, but cannot modify the global flow; when, for instance, parts with work-cycles exceeding the allowed margin are rejected, queues are speeded up provided that the shuttle assures sufficient feeding to all resources.

Downstream, after recovery of proper running situations, transients are easily damped out by the rule of prioritising transport to work-stations with smallest input buffering capabilities; parts, awaiting up stream, are dispatched according to the feedback action enabled by a priority index computed on the complement of the part completion time.

The on-process expert governor. The expert governor is basically modelled by a set of rules, written in OPS5, which yield decisions based on current situations, by acknowledging the behaviour of the section, within pre-selected contexts. The language provides a simple aid for coding rules and, since rules are individually analysed, expansion of contexts is obtained by adding new rules, only making limited consistency checks. The inference mechanism suitably grants on line operativity. The OPS5 module is interconnected with algorithmic blocks, used for coding structured command or check actions.

As explanatory example, a part transportation-service rule is shortly commented. Each time the shuttle has completed a mission, the scan through situations of all the interfaced units is performed, creating an object 'type' with situational data and priority index. The generated service request might, for instance, correspond to an idle work-station buffer; rule antecedents would consider parts that could be withdrawn for machining: the label 'connect' expresses the relation between work-station and work-cycle (the check considers 'subsequent unconstrained operations'). The context then moves to the path selection (a couple of rules check the distances between any two plant locations), and to the object 'type' priority assessment (ranged from: part-priority, buffer filling ratio, shuttle travel) orderly verifying the choice. Once the destination link is fixed, the shuttle state switches from 'waiting' to 'moving', wipes out the object 'type' and executes the selected job.

During the restoring-operation range, production management is brought down on-process and step-wise control is adapted to local situations until emergency is removed. Thereafter only, strategic planning is resumed, by exploiting off-process management with optimal schedules and balanced flows coordinated by distributed control. Details on the actually enabled recovery procedures give further hints. The section governor initially selects the part, F05, to be forwarded to the work-station CEMC1; the related sequence of jobs is: · to fetch the part within the main storage; · to load the part on the shuttle; · to transport the part to the station CEMC1, with

Time	Occurrence
fr. 1 —>	*FEEDER USED AS INTERMEDIATE BUFFER*
662	Store CEMC1 unloaded onto shuttle part F05
662	Shuttle starts moving to station FEEDER
683	Shuttle's position is FEEDER
683	OUTPUT store FEEDER starts rotating
683	Shuttle started unloading onto store FEEDER
691	Shuttle unloaded onto OQ store part F05
691	Shuttle's position is FEEDER
691	OUTPUT store of station FEEDER starts rotating
795	Next work for part F04 on same station
795	Station CEMC1 started working part F04
fr. 2 —>	*EMPTY STATION DETECTION*
3905	Control has indicated an empty station: CEMA2
3905	Store CEMA1 started rotating
3915	Shuttle's position is CEMC2
3915	OUTPUT store of station CEMC2 starts rotating
3920	Shuttle starts loading from store CEMC2
3926	Next work for part A02 on same station
3926	Station CECA1 started working part A02
3930	Store CEMC2 unloaded onto shuttle part E01
3930	Shuttle starts moving to station CEMA2
3938	Shuttle's position is CEMA2
3938	INPUT store CEMA2 starts rotating
3938	Shuttle started unloading onto store CEMA2
fr. 3 —>	*CECA1 FAILURE*
6209	Station CEMC2 started working part B03
6211	OUTPUT store of station CEMC2 starts rotating
6216	Shuttle starts loading from store CEMC2
6226	Store CEMC2 unloaded onto shuttle part B05
6226	Shuttle starts moving to station CECA2
6243	Shuttle's position is CECA2
6243	INPUT store CECA2 starts rotating
fr. 4 —>	*CECA1 RECOVER*
7629	Shuttle's position is CEMC1
7629	INPUT store CEMC1 starts rotating
7629	Shuttle started unloading onto store CEMC1
7633	Store CECA1 started rotating
7633	Store CECA1 started unloading onto station
7641	Shuttle unloaded onto store CEMC1 part B03
fr. 5 —>	*CEMA2 FAILURE*
11120	Store CEMA2 unloaded onto station part E07
11120	Next work for part E07 on same station
11120	Station CEMA2 started working part E07
11127	Store CECA2 unloaded onto station part E05
11127	Next work for part E05 on same station
11127	Station CECA2 started working part E05
11140	Shuttle's position is CECA2
11140	OUTPUT Store of station CECA2 starts rotating
11140	Shuttle starts loading from store CECA2

Fig. 11. Chunks of the monitoring log-file

travelling time computed through the averaged speed of the shuttle on actual path; · pre-setting of input buffer, performed within masked time; · unloading of the part. The shuttle moves back to the main storage, unless an enabled job sequence requires to serve a different location. The selection of part F05 follows from the fact that the first scheduled operation of work-cycle is the shortest, assuring quickest delay to re-enable programmed schedules and, within family F, part F05 resulted to have best location.

Fig. 11 shows some example sequences out coming from plant monitoring, with jobs orderly shown referring to real time. For recovery flexibility, the main storage is used as intermediate buffer (just-in-time policies are suspended at emergency), to avoid bottle-necks at stations (at t = 662 s, thus, a part F05 is temporarily retained). The adaptive scheduling faces many situations: at time t = 3 905 s, for instance, the supervisor reports that station CEMA2 could turn to become idle having an empty input buffer; at the same time, on the output buffer of station CEMC2 the part E01, that might be machined by CEMA2, is available; the shuttle is dispatched, once acknowledged data consistency. As soon as a machining sequence is attributed, the supervisor looks whether the subsequent sequence might also be performed by the same station; in this case, the service queue is built with the refixturing time (but without dispatching time); the information is used for the balancing of the material flows across available resources. At time t = 6 221 s, station CECA1 switches to unexpected interrupt (failure): · work-sequence M26 on part C03 is stopped; · the part needs be cleared out; · the scheduled sequences are inhibited; · the selection of alternatives starts be checked. At time t = 7 633 s, the station is announced again available, and supervisor performs adaptations to resume standard scheduling. At time t = 11 123 s, the supervisor faces a warning interrupt of station CEMA2 working on part E07; the machining sequence continues, while dispatching of new parts is stopped and unloading of the input buffer starts; this is a typical situation in predictive maintenance based trend monitoring diagnostic, that allows life-long resources' utilisation.

4.2 Robotic-assembly to exploit efficient product-and-process design

Assembly tasks of customer oriented manufacts are critical aspect, often representing more than one third of production costs. The benefits of knowledge intensive settings for intelligent manufacturing is already recognised; the assembly sections maximally profit of making use of such an option, to support factory automation. Assembly facilities need, accordingly, the careful selection of CFC frames; computer-simulation becomes the basic reference for fully enabling the advantages by testing alternatives before the shot off of the lay-outs. Most manufacturers are concerned by diversified production, as option to be used to satisfy clients' needs by combining standard items, perhaps, suitably adapted, but minimising additions and modifications. The complexity of the assembling sequences makes computer simulation a cheap short cut, to face the planned adaptations and to obtain useful data with little costs.

The assembly problem of standard devices, purposely developed as special parts of customer's driven products, is discussed as second case development. The example refers to the batch production of items, supplied for the realisation of air-conditioning systems tailored to individual applications. For efficiency, the fabrication of such components might include fully robotised assembly cells; the related **CFC** frame should be designed, while performing the redesign of the devices to be assembled. The development is, thus, based on simultaneous engineering, extensively exploiting design-for-assembly (DFA) procedures for flexible specialisation, and relating plant performance characteristics by quality-function deployment (QFD) policy.

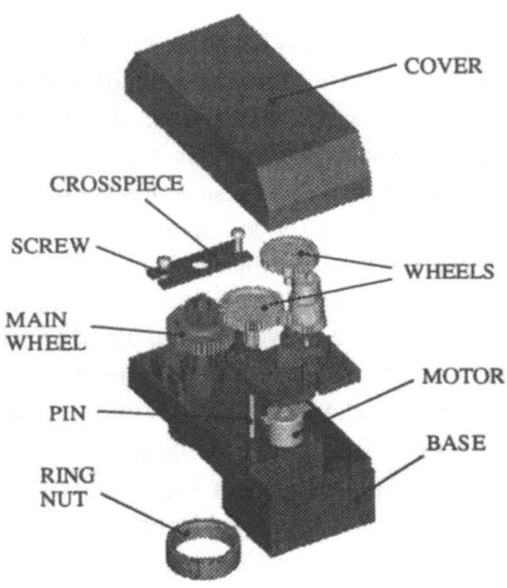

Fig. 12. Exploded view of the assembled device

Robotisation assures the automatic assembly and testing of the delivered components, with transparent access to the preparation of the assembly sequences and to the editing of the quality figures. Let refer to an example family of devices (flap-valves drivers, see Fig. 12), and consider production plans for supplying sets of devices ranged by size and differing according to requirements of (actual or future) buyers, that use them as (mass-production) components of customers' adapted air-conditioning equipment. The enterprise we are dealing with, will thereafter be placed in the position of operating as an 'active' supplier, cooperatively assuring the design of new devices, as case arises, and the manufacturing with transparent certified quality figures.

The automated assembly set-up. The cell is composed by an assembly SCARA robot, five auxiliary fixtures and a test unit that measures item performance while it is assembled, Fig. 13. The assembly and testing sequences are performed by a hierarchic supervisor, combining monitoring, decisional and command functions. The implementation of the supervisor is based on a graphical programming package, supplying the display of on-progress situations, with emergency warnings whenever they arise. The assembly robot selects the appropriate item from the supply station (automatically fed by a shuttle), the cell being designed to be operative in unattended mode for a full work-shift (8 hours).

Final devices are assembled on pallets, by putting together more than 20 elemental or preassembled parts. For the case application, the operation sequence splits into the following steps: · the actuation screw is placed in the appropriate

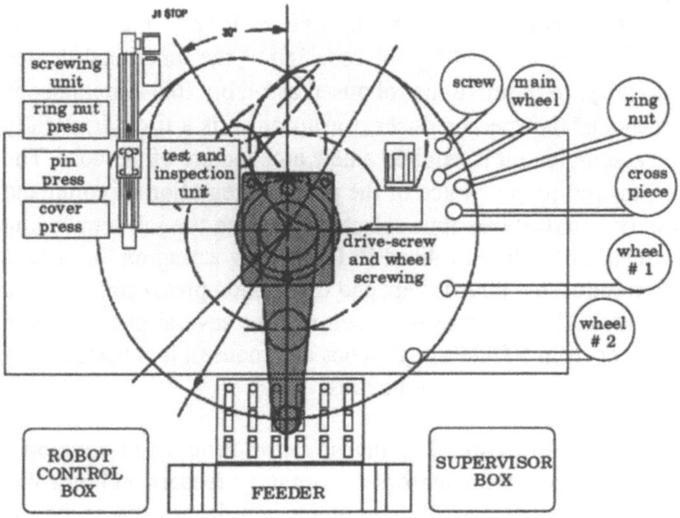

Fig. 13. The automated assembly cell

location of the pallet; · the base (device front-end) is next located on it; · pegs are fastened in the base and, parallely, the motor carrier is inserted; · the motor carrier box is fixed to the base; · the screw is fit into the driver, and the nut bodies are pressed against it; · the crosspiece is withdrawn and oriented; · the nut-screw component is located and the cross-piece positioned; · while the special rig fasten the cross-piece, the reducer two gears are prepared; · the two gears are positioned and fixed on the motor spindle, and the box cover drawn; · the cover is pressed and fixed to the motor carrier box. Fig. 14 shows the diagram of the operation timing.

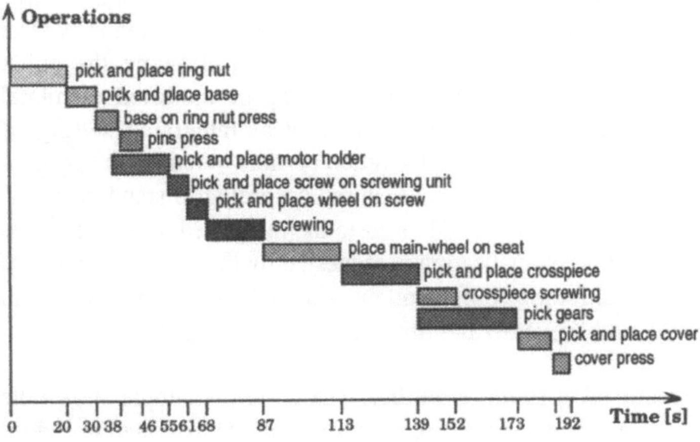

Fig. 14. Diagram of the operations timing

The supervisor enables the assembly sequences during the unattended work-shifts, by synchronising the robot with the auxiliary fixtures for peg, pin and screw connections. While performing cell setting, the redesign of the functional device appeared profitable, in order to improve the robotised assembly (the simultaneous insertion of the two gears of the speed reducer, for instance, is a trick drastically simplifying automatic assembly, but it required a new architecture of the box). The cell fitting, thereafter, moved to the choice of the proper integration of command and monitoring tasks. The control was thus selected to assure the job sequencing and coordination, by means of a decentralised PLC (directly acting on the pallets supplier, the screwing fixtures, the pins drivers and the friction press) connected to the robot governing module. The module is chosen to have several programming options (from keyboard, or from VDU through icons and mouse); it is managed by a supervisor, with monitoring and steering functions.

The supervisory process monitoring. The investigation brought to develop a fully instrumented assembly cell, with sensors on vibratory feeders and on robot grippers. The feeders are monitored through fiber optics transducers, that recognise component type and orientation; the process stops in case of unavailability of the correct piece. The check of the component on the gripper is performed by an optical gate; the robot repeats the fetching task at a 'warning' message. The testing equipment operates according to a package (based on the 'Labview' software, of National Instruments), which includes standard signal acquisition and processing options with the possibility of generating virtual instruments for speeding up, both, programming tasks and data restitution. Tests are initialised by the robot, at the moment the assembled device is located in the stand, and consider the functionality and the operative performance. Each measurement set is reported into frames, corresponding to steps included by the approval cycle; the frames display results (thrust of the screw-driver, while loop for fan opening, sequencing logic, runtime schedules, etc.) and provide full assessment for device specification.

The cell supervisor executes assembly and testing operations, and assures ordered execution of the planned agendas, according to bottom up requests. The user interface avails itself of suitably arranged displays, allowing the quick acknowledgement of new products and simple fitting of modified technical specifications. The resulting lay-outs represent typical integrated setting of strategic and execution ranges. This kind of solution permits to reduce the costs of individual items (flap-valves drivers), manufactured according to standard mass-production plans. The diversification is deferred to the integrated development of product and process. Flexible specialisation is requested by the implementation of customers' driven air-conditioning equipment, outcoming by the coordinated skill of components supplier with (several) and final equipment manufacturers.

Off-process assessment of optimal assembly. The present example shows the relevance of process monitoring and how automation could be exploited to reduce costs and increase quality. Business market share would widen, once reached this combined goal, requiring capability of further diversifying family types as new buyers are won. The set-up of a convenient assembly cell needs, therefore, the beforehand check of the technological versatility options, that support future potentialities, preserving leanness, with return on investments. The objective can be assessed with appropriate simulation supports for the setting of the **CFC** frames adapted to the manufacturing of families of standardised devices, that can readily be incorporated into personalised air-conditioning equipment. Computer simulation is used as consultation reference for assessing and testing, specially for simple production programmes, the control logics, with special attention to the on-line automation of the handling service.

The assembly cell has been developed referring to standard fixtures (robot, feeding equipment, cooperative rigs, testing instrumentation, etc.) and has already proved suitable capability of reacting to customers diversified requirements. The set-ups are selected to face individual orders with optimal, completely automated, assembly schedules. Unattended work-shifts are commonly enabled with suitably pre-arranged fabrication agendas, that (at the moment) do not require re-fixturing operations during the shift. The switching to new orders presumes a check on the **CFC** frames with the off-process consultation aid of simulation; then the highest productivity is transferred on-process and enabled all over the scheduled tactical horizons.

4.3 Adaptive-assembly to reach customers' driven requests

The third case investigation, Fig. 15, deals with an integrated section having multi-purpose automated assembly stations that work in parallel, supplied by a cartesian handling robot. The individual station delivers finished manufacts, from the fed parts. These final products evolve according to purchasers' requests, and need be assembled in parallel according to varying mixes agendas. For the case application, three families of final manufacts are dealt with simultaneously, each one presenting two intermediate assemblies. As first instance, the association 'station-and-family' is done. The discussion that follows is confined to methodological aspects, and, for confidentiality reasons, details involving final products identification are omitted.

The govern-for-flexibility fitting. Efficiency is achieved by adapting schedules to on-going situations, with integration of tactical and operational ranges, by enabling: · high productivity with decentralised control, for steady spans; and: · recovery ability through supervisory management, at discontinuities. Parts, intermediate assemblies and final products are fixed on pallets, with standard dimensions. The handling robot can withdraw individual components to further perform station loading/unloading. The assembly govern is given in charge to an expert module. This is obtained through computer simulation, performed by a modularly arranged package, whose development has been based on the shell G2 by Gensym; it allows: · the assessment of the process productivity figures, for assigned strategic choices; · the confrontation of govern-for-flexibility options, for

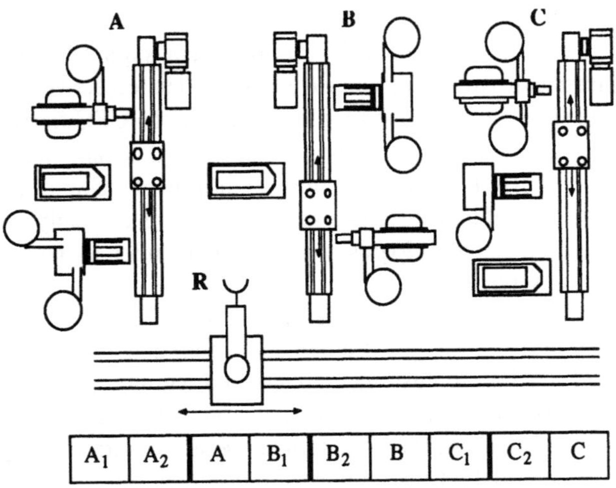

Fig. 15. Layout of the automated assembly section

selecting optimal **MDM** fittings; · the experimentation on the control-and-management module to be incorporated for factory unsupervised-mode automation. The package uses object-programming, to describe physical and logic resources. The governing module includes rules and procedures for encoding heuristic or, respectively, causal conditioning knowledge.

The rules are introduced, typically, for activating sets of tasks which create inquiries with assigned values and connections to objects, and enable the related procedures each time situations arise for the instanciation of actions dependent on the acknowledged occurrences. A different group of rules performs the feasibility test of the different requests, in relation with the current plant situation; checks make use of predefined service procedures, having in charge the instanciation of the objects to be addressed by the individual request. Several rule patterns are considered: · 'initial' rules, to activate the knowledge base; · 'whenever' rules, for rule connection; · 'if' rules, for conditioning concatenation; · etc. The procedures are used for the execution of search and re-order algorithms of lists each time the set of actions to be performed is logically assessed (e.g.: data sorting, display modification, simulation time up-dating, etc.). Specialised procedures are appended to modify the requests or to assess the request feasibility.

The assessment of flexibility effects. The time evolution of the simulation is event driven. The up-dating is driven by a strategy that acknowledges new occurrences and selects the prosecution fixed by the given logic. The occurrences are positioned in a list; and the state is continuously up-dated in relation with the acknowledged occurrence. The simulation cycle starts with the occurrence at the top of the list, performs the requested tasks that might generate new occurrences to be inserted in the list, withdraws the processed occurrence and gets ready for a new cycle. This logical timing has been programmed, as shown by Fig. 16.

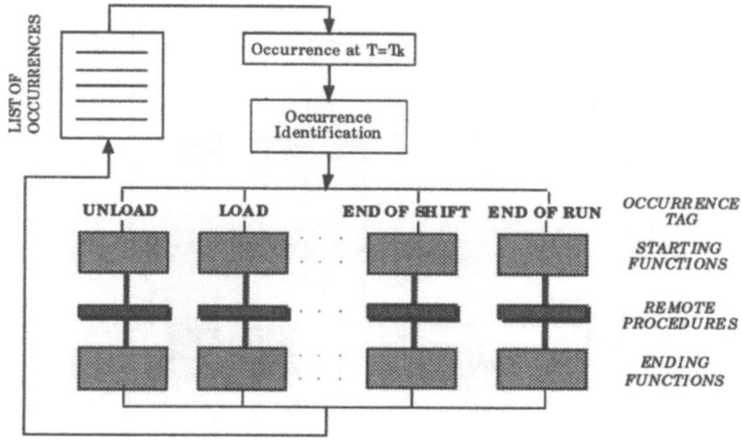

Fig. 16. Occurrences scheduler and mechanism for calling the remote procedures

The inference motor is based on several options. Rules are evoked: · with given scan interval, via forward or backward chaining as fixed by the appended attribute; · in function of the assigned focal object or focal class; · in relation with chosen rule pattern; · each time the occurrence mentioned by the antecedent verifies; · whenever a reference variable reaches the preassigned threshold. The internal branching enables, by default, the actions evoked by each rule; to suspend the activation (at the end of the rule, or to any given time) suitable 'in order' procedures need be specified. A procedure recalled by a rule, moreover, originates tasks considered by the inference motor at the end of the rule; it follows that the simulator could be requested to face sets of different tasks, and ordering requirements are necessary to grant evolution consistency.

The knowledge base of the package has been dislocated into several workspaces by grouping objects according to types and functions. Appropriate sub-workspaces gather specialised objects to organise the commands or to assemble variables and rules. The knowledge partition, with the possibility of dynamically activating the individual workspaces and sub-workspaces, assures modularity, for easy modifying or extending the structural **CFC** and behavioural **MDM** frames. A pertinent workspace is defined for the graphical restitution. The assembly section lay-out is represented by icons, for the: · I/O units, · workstations , · work-in-progress, · etc., as it is shown by the Fig. 17. The simulation outputs are further provided by several readout-tables distributed, in relation to each physical resource, into the pertinent sub-workspace.

The experimentation of control-and-management issues. The integrated assembly section, Fig. 15, is fed by pallet-grouped components that rationalise transfer tasks. Every workstation delivers finished manufacts belonging to any of the t hree scheduled families; for efficiency, the specialisation is planned, but not

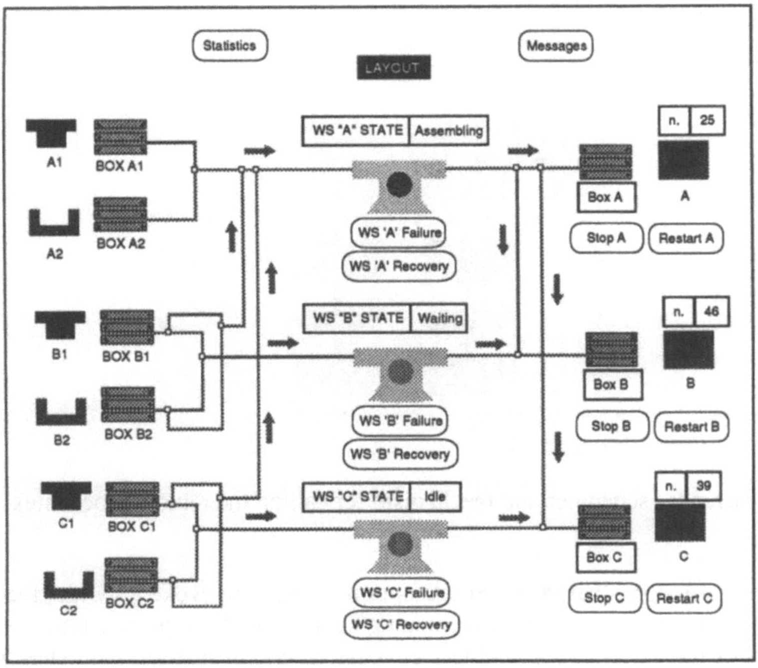

Fig. 17. Synoptic diagram of the plant by means of read-out tables

strictly obliged: the handling robot can freely perform feeding and retrieval operations. With object-programming, each resource is specified by identifiers (selected icons), structural attributes (topological location, station geometry, etc.), functional characteristics (robot performances, buffer bedspace, assembly times, etc.), operation state (free, assembling, failure, busy, idle, etc.), with colour-code allegation, and logic updating to state the situational progression. For instance, the manufact to be assembled is defined: 'not-existing' when components are still separated; 'waiting' when completed within the workstation; 'in transport' during the robot handling; 'available' when ready on the delivering pallet.

The efficiency fitting is concerned by the behavioural **MDM** frames to enable proper govern-for-flexibility issues. The assembly section presents itself as a set of concurrent, parallel processes, with connection constraints and rhythm marked by the robot services. The handling operations (pick, transfer, place) are supervised by heuristic rules to grant the scheduled delivering maximising workstations operation time (or minimising the work-in-progress). The 'occurrences' are transient objects, since created according to situational acknowledgements and destroyed once processed; the current time assures sequencing and delivering description; the identifying label specifies the actor type and the resource attribution. The simulation is, also, based on relational links, to permit indirect object addressing.

The governing policy avails itself of priority criteria, readily modified depending on delivering schedules requests. Productivity is maximised setting components feeding strategies with highest assembling stations operativity. Failure and repair occurrences are statistically driven; the switching to recovery strategies is performed after checks on the delivering agendas (and not only on the work-in-progress distribution). Similar checks are enabled when a unit is again ready to reassume operativity.

Batches	Mix M1: batch size	Mix M2: batch size	Assembly time
1	17	10	35
2	9	10	200
3	13	10	19

Robot	Speed & acceleration rates
R1	50 %
R2	100 %
R3	25 %

Fig. 18. Example parameters used during simulation runs

Efficiency through tactical and recovery flexibility. This third example shows how to deal with integrated control and management strategies for plant effectiveness. The simulation code, in this case, is not only a consultation reference, it also provides the decisional support to the govern-for-flexibility block. Assembly section performances are measured by the continuous monitoring of work progression. Specific attention is focused on the effects induced by:

- robot characteristics: the tasks are analysed and over-all duty times are acknowledged as averaged figures, since more details can be omitted once the teaching is completed;
- failure patterns: the recovery strategies are investigated for different breakdown or damage situations and modified workstations priorities;
- compositions of mixes to be assembled: the (three) manufacts have different assembly times, thus, effectiveness depends on the delivery assortment agendas.

Some typical sample results obtained by several simulation runs, whose main parameters are presented in Fig. 18, are shown by: · Fig. 19, aiming at investigating the opportunities of recovery flexibility; · Fig. 20, aiming at assessing the issues of tactical flexibility. A fair allotment of resources is reached, with low waiting intervals at workstation B and convenient flow balancing between the workstations, making use of the shorter commitment of A and C stations. As a unit switches to a failure state, the manufacturing of the related item is temporarily stopped with higher productivity of the other stations (better served by the robot), until the recovery strategy enables the alternative scheduling. The tasks transfer from station C to station A, for instance, moreover requires priority resetting, to speed up feeding of the most loaded station. The steering logic should remain in any case transparent, so that the effects of flexibility could be properly measured.

The introduction of a knowledge-based governor operating on-process and on-line needs be assessed with direct experimentation on actual situational settlements; for practical purposes this can only be achieved by expert simulation. Object programming provides a powerful support, allowing easy expandability and efficient specialisation. The sample discussions introduce to planning and scheduling solutions based on the performance evaluation of flexible manufacturing systems obtained through discrete event simulation.

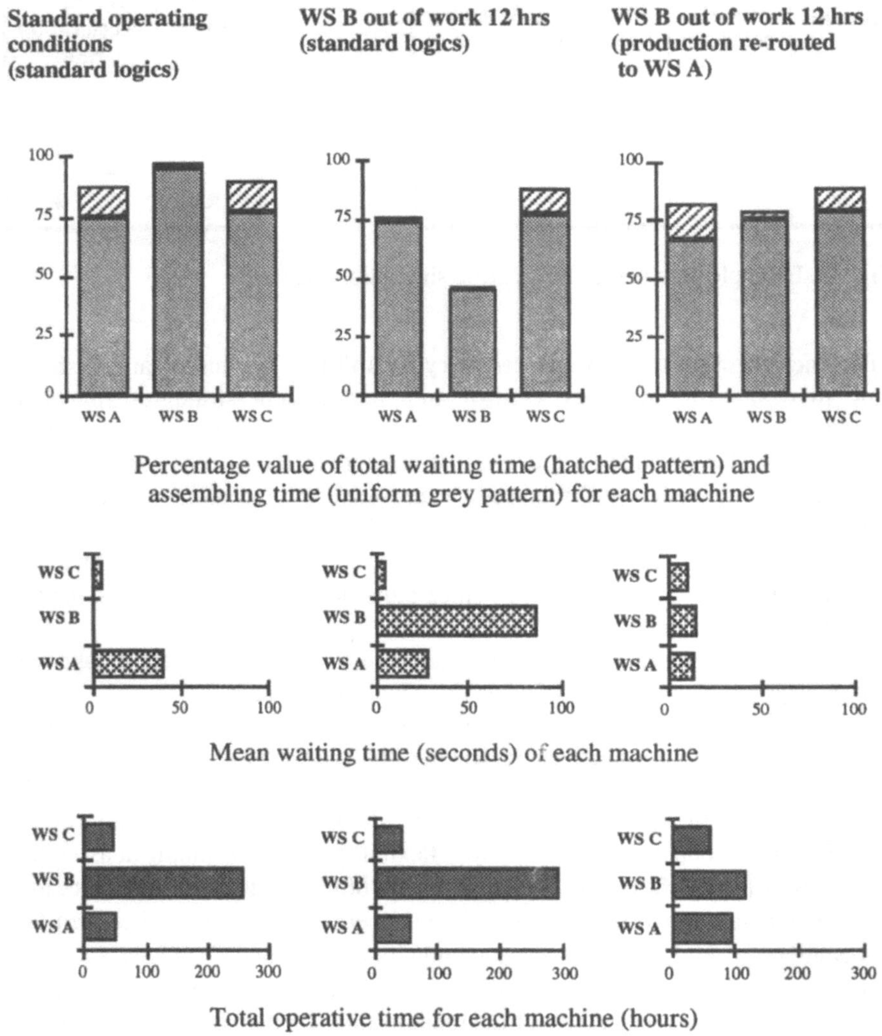

Fig. 19. Comparison among different operating conditions (mix M1 - robot R1)

The modelling encompasses products and processes and requires the joint assessment of structural and behavioural frames. However specialised decision supports are needed depending on the area application; the two typical conditions are faced in the paper, namely: when customer's driven manufacturing requests integrated product and process design, and simulation is the validation instrument for the plant set-up; when flexible specialisation needs be joined to lean engineering effectiveness based on integrated control and management, and simulation is the validation support of the governing logic.

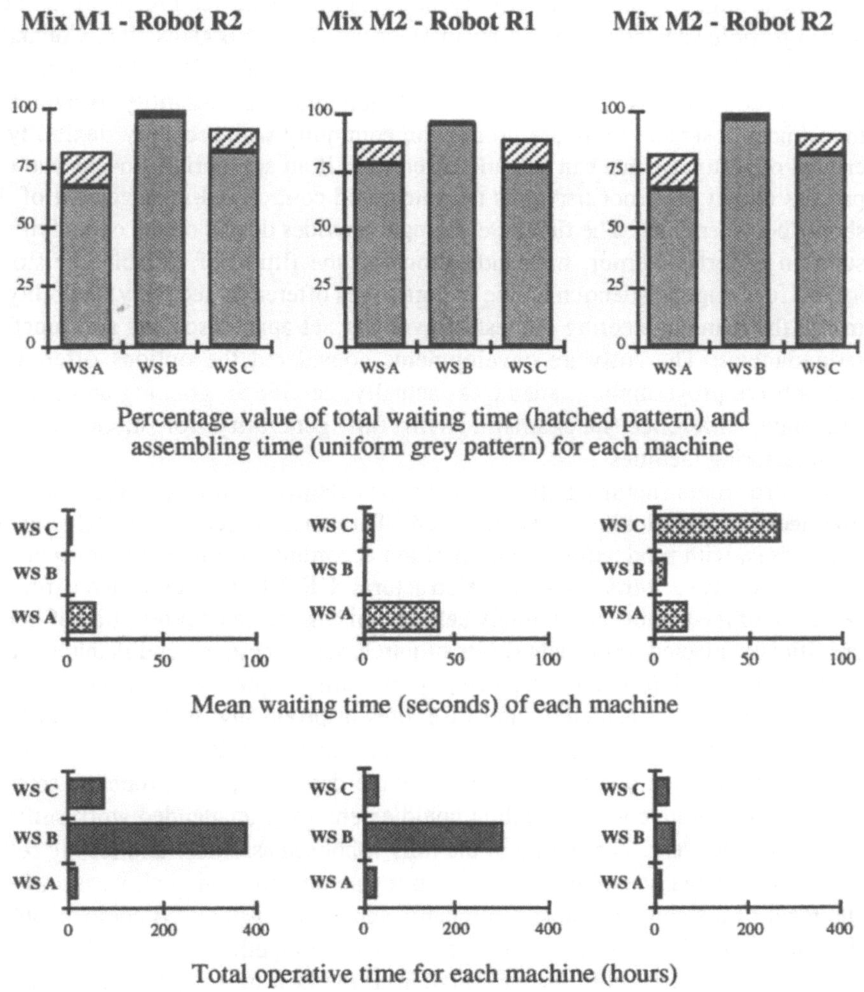

Fig. 20. Comparison among different robot rates and production mixes (standard operating conditions)

5 Concluding Comments and Remarks

The financial enterprise rentability is a mix of market-driven and technology-driven factors and the assessment of returns on investments requires a careful foundation of the enabling instruments (such as cooperative knowledge processing) in relation to the 'economy of scope' goal, namely: to fully exploit flexibility while preserve leanness; and: to maximise returns on investments by marketing extended mixes of products with customers' adapted quality.

Information technology has supplied reference supports for intelligent manufacturing. Productive flexibility is a characterising feature, with market-driven *flexible* specialisation and business-driven *flexible* organisation. Computer aids are basic *instrumental* options, and they need be tailored to the application specific requirements. The exploitation of technological versatility to minimise investment costs and to assure production continuity with recovery flexibility is critical opportunity, that can actually be enabled if an appropriate on-line and on-process expert governor performs the integrated control-and-management of the shop-floor operations. The first case example provides details on the operativity of such an expert governor, with indications on the fitting of suitable decisional logics, focusing the attention on the opportunities offered by recovery flexibility to match the lean engineering request of avoiding set-apart resources and function redundancies. The software developments considered the options offered by inexpensive programming languages (actually the OPS5), granting an efficient, real-time, inference mechanism, with full generality for modelling the manufacturing facilities.

Modern manufacturing looks after profitability through simultaneous engineering and flexible specialisation. This means network interlacing of enterprises, with productive decentralisation according to skill and competencies. The second case introduces to the structural **CFC** frames considered for the selection of the automated assembly cell, optimising the batch production of items, specifically adapted to (growing) sets of purchasers. These are end manufacturers of one-of-a-kind products, requiring personalised quality from the level of components. The considered enterprise aims at proposing itself as *cooperative supplier*, supporting the design of new components according to client's wishes, and performing the assembly cell reset while developing the products, so that, occasionally, production scheduling could extend over unattended work-shifts to face urgent delivery. The setting of the fully-robotised assembly-and-testing cell is reviewed as actually exploited option to bring automation and quality assurance at the level of a small, specialised company, supplying customised components re-designed according to the wishes of series of competing business firms. The software was restricted to the exploitation to the best of an application-oriented tool (actually the code Labview), since facility re-set needs be directly integrated with niche-specialised product-quality and process-efficiency requirements.

The third sample case discusses the situation of an enterprise that exploits intelligent manufacturing for assembling market-driven devices with just-in-time schedules, on the tactical horizons, made operative with job-floor facilities. Resources' technological versatility provides options for flexibility, enabled by a 'knowledge-intensive' outfit; efficiency requires the constraint of 'leanness', that is: additions of resources or functions needs be monitored on their ability of expanding value to actually sold products. The presentation deals with the selection of the most efficient behavioural **MDM** frames, based on the assessment of the actual net-production reached by alternative schedules and tested via expert-simulation. The software, in this case, considered a sophisticated expert environment (the shell G2), to simplify the incorporation of complex assembly sequences and recovery alternatives, while assuring the effectiveness of real-time.

The inclusion of knowledge-based modules, which support the govern-for-flexibility capabilities of factory automation, has been referred to the expected return on investments based on computer integration up to proper levels. The option is illustrated by means of cases developments, outcoming from practical applications that have concerned small to medium enterprises, with explicit attention on lean-engineering check-up assessments. Expert-simulation resulted to be an effective reference for cutting out the extension of the innovation at levels properly suited each individual situation. Small enterprises are, sometimes, still quite unaware of the opportunities offered by on-process controllers-and-managers; a somewhat smoother approach is, possibly, requiring the pervious demonstration of sub-classes of computer-integration capabilities by means of a pilot-plant implementation [12]. Even when such facility exists, however, expert-simulation cannot be left out, to provide quantitative assessments for selecting appropriate set-ups (**CFC** frame) and for enabling effective out-fits (**MDM** frame), since the pilot-plants cannot be extended to face the peculiarities of real manufacturing environments. The present chapter has, therefore, purposely summarised the characterising features of the knowledge architecture generally required for the govern-for-flexibility of actual plants, by relating three example case developments with the rules generally requested for aiming at the economy of scope issues through intelligent manufacturing.

References

[1] R.C. Michelini, R.M. Molfino, G.M. Acaccia: "Automazione di fabbrica: Robotizzazione dei processi manifatturieri", *L'Industria Meccanica*, n° 387, sett. 1984, pp. 589-600.

[2] G.M. Acaccia, R.C. Michelini, R.M. Molfino: "Automazione di fabbrica: Sistemi integrati di lavorazione", *Il Progettista Industriale*, n° 7, sett. 1984, pp. 68-79.

[3] R.C. Michelini: "Decision anthropocentric manifold for flexible manufacturing", 4th ASME Intl. Symp. on Flexible Automation, San Francisco, July 12th-15th, 1992, pp. 467-474.

[4] R.C. Michelini, G. Kovacs: "Knowledge organisation and govern-for-flexibility in manufacturing", 3rd Intl. Conf. on Cooperative Knowledge Processing for Engineering Problem Solving, Rovereto (Italy), May 29th-June 1st, 1994, pp. 801-826.

[5] R.C. Michelini, F. Pampagnin, R.P. Razzoli: "Capire e gestire il controllo di qualità", IX Conv. Naz. ADM, Caserta, 27-29 sett. 1995, pp. 273-283.

[6] R.C. Michelini, F. Pampagnin, R.P. Razzoli: "Lean engineering and performance measurement in flexible manufacturing", 27th ISATA Lean/Agile Manufacturing, Aachen, Oct. 31st-Nov. 4th, 1994, pp. 279-286.

[7] R.C. Michelini, G.M. Acaccia, M. Callegari, R.M. Molfino, R.P. Razzoli: "Knowledge-based emulation-simulation for flexible automation manufacturing", Eurosim Congress '95, TU Vienna, Sept. 11th-15th, 1995, pp. 1259-1264.

[8] G.M. Acaccia, R.C. Michelini, R.M. Molfino: "Knowledge-based simulators in flexible manufacturing", 2nd Intl. Conf. Applications of Artificial Intelligence in Engineering, Boston, Aug. 4th-7th, 1987.

[9] G.M. Acaccia, R.C. Michelini, R.M. Molfino, P.A. Piaggio: "X-SIFIP: a knowledge-based special-purpose simulator for the development of flexible manufacturing cells", IEEE Intl. Conf. Robotics & Automation, San Francisco, April 7th-10th, 1986, pp. 645-653.

[10] R.C. Michelini, G.M. Acaccia, M. Callegari, R.M. Molfino: "Flexible manufacturing by integrated control and management", in: M.B. Zaremba, B. Prasad Eds.: 'Modern Manufacturing: Control & Technology', Springer Verlag, London, 1994, pp. 225-253.

[11] G.M. Acaccia, R.C. Michelini, R.M. Molfino, A. Ragonese: "Maintenance model and failure analysis for a flexible manufacturing cell", 7th IASTED Intl. Conf. Modelling, Identification & Control, Grindelwald (Switzerland), Feb. 16th-18th, 1988, pp. 80-86.

[12] G.M. Acaccia, M. Callegari, R.C. Michelini, R. Milanesio, R.M. Molfino, A. Rossi: "Pilot CIM implementation for lean engineering experimentation", *Intl. J. Computer-Integrated Manufacturing Systems* (Elsevier Science), vol.8 n°3, Aug. 1995, pp.185-192.

9 A CIM Designed According to Ward and Mellor

W. A. Vervoort

1 The Ward and Mellor methodology

1.1 Introduction

As a result of the software crisis methodologies for the design of large software systems were developed. At first the emphases was on business applications. DeMacro [1] introduced symbolic tools to model data flows, stores and transformations, which were worked out by Yourdon [13, 14], Gane and Sarson [3] and Page-Jones [9] for administrative systems. Later on design methodologies for real-time systems were developed by Ward and Mellor [10], based on Yourdons work.

This chapter gives a summary of the methodology for structured design of real-time systems as described by Ward and Mellor. It is based on the separation of analysis and implementation. Symbolic tools are introduced to model data- and control-flows, -stores and -transformations.

1.2 Separation of analysis and implementation

Like most design strategies the Ward and Mellor methodology separates the analysis activity (**what** to solve) from the implementation activity (**how** to solve it). The advantage of this separation is:

- the analysis can be given in terms of the problem
- it is easier to keep an overview of the analysis
- errors are found earlier by application experts
- it gives most implementation freedom

Separation of these two stages does not mean that they have to be worked out one after another. An interleaved development will be more likely.

1.3 Essential modeling

The output of the Ward and Mellor analysis stage is a couple of models, together called the *essential model.* The essential model is an implementation-free description of the system behaviour and the interactions of the system with its environment. It is an outside-in rather than a top-down analysis of the problem. That is: the environment in which the system is to be embedded has a strong influence on the structure of the system under design. The essential model consists of two parts: the *environmental* model and the *behavioural* model (fig. 1).

model	purpose	documents	contents
environmental model	description of the environment in which the system operates	context diagram	description of the information (data and control) crossing the boundary that separates the system from its environment
		event list	description of events to which the system must respond
behavioural model	description of the behaviour of the system in response to the events	dataflow-diagrams	hierarchical description of the response to the events by control- and data-transformations, -flows and -stores
		data dictionary	description of the information in the dataflow diagrams

Fig. 1 Essential model

The environmental model is a description of the environment in which the system operates. It specifies the events to which the system must respond and the boundary between the system and its environment.

The behavioural model is a description of the required behaviour of the system. It specifies the way the system responds to the events and it specifies the datastructures of the system.

1.4 Implementation modeling

In the implementation stage the implementation-free essential model is mapped on a hardware configuration in such a way that all timeconstraints can be guaranteed. This results in three models (fig. 2): the *processor* model, the *task* model and the *hierarchical* model.

The first specifies which parts of the essential model will be executed on which processor. It specifies real concurrency, based on available hardware and performance specifications. In the task model, all transformations that belong to one processor are grouped into semi-parallel tasks. On a single processor these tasks simulate concurrency by multitasking. At this point is also decided at what frequency the sampling processes will be executed. Finally in the hierarchical model the procedure hierarchy of each task is specified.

model	purpose	document
processor model	allocation of dataflowdiagram elements of the essential model to processors	DFD per processor with specifications
task model	allocation of dataflowdiagram elements of one processor to tasks	DFD per task with specifications
hierarchical model	allocation of dataflowdiagram elements of one task to a hierarchy of procedures	structure chart per task with control- and dataflow between procedures

Fig. 2 Implementation model

1.5 Tools to model real-time systems

The Ward and Mellor methodology makes use of tools provided by a workbench. It contains a graphical editor connected to a data-definition facility and assists in the creation of the models described above.

With the graphical editor the context diagram (CD) and the hierarchy of data flow diagrams (DFD), and with the data-definition facility the data dictionary (DD) are created. The workbench checks the consistency of all parts.

1.6 **Context diagram (CD)**

Real-time embedded systems have the following characteristics:

- the environment is measured by sensors
- the environment is controlled by actuators
- both sensors and actuators should be served "in time"

What "in time" means depends on the application. In some robot control applications for instance robotarm positions should be measured in time intervals of equal length to let the control algorithm work properly. The real-time operating system should be able to serve sensors and actuators in parallel to meet time constraints.

In the context diagram the embedded system is drawn as a circle surrounded by entities from its environment. The entities from the environment are terminators, drawn as rectangles. They all have some relationship with the embedded system. Some could be sensors and others could be actuators. But also human operators are entities in this sense. This relationship between entities and the embedded system is shown in the CD by arrows representing flows of information and control. In some cases these terminators may chosen to be virtual (especially when the real interface is still unknown). Later on the virtual terminators can then be connected to the real ones through interface transformations. An eventlist completes this specification. Ideally it contains a description of the desired response of the embedded system to the events mentioned, where needed with time constraints.

1.7 Dataflow diagram (DFD)

The embedded system in the context diagram is in fact also the top-level dataflow diagram. It is worked out as a directed graph of selectively activated transformations and stores connected by information flows.

1.8 Flows

Information flows can have data or not. The first are called *data flows*, the second *control flows*. Data flows are time-continuous when the data is present at any moment, (it is "on" the flow at any time), and time-discrete when the data is transported over the flow only at some points in time, (in between, the flow is undefined). In the time discrete case the availability of data can be interpreted as the occurrence of a real-time event. These activation signals offer the possibility to specify time constrains (in the Data Dictionary). Control flows can only be time-discrete: their only flow is the event itself. Three special control flows are frequently used: *trigger, enable* and *disable*. They are used to start and stop the activity of transformations, and are therefore called activation signals. Time continuous data flows will mostly be sampled at a certain frequency. Note however that this will not make the dataflow a time *discrete* one in the analysis stage. Sampling is just a matter of implementation.

flow	symbol	examples
activation signal	··········>	enable gripper
control flow	··········>	gripper is closed
time discrete data flow	———→	sensed shape of object
time continuous data flow	———→>	force on gripper (sensor)
		angle of gripper (actuator)

1.9 Transformations

There are two kinds of transformations: *data-* and *control transformations*. Data transformations transform input data flows into output data flows. They are supposed to be always active unless they input one or more of the following activation signals:

- enable, which makes the transformation active explicitly
- disable, which stops the activity of a transformation
- trigger, which activates the (cyclic) transformation for one cycle

These signals offer the possibility to specify time constrains, which is important in real-time environments. In control transformations the real-time activation logic of a data flow diagram is located.

A control transformation transforms input control flows into output control flows. Control transformations should only have control flows and activation signals as in- and output, whereas data transformations may have data flows and activation signals as input and data-flows as output.

The control transformations of a DFD contain the activation information. They react on incoming control flows sent by transformations, and selectively activate other transformations in the data flow diagram.

1.10 Stores

There are two kind of stores: *data stores* and *control stores*. In a data flow diagram data can be stored in data stores. A data store is empty or contains the last written data item of a time discrete data flow. No history of values is stored, and the time of occurrence on the flow is lost. Only data transformations are allowed to read and write data stores. A time discrete data flow from a data store to a data transformation means that the data is available on the flow whenever the transformation needs it. This means that data stores offer the opportunity to change the point of time of appearance from the original one to the one desired by the receiving transformation. Control stores are only used for synchronisation purposes and are described below.

1.11 Synchronisation

It is a good custom to have only one time discrete input flow (except those that come from a data store) for each data transformation. If there are more they should be merged somewhere outside the transformation and enter the transformation as one flow. Such a merge specifies a synchronisation constraint on both flows, which has to be realized in the implementation stage. As an example consider a transformation that transforms a text into another by means of a key. Two dataflows enter the transformation: the text and the key, and one exits it: the transformed text. Clearly both input flows should arrive at the same time. In order to ensure this the two are joined together outside the transformation. Instead of merging the use of stores is also possible. The transformation is then replaced by two transformations, one that transforms an arriving text with a stored key (if the key arrived earlier), and the second that uses an arriving key to transform a stored text (if that arrived earlier). As a side effect the first stores the text while the second stores the key (draw a dataflow diagram as an exercise).

If a data transformation has more than one time discrete output data flow it is a good custom to consider those as alternatives, and have only one alternative produced at a time.

The last occurrence of a control flow can be stored in a control store. In this case the time of occurrence is lost. The occurrence is stored as a semaphore-like number (say 'x'). Three operations are defined on that number.

operation	effect
signal	x:=x+1
wait	x:=x-1; if x>=0 then 'pass' else wait till x>=0
reset	x:=init (specified in DD); eliminate possible waits

1.12 Specification

The data- and control transformations and the data- and control flows of each data flow diagram must be specified. Data transformations can be specified by means of a lower level data flow diagram (by "zooming in"), and at the lowest level by means of a *transformation specification*. This can be done in an operational way (using a high level language or structured English), or in a declarative way (with pre- and post conditions). At the lowest level a control transformation is specified by means of a *state transition diagram* (STD). Data- and control flows and -stores are specified in a *data dictionary* (DD).

1.13 State transition diagram (STD)

A state transition diagram describes the behaviour of a control transformation as a finite state machine. It consists of states (shown as rectangles) and transitions between states (shown as arrows). States should show a clear relationship with externally observable situations.

Associated with a transition are a condition and possibly one or more actions. Close to the arrow the condition is written above a horizontal line, and the actions are written below that line. A system goes from its current state to another when the condition associated with the transition between these states is satisfied. The corresponding actions are then carried out. A condition may consist of input control flows, activation signals, Boolean expressions and wait operations on control stores. Actions correspond to the output control flows and activation signals of the control transformation.

If a control transformation is disabled its state becomes undefined. If it is enabled its state will be the initial state. The initial state is the state with an arrow pointing to it which does not stem from another state. An STD may have unnamed intermediate states that do not correspond with an external situation. Their purpose is to act like a conditional state from where different states can be reached dependent on subsequently observed conditions.

A transition is supposed to be instantaneous. It takes place at the moment the condition becomes 'true', mostly at the arrival of the flow.

An STD may contain variables, to count events, and be used in Boolean expressions.

Two or more control transformation may work together, as in the following example.

Consider a critical path in a factory that does not allow two mobile robots from different sides to pass through at the same time. Sensors detect an approaching robot at both sides, and some hold mechanism (a traffic light) stops the robot when necessary. The rule to be followed is that as long as a side has green light, an infinite sequence of robots at that side may enter the critical path. Other sensors at both sides detect a robot that leaves the critical path.

The DFD shows two control transformations, one for each side of the path, called: *control left* and *control right*. Further a control store, called *path*. Control right has two input and two output control flows. Input: *robot approaches from right,* and *robot leaves at left side;* output: *right light green, and right light red.*

The other one trivially shows the opposite situation. Both inspect and write the control store. The state transition diagram for both control transformations is shown in figure 3.

1.14 Data dictionary (DD)

Data- and control flows and -stores in data flow diagrams are labeled with meaningful identifiers. They may represent single or composite data. The identifiers should be specified in the Data Dictionary. Composite data is composed of other data, which in turn may also be composite. Single data cannot be decomposed further. This composition is specified as well as the meaning (together with range, limits, precision, units, and time constraints) of the data. For control flows only time constraints can be specified.

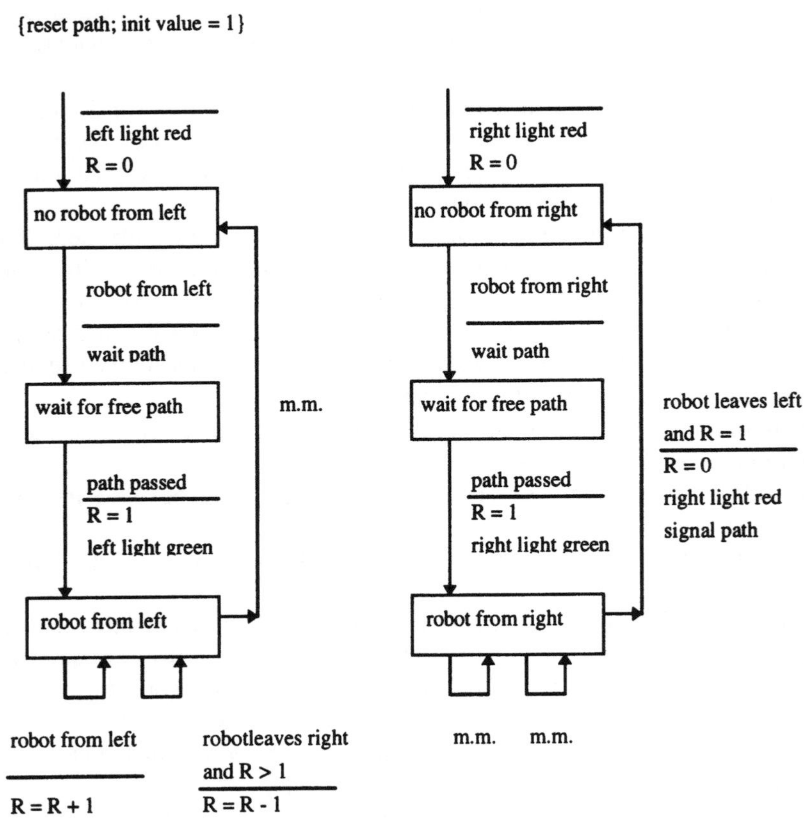

Fig. 3 Robot path control state transition diagram

2 The Ward and Mellor methodology applied to CIM

In this chapter we will discuss the application of the Ward and Mellor design methodology to the design and implementation of a part of a CIM system which has been carried out in an ESPRIT project.

The CIM part which was designed and implemented is the Cell Management And Control System (CMACS) of a factory producing series of discrete parts. It gets production information from the global factory level and schedules the production on a couple of machine-tools with the aid of an expert system. It runs on a local area network consisting of one central Cell Controller (CC) and one Machine Interface Unit (MIU) microprocessor for each machine-tool to be controlled. The MIU is the interface between the CC and the machine-tool controller. Its task is to control the job steps at the machine-tool; acquire machine status; and obtain NC part programs, tool offsets and operator instructions from the CC. In a separate report [6] details of design and implementation can be found. Below the most important experiences will be exposed.

2.1 Essential Modeling

The essential model is a model of the system to be built: the Cell Management And Control System.

The essential model consists of the environmental model and the behavioural model. The environmental model consists of the context diagram and the event list, and the behavioural model of data flow diagrams and the data dictionary. The construction of these parts in relation to the CIM design will now be discussed. A simplified version of the result of the analysis stage is shown in figures 4 and 5.

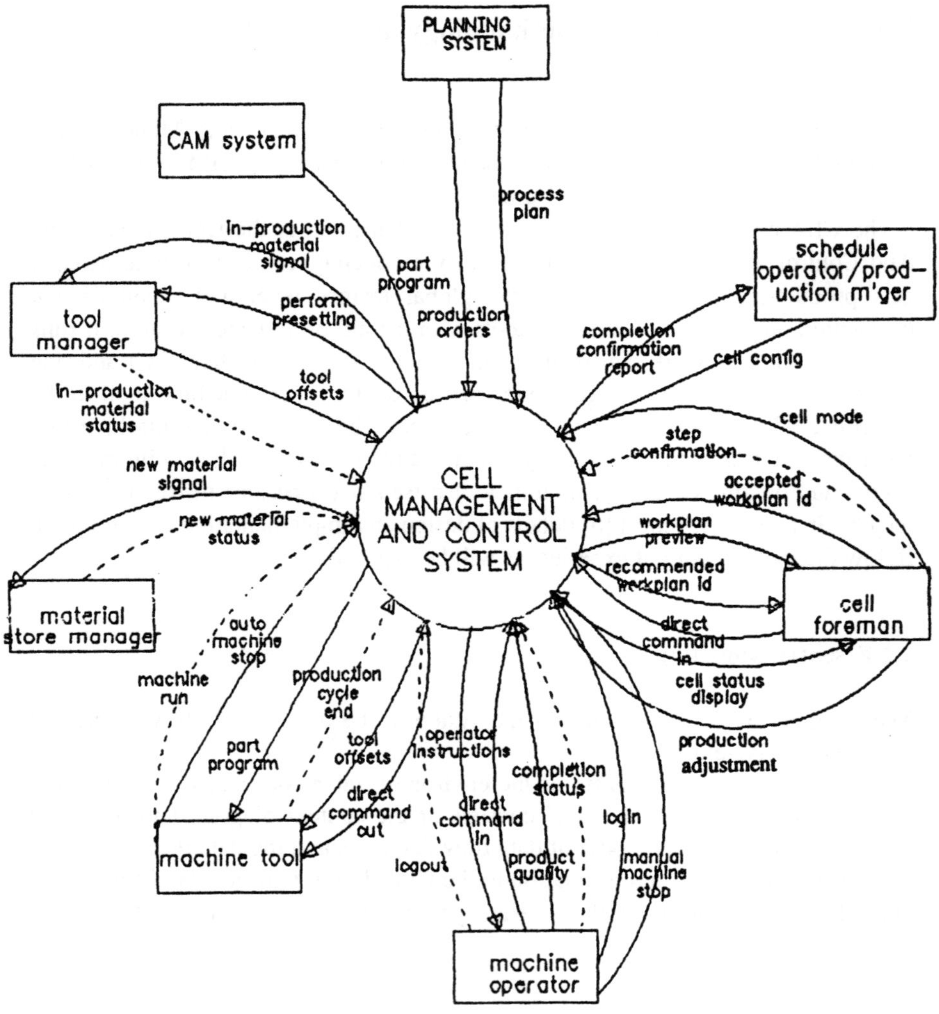

event	response
production planning system sends production orders	store production orders
material is mounted	machine operator may start if other requirements are met
time to display current cell status	update cell status display

Fig. 4 Context diagram and part of the eventlist

Context diagram: The context diagram (fig. 4) identifies the system boundaries and the interactions between the system to be built and its environment. The environment is represented by the terminators cell foreman, machine operator, machine-tool, material store manager, tool manager, CAM system, planning system and schedule operator/ production manager.

The terminators are connected to the CMACS by means of data- and control flows.

The flows of the context diagram do not show implementation details. For example the flow *production adjustment* from the cell foreman to the CMACS will be implemented as a two-way dialogue. The extra flows needed to model this conversation are not shown.

Event list: The eventlist shows the events that occur at certain times in the systems environment and to which the system must respond with some planned reaction. There are three types of events: external, indirect and temporal.

An external event is generated by the environment, possibly at specific moments in time. The embedded system should respond on it. An indirect event is generated by the embedded system when for instance a continuous dataflow value exceeds a threshold value. The system itself responds to that event by starting or stopping certain functions. A temporal event is generated by the embedded system at a predefined point in time, for example to generate a periodic report.

Some of the events are represented by control flows entering the system, others are generated in the system. There is no one to one relationship between the flows of the context diagram and the events in the event list.

The eventlist (fig. 4) was constructed partly based on the *incoming* flows in the context diagram described above (like: 'Production planning system sends production orders'), partly on knowledge of the production cycle described in a problem statement document (like: 'Material is mounted') and partly on other events that came up by brainstorming about the system behaviour (like: 'Time to display current cell status').

A response-part was added to the event list to define the systems response. Defining the eventlist left us with some feelings of uncertainty.

Datflowdiagram-0: In the level zero dataflowdiagram of the behavioural model (fig. 5) the responses of the system to the events is specified. First a preliminary one-level data flow diagram is drawn to collect all system responses to the events. This should make clear how the system behaves when it consumes input flows from the environment and produces output flows. It should then be transformed into a hierarchical one by grouping together transformations and stores.

In CMACS five functional parts were recognized and a transformation was created for each of those. The model was further constructed in a hierarchical way. Details can be found in [6]. Many discussions were needed before the partners could agree on the model. There were no textual explanations to the data flow

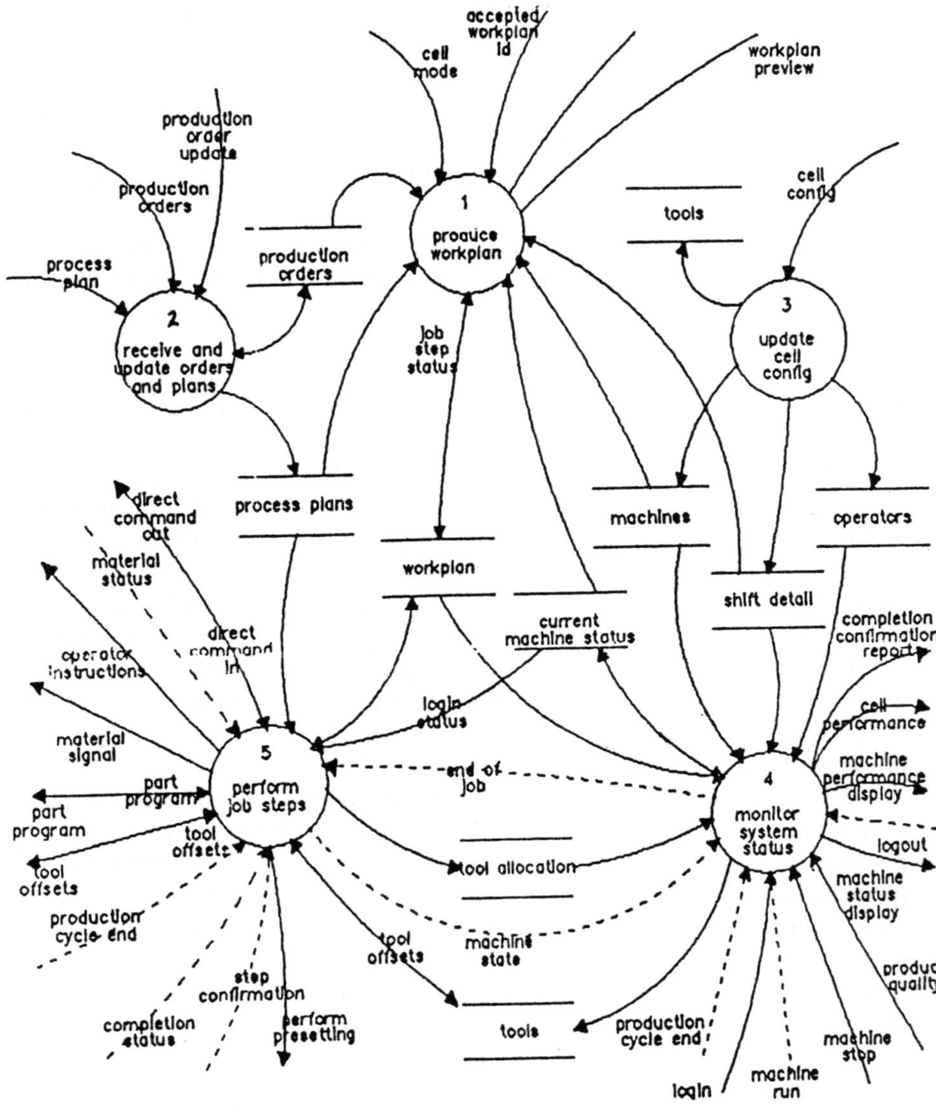


```
cell = { machine }
machine = @machineID + machine name + machine type +
          machine availability
machine state = [ idle | setting up | loading material |
                  machining | unloading part | cleaning
                  up ]
```

Fig. 5 Behavioural model and part of the datadictionary

diagrams, and the decisions taken during the modeling process were not documented. Much time would have been saved if we had done so.

The DFD's have been of great importance for the group of designers, because they were used as a common blackboard.

Datadictionary: For the interpretation of data flow diagrams the data dictionary (fig. 5) is of extreme importance. It makes the specification complete and consistent and it helps in understanding each other.

Discussion: Essential modeling was carried out by one partner in the ESPRIT project. In the meetings the model was walked through, i.e. the builder presented it to the other partners and explained his decisions. A lot of time would have been saved, if the DFD's of the model had been extended with these explanations. At some stage the partners felt that the model was close enough to describe the top level solution to the problem. However there always remained a feeling of uncertainty.

When the partners finally agreed upon the context diagram, the event list, the functional parts in the top level of the behavioural model and the data dictionary, details were worked out in a hierarchical way. From all the problems we encountered during this exercise three are worth to be mentioned in particular.

First we did not use activation signals in the models. This means that we assumed all transformations to be active all the time. As a result the (parallel) sequence of activity of the transformations then depend on the occurrences of the time-discrete data flows. It appeared hard to understand these sequences. A second problem arose when a control transformation had to send commands over a control flow to a data transformation. We allowed a control flow to have data indicating the command. Finally it was found hard to detect the level of detail at which to stop the top down development. One often goes down far beyond the point where simple programming would do the job. At some point the lower level diagram could better be expressed in a transformation specification directly.

2.2 Implementation modeling

The essential model is based on ideal implementation technology: no time delays and infinite processing and storage capacity. In the implementation stage one has to face reality. Implementation modeling consists of three stages: the processor-, task- and hierarchical modeling stage.

For a simplified version of the CMACS these were tried out precisely according to Ward and Mellor. After that experience we decided to leave Ward and Mellor after the processor allocation stage. The task allocation stage and the hierarchical stage were replaced by other design methods.

Processor modeling: In the processor modeling stage, the essential model must be mapped on one or more processors. The decision to allocate transformations and stores to a particular processor can be based on several criteria. It may be qualitative: for instance when a processor has an extended arithmetic instruction set or a suitable interface to a certain device, or quantitative: for instance when more processors are needed to meet time and storage requirements. Often separate processors are used for data acquisition, process control, performing calculations, operator interfacing and long-term datastorage. Smart interfacing instruments should also be modeled as separate processors. Also the environment might influence the allocation choice, for example when a system is to be implemented on geographically separated processors.

The last one holds in our case: one processor is used for the cell control computer and between every machine-tool controller and the cell control computer a Machine Interface Unit (MIU) is allocated. Each MIU is implemented on a separate PC.

When a transformation is assigned to a processor in principle all its internal lower level transformations, which together build a tree, will also be allocated there. In cases where this does not seem appropriate the tree should be split up into separate sub-trees and divided over the different processors.

A small number of transformations of the behavioural model had to be divided over the cell control computer and the MIU's. Transformations to be located on the MIU's are duplicated, so that they can serve as a data buffer between the machine-tool and the cell control computer in case the latter is temporary off line.

The three transformations (1,2 and 3) at the top of the behavioural model (fig 5) were allocated to the cell controller and the other two were split into parts to be allocated to the machine interface units and the cell controller, based on real-time requirements of the CMACS.

Task modeling: In this stage the transformations of each processor are grouped into sequential tasks, taking into account possibilities of concurrency. For this allocation no criterions are given by Ward and Mellor. We found some by experience:

Degree of parallelism: Most important is the desired degree of parallelism which may vary from high to low. It depends more on the designers taste than on the particular cases that are investigated.

Sequence of activation: When a number of transformations is always executed in the same sequence, they can be allocated in one task.

Store consistency: Transformations updating and reading the same store can be allocated in one (sequential) task to avoid concurrent access to the store. A special case is a group of transformations that display information on one screen.

Low level considerations: When it is known that a transformation will be implemented by a (hardware) process (for instance a device driver) that runs in parallel with the rest of the system the transformation can be allocated to a separate task.

All allocation decisions should be (but were not) documented.

The portion of the essential model allocated to the cell control computer and the MIU's was subdivided into sequential tasks. The transformations allocated in one task can only run sequential. It was therefore important to decide which transformations must or may run concurrently.

Transformations acting on time-continuous data flows can be collected in tasks that are triggered periodically. The trigger frequency should be high enough to sample the incoming data flow appropriately. Transformations acting on time-discrete data flows are allocated to tasks that are scheduled to run until they have to wait for input.

Transformation 4 *monitor system status* is an example of such a periodically triggered task. It displays the state of the machine-tool on a screen. This state changes once in a while, and when the task is triggered the screen will be refreshed. If the sample frequency is high enough this seemed to be acceptable.

Hierarchical modeling: In this final implementation stage the structure of each task is modeled in a structure chart: a hierarchy of modules (procedures and functions). Mostly a control transformation appears at the top of the hierarchy and controls the activation of the lower level data transformation procedures. How this is done is not visible in the structure chart however. Only the relations between procedures and not the sequences of their calls are shown. This control information is given by the state-transition diagram of the control transformation at the top of the structure chart. Coding large and complex state transition diagrams may encourage the old-fashioned goto-programming style, which of course should be avoided.

There are different ways of translating the state transition diagram of a control transformation into a program part calling other modules. A straight forward way is a module with a CASE statement containing an entry for every state in the STD that checks for the conditions that may occur in this state and determines the new state. This usually results in a large module that is hard to prove correct. It is also hard to change such a module in case the STD is changed because the STD is coded and not put in an easily changeable data structure. A more systematic way of translation makes use of a state transition table. Every combination of a condition and a state has a unique entry in this table which specifies the new state and the action to be performed in this state.

To keep the system easy to maintain it is necessary during hierarchical modeling to keep the control logic separated from the data transformation. The other modules contain the data transformation level. When the STD is changed this only affects the control logic modules, and when the data transformation is changed only the data transformation parts are affected.

3 Evaluation of the Ward and Mellor Methodology

During the modeling process we got a good impression of the applicability of Ward and Mellor. In this chapter we will relate the Ward and Mellor methodology to a combination of Dijkstra's structured programming methodology [2] and an operating system design method based on communicating sequential processes [4], used to design and implement real-time systems. Also some suggestions to extend the Ward and Mellor methodology and improve its use will be given.

The ultimate goal of every analysis, design and implementation activity is to build a bridge between the problem statement and a working product. Traversing in the above mentioned direction is called *top down* design. One starts with a global description of the solution to the problem and by descending towards implementation one fills in more detail. This is called *stepwise refinement*. At last the detailed solution should fit with the available environment (mostly hardware) in which the product has to be embedded. It is also possible to start with the hardware environment in order to ensue fitting and to abstract from its details by building so called *abstract datatypes* on top of it. This approach is called *bottom up* design. In general the best method is to alternate both directions and to build the bridge starting from both sides. These are well known methods. The *process design method* is supported by languages like CSP [4], Occam [5], Modula [12] and ADA, and is best explained by an example.

Consider the Machine Interface Unit (MIU). It is surrounded by an environment consisting of the Cell Controller (CC), the operators console and the machine-tool logic. As this environment asks for parallel service, the process design method fills the empty MIU circle with communicating processes instead of dataflows and -transformations, according to Ward and Mellor. First interface processes are defined at the border. They abstract from the environmental details and differences, and communicate with the environment. They offer a standard interface to internal application processes which communicate about the systems global state and perform the functional tasks of the machine interface unit. These processes run on virtual processors. This leaves the decision open to implement the MIU on a single processor, that simulates virtual processors, or on a multi-processor or transputer network, with real parallelism in order to meet real-time constraints.

3.1 Ward and Mellor related to other design methods

So far four design methods have been mentioned: the top-down, bottom-up, process design and Ward and Mellor. What is the influence of each of these in the development of a system, and which can best be used where?

A system, together with a (network of) machine(s) on which it runs, can be seen as an abstract machine especially designed to solve a given problem. According to

Koster [7] such an abstract machine is recursively composed of systems running on abstract machines. At the lowest level the abstract machine is the real one. Each (abstract) machine offers an interface (its "instruction set") to the system "running" directly on it, and together with this system it builds a new one with a more abstract interface. So an abstract machine has layers of abstraction connected by language interfaces. In an abstract machine the vertical direction is the direction of *abstraction*. In each layer the language can be extended by adding parts which introduce new instructions. The horizontal direction is the direction of *extension*.

Apart from the horizontal division in abstract layers the abstract machine we want to build can also be divided vertically in functional parts. This is called *modularity*. It is orthogonal to the above *hierarchy*: each module has a hierarchy of units together performing a certain function, and each hierarchical layer is divided in functional units which all have the same level of abstraction.

3.2 Characteristics of the four methods

The method of Ward and Mellor starts at the problem statement side (the top level) with a precise description of the environment and its events, to which the system must respond. The result (the context diagram and event list) is like a giant empty circle which is the systems border. In this circle data transformations and -stores are placed, beginning at the border, to describe the reactions to the events. This proceeds until the whole behaviour is visible inside the circle (the behavioural model) as a directed graph of separate functional units. It is a *one level functional decomposition in dataflows*, in which the order of execution is not visible. Only for the sake of overview this behavioural model may be presented in a hierarchical way. We may conclude that the analysis stage of Ward and Mellor is used in the first place for functional decomposition (vertical cuts in the abstract machine), and depending on the designer this decomposition may be carried out at one level or (as we did) hierarchically.

The top-down (stepwise refinement) method is based on *functional decomposition in program parts and anti-abstraction*. Going down the layers of abstraction is the implementation of an abstract action by a sequence of less abstract actions. These actions may but need not be separate functions. Important is the visibility of the control structure. Functional refinement in a hierarchy of data flow diagrams can be compared with stepwise refinement. The main difference is that the top-down method does not give strict rules to follow, nor (graphical) tools to use. (The reason we preferred the hierarchical approach could have been our custom to use stepwise refinement).

The bottom-up method is based on the construction of abstract machines and abstract data types. It is used to hide machine dependencies and pragmatic representation decisions. Extension of the functionality of a layer is an important activity, so *functional decomposition* is as important as *abstraction*.

The process design method starts at the systems border where the environment demanding parallel service is abstracted by parallel processes communicating with each other about the systems global state. It is based on *functional decomposition in parallel processes*. Processes may also have hierarchical family relationships. Processes and their communication look alike Ward and Mellor's data transformations and -flows. They are quite different however. The process design method allocates functionality directly to parallel processes. These are supposed to run on *virtual* processors, which are mapped on real processors in lower abstraction layers. Ward and Mellor's methodology leaves the allocation of transformations to the implementation stage. There they are split and recombined into different transformations and then allocated by the first two implementation stages: *first* to (real) processors and *then* to (virtual) processes (tasks).

3.3 Three phases in system design

To design a system Koster describes a three-phase model. Generally one starts at the top of the abstraction hierarchy, inventing and refining useful abstractions. If a large problem has to be solved it is unlikely that all decisions taken will finally fit on the available lowest level machine(s). When taking the right decisions is becoming difficult, one should carefully define the interface just reached. At this stage the problem has been analysed and replaced by a "set of instructions" in terms of which the problem is solved. It is called the *application phase*.

One can now start at the bottom, building layers on top of each other (nested abstract machines) keeping the above mentioned necessary 'set of instructions' in mind. Extension however may lead to interfaces which become too rich. Then again it is time to stop and define the interface just reached. At this stage the available machine has been made more friendly by abstraction from machine dependencies and pragmatic representation decisions. As a result the system has now become more portable. It is called the *representation phase*.

From this point on the gap between the two interfaces can easily be bridged most of the time by the process design method described below, supported by an alternation of top-town and bottom-up design. This phase is called the *function phase*.

Which of the methods can best be used in each of the above phases? From our experience with operating system design and the use of Ward and Mellor in the Esprit project we may conclude that each of the three phases seems to have its favorite design method(s). These are:

In the *application phase* Ward and Mellors analysis and the top-down method are the only candidates. There is a great resemblance between the two. The main difference is that the latter does not give strict rules to follow, nor tools to use. The idea behind both is the same: walking down a DFD-tree is the same as following stepwise refinement. (In the commercial plugger experiment [11] both methods were used to design the same product. There were no significant differences in the structure of the two results). The difference between both methods may become more important when the problem is larger. For relatively small problems also the difference between the essential- and implementation models is small, especially when one processor is used. The border between the two stages becomes vague and the first is gradually converted into the second by working out details. Ward and Mellor can be preferred because of its graphical tools, but only if used in a top-down manner, by zooming-in. These methods never replace human creativity.

In the *function phase* the process design method is better than Ward and Mellor's. The problem is analyzed far enough to be functionally decomposed into parallel processes with a communication structure, instead of datatransformations that still have to be reorganized before being mapped on processors and tasks. Allocation of processes remains flexible because they are mapped on virtual processors. Secondly an alternation of the top-down and bottom-up methods can be used.

In the *representation phase* the bottom-up and process design methods are candidates in that order. Ward and Mellor does not give any bottom-up facilities. A strict top-down design strategy may lead to mismatches at the lowest implementation level.

3.4 Evaluation of the methodology

During the modeling process we got a good impression of the applicability of Ward and Mellor. In this chapter we will present some suggestions to extend the Ward and Mellor methodology and to improve its use.

Remarks on the methodology: The use of Ward and Mellor (or any comparable design discipline - as long as it is used as a *discipline*) in a project serves as a communication vehicle. This becomes more important the larger the project is and also when the project members stem from diverse disciplines (as is the case in Esprit projects). For relatively small projects however, the method may give to much overhead in time and paperwork.

Using Ward and Mellor ensures the availability of good documentation material for later maintenance, provided all chosen and not chosen alternatives are explained in a text accompanying the models.

Using Ward and Mellor gives no assurance that the solution found is the right one, nor can it be proved. The same holds for the implementation: Ward and Mellor does not support any form of testing. The implementor should be aware of that and provide for testability.

During the design of a system the model will be updated frequently. If a good overall view of the development of a system is required, it is necessary to store dated versions of the model, each time it is updated.

Remarks on the data flow diagrams: One must be careful not just to concentrate on the Data flow diagrams, because they do not contain all the information. The data dictionary and the transformation specifications contain information that is necessary to understand the data flow diagrams. Also their choice should be documented. In particular activity and timing are not shown explicitly. One transformation may activate another by activation signals but this does not say anything about when these signals will be given. That information is contained in the transformation specifications.

Another problem related to DFD's is to detect when to stop zooming in on lower level DFD's. At some point the lower level DFD could better be expressed in a transformation specification directly. It may be hard to find that point.

The use of time-continuous data flows and data transformations is restricted. A data flow is only continuous when it is connected to a device that produces or requires an analogue signal. A transformation is only continuous when it has an incoming or outgoing continuous data flow.

State transition diagrams, especially when they are large and complex encourage the old-fashioned GOTO programming style. This of course should be avoided.

With respect to multiple instances of transformations one has to define what is meant by their input and output flows. The flows can be parallel multiple instances or one input flow can spread out to reach all transformations and / or all output flows can join into one output flow. A store connected to a multiple instance flow has also multiple instances.

Remarks on the event list: It is questionable to what use the event list is to the reader of an essential model when he did not participate in the brainstorm process that creates it. After reading a context diagram and an event list one tries to find a one to one relationship between the flows in the context diagram and the events stated in the event list and finds out that there are more events than data flows. Then one starts to question where the other events come from, and finally one starts to invent new ones.

It is difficult to check the completeness of an event list, and Ward and Mellor admits that there is no analytical way of building it.

We might conclude that although the creation of an event list is essential in the analysis stage of the design, it is rather confusing to the reader of an essential model. But in case enough information on the definition of events is available, the event list gives an impression of the co-operative behaviour of the system and its environment.

Remarks on the data dictionary: While building the essential model, a lot of time and effort has to be put in the realization of the Data dictionary (DD). This should be done with quite a lot of details. The resulting DD must be consistent with the data flow diagrams, and complete.

Completeness also incorporates the definition of a key-element of a structure through which particular parts can be searched. If the DD is detailed far enough and consistent, it will be quite easy to design the data structures and databases during implementation modeling.

Remarks on processor modeling: It is questionable whether the allocation of parts of the essential model to more processors because of lack of processing power is important at this stage of implementation modeling. In other design disciplines it is quite usual to assume enough virtual processors on which the whole system can run and then simply go on with the task modeling. If at a later stage of implementation the system appears to be too large for one processor, portions of it can still be allocated to different real processors.

Remarks on task modeling: The Ward and Mellor methodology does not give much support with respect to task modeling. Criterions for allocating parts of the processor model to tasks depend heavily on the builder of the model. Some criterions we found were mentioned above.

Remarks on hierarchical modeling: Hierarchical modeling is only useful when the system is implemented making use of a block structured language. The hierarchy of a structure chart typically reflects the structure of such languages. At this point the structured programming methodology mentioned earlier is much more appropriate.

3.5 Extensions to the methodology

A response part added to the event list to specify shortly the proper response of the system is of great help when building the behavioural model.

To the DFD's and STD's textual explanations should be added, to elucidate the decisions taken during the modeling process.

Also during task modeling some explanation of the decisions taken is desirable. Therefore allocation decisions and task descriptions are added. The first explain the criterions used and the decisions taken during essential modeling. The latter describe the behaviour of a task once it is activated.

To processor modeling and hierarchical modeling similar explanations should be added with respect to allocating portions of the essential model to processors and the translation of networks of transformations into hierarchies of modules.

For some unknown reason Ward and Mellor's method forbids a data transformation to have data- and control flows at the same time. We found that this rule should not be followed. Both are needed for instance when a data transformation has to tell by means of a control flow whether its output data is correct or not.

With respect to error situations we found that in a state transition diagram it often happened that we had to draw a state transition from all states to one error state. To avoid numerous arrows one general state called 'all states' was drawn with only one transition to the error state. This general state is assumed to be overlaid with all states.

Another case where too many arrows in a DFD should be avoided is when many transformations output data to one receiver transformation (for instance screen display). In those cases we only draw a short output flow at each of the senders and one input flow at the receiver side, labeled with the identifications of the sending transformations.

Another problem arose when a control transformations had to send commands over a control flow to a data transformation. A control flow can have no information, not even a command indication. To avoid splitting up the data transformation into one for every command, and one trigger signal to each of them, we allowed trigger signals to indicate the data transformation part to be triggered.

4 Conclusions

The Ward and Mellor methodology was applied to the analysis and implementation stages of the design of the CMACS-part of a CIM system. In the Esprit project we had to deal with different disciplines as an expert system for scheduling, a local area network, an open system interconnection protocol, and a distributed real-time operating system to control the machine tools. We had to go through a real application phase for which Ward and Mellor's analysis stage was useful .

Related to Kosters three-phase-model for programming large systems the Ward and Mellor methodology is almost only useful in the first application phase as a tool to specify and document large real-time systems. It bridges the gap between designer, implementor and user. Used as a *discipline* in a project like ESPRIT with partners stemming from different disciplines it also serves as a communication vehicle and as a tool to plan the project. It ensures the availability of good documentation material for later maintenance, provided all alternatives are explained in a text accompanying the models. However using Ward and Mellor gives no assurance that the solution found is the right one, nor can it be proved.

In the implementation stage Ward and Mellor is, if at all, only useful for communication and documentation purposes. The use of processor and task stages in the representation phase are questionable, and the module stage is of little use in the function phase because it is less more than a one level relation diagram showing which procedures call each other, not showing their calling sequences. For the implementation almost no criterions or heuristics are given by Ward and Mellor's method, nor does it support any form of testing.

The processor- and task modeling stages overlap with the design methods based on parallel processes. These assume parallel processes to be allocated to (virtual) processors, which is to be preferred in this stage of implementation. At the lowest implementation level some mechanism (a real-time operating system kernel with process scheduling) is then assumed to implement these virtual processors on a number of real ones, to meet the real-time requirements.

When the system to be designed is relatively small Ward and Mellor will give too much overhead. The other designmethods (process design and an alternation of top-down and bottom-up) are then more appropriate.

5 References

[1] T.DeMacro, Structured Analysis and System Specification.
[2] E.W.Dijkstra, A discipline of programming.
[3] C.Gane and T.Sarson, Structured System Analysis, Tools & Techniques.
[4] C.A.R.Hoare, Communicating Sequential Processes, CACM, vol 21, 8.
[5] Inmos Ltd., OCCAM programming.
[6] J.Johnson, the CMACS Behavioural Model, Esprit internal publication.
[7] C.A.H. Koster, Modules and hierarchy, University Nijmegen.
[8] J.Peterson and A.Silberschatz, Operating System Concepts.
[9] M. Page-Jones, Practical Guide to Structured System Design.
[10] P.T.Ward and S.J.Mellor, Structured Development for Real Time Systems.
[11] A.J.De Weerd, Evaluation of Real-time design and specification methods applied to a commercial plugger, doctoral thesis, Twente University.
[12] N.Wirth, Programming in Modula-2.
[13] E.Yourdon, L.Constantine, Structured Design, Fundamentals of Computer program and System design.
[14] E. Yourdon, Modern Structured Analysis.

10 Monitoring and Automatic Supervision in Manufacturing Systems

M. Szafarczyk

1. Introduction

The market for manufactured goods is more and more world-wide and competition on it is becoming exceptionally fierce. In modern manufacturing, production obeys the dictates of consumers' demand. This, in turn, leads to small batches, short lead times and reliable dates of delivery. High reliability of manufacturing systems, together with the proper organization of production, is a very important factor in gaining a good position on the market. For producers, lower reliability may mean losing their position on thr market.

Generally speaking, there are two ways of achieving high reliability of manufacturing systems. One is: to learn very well the system and its working conditions, use the safe parameters of operation and in planned time intervals perform the proper maintenance and preventive repairs. Such attitude is possible in mass production of relatively simple products, when the same operations are repeated many times, material or prefabricated parts to be machined are uniform as well as the tools used for machining. After a long period of repetitive operations it is possible to adjust the safe parameters, and knowing working conditions of machine tools and manufacturing equipment, to plan properly preventive repairs. But because of changing market demand the scope of mass production is diminishing and such aproach is rarely possible..

In the changeable conditions of flexible manufacturing, the only way of increasing reliability is to use sensors which measure the chosen features of manufacturing systems and manufacturing processes. The results of the measurements allow assesing actual disturbances and detecting a failure, or even better detecting growing possibility of a failure. This leads to monitoring or automatic supervision of manufacturing systems. Monitoring alone is used in manufacturing with a steady presence of operators - it signals to the operators the occurrence of disturbances which require some action. *Automatic Supervision in Manufacturing* (ASM) is very useful in manually operated manufacturing systems and is necessary in automatic ones. Apart from detecting disturbances (monitoring) an ASM system acts in such a way that the influences of disturbances on the manufacturing system are eliminated or at least diminished.

The term *disturbance* has a very broad meaning in automatic supervision. It is connected with quality (quality of a product and quality of manufacturing process) and with a model of manufacturing system used during process planning. The manufacturing production is in every case preceded by the process planning during which the sequence of operations and their parameters are established with the aim of achieving high quality. The process planning is based on a model of the manufacturing system and on models of manufacturing operations to be used. The models are based on the theoretical knowledge and on previous experience. Between models considered during the process planning and the actual conditions of manufacturing there are normally differences. Any difference which affects quality is treated in automatic supervision as a disturbance.

In modern „intelligent" manufacturing systems many, sometimes quite sophisticated, ASM systems are used [7]. Due to new developments in sensors' technique and signal processing systems as well as deeper knowledge of the manufacturing processes and machines to be supervised, there are growing possibilities of using reliable, effective and low-cost ASM systems. Moreover, there is a growing need for automatic supervision. Apart from reliability there are for other reasons the use of ASM in modern manufacturing systems.

Table 1 Reasons for the use of ASM systems

1	Full automation	ASM fights against influence of disturbances allowing manufacture without human supervision
2	Integration of manufacturing processes	Integration makes interruptions in production more costly - ASM enables to eliminate some interruptions and to shorten the rest of them
3	Accuracy of workpieces	Growing requirements concerning accuracy of workpieces in many cases can not be fulfilled without ASM
4	Quality standards	ISO and EU quality standards require registration of machining conditions for every workpiece - ASM allows automatic registration of the conditions
5	Optimization of manufacturing processes	Because of disturbances on-line optimization of machining processes is needed
6	Different products and volumes of production	There is a higher probability of disturbances when production is changeable
7	Intensification of manufacturing processes	Development in tools and machine tools leads to movements at high speed - the reaction of an operator is not quick enough
8	New manufacturing processes	Many manufacturing processes (EDM, ECM...) make direct observation of machining conditions impossible

2. Classification of Disturbances

Disturbances play a key role in the monitoring and automatic supervision. A lot of different disturbances may occur during manufacturing processes. Some kind of the classification of disturbances is needed. Different classifications of disturbances may be used (outside - inside disturbances, disturbances connected with the different elements of the manufacturing system...). The most important seems to be the classification based on the influences which disturbances have on the manufacturing system.

According to this classification [5], Fig.1, disturbances may lead to three final consequences:

1) Disturbances cause a full stop of production because of a break-down of the manufacturing equipment.
2) The manufacturing system may work but continuation of work is useless because products must be rejected for quality reasons
3) It is possible to manufacture good products but the production process is not optimal from the economic or organizational point of view.

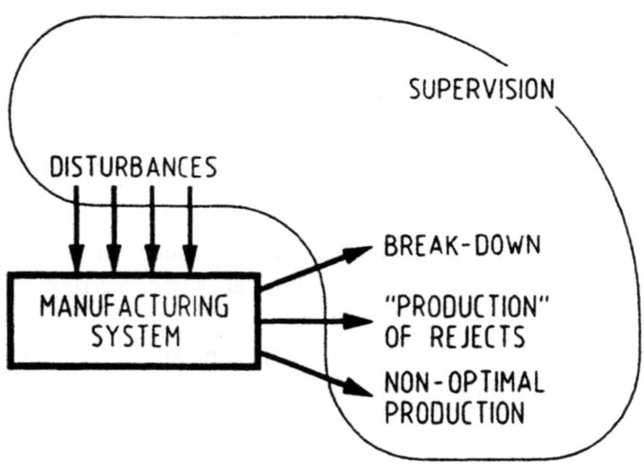

Fig.1 Three classes of consequences of disturbances during manufacturing

First two groups of disturbances deal with „good-bad" assessment of quality. The quality index q may be treated as a binary value. It is either $q=1$ and the manufacturing production may be continued or $q=0$ when production should be stopped and something must be done before its continuation.

The third group of disturbances has a different character. Assessment of economic features (production costs, profit...) or organization features (output, load of the machine tools...) are expressed by multi-valued quality indices. There is a need for optimization - the value of quality index should be as big as possible but optimization is not necessary for the continuation of production.

3. Sensors and Methods of Monitoring

Because a lot of different disturbances may be expected, there is a need for many sensors for monitoring of manufacturing systems. The example of a set of sensors needed for the purpose is shown in Fig.2 [1].

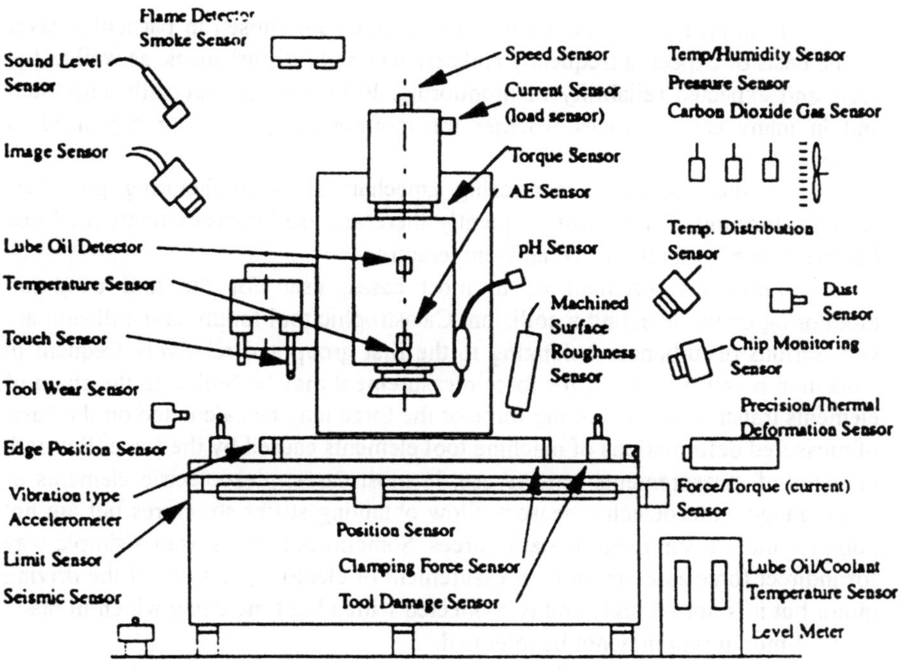

Fig.2 Extensive set of sensors for machine tool monitoring [1]

In practice much fewer sensors are used both because of costs and because of very high requirements which must be met by them in workshop conditions. For example, sensors used for machining process monitoring should meet the following requirements [1]:

- measurement as close to the machining point as possible,
- no reduction in the static and dynamic stiffness of the machine tool,
- no restriction of working space and cutting parameters,

- wear- and maintenance-free, easily changed, low costs,
- resistance to dirt, chips and mechanical, electromagnetic and thermal influences,
- function independent of tool or workpiece,
- adequate metrological characteristics,
- reliable signal transmission, e.g. from rotating to fixed machine components.

In many research works carried out on monitoring and automatic supervision in manufacturing, a lot of original sensors, based on different physical phenomena, were tested. Only some of them are used in industrial practice.

In industry, sensors and monitoring units are chosen in particular cases on the basis of expected frequency and seriousness of disturbances, as well as the costs and expected reliability of monitoring. Reliability is especially important and in many cases decides whether the monitoring system is really used in practice.

In the cases of monitoring „mechanical" manufacturing processes (cutting, forming...), the most frequently there are used measurements of: force (torque, power), vibration, AE and temperature.

Force measurement is, in most cases, used for the tool condition monitoring or for detecting a collision. Catastrophic tool failure and collision are very serious disturbances belonging to the first group and relatively frequent in workshop practice. Sensors for force measurement may be built into the chain of elements transferring the cutting force or the force may be calculated on the basis of measured deformations of machine tool elements caused by the force. For both methods of measurements sensors use, in most cases, piezoelectric elements or strain gauges. Piezoelectric sensors allow obtaining stiffer structures but are not good for the measurement of static forces. Sometimes there is used a simple way for indirect force assessment by measurement of electric quantities of the driving motor but it is not accurate and is connected with a big time delay which in many cases of monitoring may not be tolerated.

Accelerometers with piezoelectric sensors are frequently used for measurement of vibration. Vibration measurement may be used for different purposes but, first of all, for monitoring of chatter, which is an important disturbance especially in the case of machining a slender workpiece or machining with a slender tool.

Elastic stress waves which propagate through the mechanical structure, called frequently acoustic emission, AE, are generated during friction, breaks or impacts. In manufacturing, stress waves have high frequency - between 100kHz and 2Mhz. The measurement of AE may be very useful for monitoring purposes, because the signal is generated by the main mechanical phenomena in the manufacturing system. It is especially important during finishing when forces are not big enough to be used for monitoring. For measurement of AE the special piezoelectric sensors are used. The signal processing is difficult because of its high frequency - an AE sensor will generate a signal which needs between 10^6

and 10^7 data points per second during monitoring if one wishes to capture the full dynamic range of the signal [1].

Temperature in different points of machine tools is measured mostly because of thermoelastic deformations of the machine. The deformations are usually the main cause of dimensional errors of workpieces (second group of disturbances). On the basis of the results of temperature measurements (thermocouples or thermistors are used as sensors) and a thermoelastic model of the machine tool, the compensatory displacements of the tool relative to the workpiece may be calculated.

Touch trigger probes are specific sensors used in the NC machine tools against disturbances influencing dimensional accuracy of workpieces (second group). Touch trigger probes are used for many years in coordinate measuring machines, CMM. In an ASM system when a touch trigger probe is located in the working position of a tool in the NC machine tool, automatic measurement of the NC coordinates of the chosen points and calculation of the workpiece dimensions are possible. Improvement of dimensional accuracy is achieved by the insertion of proper corrections to the NC program. When a touch trigger probe is located on the workpiece side of the NC machine tool and can be reached by the tool located in the working position, it may be used for automatic checking of the position of the cutting edge in the coordinate system of the NC machine tool.

In „non-mechanical" manufacturing processes, there may be a need for measurement of quite different quantities. For example, in electro-discharge machining (EDM), monitoring of the chosen parameters of the electric pulses supplied to the working electrodes is very important. The measurement of electric quantities is easy It is much more difficult e.g. to measure the intensity of a laser beam during LBM.

Monitoring manufacturing systems is difficult both because of noisy environment and uncertain signals from the sensors and because of the complicated, uncertain models of manufacturing units or processes to be monitored. Substantial progress has been made in this field over the last decade mostly as a result of developments in the sensors and signal processing technique and because of improved modeling of manufacturing systems. The progress in the technique of measurements led to so called sensor fusion and intelligent sensors.

A new focus of much of the research on monitoring is on *sensor fusion* - integration of information from several sensors, measuring different features, for a better and more robust characterization of the process or the machine. In many cases it increases reliability of monitoring, which is crucial for practical application.

Intelligent sensor is practically a „conventional" sensor integrated with elements of signal processing and feature extraction, thanks to the developments in micro-electronic technology. Intelligent sensors are able to do some or all of the following things: self-calibration, self-diagnosing, signal processing, decision making, fusion ability, learning capability. Signals from the sensors are processed and the needed information is sent to the control system in a digital form which is more reliable for transmission in workshop environment than an analogue one. A lot of research work has been done in this field.

An intelligent sensor module has been developed [8] at the University of Hannover. It may be connected to two different analogue sensors and processes signals from the sensors in accordance with the requirements described by the software. Its processing abilities are quite big, because the module contains a digital signal processor (DSP) and memory units. Due to the modern micro-electronic technology, the entire circuit of the sensor module can be compressed to the size of a cigarette pack. Thanks to the miniaturization the module may be placed near the sensors, process the signals from them and send properly reduced information in a digital form to a monitoring system through CAN fieldbus.

Signal processing should transform information from the sensor into a form which is the most useful for monitoring purposes. Many different transformations are used e.g. RMS, frequency content, cepstrum, time series, higher order spectrum analysis... The transformation should allow feature extraction - to extract characteristic features which will allow the proper pattern recognition procedure. Decision making, a final act of monitoring, is based on a kind of pattern recognition. It may be either recognition based on similarity to the pattern typical for the proper behavior of a manufacturing system or on similarity to the pattern characteristic for a particular disturbance.

In the first case, the signal for the undisturbed operation under the same conditions should be stored in the memory of the monitoring system. As the conditions during machining are changeable, there is frequently a need for storing in the memory many patterns connected with the particular phases of machining. In Fig.3 [5] such procedure is shown on the example of turning monitored by the measurement of the cutting force. The proper forces for every NC block are recorded together with deviations which may be tolerated. A disturbance is monitored when the actual value of the force measured during turning is outside tolerances defined for this phase of machining.

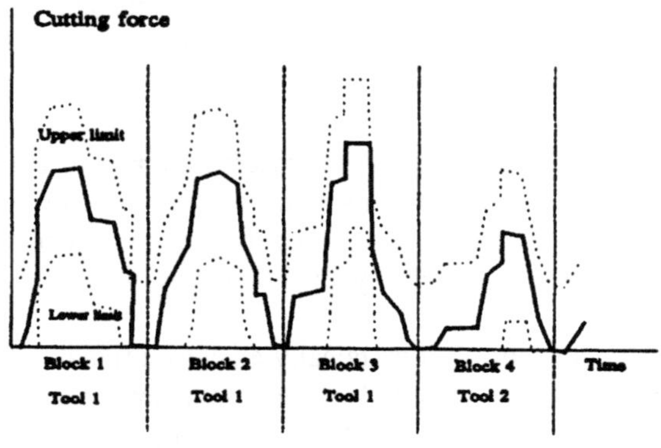

Fig.3 Cutting forces during undisturbed turning in different NC blocks

This method of monitoring may have some disadvantages. If several disturbances may increase (or decrease) the value of the measured quantity, it is impossible to distinguish which of them was the reason for the change in the value. It may be even worse when it is possible that at the same time two disturbances may appear of which one will increase and the second decrease the value of the measured quantity.

The second method of monitoring may be demonstrated on the example of monitoring a tool catastrophic failure by the measurement of cutting forces and AE during turning, Fig.4 [1]. In this case, monitoring is based on the sudden change of both AE and force signals (especially cutting force component in the feed direction F_f), repeated after one revolution of the workpiece. It is a typical pattern in the case of the cutting edge breakage.

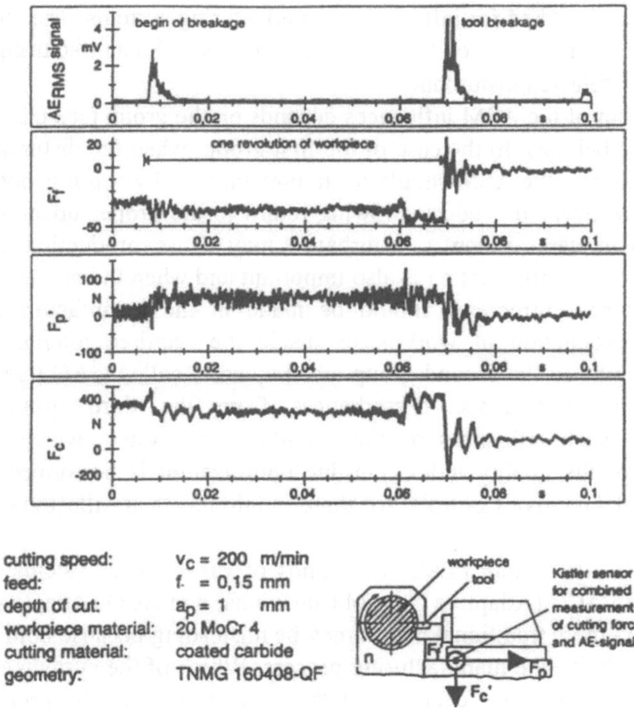

Fig.4 Tool breakage detection by measurement of cutting forces and AE

Formulation of mathematical models of manufacturing processes and machines, which are needed for monitoring purposes, is difficult both because of their complicated nature and because of multiple diversification between particular cases. This is the main reason for many attempts to use „artificial intelligence" (AI): neural networks, fuzzy logic and genetic algorithms in monitoring and in ASM systems. The aim is to build self-learning systems which will improve their performance on the basis of experience - the results of previous actions. In some cases, it is also important to have a possibility of transfering human experience in a „soft" fuzzy form into the automatic system. The big progress has been made in the field, but AI systems are still not used in industrial practice.

4. Methods of Influencing Manufacturing Systems

Monitoring is only part of ASM activity. After detection of a disturbance, the ASM system should influence the operation of the manufacturing system eliminating or at least diminishing losses which may occur. There are a lot of methods used by ASM for influencing manufacturing systems. The choice of a method depends, first of all, on the type of the supervised manufacturing process and the type of detected disturbance.

The aim of the ASM influences depends on the group (see p.2) to which the disturbance belongs. In the case of the first group, when the disturbance may cause a break-down, the ASM should try to prevent it and when it is not possible to perform a recovery procedure allowing return to the proper operation of the manufacturing system. When a disturbance may cause production of rejects (second group) preventive action is also important and when the prevention is not successful, proper corrections should be made in the manufacturing system allowing for production of workpieces inside the required tolerances. ASM systems belonging to the second group are frequently called GAC (Geometrical Adaptive Control) systems. Disturbances from the third group require optimization - such influences on the manufacturing system which lead to a bigger value of the quality index. On-line optimization is performed by TAC (Technological Adaptive Control). All these requirements are illustrated in Fig.5 [5].

In the third group (TAC) two kinds of supervisory systems are used. They are called ACC (Adaptive Control Constrains) and ACO (Adaptive Control Optimization) - these traditional names may be misleading because both types are intended to optimize the manufacturing process. Which of the systems should be used in the particular case depends on the characteristic of the process to be optimized - depends on the character of relationship between the process parameters chosen as controlled variables (parameters changed by the ASM system) and the quality index.

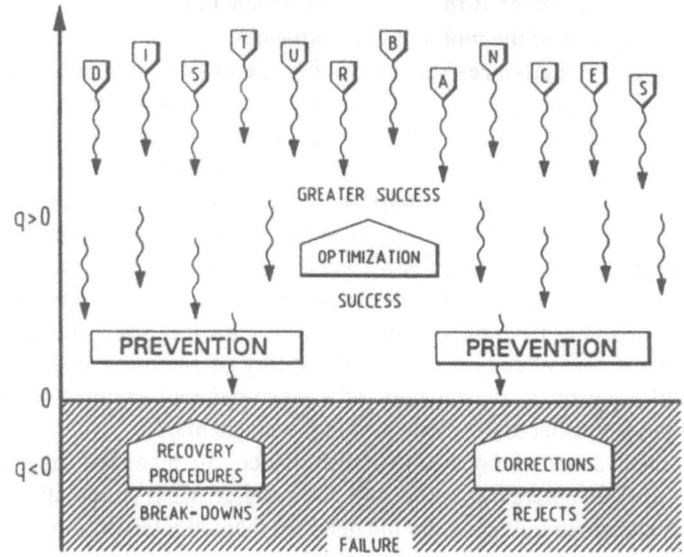

Fig.5 Supervisory functions in manufacturing

Two characteristic relationships in the simple cases, when only one process parameter C is changed in order to maximize the quality index q, are shown in Fig.6. In the case of the monotonic function of q within the range of C permissible by the constraints, between C_{min} and C_{max}, (Fig.6a) the biggest value of q at the limit value C_{max} and ACC system may be used with a simple strategy to increase C up to the constraint. In the case in which there is an extreme of q within the permissible range of C (Fig.6b), an ACC strategy would be useless. There is a need for a much more complex ACO system. The ACO strategy is based on looking for the extreme of q by small changes in the C value and analyzing the influence of these changes on the q value.

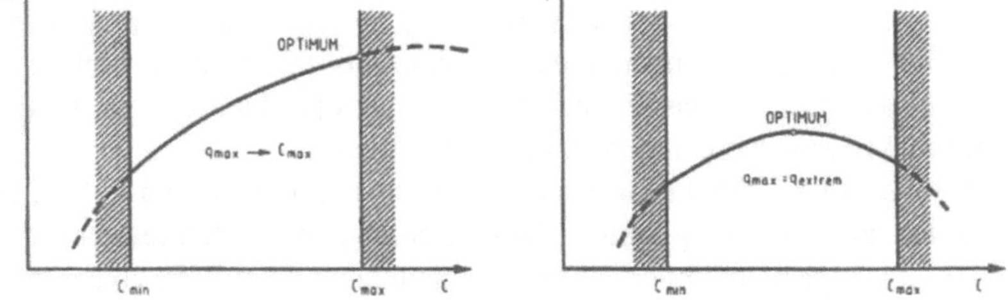

Fig.6 Quality index q as a function of controlled variable C

There are a lot of different actions which may be taken by different ASM systems because of the monitored disturbances.

Frequent disturbances in machining processes are connected with deterioration of the tool condition. Automatic condition improvement of the tool located in the working position of the machine tool is used only in grinding (redressing of a grinding wheel). In the case of a deteriorated condition of a tool in other machining processes the tool should be replaced by a sharp one. In machining centers it is possible to replace automatically the worn-out tool with a sharp one. From the theoretical point of view such operation is a very advanced activity of supervision (automatic replacement of a worn-out element of the supervised system) but from the practical point of view it is quite simple. Because machining centers are already built with the ability of automatic change of tools from the tool magazines, replacement of a worn-out tool requires only proper software in the controller and a clone tool in the magazine.

In most cases of break-down disturbances in manufacturing systems, ASM generates an emergency stop and later either starts automatic recovery procedure or signals to the operators the need for manual intervention.

Improvements of workpiece dimensions by GAC systems are relatively easy in a modern machine tool because of the high flexibility of numerical control. It is a software operation - appropriate changes of the values of coordinates in the NC program.

5. Automatic Supervisory Systems

Automatic supervisory systems used in practice are, so far, built mostly as specialized units for dealing each one with a particular disturbance. They may be divided into the same three groups as disturbances.

5.1 Safety systems dealing with break-downs

Practically in every machine tool there exist simple supervisory systems belonging to this group. It may be, for example, a limit switch which checks whether the machining area is properly closed by the cover. In a typical stand-alone machine tool, there are many feedback signals of this kind confirming the correctness of activity. In transfer lines used for mass production or in flexible manufacturing systems, there may be thousands of them.

A safety signal should cause an emergency stop in case of a failure, but apart from that, the monitoring system should also help to find the reason for the malfunction. It is especially important for big manufacturing systems like transfer lines. The detailed monitoring with precise indication of the element which should be repaired or replaced is becoming even more important because of the current changes in the organization of production, Fig.7 [4].

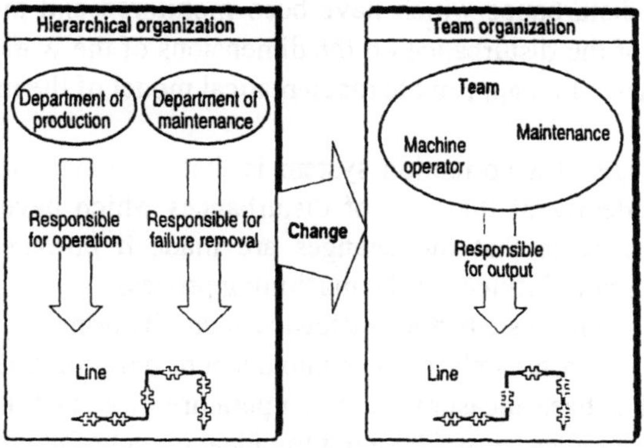

Fig.7 Change within the production organization of transfer lines

In hierarchical organization a separate department is dealing with production on the transfer line and a separate one with the line maintenance and repair in case of a break-down. There are specialists in the department of maintenance which do not need a very detailed diagnosis, because they are especially trained how to find a wrong element and how to deal with it. Such organization is criticized because it is not directed at the main aim - productivity of the line. It causes a relatively long delay in the repair process for administrative reasons.

In a team organization, the operators are responsible for output of the transfer line. High specialization of them is not possible because they have a big number of different tasks. They need well prepared detailed information before taking an action. Monitoring of the manufacturing system is very important and should contain a detailed diagnosing of malfunctions.

5.2 Systems dealing with dimensional inaccuracies, GAC

Demand for higher dimensional accuracy is an important trend in modern machining. In most cases of machining, the demanded accuracy is measured in micrometers and typical machine tools (with support of GAC systems) may be used. In cases of very high precision, accuracy is measured in nanometers and a special design of machine tools is then needed.

Many different GAC systems are used in industry. Two approaches may be used by a GAC system in the fight against disturbances spoiling the precision of machining. It may use a *correction* of dimensional errors on the basis of measurement of the workpiece after machining or it may use a *compensation* of errors caused by a disturbance which have been monitored and measured. The expected influence of the disturbance on the dimensions of the workpiece is then calculated on the basis of an appropiate mathematical model of the manufacturing system.

An advantage of a correction system is that it changes the workpiece dimensions independently of the kind of disturbances which have induced the errors. Its disadvantage is that the changes are made *a posteriori*, after the disturbances have already influenced the machining process.

A compensation system may influence manufacturing *a priori*, before the error can occur. Unfortunately it takes into account only the influence of this disturbance which has been measured (e.g. temperature of a machine tool body). It contains also inaccuracies caused by a not too ideal model of the manufacturing system used for the calculation.

In grinding, it is possible to measure dimension of a workpiece (e.g. diameter of a shaft) during machining, just after the dimension appears as a result of material removal process. Correction based on this measurement may be very effective and is inbuilt in practically every controller of a cylindrical grinding machine. Measurement of this kind is impossible in other methods of machining. In a typical case of turning or milling on an NC machine tool, a correction system uses a trigger probe located in the working position of a tool for the measurement of workpiece dimensions after machining has been interrupted. Correction of dimensional errors may be then carried out through the NC controller.

For automatic compensation of inaccurate location of the tool cutting edges in the NC coordinate system, the automatic measurements are usually made by a touch trigger probe located near the workpiece. This method of compensation is reliable and relatively simple because the model for the calculation of needed changes of coordinates in the NC program is purely geometrical.

Thermoelastic deformations of the machine tool structure are the main source of dimensional inaccuracies in machining [9]. In a typical machine tool, there are a lot of internal heat sources: motors, bearings, ball screws... They generate heat with different intensity during operation of the machine tool creating a complicated and changeable temperature distribution. There is also an influence of external heat sources: solar radiation, heaters... Thermal expansion of elements influences geometry of the machine tool and may cause serious inaccuracies in machining.

A lot of research work has been done in the field of compensation of thermoelastic deformations of machine tool structures. Some compensation systems are used in industrial practice. Temperature measurement of the machine tool body is relatively easy and in most cases it is enough to measure temperature in several points of the machine. Mathematical models describing deformations on the basis of temperature distribution are quite complicated. Because of their complexity thermoelastic models are frequently described by means of neural networks or/and fuzzy logic.

Even with a careful design of the machine tool (minimizing intensity of heat sources and their location in less sensitive places of the machine tool body) and the use of thermoelastic compensation systems, deformations are still too big for cases when high accuracy is required. In nanotechnology there is a need for stabilization of temperature. It is not only stabilization of room temperature (elimination of external heat sources) but also stabilization of the machine tool temperature. The circulation of oil with controlled temperature or even oil shower which covers the machine structure with laminar flow of oil is frequently used for this purpose. In some cases, many local stabilizing systems are used. In a grinding machine shown in Fig.8 [2], five stabilizing sub-systems are used. They are keeping temperature at 20^{0}C within +/- 0.01^{0}C of:

1 - Workhead hydrostatic bearing oil.
2 - Workhead motor coolant.
3 - Oil shower system over the machine structure.
4 - Linear and rotary hydrostatic bearing oil.
5 - Grinding coolant.

The most demanding has proven to be the workhead hydrostatic bearing oil system, but even in this case, the required temperature can be met within 6 minutes of the spindle start up for 750 r.p.m. and after 12 minutes for 1500 r.p.m.

Even with such demanding temperature control a compensation for thermal drift of the workhead spindle has been used. Two inductive probes are installed, one at each side of a diamond disc mounted on the spindle faceplate. The axial drift is stabilized at around 50 nanometers.

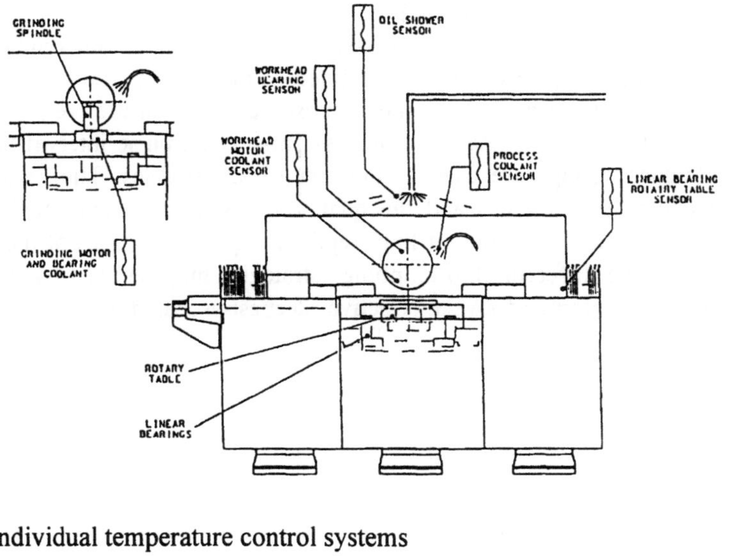

Fig.8 Individual temperature control systems

5.3 Systems dealing with on-line process optimization, TAC

In process optimization the most common approach is to maximize the value of material removal rate, which is treated as a process quality index. In many cases it leads to an ACC system.

Some years ago, ACC was very popular for optimization of rough turning. When surface roughness is not important, the typical strategy of the ACC controller is to keep feed as big as possible with the constraint resulting from a maximum permissible value of the cutting force. It may be important when the diameter of the workpiece before turning or/and its material strength is not known beforehand e.g. during turning free forging. Because of the character of turning process the feed may be increased when the measured cutting force has a low value during at least one revolution of the workpiece. When the cutting force exceeds the maximal permissible value, the feed should be diminished (or even stopped) at once. Now ACC controllers are not so popular in turning because in practice accuracy during finishing is really important and not the material removal rate during rough turning.

Maximization of material removal rate is important during milling operations in aerospace industry where a lot of material should be removed from the workpiece and the end milling cutters of small diameters are used because of shapes of pockets to be machined. The ACC systems increase feed with the constraint resulting from cutting forces. The cutting forces are limited by the strength of end milling cutters.

Technological adaptive control is frequently used in „nontraditional" machining methods like EDM. It is both because of many disturbances in the process of electro-discharge machining and because of relatively easy measurement and change of electrical parameters crucial for the process. In most cases ACC systems are used but with more sophisticated approach to the optimization an ACO may be needed.

Kruth [3] has built an ACO system for maximizing the material removal rate. The average pulse efficiency Φ, defined as the average ratio of discharge duration over total pulse time, has been chosen as the measure well correlated with the material removal rate. Two process parameters: pulse interval time and servo reference voltage are used as the controlled variables. The relationship between controlled variables and the quality index may be presented in two dimensions as a map with contour lines representing steady values of the quality index, Fig.9. The different „hill climbing" strategies may be used by the ACO controller. The example of a very effective Steepest-ascent strategy is illustrated in Fig.9.

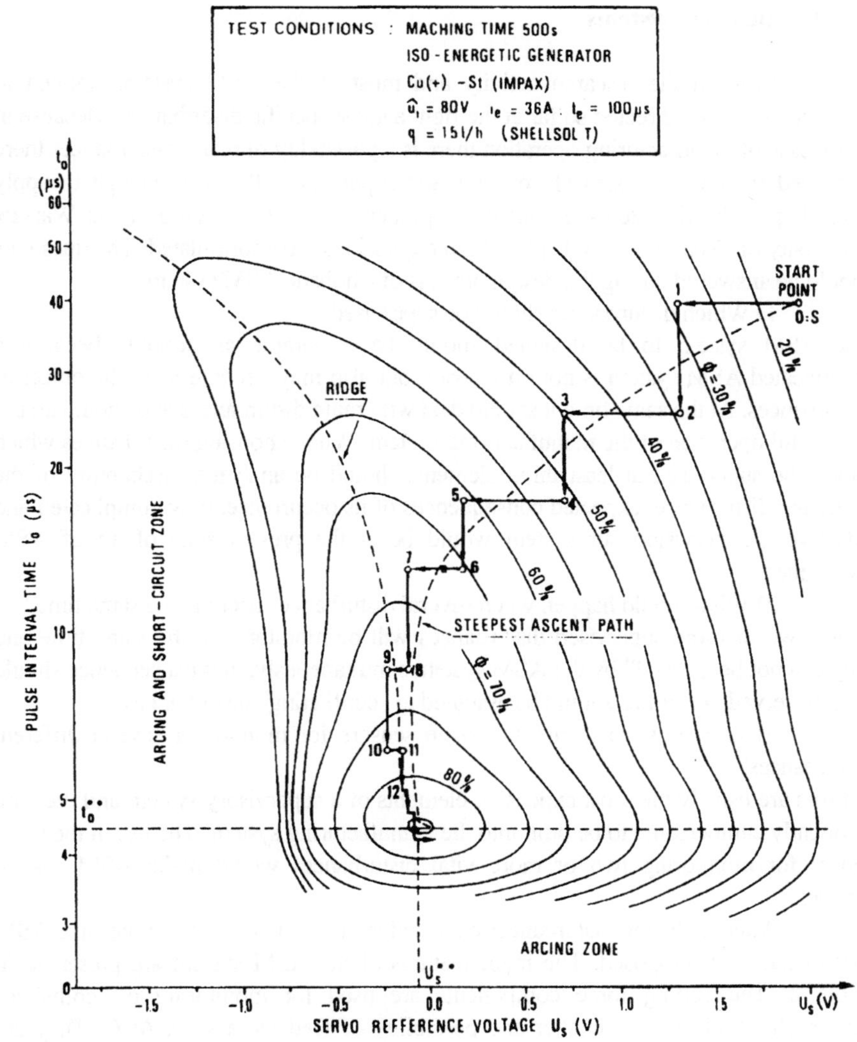

Fig.9 Organization of steepest-ascent search

5.4 Holistic ASM systems

Most of the research activity and most of the ASM systems applied in industry have been directed so far at the fight against specific disturbances. Because in every case of manufacturing operation there is a possibility of many disturbances there is a need for holistic approach to automatic supervision. It is not enough to apply several specialized systems. A holistic approach has been analyzed at the Warsaw University of Technology (WUT) and three questions were formulated. The questions should be answered during the design process of a holistic ASM system.

1) Which disturbances should be supervised.

The ASM system to be designed should be as simple as possible because a complicated ASM system is not only costly but also may become itself the object of disturbances. At the same time it should deal with main disturbances and should assure successful operation of the manufacturing system. When choosing disturbances which should be supervised at least three elements should be analyzed: probability of the particular disturbance, expected consequences of its occurrence, how complicated and effective a supervision sub-system would be at the present state of art of ASM technique.

2) What should happen when several disturbances occur at the same time?

When two or more supervised disturbances will be monitored at the same time and they can not be „served" by the ASM system simultaneously, in what sequence should they be served? Priorities should be assigned to „conflicting" disturbances.

3) Is it possible to use the same sensors for monitoring several different disturbances?

Sensors are usually the most expensive elements of a supervisory system and they are frequently cumbersome to be built into the manufacturing system. The use of the same sensor for monitoring two or more vital disturbances would make ASM system simpler.

After analyzing disturbances expected in an NC lathe a comprehensive ASM system has been proposed. The main features of the ASM system are presented in Table 2. The cutting force components are used for monitoring of: collisions, catastrophic tool failures, chatter and possibility of feed increase in ACC. They are measured with the help of strain gauges. The required accuracy of workpieces is supervised by touch trigger probes. The highest priority is assigned to collisions and catastrophic tool failures.

6. Conclusions

Monitoring and automatic supervision are crucial at the present state of art of manufacturing. ASM is a condition for full automation of production and is very important for new manufacturing processes and new requirements of product quality.

Automatic control of even very complicated cycles is possible and relatively easy with modern numerical controllers (NC), programmable logic controllers (PLC) and computers. The real problem at the present state of development is fight against

Table 2 Functions of the ASM system, priorities of the reactions and the means of monitoring

Break-downs of the manufacturing system	Blocking of malfunctions (2) (low oil pressure, open guard...)	Limit switches, pressure relay...
	Exceeding the limits of NC movements (3)	
	Formal errors in application programs (*)	Software
	Collisions (1)	Cutting force measurement
	Catastrophic tool failure (1)	
	Chatter (4)	
Production of scraps (dimensions of workpieces out of tolerances)	Tool edge offset (*)	Tool touch trigger probe
	Coordinate value of the touched point (*)	Workpiece touch trigger probe
Optimization of manufacturing process (maximization of output)	Increase in feed with limitation of the cutting force (5)	Cutting force measurement

(1), (2)... Priorities of the actions **(*)** Priority may not be assigned

disturbances during manufacturing. Comprehensive automatic supervision makes manufacturing system „intelligent" - the system can deal by itself with disturbances which occur during its operation.

Thanks to last developments in integrated circuits, computers, sensor technology and also thanks to deeper knowledge of manufacturing processes there are on the market relatively simple and reliable systems directed at specific disturbances. There is a need for holistic approach to automatic supervision in manufacturing.

Holistic approach should lead to a comprehensive supervisory system which would deal with every kind of serious disturbances during manufacturing. The holistic

ASM system should have an open architecture and should be integrated with the control system of the operation cycle (e.g. with the NC controller) [6].

References

[1] Byrne G., Dornfeld D., Inasaki I., Ketteler G., Koenig W., Teti R. „Tool Condition Monitoring (TCM) - The Status of Research and Industrial Application" CIRP Annals Vol. 44/2/1995

[2] Corbett J. „Automatic Supervision Research and Application in Precision Machining" IV International Conference on Monitoring and Automatic Supervision in Manufacturing, Miedzeszyn, Poland, 1995

[3] Kruth J.P., Van Coppenolle B. „New Trends in Automatic Control of Electro-Discharge Machining". IV International Conference on Monitoring and Automatic Supervision, Miedzeszyn, Poland, 1995

[4] Pritschow G., Ordenewitz R. „Diagnosis and Information Systems for Production Lines" IV International Conference on Monitoring and Automatic Supervision in Manufacturing, Miedzeszyn, Poland, 1995

[5] Szafarczyk M. (ed) „Automatic Supervision in Manufacturing" Springer-Verlag, 1994

[6] Szafarczyk M. „Open Architecture Controllers and Automatic Supervision in Manufacturing" 27th CIRP International Seminar on Manufacturing Systems, Ann Arbor, 1995

[7] Toenshoff H.-K., Wulfsberg J.P., Kals H.J.J., Luttervelt C.A. „Developments and Trends in Monitoring and Control of Machining Processes" CIRP Annals Vol. 37/2/1988

[8] Toenshoff H.-K., Nitidem-Ndengue E. „Intelligent Sensor Modules for Decentralized Signal Processing in Monitoring and Diagnostic Systems" IV International Conference on Monitoring and Automatic Supervision in Manufacturing, Miedzeszyn, Poland, 1995

[9] Weck M., McKeown P., Bonse R., Herbst U. „Reduction and Compensation of Thermal Errors in Machine Tools" CIRP Annals Vol. 44/2/1995

11 Petri Nets for Designing Manufacturing Systems

A. W. Colombo and R. Carelli

1 Introduction

Petri nets (PN) are a graphic-mathematical tool which can be used for modelling and analyzing manufacturing systems. The dynamic behaviour of a complex discrete-event system may be readily visualized through PN, which can simulate the system's evolution, analyze its synchronism, describe task concurrency and parallelism, and verify the reachability of a state by means of mathematical analysis. Besides having all the above characteristics as a modelling tool, the PN model of the system may be used for its qualitative - functional - and quantitative performance - analysis, thus becoming an advantageous instrument for validation during the design process of Flexible Production Systems (FPS). Indeed, through model validation and verification with PN, the user of the FMS/FAS can rearrange system specifications and objectives due to the easy descriptive characteristics of the PN. In the authors' opinion, it is very important to develop the modelling, analysis and implementation of FPS simultaneously, for none of these processes can be adequately developed independently.

This chapter presents the main concepts and definitions related to Petri nets and their interpretation when applied to manufacturing systems. Methods and concepts used for modelling FMS by means of PN and for validating PN models are overviewed. An introduction to Coloured Petri Nets for designing FMS is also presented. Quantitative analysis is considered using model simulation by means of a temporized Petri net and by defining performance indexes. Finally, a methodology for designing the PN based model of the FMS control system is

---

The authors are participant researchers of the project ECLA FLEXSYS, for scientific cooperation between the European Union and Latin-America in the area of FMS/FAS.

proposed and applied to a comprehensive example of a flexible manufacturing system.

2 Petri Nets

A Petri Net is a directed graph consisting of two types of nodes: places and transitions, where places are always followed by transitions and viceversa [Peterson, 81], [Silva, 85], [Murata, 89]. A PN is defined by a 4-tuple:

$$PN = \{ P, T, E, S \} \tag{1}$$

where $P = \{p_1, p_2, \ldots, p_m\}$ is a finite set of m places; $T = \{t_1, t_2, \ldots, t_n\}$ is a finite set of n transitions. E is the input matrix defined as $E: PxT \rightarrow \mathbf{N}$. Each element $E(p_i, t_j) = e_{ij}$ represents the positive integer weight of the input arc from place p_i to transition t_j. S is the output matrix defined as $S: TxP \rightarrow \mathbf{N}$. Each element $S(t_i, p_j) = s_{ij}$ represents the weight of the output arc from transition t_i to place p_j. Now, the *incidence matrix* A is defined as $A = S^T - E$. From above, it is clear that the incidence matrix of a PN represents the net structure. It may occur that a transition has the same place as both, input and output (self-loops). In this case the net is called a *not pure PN*, because it does not entirely represent the network structure [Silva, 85].

A *marking* M of a PN is a function $M: P \rightarrow \mathbf{N}$ which assigns markings or tokens to each place of the net. The presence or absence of tokens indicates the status of the place. Marking M can also be regarded as a vector $M = [m_1, m_2, \ldots, m_m]$ where $m_i = M(p_i)$. Therefore, a Marked PN (MPN) is a PN plus a marking, thus becoming a 5-tuple

$$MPN = \{ P, T, E, S, M \} \tag{2}$$

For a transition t, the input set of places is defined as $\cdot t = \{p / E(p, t) \geq 0\}$. Then, a transition t is said to be *enabled* for a marking M if and only if $\forall p_i \in \cdot t, M(p_i) \geq E(p_i, t)$. If a transition is enabled, it can be fired.

Graphically, a MPN is represented by circles indicating places or locations, and bars indicating transitions. Directed arcs are represented by arrows. Place markings are represented by dots inside the circle or, if the number of markings becomes excessive, its numerical value is denoted instead of dots.

The dynamical *evolution of the PN* can be described by the following evolution rule:

$$M_f = M_0 + A \,.\, \sigma \tag{3}$$

where M_0, M_f are, respectively, the initial and final markings - states - of the net evolution. σ is the firing counter vector. Each element σ_i represents how many times the transition t_i has been fired during the net evolution from M_0 to M_f.

Firing of a transition is interpreted as a change of state or the occurrence of an event. When an enabled transition t_i is fired, $E(p, t_i)$ tokens are removed from each of its input places, and $S(t_i, p)$ tokens are added to each of its output places. If there exists a firing sequence with a firing counter vector σ from M_o to M, then M is said to be *reachable* from M_0. The reachability set $R(M_o)$ is defined as the set of all markings which are reachable from M_0.

Different types of Petri Nets

Following, the main types of PN and their extensions, as regards of FMS, are given.

Ordinary Petri Nets are a type of PN similar to those defined at the begining of this section, where the elements of input matrix E and output matrix S are unit integers; therefore, all its arcs have unitary weights.

Generalized Petri Nets correspond exactly with the definition given at the beginning of this section, where arc weights admit any positive integer value.

Temporized Petri Nets (TPN) are introduced in order to extend PN modelling and analysis potentialities to evaluate system performance. To this purpose, the time factor is incorporated to the former definition of Petri nets. Several different approaches have been proposed in the literature, [Ramchandani, 74], [Sifakis, 77], [Molloy, 82], [Freedman, 91], [Marsan, 89], [Marsan, 95]. The main alternatives that characterize the different proposals are: a) The PN elements (either places or transitions) which the time is associated to. b) The semantics of the firing in case of timed transitions (either atomic firing or firing in three phases). c) The nature of temporal specifications (either deterministic or probabilistic).

The following definition is considered: A temporized PN is a pair $\langle MPN, \Theta \rangle$ so that $MPN = \{P, T, E, S, M\}$ and Θ is a function which assigns a non negative real number θ_i to each transition, $\Theta: T \rightarrow \mathbb{R}^+$. $\theta_i = \Theta(t_i)$ is called the firing time for transition t_i and can be defined as $\theta_i = du_i + rand(\delta_i)$ with parameters $du_i \in \mathbb{R}^+$ and $\delta_i \in \mathbb{R}^+$. The function $\mathrm{rand}(\delta)$ multiplies the parameter δ by a positive real number taken from interval *[0,1]*. This function introduces into the model a stochastic - random - firing delay for each temporized transition of the net. The marking evolution rule is the same as for the formerly defined PN, excepting that the firing of t_i lasts θ_i time units [Colombo, 94a].

The existence in the model of random firing delay times requires the definition of the TPN *execution policy* for each temporized transition. A discussion of the effect of firing policies is contained in [Campos, 90]. In this approach a *race policy with enabling memory* is used. Whenever a change of marking enables a set of transitions, which were not previously enabled since their last firing, an algorithm - implemented to solve conflicts among enabled transitions - must be called to assist the net evolution. Upon solving the conflicts, the execution policy is applied to each effectively enabled transition. According to the adopted execution policy, whenever a new marking is entered, each effectively enabled transition t_i samples an instance of the random firing delay from the associated $\theta_i = du_i + \mathrm{rand}(\delta_i)$.

Axiom 1: Transitions compete for firing. Among the set of enabled transitions, the competition is won by the transition that samples the shortest delay time. Because of this rule, immediate transitions have priority to fire over timed transitions. If two or more enabled transitions have identical delay time, then they fire together. *Enabling memory* indicates that the re-sampling is performed only after the transition becomes enabled. The transition which samples the minimum firing delay is the one whose firing determines the change of marking: the sojourn time in the marking is equal to the minimum sampled delay time of the enabled transitions. The new marking is obtained through the rules of the underlying untemporized PN; then, the process is started again.

Remark: The methodology, which must be adopted for modelling with the proposed TPN, does not allow to disable one enabled transition if another one (concurrently enabled) fires before. The models must verify the so called *persistence* property [Murata, 89].

Since the transitions fire in three phases, a transition immediately starts firing provided it is effectively enabled, it sets a timer at the value of its sampled delay instance, and it removes tokens from its input places. Tokens, however, are not deposited into the transition output places until the transition delay θ had elapsed, and its timer had reached zero. Tokens are *kept in the transition* during the entire transition delay. Just upon transition firing, the previously removed tokens are deposited into the output places of the transition under study. The token-game of the TPN is graphically described in Fig. 1.

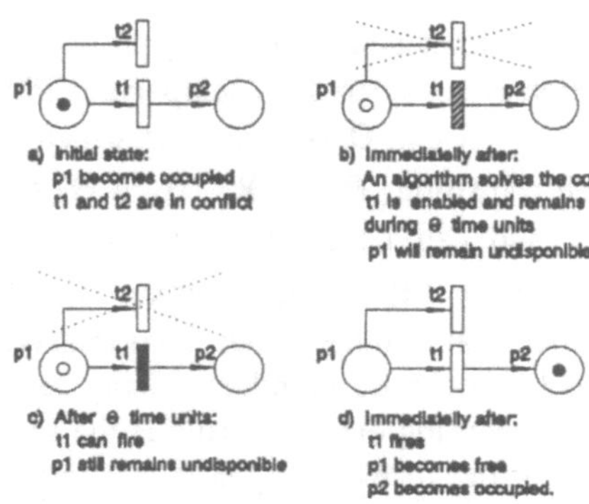

Fig. 1. Token - game of the TPN

PN with disabling arcs [Hack, 75] introduce a new element called a *disabling arc*, which connects a place with a transition and it is represented by a line ended in a small circle which links it to the transition. The functionality of the disabling arc is based on the following conditions [Murata, 89]: If an input place to the disabling arc is marked - i.e., it has at least one marking -, the transition this arc is connected to becomes disabled for firing, regardless of the fact that other input places to the same transition may try to enable it. This type of arc is used for modelling systems with priorities.

PN with predicates admit certain Boolean-type variables associated to their transitions. True or false value of these variables, called *predicates*, depends on the external conditions of the net and such values set constraints for enabling the transition. For instance, when the PN is used as a discrete controller, the net interacts with the controlled system [Villaroel, 90]. In such a case, both the enabling and firing of transitions depend on the compliance with certain system operative conditions, which modify in real-time the value of such predicates [Feldmann et al., 95].

Based on the Generalized PN and the Ordinary PN, various *PN extensions* have been developed for different applications. It is important than the extensions maintain most properties of conventional PN, therefore remaining an important tool for validating the model of any given system. *Continuous Petri Nets* have been defined, where markings are positive real numbers and the transition firing is carried out as a continuous flows [Le Bail et al., 92]. Continuous PN can model systems which cannot be modelled by ordinary PN; this new model keeps most properties of a classic PN [David and Alla, 93] , [David and Alla, 94]. *Extended Petri Nets* have also been developed as extensions of Ordinary PN, where six new places have been defined to represent different characteristics of a FMS. Extended PN use subnets for modelling parts of the system. By analyzing the reachability graphs of subnets, the properties of the entire system can be studied as well. The general model of the PN system can be generated using the smaller PN and regarding synthesis techniques which keep the properties of the nets [Ramaswamy and Valavanis, 94]. Another kind of PN, nowadays being increasingly used, are the *Fuzzy Petri Nets*, where the markings adopt some fuzzy value. The fuzzy value of a marking indicates the completion level of a process. Each transition has a weighing factor which decides the fuzzy value of the marking of the post-condition place when the transition is fired. This kind of nets are utilized, among other applications, for error recovery in robot task planning based on sensors [Cao and Sanderson, 92], planning of task sequences [Cao and Sanderson, 95], etc. The *Mark Flow Graph/Production Flow Schema* (MFG/PFS) methodology accurately describes complex characteristics of production systems, asynchronism, parallelism and concurrency of activities. In this methodology, systems are approached in two differnt levels: a conceptual model is obtained by using PFS at a first level, then it is refined into a functional model by using MFG. This MFG/PFS model specifies the control algorithm, for the production systems [Hasegawa et al., 88], [Miyagi, 88]. Structural properties and invariants in MFG/PFS have been studied in [Silva and Miyagi, 95].

Coloured Petri Nets

Coloured Petri Nets (CPN) have been introduced in reason of their power for a concise and readable representation of complex systems with a high degree of abstraction [Alla et al., 84]. These nets provide a formalism for the state representation, the activities and information flow, with a much greater graphical and mathematical expression capability than ordinary and generalized PN, thus becoming a very efficient means for modelling complex systems [Jensen, 81]. In a coloured PN, a set of attributes/colours is assigned to each place and transition. Thus, a transition can be fired with respect to a certain colour. Arcs are labelled with functions which relate the set of marking-colours with the set of firing-colours. These functions indicate which firing modes are activated, i.e. which colours are enabled to be fired in a given transition [Jensen, 81].

Let's recall here, under the definitions of CPN, the firing rule of transitions, the incidence matrix and the rules of the dynamical evolution for the Order Coloured Petri Nets (OCPN) used in this chapter [Jensen, 92], [Haddad and Couvreur, 88], [Colombo et al., 94b].

Definition 1. A CPN is a 7-tuple $CPN = \{P, T, C, I^+, I^-, G, M_o\}$ where $P = \{p_1, p_2, \ldots, p_m\}$ is a finite set of places, $T = \{t_1, t_2, \ldots, t_n\}$ is a finite set of transitions. C is the colour function defined from $P \cup T$ into Ω, where Ω is a set of finite and not empty sets. An item of $C(s)$ is called a colour of s and $C(s)$ is called the colour set of s. C attaches to each place a set of possible *token-colours* and to each transition a set of possible *occurrence-colours*. $I^+(I^-)$ are, respectively, the input function and the output function defined on PxT, such that $I^+(p,t)$ is a function from $C(t)xC(p)$ to $\mathbb{N}$ (i.e. a function from $C(t)$ to $Bag(C(p)), \forall (p,t) \in PxT$). It is noted $I^+(I^-)$, $I^+(p,t,c_t)$, where c_t belongs to $C(t)$. The *incidence matrix* I of a CPN is defined by $I(p,t,c_t) = I^+(p,t,c_t) - I^-(p,t,c_t)$, then I is a function from $C(p)xC(t)$ into $\mathbb{Z}$ (set of integers). G is a *guard function*, defined from T into expresions of Boolean type, (i.e., a predicate), such that $\forall t \in T$: $[Typ(G(t)) = Bool \wedge Typ(Var(G(t))) \subseteq C]$. According to this definition, a standard form of a guard is described as follows: $\forall\, c_{t_i}, c_{t_j} \in C(t) \rightarrow GS(t) = (c_{t_i} \wedge c_{t_j} \wedge \ldots),\ i \neq j$, $G(t) = GS1(t) \vee GS2(t) \vee \ldots$. Each $GS(t)$ is a Boolean function of occurrence-colours related to transition t. M_o is the initial marking of the net. It is a function defined on P, where $M_o(p)$ is a function from $C(p)$ to $\mathbb{N}$ (i.e. an item of $Bag(C(p)), \forall\, p \in P$). Elements of $C(p)$ and $C(t)$ are called *colours* (marking- and firing-colours). $M(p)$ and $M_o(p)$ respectively indicate the number of tokens of each colour in the place p for the current and the initial markings.

Note: The set of *colour sets* determines the types, operations and functions that can be used in the net descriptions (i.e., arcs expressions, guards, initialization expressions, etc.). All colour sets must be finite, although they may have a very large cardinality. Without loss of generality, it is supposed that the basic - standard - colour domains Ω_j are circularly ordered by the enumeration of their items.

Definition 2. A transition t is *marking-enabled* in a marking $M(p)$ for the occurrence-colour $c_t \in C(t)$ iff the following property is satisfied: $\forall p \in P$: $[\sum I^-(p,t,c_t) \leq M(p)$.

Definition 3. A transition t is *enabled* in a marking $M(p)$ iff t is marking-enabled and the corresponding guard $G(t)$ has true value (i.e. 1).

Definition 4. Two or more occurrence-colours in a transition t are concurrently enabled to fire iff the corresponding $GS(t)$ is true.

Definition 5. The firing of transition t for a marking M and a colour $c_t \in C(t)$ gives a new marking M', defined by:

$$M'(p) = M(p) + I(p,t,c_t) \,.\, \Sigma, \quad \forall p \in P \tag{4}$$

Here Σ is a column vector of colour functions and is called the *firing count vector* corresponding to the occurrence sequence $\sigma_1, \sigma_2, \sigma_3, \sigma_4 \ldots \sigma_q$. The *ith* entry of Σ denotes the number that transition t_i must be fired as regards of the respective colour to transform $M(p)$ into $M'(p)$.

Definition 6. A marking $M'(p)$ is reachable from a marking $M(p)$ iff there exists a finite occurrence sequence having $M(p)$ as start marking and $M'(p)$ as end marking, i.e. iff there exists, for some $n \in \mathbf{N}$, a finite firing/occurrence sequence of steps such that: $M[\sigma_1 \sigma_2 \sigma_3 \sigma_4 \ldots \sigma_q > M'$.

Notation: Operator "." is the generalization of matrix multiplication, substituting each product by a function composition.

Definition 7. Consider the equations system $x^T . I = 0$ with $x = v$ a solution. If Eq. (4) is multiplied on the left by v^T, for all reachable markings M and M', $v^T . M' = v^T . M + v^T . I . \Sigma$, where Σ is a vector of the positive weighed set of transitions with dimension $card(T)$. Then $v^T . M' = v^T . M$ because $v^T . I = 0$. This equation is called a linear invariant of markings (place-invariants). The set of place-flows (p-flows) E of a CPN is defined by

$$E = \{v: \mathit{vector\ of\ coulored\ places}\ /\ v^T . I = 0\} \tag{5}$$

Definition 8. Consider the equations system $I . y = 0$ with $y = v$ a solution, where v is a vector of the positive weighed set of transitions with dimension $Card(T)$. Equation (4) can be written as $M' = M + I . v$ for all reachable markings M and M', then $M' = M$ because $I . v = 0$. This last equation is called a linear invariant of firing (transition-invariants). The set of transition-flows (t-flows) D of a CPN is defined by

$$D = \{v: \mathit{vect.\ of\ firing\ modes\ for\ net\ trans.}\ /\ I . v = 0\} \tag{6}$$

Remark: These relationships allow to verify specific properties of CPN models and, particularly, allow to show that the CPN is bounded, deadlock-free and reversible.

Standard -basic- colour domains: The objects of manufacturing systems can be decomposed into classes according to their nature. For instance, there are two or more classes of parts to be assembled, conveying systems with two or more places, etc. The classes of objects modelled by the Petri net will be the standard colour domains $\Omega_j, j \in [1:n]$ [Colombo et al., 94d].

Universal colour domain: The universal coloured domain Ω of the net will be the union of standard colour domains Ω_j and the Cartesian product Ω_j^* of the basic domains above defined, $\Omega_j^* = \Omega_1 x \Omega_2 x \Omega_3 x \ldots x \Omega_n$.

Standard colour functions: These functions are built from the following functions or their linear combination [Colombo et al., 94b]: Projection functions, which select a component of an item (color); identity functions, which select all the items of a basic colour domain; successor functions, which select some successor of an item; predecessor functions, which select some predecessor of an item; decoloured function, which transforms a coloured marking in the uncoloured token.

According to the above description a CPN consists of three different parts: *net structure* (i.e., places, transitions and arcs), *declarations* (functions, sets of colours and guards) and *net inscriptions* (i.e., the various text strings which are attached to the elements of the net structure). The complexity of a CPN description is then distributed among the *net structure*, *declarations* and *net inscriptions* [Jensen, 92].

3 Modelling FMS using Petri nets

Manufacturing Systems (MS) can be straightforwardly defined as a set of resources which cooperate or compete for performing a set of activities in order to obtain a set of products. MS resources are: the plant personnel, machines and equipment, raw materials, which are required to carry out the tasks within a workplan for manufacturing a product. MS activities are: the manufacturing process itself, including machining, materials handling and storing, and information processing necessary for production. The workplan also incorporates the precedence relationships among activities, including possible concurrencies when necessary resources are available. The process plan may specify several alternative resources for any given activity, and it may asign some priority level to the preferred resource [Zhou and DiCesare, 93].

A highly important consideration is to carry out the modelling, analysis and implementation processes of the FMS simultaneously, for none of such processes can be adequately developed independently. From a brief revision of referenced works, it is clear that, from modelling to real-time control, the theory of Petri Nets presents interesting solutions for FMS design, qualitative validation, performance evaluation and implementation, due to their capacity to graphically present certain relationships and to visualize certain concepts, such as: parallelism and concurrency, synchronization, resource sharing, memorizing, monitoring. The main goal of modelling FMS with PN is to obtain an error-free description of system specifications. Starting from the evaluation of possible flows of materials, a process model is built, which already incorporates buffer capacities and deadlock avoidance. These are necessary conditions when PN are intended to be used as real-time control systems of FMS. Petri nets, as defined in graphical and mathematical form, have no physical meaning. For using PN in practical applications, the designer must provide the domain interpretation by labelling PN places and transitions. In this chapter, the following interpretations for places, transitions and tokens are regarded, which have already been adopted by many researchers [Sifakis, 77], [Martinez et al., 87], [Zhou and DiCesare, 93], [Colombo et al., 94b]. Actions ocurring in the system are modelled by transitions. Transition enabling depends on the PN marking and the values of the predicates associated to the transition. The firing of a transition changes the PN marking and, hence, changes the state of the system as well. PN places and their respective markings represent the various states of resources of the modelled system. They are also used as pre-conditions for a posible firing of transitions. The initial number of tokens can be either a constant, e.g., the number of robots-machines (assuming the plant is fixed), or a variable, e.g., the number of tasks or pallets within the system. For modelling many physical systems, an upper limit is generally considered for the number of tokens that each place can hold. Such a PN is referred to as a finite capacity net. For the above defined PN, each place p has an associated monitor place [Campos, 90], [Colombo et al., 94b].

A Petri net model can be obtained by using top-down methods in the initial stage, and bottom-up methods in the final stage [Zhou and DiCesare, 93] of a design procedure. Two major benefits are reapped. First, the details can be introduced incrementally, thus alleviating model complexity and, second, important properties of the resulting Petri nets are safeguarded, thus avoiding costly analysis for some qualitative properties. *Top-down design* of PN refers to a descending approach using stepwise refinement of places, transitions or substructures. This is accomplished by replacing these elements with more complicated subnets. Each successive step grows in detail until implementation level is reached. Refinement theory is used to guarantee that each step of the refinement procedure preserves the desirable system properties of PN. Since those refinements are often local and required not to originate any side-effect, interaction among subnets is very difficult to handle using such a strategy. The possibility of a progressive modelling is absolutely necessary for FMS because they are usually large and complex systems. The refinement mechanism allows for building hierarchically structured net models. If the refinement-transformation rule catalog is adequately selected and the model is well designed, the design can be easily or trivially proven correct with respect to system specifications [Silva and Valette, 89].

Whenever the top-down approach is not feasible, the system can be specified by using partial descriptions and the overall net obtained by compounding the various subnets [David and Alla, 92]. Therefore, *bottom-up design* is also adopted for solving this problem. Bottom-up design focuses on a correct construction of interactions among subsystems or detailed operation processes. All shared resources are considered at this stage and become key elements for constructing a PN with desirable qualitative properties [Jeng and DiCesare, 92], [Zhou and DiCesare, 93], [Colombo et al., 94b]. By applying a *hybrid synthesis* methodology as presented in [Zhou, 90a] ,[Zhou and DiCesare, 90b], [Zhou et al., 92a], [Zhou et al., 92b], [Zhou, 92c], [Chao et al., 92], [Jeng and DiCesare, 92], [Zhou, 92c], [Zurawsky, 92], [Zhou and DiCesare, 93], two stages arise, namely, top-down design and bottom-up design which include six steps, as follows: Design of a first-level PN; decomposition and refinement of operations; addition of non-shared resources; addition of monitor-places; addition of shared resources to form GPMEs [DiCesare et al. 93]; addition of shared resources to form GSMEs [Zhou and DiCesare, 93].

In the work of [Ezpeleta and Martinez, 91], a system modelling using PN is presented from a methodological viewpoint, which can be basically attained in two main stages.

a) System operation is broken down into small tasks. Two criteria may be used: each part to be processed has an associated workplan (sequence of operations to be carried out on this part) or, the system is considered as a set of small independent subsystems. From this first stage a set of independent specifications is obtained; each of them is then modelled by means of a PN. Therefore, a PN will model either the workplan for each part to be processed in the system, or the material flow in each subsystem of the plant.

b) Integration of specifications regarding the following relationships among subsystems: Competence for shared-resources and cooperation between two or more subsystems in order to perform some task of the workplan.

The following steps are now proposed for *modelling the FMS control system*: 1) Identify activities and resources required for manufacturing one item of each product. 2) Arrange activities according to the precedence relationships as assigned in the process plan. 3) Sketch a PN model transition for each activity, or, define the firing-modes (occurrence-colours) for a transition. In Section 6, figure 5 shows a first model of the process to be developed in a FPS for manufacturing solar heating equipments. 4) For each resource of the system: create and label places to represent the busy states (before and after performing a task) of such a resource; create and label a place to represent the free-status of that resource - monitor place - [Campos, 90], [Colombo et al., 94b]. 5) Link each place with an arc to and from the corresponding transitions, according to the above defined precedence relationships. A corresponding function must be associated to each arc if the model is built with CPN. 6) Specify the initial marking of the net according to the initial state of the modelled system. 7) When the net is executed, a token in a place (busy state of a resource) will indicate the resource is already busy. 8) When the net is executed, a token in a place (free state of a resource) will indicate the resource is already free. 9) During the net evolution, the marking of the free-state and busy-state places of a resource will be in mutual exclusion [Colombo, 94a]. 10) Multiple tokens in a place will indicate a multiplicity of resources or parts; for example, in a buffer-free place, three tokens might represent three positions of a buffer which are free at the same time. If the same model is built with CPN, three tokens with different attributes will model three positions of a buffer [Colombo et al., 94b]. Figures 6 and 7 in Section 6 depict the PN and CPN models of a FPS resource (machine) and illustrate the application of the above steps. 11) A bottom-up methodology is applied for composing the models of FPS resources, regarding both the competence and cooperation relationships, in order to obtain a complete model of the FPS coordination level, whose behaviour corresponds to the process modelled in the first three steps above presented. This approach needs an overall synchronization of subnets by using, for example, a kind of rendez-vouz mechanism [David and Alla, 92]. Figure 8 of Section 6 shows the results after applying this step to the FPS under consideration. 12) At this point of the design procedure, a well-defined PN or CPN model must comply with some basic properties such as liveness, boundedness and reversibility. A validation stage is then very important to prove the system specifications related to the net's properties [Colombo et al., 94b], see Section 4. 13) By means of a Top-Down design method, the transitions of the PN model, or their firing-mode (occurrence-colours) in a CPN model are now refined by assigning an underlying PN to each of them. As an example, Fig. 9 in Section 6 presents the results of this step in relation to transition t_{10} of Fig. 8, where the newly obtained set of transitions models both the technical actions and the handling of real signals to be exchanged between the control model and the sensor/actuator interface of the production environment [Villaroel, 90], [David and Alla, 92], [Feldmann et al., 95].

4 Validation of PN models

Validation analysis aims at verifying the correctness of system design and at verifying all initial specifications. This analysis is achieved through a description, namely, a model of the system. This model needs to include all significant information on system operation, processing plans, machines, workcenters, tools, materials, commands, etc. and their interrelationships. The model may result in a computation model practical for analytic validation, or in a simulation model which allows experiments to be performed onto the system model. Basically, two kinds of analysis of system model can be made: the qualitative and the quantitative analysis. The first one verifies the compliance of certain characteristics such as the absence of deadlock, cyclicity, finite number of states, finite capacity of buffers and storage areas, etc. The quantitative analysis is often called performance evaluation, and it takes into account system specifications, therewith checking the system's compliance with specified performance indexes.

As expressed in the above paragraph, *qualitative analysis* aims at proving some desirable properties of the system's model. Considering PN models, these properties can be classified in properties dependent on the initial marking of the net like liveness, cyclicity, conservativity, mutual exclusion of places, and structural properties or properties independent of the initial marking of the net like repeativity, boundedness and conservativity. For a definition and understanding of these properties, the following references can be consulted: [Peterson, 81], [Silva, 85], [Murata, 89]. PN properties are related to system specifications. For example: the boundedness of a PN place may be related to the maximum number of pallets in a storage or the items in a machine; mutual exclusion of the marking of two places implies complementary states for the resource being modelled by these places (as busy-free states); cyclicity of the PN is related to production cycles [Colombo, 94a], [Silva and Valette, 89].

There are many *methods* for PN validation analysis, which can be classified into the following groups: 1) reachability graph, 2) transformation analysis, 3) structural analysis, which includes the method based on Linear Algebra and the method of latches and traps. 4) simulation analysis [Silva, 85], [Murata, 89]. The *method based on Linear Algebra* exploits the fact that, regardless of place markings, the PN structure contains a good deal of information that can be used to verify some desirable properties that correspond to functional specifications of the modelled system. Provided the PN is pure, its structure is explicily represented by the incidence matrix. Therefore, from this matrix, many a worthy conclusion can be drawn on net performance without resorting to simulate its evolution. The analysis is made by obtaining invariant relationships of the incidence matrix, which provide conditions for fullfilment of a number of good performance characteristics of the net, such as liveliness, deadlock free condition and boundedness [Murata, 89], [Silva, 85]. The analysis by means of *simulation* allows to gain confidence on the system description, as well as to detect certain wrongly described situations, such as incompatibilities, omission errors, etc. The simulation of a system described through PN is performed by using CAD systems. A properly designed simulator is useful for operational and performance studies with corresponding visualization of results in the token-game of the net. Some of these simulators are: GreatSPN [Chiola, 87], PETSY [Jarschel et al., 92], SIMRdP [Colombo et al., 94c]. In [Carelli et al., 95] a combined discrete-event and continuous motion simulation of FMS is proposed.

Quantitative validation is related to performance analysis including such elements as capacity, cycle time, inventory, utilization, service level and costs. Some research lines in quantitative validation are given in [Chiola, 87], [Marsan, 89], [IEEE, 89], [Campos, 90], [Freedman, 91], [Marsan et al., 95]. The high operative cost of manufacturing systems is a critical reason for carrying out a quantitative analysis of them previously to their implementation. This is a stage where an intensive computational assistance is required, in order to obtain the fundamental ideas on the number of machines to use, the kind of conveyance system, and the space distribution of resources on the shopfloor and all aspects necessary to achieve the specific production objectives.

The use of temporized PN (TPN) for modelling a given system, allows for carrying out a faithful study on the system's behaviour. A number of specifications of the modelled system can be thus verified from the TPN model, based on the values of the pre-defined performance indexes. This stage of the analysis is reached by previously qualitatively validating the original non-timed PN. This is so because the PN temporizing does not modify its basic structure, therefore maintaining the t-invariants and p-invariants [Freedman, 91]. Also, the set of firing sequences of a temporal PN is a subset of the firing sequences of its non-timed PN, since the timing information may inhibit certain combinations of firing transitions. It can be ascertained that a PN boundedness is sufficient condition for the boundedness of the corresponding TPN [Freedman, 91], [Silva, 85].

In order to attain a measurable-quantifiable representation of the performance of the plant, there arises the concept of performance indexes. *Global indexes* are obtained by regarding the evolution of a TPN which models the operations of a set of resources competing and cooperating in the production of a given product following a pre-determined processing plan. Some global indexes are [Colombo, 94a]: cycle time of the net, output capacity, production time. *Local indexes* are obtained by separately considering the various modelled resources and their operations. Generally, local indexes are referred to the processing time of the modelled plant with the worktime for each resource involved in the modelled manufacturing process. Some local indexes [Colombo, 94a] are: throughput of a temporized transition, processing time, percentage of resource-product utilization, percentage of deadtime.

Basically, at present the quantitative analysis of a TPN is done for steady state of the net evolution [Marsan, 89], because there exists very little literature as regards of transitory state of the modelled plant. Taking this aspect into account, the starting conditions for a quantitiative analysis through simulation of a TPN are the following: 1) A sufficient number of evolution simulations have to be carried out so as to ensure it evolves in steady state. This is a necessary and sufficient condition to ensure a correct computation of the performance indexes. 2) Some sort of information is needed to be available beforehand on the elementary operations which are being modelled, for instance, the main operative characteristics of the resources. From this previous knowledge there arise the initial values for temporizing the transitions of the originally developed PN. 3) Considering the modelled system, several simulations are carried out, suggesting for each one various resource models to perform the given task. Some indications will arise as regards of the most favorable model for designing the system, so as to comply with the performance specifications. 4) When several simulations of one TPN are carried out, the changes proposed in each simulation have to be performed, whenever possible, on only one resource at a time.

Validation represents a very important stage when designing FMS using PN. In fact, PN appears as a powerful tool for validation, as it was presented above. It must be emphasized that modelling and validation should be carried out interactively, in this way models are optimized and the designer can guarantee the compliance with system specifications.

5 Coordination Control of Production Systems by Means of PN

Due to its complexity, the FMS control problem is generally dealt with by considering the following hierarchical control levels: production planning, task scheduling, coordination control and local control. At coordination level, the various resources of the production plant are coordinated to cooperate in the execution of the operations corresponding to the task schedule. The coordinator control system is linked to the production plant through a set of signals [Villaroel, 90]: a) Controller-related signals: commands from the coordinator to the local controllers for executing a certain task; end-of-activity messages from the local controllers to the coordinator in response to the above cited commands; exception messages sent by the controllers to the coordinator whenever an error occurs in the execution of a command. b) Signals from sensors, which complement the information on plant evolution received by the coordinator. c) Alarm signals for breakdown and failure of pieces, machines, etc.

Petri Nets have been used mainly for modelling, simulating and validating Discrete Event Systems (DES) [Peterson, 81], [Silva, 85], [Murata, 89]. As regards to specification and validation of DES controllers, some works have been developed, mainly oriented to the application of Programmable Logic Controllers (PLC) [Furukawa et al., 90], [Silva and Velilla, 82], [Silva, 85], [Ferrarini, 92], [Gomes et al., 93]. Some references are available on implementation of DES coordination control using PN [Brand and Kpoansky, 88], [Zhou, 90c], [Cofer and Garg, 92], [Zhou and DiCesare, 93]. In [Miyagi, 93] the Mark Flow Graph/Production Flow Schema (MFG/PFS) for DES control is used. The MFG, which is derived from PN, focuses on modelling the DES local controller.

The implementation of *DES coordination control* by means of PN necessitates a PN structure capable of sending and receiving signals to and from the external world, namely the local controllers, and of solving the conflicts arising in the PN which models the coordinator. The implementation also needs that the control be continuously monitored and that its safe operation be warranted. For allowing a PN to represent a system, it needs to be given an *interpretation*. To interpret a GPN means to set an agreement to define the following aspects [Silva, 85]:

- A physical significance to the necessary conditions for firing a transition. The evolution rules are slightly modified by the interpretation, which also turns to be an evolution function of the process to be controlled.
- The actions resulting from the marking evolution.

The autonomous PN (Petri nets which lack interpretation) cannot be used for describing the operation of system controllers. These nets lack the necessary information for taking into account the present state of the plant, and for determining the necessary operations for modifying the state of the system. For specifying the control function on a DES by means of a PN, three new functions should be introduced. One is in charge of reading the present state of the plant represented in the model by the marking; another function modifies such plant state corresponding to the firing of transitions. The third function is assigned to solving conflicts [Treviño, 90], [Feldmann et al., 95]. Consequently, the characteristics for such a controller model should regard the following aspects [Miyagi, 93]:

- To prevent the occurrence of conflicts.

- To warrant the system will not get blocked. There always has to be an activated event about to occur.
- To be capable of sending the model state signals to the external devices.
- To be capable of receiving the signals generated from the external devices and, based on these signals, to control the occurrence of model events.

The next section presents an example which includes the design of the PN-based model for the coordination controller of a FMS, following the steps proposed in Section 3.

6 Design, Modelling and Analysis of a FMS

Product to be manufactured
As an example, a Flexible Manufacturing System with two flexible manufacturing cells will be considered, for producing solar energy collector panels of solar heating equipments [Tomeco, 95].

Configuration of the FMS
Considering the characteristics of the involved processes for performing the cited product, one possible configuration of the FMS for carrying out the set of production tasks is shown in Figure 2.

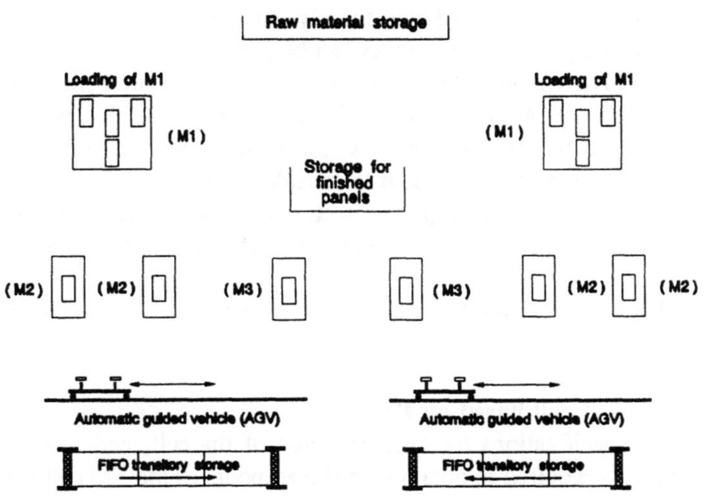

Fig. 2. Layout of the flexible manufacturing system

The system has two identical workcells, each one composed of the following resources:

- One raw material storage area with capacity for storing twenty-four stainless steel plates.
- One machine (M1) for the alignment and first welding of two plates per panel.
- Two machines for welding the water convection channels (M2).
- One discretized three-position FIFO for transitory storage of panels.
- One machine for welding panel contours (M3).
- One automatic guided vehicle (AGV) for transferring the panels from M2 to M3.
- One storage area for finished panels, capable of storing a finite number of processed panels.

First, to manufacture one collector panel, a worker loads two stainless steel plates from the raw material storage onto the alignment sector of M1. Then, the panel comes into the welding sector of the same machine, where the plates are welded. Upon completing this task, a magnetic code is written on the pallet, specifying the information about solar collector model and the next workplan task to be performed on the panel. According to this information, an automatic mechanism unloads M1 and loads one machine M2 (in each cell there are two identical machines M2 which make the channels for water convection). After completion of this task, a magnetic code is written on the pallet, giving information about collector model and the next workplan task to be carried out on the panel. Then, the panel is transferred by the AGV to the transitory buffer storage (the three-position FIFO). The same AGV unloads the FIFO and loads machine M3, where panel contours are welded. Finally, a worker unloads M3 and delivers the final product into the output storage. This task sequence is repeated for each panel.

Note that this application has all the ingredients of a typical FMS. Some operations should be performed sequentially, such as alignment and welding on machine M1, though many other tasks can be performed in parallel, such as the operations on M1, M2 and M3. There are instances of synchronization as well, such as that between AGV hauling and FIFO storage. At this point it is very important to recall that the same process can be developed in both cells.

In the configuration of the FMS, it is considered that:
- only one worker loads the alignment sectors of machine M1 of both cells;
- only one worker unloads both cells and carries the final products into the output storage;
- the FMS has only one area for raw material storage and only one output storage area.

Model of one FM cell based on PN

According to specifications for each resource of the cell, and to the cited task plan, synthesis is required to obtain a Petri Net model of the FM cell coordination controller. Figure 3 shows an approach using the Temporized Petri Net defined in Section 2, where timed transitions are denoted by a thick bar.

Transitions t4 and t5 present a structural conflict. To assist the TPN token-game, the model has a decision-free choice structure comprising places p20 and p21 [Zhou et al., 92a].

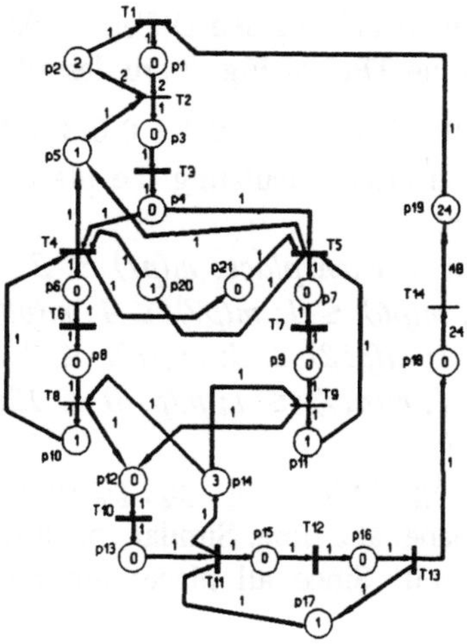

Fig. 3. Temporized Petri net model of the Flexible Manufacturing Cell

Table 1. Description of places and transitions of figure 3

p1	Alignment sector of M1 busy	t1	Loading of alignment sector of M1 from raw material storage
p2	Capacity monitor for p1		
p3	Welding sector of M1 busy	t2	Loading of welding sector of M1
p4	M1 finished welding of panel	t3	M1 performs welding operation
p5	Capacity monitor for p3 and p4	t4	Loading of first M2 from M1
p6	Welding sector of first M2 busy	t5	Loading of second M2 from M1
p7	Welding sector of second M2 busy	t6	First M2 welds a panel
p8	M2 finished welding of panel	t7	Second M2 welds a panel
p9	M3 finished welding of panel	t8	Unloading of first M2 into the FIFO storage
p10	Capacity monitor for p6 and p8		
p11	Capacity monitor for p7 and p9	t9	Unloading of second M2 into the FIFO storage
p12	FIFO loaded with at least one panel		
p13	Panel transferred into FIFO storage	t10	Transfer operation in FIFO storage
p14	Capacity monitor for p12 and p13	t11	Loading of M3 from FIFO storage
p15	Welding sector of M3 busy	t12	M3 welds a panel
p16	M3 finished welding of panel	t13	Unloading of M3 into the storage for finished products
p17	Capacity monitor for p15 and p16		
p18	Storage of finished products	t14	Re-initialization of TPN evolution
p19	Raw material storage		

Functional analysis of the TPN model of a cell

The TPN model was developed by separating the Flexible Cell into its several functional parts such as M1, M2, M3, FIFO transitory storage, etc. Therefore, a TPN model was constructed for each part and the final net was obtained by merging the constructed models [Jeng and DiCesare, 93]. The results of applying the structural analysis to the TPN of Fig. 3 are described below.

By considering $M_0^T = [0\ 2\ 0\ 0\ 1\ 0\ 0\ 0\ 0\ 1\ 1\ 0\ 0\ 3\ 0\ 0\ 1\ 0\ 24\ 1\ 0]$ as the initial marking of the TPN, and after calculating the place invariants, the following conclusions can be drawn.

Bounds for the capacity of each place: $m(p1) \leq 2$; $m(p2) \leq 2$; $m(p3) \leq 1$; $m(p4) \leq 1$; $m(p5) \leq 1$; $m(p6) \leq 1$; $m(p7) \leq 1$; $m(p8) \leq 1$; $m(p9) \leq 1$; $m(p10) \leq 1$; $m(p11) \leq 1$; $m(p12) \leq 3$; $m(p13) \leq 3$; $m(p14) \leq 3$; $m(p15) \leq 1$; $m(p16) \leq 1$; $m(p17) \leq 1$; $m(p18) \leq 12$; $m(p19) \leq 24$; $m(p20) = 1$, $m(p21) = 1$.

The bound of the marking of places p_3, p_4, p_5 shows the capacity of machine M1 to process only one panel at a time. Similar conclusion can be obtained for all other resources of the cell. Since all places are bounded, then the TPN is structurally bounded.

Mutual exclusion of markings: The following relationships establish the mutual exclusion of markings:

$m(p3) + m(p4) + m(p5) = 1$; $m(p6) + m(p8) + m(p10) = 1$;
$m(p7) + m(p9) + m(p11) = 1$; $m(p15) + m(p16) + m(p17) = 1$; $m(p20) + m(p21) = 1$.

Places p_3, p_4 and p_5 build a p-flow [Colombo, 94a] and the mutual exclusion among their marking shows the three possible states in which M1 can be found: free, busy by a non-processed panel, and busy by a processed panel. Similar conclusions can be obtained for all other resources of the cell.

In addition, the following relationships are valid, indicating mutual exclusion related to the total marking:

$m(p1) + m(p2) = 2$; $m(p12) + m(p13) + m(p14) = 3$;
$m(p1) + 2.[m(p3) + m(p4) + m(p6) + m(p7) + m(p8) + m(p9) + m(p12) + m(p13) +$
$+ m(p15) + m(p16) + m(p18)] + m(p19) = 24$.

The net is live (necessary condition [Murata, 89]) and deadlockfree. There is one t-flow:

$$y^T = [4t1\ 2t2\ 2t3\ t4\ t5\ t6\ t7\ t8\ t9\ 2t10\ 2t11\ 2t12\ 2t13\ t14]$$

The net is reversible and it has at least one home-state.

Performance analysis of the FM cell

At this design stage, the use of timing values associated to each task to be performed in the cell is of great importance. Of course, the main specification in this study must be an estimation of the duration of all elementary operations developed by each resource of the cell. To better understand the results from the quantitative analysis, deterministic firing delay times are proposed for each task of the FM cell. The duration of different operations vary widely from a short time (immediate re-initialization operation) to many time-unit lapse (welding of panels). Table 2 gives the duration of each operation and the transition which models it, according to TPN of figure 3. After the temporized evolution of the TPN, by simulating the manufacturing of a batch of 12 solar energy collector panels, a

Gantt diagram is obtained. For illustrative purposes, figure 4 shows the first part of the complete diagram.

Table 2. Duration of each operation

	Operation	Duration [time units]	Transition
op1	Loading of alignment sector of M1	0.2	t1
op2	Loading of welding sector of M1	0.0	t2
op3	M1 welds a panel	3.5	t3
op4	Loading of first M2 from M1	0.2	t4
op5	Loading of second M2 from M1	0.2	t5
op6	First M2 welds a panel	7.0	t6
op7	Second M2 welds a panel	7.0	t7
op8	Unloading of first M2 into the FIFO storage	0.0	t8
op9	Unloading of second M2 into the FIFO storage	0.0	t9
op10	Transfer operation in the FIFO storage	3.0	t10
op11	Loading of M3 from FIFO storage	0.0	t11
op12	M3 welds a panel	5.0	t12
op13	Unloading of M3 into the product storage.	0.2	t13
op14	Re-initialization of the TPN evolution	0.0	t14

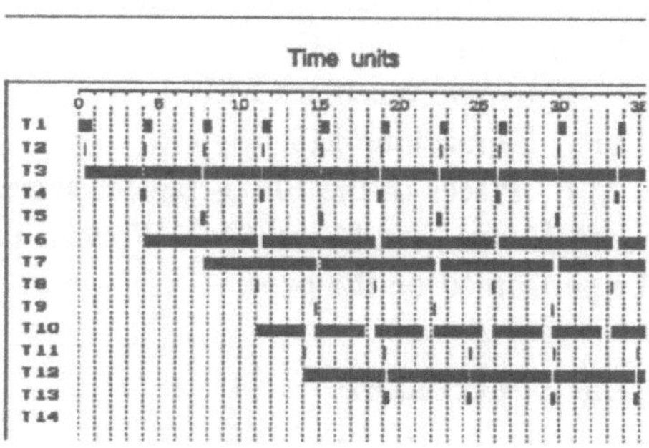

Fig. 4. Gantt diagram

From the Gantt diagram, simulation results of the performance analysis for the FM cell are:

Production period for a part (PPfP) [Colombo, 94a]: 19.30 time units. Total time needed by the cell to produce a solar collector panel.

Production period for a batch (PPfB) [Colombo, 94a]: 76.50 time units. Total time needed by the cell to produce a batch of 12 solar collector panels.

Percental use of a resource related to a part (PURP) [Colombo, 94a]: Time lapse each resource of the cell is working when parameter *PPfP* is being considered. From the diagram of figure 4 and considering only the main operations performed by each resource, the *PURP* for each operation is: M1 (process): 18.13%; M2 (process): 36.27%; M3 (process): 25.90%; FIFO storage: 15.54%.

Percental use of a resource related to a batch (PURB) [Colombo, 94a]: Time lapse each resource of the cell is working when parameter *PPfB* is being considered. From the diagram of Fig. 4 and considering only the main operations performed by each resource, the relationship among resource utilization and the number of processed panels can be exactly calculated: M1 (process): 54.90%; M2 (process): 54.90%; M3 (process): 78.43%; FIFO storage: 47.05%.

Manufactured Parts per Time Units (MPpTU) [Colombo, 94a]: 0.1923 [solar panels/time units].

Modelling and analysis of the FMS by means of CPN

Since both cells are identical, a CPN modelling and analysis of the FMS is now proposed. According to the concepts introduced in Section 2, the main steps for designing a CPN control structure for the FMS are next described.

- *Definition of the standard - basic - colour domains:*

In this step, colours are mapped to each of the relevant characteristics of the modelled resources, e.g. places of a buffer, sequence of tasks, classes of parts to be processed, different types of machines, etc. The elements of each standard colour domain are sequentially ordered.

$Cell = \{c_j, j \in [1:2]\}$ - cell 1 or cell 2

$Al = \{a_y, y \in [1:2]\}$ - alignment sector 1 or alignment sector 2

$M1 = \{m1_{k1}, k1 \in [1:1]\}$ - machine M1

$M2 = \{m2_{k2}, k2 \in [1:2]\}$ - two machines M2

$M3 = \{m3_{k3}, k3 \in [1:1]\}$ - machine M3

$TS = \{ts_x, x \in [1:3]\}$ - Places of the FIFO-transfer system

$WP = \{\langle pa_s, wp_r\rangle, s \in [1:n], r \in [1:m]\}$ - Workplan: panel-type and workplan task to be performed on the panel

- *Cartesian product colour domain:*

$$\Omega_j^* = Cell x Al x M1 x M2 x M3 x TS x WP$$

$$\rightarrow \omega_j^* = \langle c_j, a_y, m1_{k1}, m2_{k2}, m3_{k3}, ts_x, \langle pa_s, wp_r\rangle\rangle \in \Omega_j^*$$

- *Universal colour domain:*

$$\Omega = Cell \cup Al \cup M1 \cup M2 \cup M3 \cup TS \cup WP \cup \Omega_j^*$$

- *Definition of functions:*

Identity $Id(\omega_j^*) = \omega_j^*$; Sucessor

$$Suc72_1(\omega_j^*) = \langle c_j, a_y, m1_{k1}, m2_{k2}, m3_{k3}, ts_x, \langle pa_s, wp_{(r+1)}\rangle\rangle \quad if\ r < m$$

$$Suc72_1(\omega_j^*) = \langle c_j, a_y, m1_{k1}, m2_{k2}, m3_{k3}, ts_x, , \langle pa_s, wp_1\rangle\rangle \quad if\ r = m$$

Projection $Proj12(\omega_j^*) = \langle c_j, a_y\rangle$; Projection $Proj13(\omega_j^*) = \langle c_j, m1_{k1}\rangle$;

Projection $Proj14(\omega_j^*) = \langle c_j, m2_{k2}\rangle$;

Projection $Proj15(\omega_j^*) = \langle c_j, m3_{k3}\rangle$; Projection $Proj16(\omega_j^*) = \langle c_j, ts_x\rangle$

- *Definition of guards:* Guards are Boolean functions built from the colours of the standard colour domains.

According to the specifications for each resource, and the production task plan above described, synthesis is required to obtain the CPN model for the *coordination controller* of the proposed FMS. The problem can be stated as follows [Zhou et al., 92a]: Given the specifications for the system, model it as a Petri net such that its structure and initial marking of places make it bounded, live, and reversible.

Figures 5, 6, 7 and 8 illustrate the synthesis design following the sequence of steps proposed in Section 4 for modelling the control system of a flexible production system. Fig. 8 represents the complete model obtained after a bottom-up design considering PN modules of resources like those of figures 6 and 7.

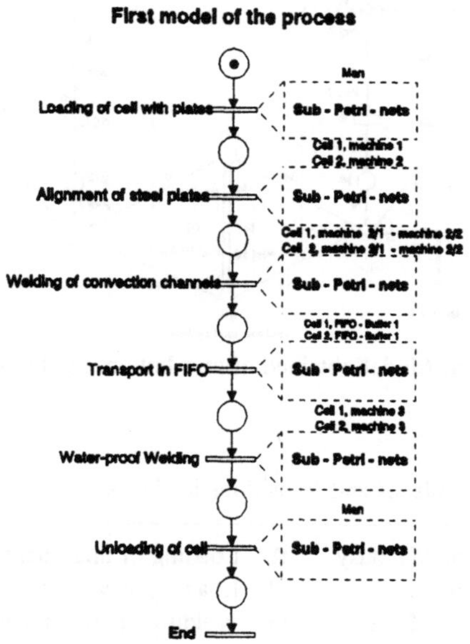

Fig. 5. First model

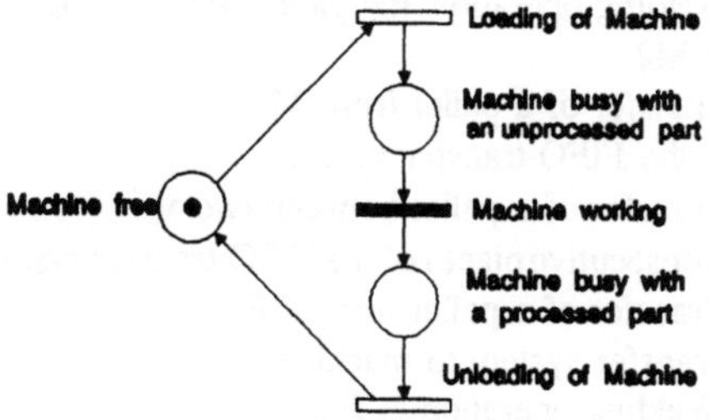

Fig. 6. PN-Module for describing a machine operation

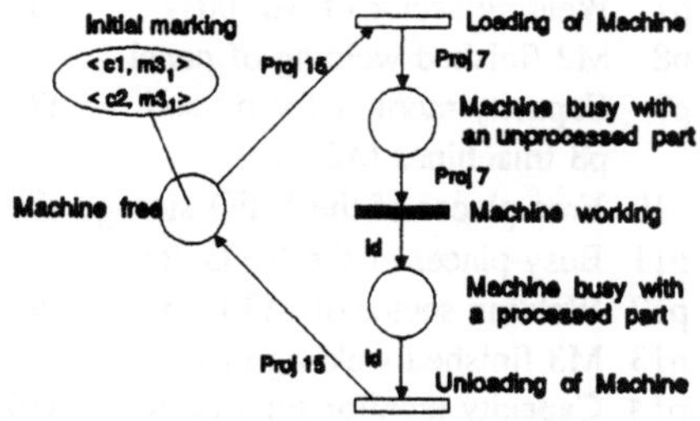

Fig. 7. CPN-Module for describing the operation of machine M3 (both cells)

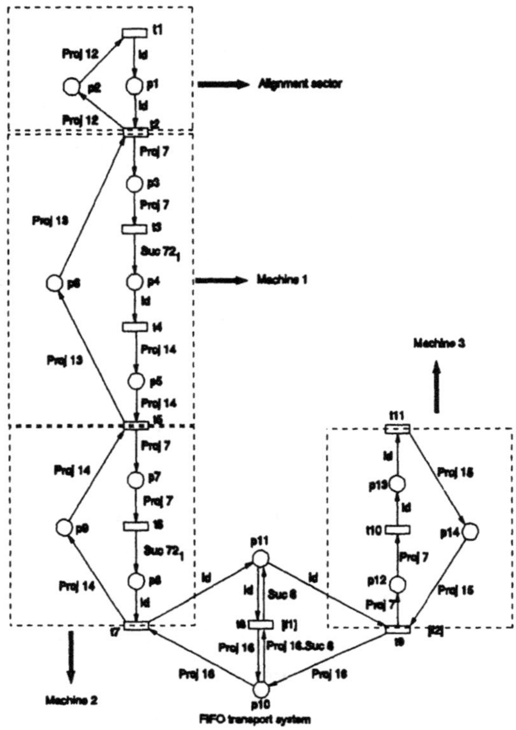

Fig. 8. Model obtained after a bottom-up design

Table 3. Description of places and transitions in figure 8

p1	Alignment sector of M1 busy	t1	Loading of alignment sector of M1
p2	Capacity monitor for p1 (Alignment sector of M1 free)	t2	Loading of welding sector of M1
p3	Welding sector of M1 busy	t3	Welding operation and pallet code-writing in M1
p4	M1 finished welding of panel	t4	Reading of the following task and which machine M2 (in which cell) the pallet is forwarded to
p5	M1 finished welding of panel		
p6	Capacity monitor for p3, p4 and p5 (machine M1 free)	t5	Transfer of a pallet from M1 to M2
p7	Welding sector of M2 busy	t6	Welding operation and pallet code-writing in M2
p8	M2 finished welding of panel		
p9	Capacity monitor for p7 and p8 (machines M2 free)	t7	Transfer of a pallet from M2 to the FIFO-transfer system
p10	Free-places of the FIFO storage	t8	Transfer of a pallet between two consecutive places of the FIFO-transfer system
p11	Busy-places of the FIFO storage		
p12	Welding sector of M3 busy	t9	Transfer of a pallet from FIFO-transfer system to machine M3
p13	M3 finished welding of panel		
p14	Capacity monitor for p12 and p13 (machine M3 free)	t10	Welding operation in M3
		t11	Unloading of machine M3

The initial marking of the CPN is: p1 = 0; p2 = $<c_1,al_1>, <c_2,al_2>, <c_1,al_2>, <c_2,al_2>$; p3=0; p4=0; p5=0; p6 = $<c_1,m1_1>, <c_2,m1_1>$; p7 = 0; p8 = 0; p9 = $<c_1,m2_1>, <c_2,m2_1>, <c_1,m2_2>, <c_2,m2_2>$; p10 = $<c_1,ts_1>, <c_2,ts_1>, <c_1,ts_2>, <c_2,ts_2>, <c_1,ts_3>, <c_2,ts_3>$; p11=0; p12=0; p13=0; p14 = $<c_1,m3_1>, <c_2,m3_1>$.

- *Structural analysis of the CPN model:* The set of p- and t-flows of the CPN obtained from its incidence-matrix determines the following p- and t-invariant relationships:

PI1: $Proj12.m(p2)+m(p1) = Proj12.mo(p2)+mo(p1) = (<c_1,al_1>, <c_1,al_2>, <c_2,al_1>, <c_2,al_2>)$

PI2: $[m(p3)+m(p4)+m(p5)]+Proj13.m(p6) = [mo(p3)+mo(p4)+mo(p5)]+Proj13.mo(p6) = (<c_1,m1_1>, <c_2,m1_1>)$

PI3: $[m(p7)+m(p8)]+Proj14.m(p9) = [mo(p7)+mo(p8)]+Proj14.mo(p9) = (<c_1,m2_1>, <c_1,m2_2>, <c_2,m2_1>, <c_2,m2_2>)$

PI4: $[m(p12)+m(p13)]+Proj15.m(p14) = [mo(p12)+mo(p13)]+Proj15.mo(p14) = (<c_1,m3_1>, <c_2,m3_1>)$

PI5: $m(p11)+Proj16.m(p10) = mo(p11)+Proj16.mo(p10) = <ts1>+<ts2>+<ts3> = \sum_{1\le i\le 3} <tsi>$

PI6: $m(p1)+Proj7.m(p3)+Suc72_1.m(p4)+Proj14.m(p5)+Proj7.m(p7)+ Suc72_1.mo(p8)+m(p11)+Proj7.m(p12)+m(p13) = mo(p1)+Proj7.mo(p3)+Suc72_1.mo(p4)+Proj14.mo(p5)+Proj7.mo(p7)+ +Suc72_1.mo(p8)+mo(p11)+Proj7.mo(p12)+mo(p13) = 0$

TI1: $I(p,t,c_j).tf1 = I(p,t,c_j).[Id\ Proj7\ Suc72_1\ Proj14\ Proj7\ Suc72_1\ Id\ S<X>\ Proj7\ Id\ Id\ ]^T$

Remark: $S<X> = <X>+<X+1>+..+<X+i>+..+<X-1>$, where $<X>$ represents a particular marking- or occurrence-colour.

- *Formal specification of the modelled FMS by means of the invariant relationships:* Validation of the cell specifications.

From the results obtained in the last item, it is possible to ensure that the proposed OCPN model correctly represents the priorly stated specifications. For example, the structure of PI3 shows that machines M2 of cell 1 can be found in one of the following states: free (marking $<c_1,m2_1>$, $<c_1,m2_2>$ in place p9), busy by a non-processed part (marking $<c_1,m2_1>$, $<c_1,m2_2>$ in place p7), busy by a processed part (marking $<c_1,m2_1>$, $<c_1,m2_2>$ in place p8).

From the structure of PI3, it can also be concluded that places p7, p8 and p9 have a marking-bound given by the initial marking of place p9. These places are in mutual exclusion with respect to colours ($<c_1,m2_1>$, $<c_1,m2_2>$, $<c_2,m2_1>$, $<c_2,m2_2>$).

A similar analysis applied to TI1 allows to prove that the enabling-condition and the firing of transition t3 for the firing-mode (occurrence-colour) $<c_1,m1_1>$, turns out the following set of actions: In machine M1 (marking $<pa_s, wp_r>$ in p3) runs a program (marking $<wp_r>$); M1 works on a part (markings $<pa_s>$), and the next task to be performed on the panel ($Suc72_1$) is written onto it.

The validation of the FIFO system has already been presented in [Colombo et al., 94d]

Finally, and referring to step 13 of the procedure suggested in Section 3 for designing the control system of a flexible production system, by means of a Top-Down design method, the firing-modes (occurrence-colours) in the CPN model are now refined by assigning an underlying PN to each of them. As an example, figure 9 presents the results of this step in relation to transition t10 of figure 8, where the newly obtained set of transitions models both the technical actions and the handling of real signals to be exchanged between the control model and the sensor/actuator interface of the production environment [Villaroel, 90], [David and Alla, 92], [Feldmann et al., 95].

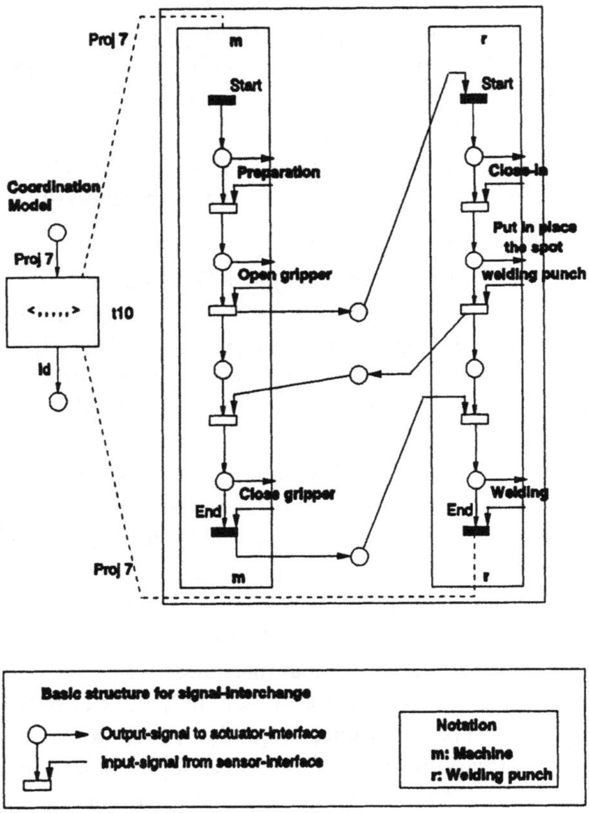

Fig. 9. Model obtained after a second top-down stage for refinement of transitions.

In the CPN of Fig. 9, when transition t_{10} is effectively enabled, the occurrence-colours Proj7 corresponds to an underlying PN where the workplan <wpr> is performed by enabling and firing a new set of place-transitions mapped into the elements composing the machine 3 of each cell.

7 Conclusions

In this chapter, main concepts related to Petri nets for the design, modelling and analysis of manufacturing systems have been overviewed. A methodology for obtaining the PN based model of the FMS control system is proposed. It is emphasized the importance of interactive modelling and validation for a successful implementation of the FMS. A design example of a FMS using TPN and CPN illustrates the main concepts.

References

Alla H., Ladet P., Martínez J. and Silva M. (1984). Modelling and Validation of Complex Systems by Coloured Petri Nets. Lecture Notes on Computer Science 188. Advances in Petri Nets 1984. Springer-Verlag, pp.15-29.

Brand K.P. and Kpoansky J. (1988). Principles and Engineering of Process Control with Petri Nets. IEEE Trans. on Automatic Control. V.33, no.2, pp.138-149.

Campos J. (1990). Performance Bounds for Synchronized Queueing Networks. (Doctoral Thesis). Dept. de Ingeniería Eléctrica e Informática. Univ. de Zaragoza. España.

Cao T. and Sanderson A.C. (1992). Sensor-based Error Recovery for Robotic Task Sequences Using Fuzzy Petri Nets. IEEE International Conference on Robotics and Automation. Vol. 2. pp.1063-1069. Nice, France.

Cao T. and Sanderson A.C. (1995). Task sequence Planning Using Fuzzy Petri Nets. IEEE Transactions on Systems, Man, and Cybernetics. V. 25. Nº5. pp.755-768.

Carelli R., Colombo A.W., Bernhard R. and Schreck G. (1995). Discrete Event and Motion Oriented Simulation for FMS. BASYS'95, Vitoria, Brasil. Balanced Automation Systems, Architectures and Design Methods, Ed. Chapman and Hall, pp.107-120.

Chao D.Y., Zhou M.C. and Wang D.T. (1992). Extending Knitting Technique to Petri Net Synthesis of Automated Manufacturing Systems. Proc. Rensselaer's Third Conf. on Computer Integrated Manufacturing. Troy NY, pp.56-63.

Chiola G. (1987). GreatSPN, User's Manual. Versión 1.3. Dipartamento di Informática. Universita degli Studi di Torino. Italy.

Cofer D. and Garg V. (1992). A Timed Model for the Control of Discrete Event Systems Involving Decisions in the Max/Plus Algebra. Proc. of the 31st. Conference on Decision and Control, pp.3363-3368. Arizona, USA.

Colombo A.W. (1994a). Modelling and Analysis of Flexible Production Systems (in Spanish). MsC. Thesis, Univ. Nac. San Juan, Ed. Fund. UNSJ, Argentina.

Colombo A.W., Martinez J. and Carelli R. (1994b). Formal Specification and Validation of Complex Production System using Coloured Petri Nets. Proceedings of the IEEE - International Conference on Robotics and Automation ICRA '94. Vol. 2, S. 1713-1718. California, USA.

Colombo A.W., Pellicer J., Martín M. and Kuchen B. (1994c). FMS Simulator by using Temporized Petri Nets (in Spanish). 6º Congreso Latino Americano de Control Automático, vol. 1, pp.100-102. Rio de Janeiro. Brasil.

Colombo A.W., Martínez J. and Kuchen B. (1994d). Formal Specification and Validation of a FIFO Module using Coloured Petri Nets. Proc. IEEE Int. Symp. on Industrial

Electronics, ISIE'94, pp.231-235, Santiago, Chile.
David R. and Alla H. (1992). Petri nets & Grafcet: Tools for modelling discrete event systems. Prentice Hall.
David R. and Alla H. (1993). Autonomous and Timed Continuous Petri Nets. Lectures on Computer Science. Advances in Petri Nets. pp. 71-90. Springer-Verlag.
David R. and Alla H. (1994). Petri Nets for Modeling of Dynamic Systems - A Survey. Automatica. Vol. 30, Nº2, pp.175-202.
Ezpeleta J. and Martínez J. (1991). Petri Nets as a Specification Language for Manufacturing Systems. 13th. IMACS World Congress on Computation and Applied Mathematics. Dublin.
Feldmann K., Colombo A.W. and Schnur C. (1995). An Approach for Modelling, Analysis and Real-Time Control of Flexible Manufacturing Systems using Petri Nets. Proceedings of the 7th European Simulation Symposium. ESS'95, pp.661-665. Erlangen, Germany.
Ferrarini L. (1992). An Incremental Approach to Logic Controller Design with Petri Nets. IEEE Trans. on Syst., Man, and Cybernetics. 22 (3), pp.461-473.
Freedman P. (1991). Time, Petri Nets and Robotics. IEEE Trans. on Rob. and Autom., vol.7, no.4, pp.417-433.
Furukawa C.M., Marchese M. and Miyagi P. (1990). Methods and Techniques for PLCs Efficient Programming in Complex Applications (in Portuguese). Proceedings of the IV Latinamerican Congress of Automatic Control, pp.250-255. Puebla, Mexico.
Gomes L., Steiger-Garcao A., Gama L., and Correia N. (1993). Controllers Programming by Means of Petri Nets (in Portuguese). Faculdade de Ciencias e Tecnologia, Universidade Nova de Lisboa.
Hack M. (1975). Petri Nets Languages. Comput. Struct. Gr. Memo 124. Project HAC. M.I.T. Cambridge Mass.
Haddad S. and Couvreur J.M. (1988). Towards a General and Powerful Computation of Flows for Parametrized Coloured Nets. 9th. European Workshop on Applications and Theory of Petri Nets. Venice, Italy.
Hasegawa K., Takahashi K. and Miyagi P. (1988). Application of the Mark Flow Graph to Represent Discrete Event Production Systems and System Control. Trans. of SICE, Vol. 24, Nº1. Japan.
IEEE (1989). Proceedings of the 3rd. International Workshop on Petri Nets and Performance Models. Kyoto. Japan.
Jarschel W., Drebinger A. and Bolch G. (1992). Modellierung von Fertigungssystemen mit dem Petri-Netz Simulator PETSY. Witschaftsinformatik, 34, Jahrgang, Heft 5,S.535-545.
Jeng M. and DiCesare F. (1992). A Modular Petri Net Synthesis Approach for Automated Manufacturing Systems. Proc. Japan USA Symp. on Flexible Automation, San Fransisco CA, pp.1163-1170.
Jeng M. and DiCesare F. (1993). A Review of Synthesis Techniques for Petri Nets with Applications to Manufacturing Systems. IEEE Trans. Syst., Man, and Cyber., vol.23, no.1, pp.301-312.
Jensen K. (1981). Coloured Petri Nets and the Invariant Method. Theoretical Computer Science 14. North-Holland, Amsterdam, pp.317-336.
Jensen K. (1992). Coloured Petri Nets. Basic Concepts, Analysis Methods and Practical Use. Vol.1, Monographs and Theoretical Computer Sciences. Springer-Verlag.
Le Bail J., Alla, H. and David R. (1992). Asymptotic Continuous Petri Nets: An Efficient Approximation of Discrete Event Systems. Proceedings IEEE International Conference on Robotics and Automation. V.2. pp.1050-1056. Nice, France.
Marsan M.A. (1989). Stochastic Petri Nets: An Elementary Introduction. Advances in Petri

Nets 1989, pp.1-29. Springer-Verlag.

Marsan M.A., Balbo G., Conte G., Donatelli S. and Franceschinis G. (1995). Modelling with Generalized Stochastic Petri Nets. Wiley Series on Parallel Computing.

Martinez J., Muro P. and Silva M. (1987). Modelling, Validation and Software Implementation of Production Systems using High-Level Petri Nets. Proc. IEEE Int. Conf. on Robotics and Autom., Raleigh, North Carolina, pp.1180-1185.

Miyagi P. E. (1988). Control System Design, Programming and Implementation for Discrete Event System by Using Mark Flow Graphs. Doctoral Thesis, Tokyo Institute of Technology, Japan.

Miyagi P. E. (1993). Discret Event Systems Control (in Portuguese). VI Escuela de Robótica. CINVESTAV. 1993.

Molloy M. (1982). Performance Analysis using Stochastic Petri Nets. IEEE Trans. on Comp., vol. c-31, no.9, pp.913-917.

Murata T. (1989). Petri Nets: Properties, Analysis and Applications. Proceedings of the IEEE, vol.77, no.4, pp.541-580.

Peterson J.L. (1981). Petri Net Theory and the Modelling of Systems. Prentice-Hall Inc., Englewood Cliffs, N.J.

Ramaswamy, S. and Valavanis, K. P. (1994). Modeling, Analysis and Simulation of Failures in a Materials Handling System With Extended Petri Nets. IEEE Transactions on Systems, Man, and Cybernetics. V. 24. Nº9. pp. 1358-1373.

Ramchandani C. (1974). Analysis of Asynchronous Concurrent Systems by Petri Nets. PhD dissertation, L.C.M./MIT, Cambridge, USA.

Sifakis J. (1977). Use of Petri Nets for Performance Evaluation. H. Beilner and E. Gelenbe, Eds. Measuring, Modelling and Evaluation of Computer Systems. New York: North-Holland, pp.75-93.

Silva M. and Velilla. (1982). Programmable Logic Controllers and Petri Nets: A Comparative Study. IFAC Software for Computer Control. Madrid, Spain.

Silva M. (1985). Las Redes de Petri en la Automática y la Informática. Editorial AC, Madrid, Spain.

Silva M. and Valette R. (1989). Petri Nets and Flexible Manufacturing Systems. Advances in Petri Nets, Springer-Verlag, pp.374-417.

Silva R. and Miyagi P. (1995). PFS/MFG: A High Level Net for the Modeling of Discrete Manufacturing Systems. BASYS'95, Vitoria, Brasil. Balanced Automation Systems, Architectures and Design Methods. Ed. Chapman and Hall. pp.107-120.

Tomeco, Ingeniería. Solar Heating Equipments. Product catalog, Argentina, 1995.

Treviño A.R. (1990). Task Modelling and Controllers Specifications in Robotic Cells (in Spanish). MsC Thesis, CINVESTAV, México.

Villaroel J. (1990). Integración Informática del Control en Sistemas Flexibles de Fabricación. Doctoral thesis, Informatics and Electrical Eng. Dept., University of Zaragoza, Spain.

Zhou M. (1990a). A Theory for the Synthesis and Augmentation of Petri Nets in Automation. Doctoral Dissertation, ECSE, Rensselaer Polytechnic Institute, Troy NY.

Zhou M. and DiCesare F.A. (1990b). Petri Net Design Method for Automated Manufacturing Systems with Shared Resources. Proc. of 1990 IEEE Int. Conf. on Robotics and Autom, Cincinnati OH, pp.526-531.

Zhou M.C., DiCesare F. and Rudolph D. (1990c). Control of Flexible Manufacturing System using Petri Nets. 11th. IFAC World Congress, Tallinn, Estonia.

Zhou M., DiCesare F. and Desrochers A. (1992a). A Hybrid Methodology for Synthesis of Petri Net Models for Manufacturing Systems. IEEE Trans. on Robotics and Autom., vol.8, no.3, pp.350-361.

Zhou M., DiCesare F. and Rudolph D. (1992b). Design and Implementation of a Petri Net Supervisor for a Flexible Manufacturing System. Automatica, vol. 28, Nº67, pp.1199-1208.

Zho M.C. (1992c). Computer-aided Modeling, Analysis and Design of Discrete Event Systems using Petri Net. Proc. 1992 IEEE Symp. on Computer Aided Control System Design (CACSD '92). Napa CA, pp.255-261.

Zhou M.C. and DiCesare F. (1993). Petri Nets Synthesis for Discrete Event Control of Manufacturing Systems. Kluwer Academic Publishers, Boston, MA.

Zurawsky R. (1992). A Class of Petri Net Models used to Represent Primary Components of Flexible Manufacturing Systems. Proc. Japan USA Symp. on Flexible Automation, San Francisco CA, pp.1695-1702.

12 Petri Net-Based Approach to Synthesis of Intelligent Control Systems for DEDS

Spyros Tzafestas and František Čapkovič

1. Introduction

In order to control a system, especially a kind of discrete event dynamic systems (DEDS) like flexible manufacturing systems (FMS), transport systems, communication systems, etc., an additional amount of knowledge (besides the system model) is needed in the process of the control synthesis. Consequently, a suitable knowledge representation scheme must be used in addition to the standard mathematical model of the system to be controlled. Such a knowledge representation is necessary in order to express additional information concerning the particulars of the control task specifications - e.g. control aims, external circumstances and influences, different limitations, human experience, etc. Sometimes, the knowledge in question may have a logical character. However, many times its character is fuzzy. Usually, such knowledge can be expressed by means of implications in the form of the well known *IF-THEN* rules. A set of rules creates a knowledge base (KB) which can be utilized *"off line"* - i.e. during the process of the control systems synthesis - or *"on line"* - i.e. directly in the feedback control process of the real kind of the DEDS. The former application is illustrated by Fig. 1 and the latter one by Fig. 2. The meaning of the variables used in both figures are given

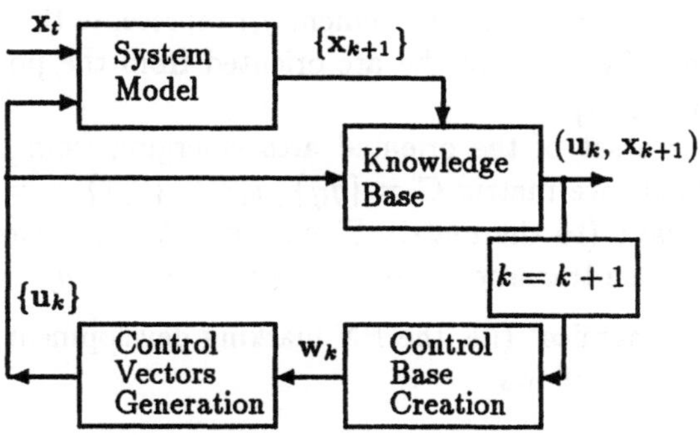

Fig. 1. The principal scheme of the control synthesis (off line approach)

in Part 3.

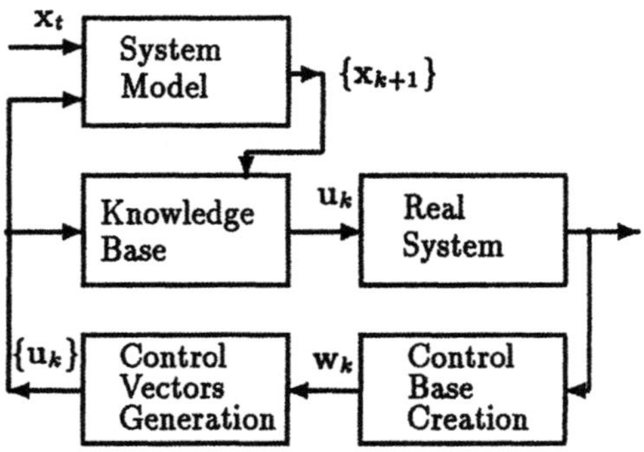

Fig. 2. The principal scheme of the control synthesis (on line approach)

2. Petri net-based modelling of DEDS

The DEDS are often described by different types of Petri nets (PN). The ordinary PNs are chosen here. They are understood to be the directed bipartite graphs. Consequently, the PN structure can be represented by the quadruplet

$$\langle P,T,F,G \rangle \quad ; \quad P \cap T = \emptyset \quad ; \quad F \cap G = \emptyset \tag{1}$$

where

P $= \{p_1, ..., p_n\}$ is a finite set of the PN positions with p_i, $i = 1, n$, being the elementary positions

T $= \{t_1, ..., t_m\}$ is a finite set of the PN transitions with t_j, $j = 1, m$, being the elementary transitions

F $\subseteq P \times T$ is a set of the oriented arcs enterig the transitions. It can be expressed by means of the arcs incidence matrix $\mathbf{F} = \{f_{ij}\}$, $f_{ij} \in \{0,1\}$, $i = 1, n$; $j = 1, m$. The element f_{ij} represents the absence (when 0) or presence (when 1) of the arc oriented from the position p_i to its output transition t_j

G $\subseteq R \times S$ is a set of the oriented arcs emerging from the transitions where the incidence matrix $\mathbf{G} = \{g_{ij}\}$, $g_{ij} \in \{0,1\}$, $i = 1, m$; $j = 1, n$ very analogically (to the matrix F) expresses the occurrence of the arc oriented from the transition t_i to its output position p_j

As to the PN "dynamics" (i.e. the PN marking development) it can be formally understood as follows

$$\langle X, U, \delta, \mathbf{x}_0 \rangle \quad ; \quad X \cap U = \emptyset \tag{2}$$

where

$X = \{\mathbf{x}_0, ..., \mathbf{x}_N\}$ is a set of the state vectors of the PN (the marking) with $\mathbf{x}_k = (\sigma_{p_1}^k, ..., \sigma_{p_n}^k)^T$; $k = 0, N$ being the state vector of the PN (the state of marking) in the step k, where k is the discrete step of the PN dynamics development, and $\sigma_{p_i}^k$, $i = 1, n$ is the state of the marking of the elementary positions p_i in the step k. T symbolizes the vector or matrix transposition and N is an integer representing formally the number of different situations during the PN dynamics development

$U = \{\mathbf{u}_0, ..., \mathbf{u}_N\}$ is a set of the "control" vectors of the PN (i.e. the state vectors of enabling the transitions) with $\mathbf{u}_k = (\gamma_{t_1}^k, ..., \gamma_{t_m}^k)^T$; $k = 0, N$ being the control vector of the PN in the step k, where $\gamma_{t_j}^k$, $j = 1, m$ is the state of enabling the elementary transition t_j in the step k

$\delta : X \times U \longmapsto X$ is a transition function of the PN

$\mathbf{x}_0$ is the initial state vector of the PN

Consequently, the simplest form of the PN-based model of the FMS can be expressed in analytical terms by means of the linear discrete system (with the coefficient matrix $\mathbf{A}$ being the identity matrix) as follows

$$\mathbf{x}_{k+1} = \mathbf{x}_k + \mathbf{B}.\mathbf{u}_k \quad , \quad k = 0, N \tag{3}$$

$$\mathbf{B} = \mathbf{G}^T - \mathbf{F} \tag{4}$$

$$\mathbf{F}.\mathbf{u}_k \leq \mathbf{x}_k \tag{5}$$

where

k $= 0, 1, ..., N$ is the disrete step of the system dynamics development

$\mathbf{x}_k$ is the n-dimensional state vector of the system in the step k. Its components express the states of the FMS elementary subprocesses. They acquire their values from the set $\{0, 1\}$ (0 - passivity or 1 - activity).

$\mathbf{u}_k$ is the m-dimensional control vector of the system in the step k. Its components represent the occurence of the FMS elementary discrete events (e.g. starting or ending the elementary subprocesses or other activities). They acquire their values from the set $\{0, 1\}$ (1 - presence, 0 - absence).

$\mathbf{B}$, $\mathbf{F}$, $\mathbf{G}$ are respectively, $(n \times m)$, $(n \times m)$ and $(m \times n)$- dimensional structural matrices of constant elements. The matrix $\mathbf{F}$ expresses the causal relations among the states of the FMS and the discrete events occuring during the FMS operation, where the states are the causes and the events are the consequences. The matrix $\mathbf{G}$ expresses the causal relation among the discrete events (the causes) and the FMS states (the consequences). Both of these matrices are the oriented arcs incidence matrices. The elements of them acquire their values from the set $\{0, 1\}$, 1 - existence, 0 - nonexistence of the corresponding causal relations.

$(.)^T$ symbolizes the matrix or vector transposition

More details about such an understanding of the PN can be found e.g. in [1], where the corresponding formalism is also presented.

3. Analysis of the DEDS control possibilities

In order to find a suitable control vector $\mathbf{u}_k$ the following procedure will be used:

$$\overline{\mathbf{x}}_k = \underline{\text{neg}}\, \mathbf{x}_k = \mathbf{1}_n - \mathbf{x}_k \tag{6}$$

$$\mathbf{v}_k = \mathbf{F}^T \underline{\text{and}}\, \overline{\mathbf{x}}_k \tag{7}$$

$$\mathbf{w}_k = \underline{\text{neg}}\, \mathbf{v}_k = \mathbf{1}_m - \mathbf{v}_k = \mathbf{1}_m - (\mathbf{F}^T \underline{\text{and}}\, (\mathbf{1}_n - \mathbf{x}_k)) \tag{8}$$

where
$\underline{\text{neg}}$ is the operator of logical negation
$\underline{\text{and}}$ is the operator of logical multiplying
$\mathbf{1}_n$ is n-dimensional constant vector with elements equaled to 1
$\mathbf{v}_k$, $\mathbf{w}_k$ are, respectively, m-dimensional auxiliary vector and m-dimensional vector of the base for the control vector choice

The vector $\mathbf{w}_k$ expresses the possible candidates for the control vector $\mathbf{u}_k$ in the step k. When only one of its components is different from zero, it can be used to be the control vector, i.e. $\mathbf{u}_k = \mathbf{w}_k$. When there are several components of the $\mathbf{w}_K$ different from zero, the control vector $\mathbf{u}_k$ has to be chosen on the basis of additional information about the actual control task. The choice of the control vector can be made either by an expert in the corresponding domain or automatically on the basis of the rules predefined by means of the constraints imposed upon the task in question. The conditions which must be satisfied during the actual control problem solving (including the final aim of the control process) are understood to be such constraints. In Fig. 1 the symbol $\{.\}$ representss a set of alternative vectors possible in the step k, $\mathbf{x}_t$ is a terminal state of the system to be reached and $(\mathbf{u}_k, \mathbf{x}_{k+1})$ represents the pairs (control vector; corresponding response of the system). The vector $\mathbf{w}_k$ represents a base for creation of a set $\mathbf{u}_k$ of different possible control vectors in the step k.

More details concerning the DEDS modelling and control synthesis can be found in [1, 2].

4. Petri nets in the rule-based knowledge representation

As we can see on Fig. 1 or/and on Fig. 2 a suitable form of the problem oriented rule-based KB is needed within the frame of the control synthesis procedure in order to choose the most suitable control vector at each step k of the system dynamics development when there are more possibilities at our disposl. To give a sufficiently general possibility how to represent such a KB we shall use the Petri nets-based approach. The kinds of the PN we shall use are the logical PN (LPN) or/and fuzzylogical PN (FPN) - see e.g. [8].
Some pieces of knowledge (e.g. some statements) can be expresed by means

of the PN positions and the rules can be expressed by means of the PN transitions (taken together with their input and output positions). The mutual causality interconnections among the statements and rules can be expressed by means of the analogy with the oriented arcs among the PN positions and transitions. An example of such a rule is given in Fig. 3. The meanings of sym-

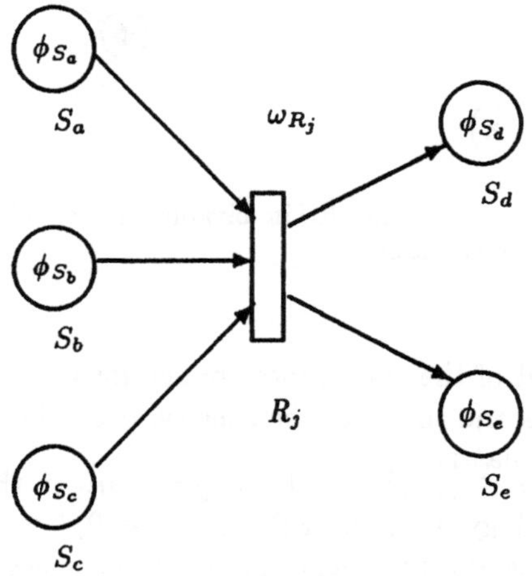

Fig. 3. The rule R_j with the input and output statements

bols are explained below. The KB truth propagation (i.e. the KB "dynamics" development) is clear from Fig. 4, where the tokens mean the true statements - in case of bivalued logic of course. In case of fuzzy logic the situation is a little different. As we can see in Fig. 5, no tokens can be used, however, the truth of the statements can be expressed by the real fuzzy values. Hence, the KB structure can be formally expressed as

$$\langle S, R, \Psi, \Gamma \rangle \quad ; \quad S \cap R = \emptyset \quad ; \quad \Psi \cap \Gamma = \emptyset \tag{9}$$

where

$S \ = \{S_1, ..., S_{n_1}\}$ is a finite set of the statements.

S_i , $i = 1, n_1$, are the pieces of knowledge (the elementary statements).

$R \ = \{R_1, ..., R_{m_1}\}$ is a finite set of the rules.

R_j , $j = 1, m_1$, are the rules either in the form of implications:

$R_j \ : \ (S_a \, and \, S_b \, and \, ... \, and \, S_c) \Rightarrow (S_d \, and \, S_e)$

or in the form of *IF-THEN* structures:

$R_j \ : \ \textbf{IF} \, (S_a \, and \, S_b \, and \, ... \, and \, S_c) \, \textbf{THEN} \, (S_d \, and \, S_e)$.

where S_a, S_b, ..., S_c are the input statements of the rule R_j, and the S_d, S_e are the output statements of the rule R_j.

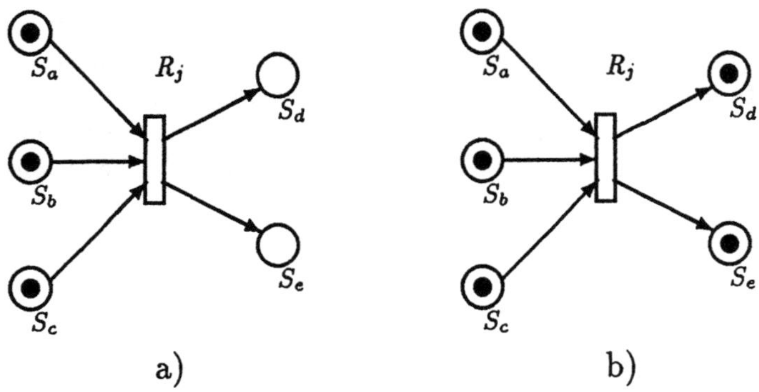

Fig. 4. The example of the rule R_j in bivalued logic. The state: a) before its evaluation; b) after its evaluation

$\Psi \subseteq S \times R$ is a set of the causal interconnections between the statements entering the rules and the rules themselves. It can be expressed by means of the incidence matrix
$\boldsymbol{\Psi} = \{\psi_{ij}\}$, $\psi_{ij} \in \{0,1\}$, $i = 1, n_1$; $j = 1, m_1$ in the analogy with the LPN. In the analogy with the FPN $\psi_{ij} \in <0,1>$. It means that the element ψ_{ij} represents the absence (when 0), presence (when 1) or a fuzzy measure of existence (when the value is between these boundary values) of the causal relation between the input statement S_i and the rule R_j.

$\Gamma \subseteq R \times S$ is a set of the causal interconnections between the rules and the statements emerging from them. It can be expressed by means of the incidence matrix
$\boldsymbol{\Gamma} = \{\gamma_{ij}\}$, $\gamma_{ij} \in \{0,1\}$, in analogy with LPN or $\gamma_{ij} \in <0,1>$ in analogy with FPN , $i = 1, m_1$; $j = 1, n_1$ very analogically (to the matrix $\boldsymbol{\Psi}$) expressing the occurrence of the causal relation between the rule R_i and its output statement S_j.

$\emptyset$ is an empty set.

The KB "dynamics" development (i.e. the statements truth propagation) can be formally expressed as follows

$$\langle \Phi, \Omega, \delta_1, \boldsymbol{\Phi}_0 \rangle \quad ; \quad \Phi \cap \Omega = \emptyset \tag{10}$$

where

$\Phi = \{\boldsymbol{\Phi}_0, ..., \boldsymbol{\Phi}_{N_1}\}$ is a set of the state vectors of the KB.
$\boldsymbol{\Phi}_K = (\phi^K_{S_1}, ..., \phi^K_{S_{n_1}})^T$; $K = 0, N_1$ is the state vector of the KB (the state of the statements truth propagation) in the step K.
K is the discrete step of the KB dynamics development
T symbolizes the vector or matrix transposition
N_1 is an integer representing the number of different situations during the KB dynamics development (i.e. during the statements truth propagation)

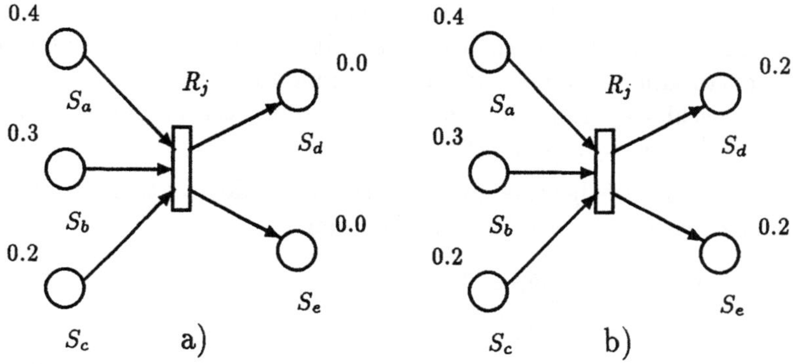

Fig. 5. The example of the rule R_j in fuzzy logic. The state: a) before its evaluation; b) after its evaluation

$\phi_{S_i}^K$, $i = 1, n_1$ is the state of the truth of the elementary statement S_i in the step K. It means that the statement is false (when 0), true (when 1) or that the statement is true with a fuzzy measure (when this parameter acquires its value from the real interval between these two boundary values).

Ω $= \{\Omega_0, ..., \Omega_{N_1}\}$ is a set of the "control" vectors of the KB.

$\Omega_K = (\omega_{R_1}^K, ..., \omega_{R_{m_1}}^K)^T$; $K = 0, N_1$ is the "control" vector of the KB (i.e. the state of the rules evaluability) in the step K.

$\omega_{R_j}^K$, $j = 1, m_1$ is the state of the rule R_j evaluability in the step K. It means that the rule is not able to be evaluated (when 0), the rule is able to be evaluated (when 1) or that the rule is able to be evaluated with a fuzzy measure (when this parameter acquires its value from the interval between these two boundary values).

δ_1 $: \Phi \times \Omega \longmapsto \Phi$ is a transition function of the KB.

Φ_0 is the initial state vector of the KB.

Consequently, the KB dynamics development - i.e. the KB truth propagation - can be expressed in analytical terms as follows

$$\Phi_{K+1} = \Phi_K \; \underline{or} \; \Delta \; \underline{and} \; \Omega_K \quad , \quad K = 0, N_1 \quad , \quad \Phi_{K_{|K=0}} = \Phi_0 \tag{11}$$

$$\Delta = \Gamma^T \; \underline{or} \; \Psi \tag{12}$$

$$\Psi \; \underline{and} \; \Omega_K \leq \Phi_K \tag{13}$$

where

$\underline{and}$ is the operator of logical multiplying in general. For both the bivalued logic and the fuzzy one it can be defined (for scalar operands) to be the minimum of its operands. For example the result of its application on the scalar operands a, b is a scalar c which can be obtained as follows: $a \, \underline{and} \, b = c = min\{a, b\}$.

$\underline{or}$ is the operator of logical additioning in general. For both the bivalued logic and the fuzzy one it can be defined (for scalar operands) to be the maximum of its operands. For example the result of its application on the scalar operands a, b is a scalar c which can be obtained as follows: $a\, \underline{or}\, b = c = max\,\{a, b\}$.

Hence, the knowledge inference will be very analogical to obtaining the control base vector $\mathbf{w}_k$ in 8. It is the following

$$\overline{\boldsymbol{\Phi}_K} = \underline{neg}\, \boldsymbol{\Phi}_K = \mathbf{1}_{n_1} - \boldsymbol{\Phi}_K \tag{14}$$

$$\mathbf{v}_K = \boldsymbol{\Psi}^T \underline{and}\, \overline{\boldsymbol{\Phi}_K} \tag{15}$$

$$\boldsymbol{\Omega}_K = \underline{neg}\, \mathbf{v}_K = \mathbf{1}_{m_1} - \mathbf{v}_K = \mathbf{1}_{m_1} - (\boldsymbol{\Psi}^T \underline{and}\, (\mathbf{1}_{n_1} - \boldsymbol{\Phi}_K)) \tag{16}$$

$$= \underline{neg}(\boldsymbol{\Psi}^T\, \underline{and}(\underline{neg}\, \boldsymbol{\Phi}_K)) \tag{17}$$

where the meaning of quantities is also very analogical to those utilized at the vector $\mathbf{w}_k$ deriving, i.e.

$\mathbf{v}_K$ is a m_1-dimensional auxiliary vector pointing out (by its nonzero elements) the rules that cannot be evaluated, because there is at least one false (of course in the LPN analogy) statement among its input statements

$\boldsymbol{\Omega}_K$ is a m_1-dimensional "control" vector pointing out the rules that have all their input statements true and, consequently, they can be evaluated in the step K of the KB dynamics development. This vector is a base of the inference, because it contains information about the rules that can contribute to obtaining the new knowledge - i.e. to transfer the KB from the state $\boldsymbol{\Phi}_K$ of the truth propagation into another state $\boldsymbol{\Phi}_{K+1}$. These rules correspond to the nonzero elements of the vector $\boldsymbol{\Omega}_K$.

$\underline{neg}$ is the operator of logical negation in general. For both the bivalued logic and the fuzzy one it can be defined (for scalar operands) to be the complement of its operand. For example the result of its application on the scalar operands a is a scalar b which can be obtained as follows: $\underline{neg}\, a = b = 1 - a$.

The equations (15) - (17) can be understood to be an expression of a very simple inference mechanism of an automatic reasoning procedure in analytical terms.

5. Utilizing the knowledge base in the control synthesis

In order to give more particulars (in comparison with Fig. 1) as to utilizing the above introduced KB in the control synthesis let us write the control synthesis procedure in the form of the following "flow chart"

START

– $k = 0$
– defining the initial state of the system $\mathbf{x}_k = \mathbf{x}_0$
– defining the terminal state of the system $\mathbf{x}_t$

LABEL:

– generation of the control base $\mathbf{w}_k$ utilizing the system model
– generation of the set $\{\mathbf{u}_k\}$ of the possible control vectors $\mathbf{u}_k \in \mathbf{w}_k$
– generation of the set of the corresponding model responses $\{\mathbf{x}_{k+1}\}$
– **confrontation** of such information with the *knowledge base*
– choice of a suitable control possibility by means of the actual knowledge base
– *if* (the solution of the control synthesis task was found) *then (goto* END) *else (begin* $k = k + 1$; *goto* LABEL; *end*)

END
In spite of the fact that this scheme is only verbal and very rough, we can see the actual place and the principal importance of the KB in the process of the DEDS control synthesis.

6. Illustrative examples

6.1 Flexible manufacturing systems

To illustrate the above problems let us introduce the following examples published in [2 - 5], [7].

6.1.1 A simple kind of the FMS. Consider the flexible manufacturing system given on Fig. 6. As we can see it consists of two robots serving five

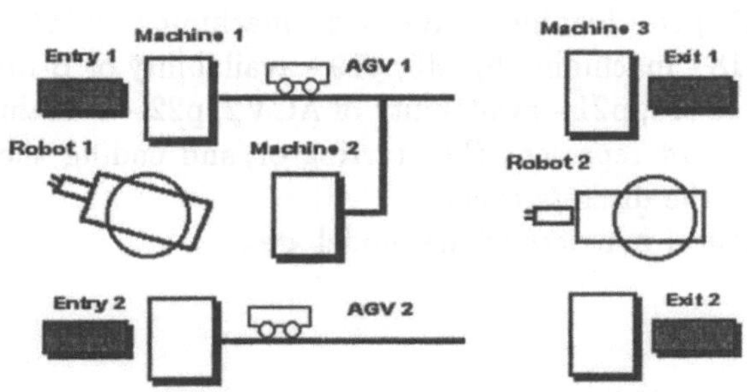

Fig. 6. Flexible manufacturing system

machine tools. The AGVs are automatic guided vehicles. Two entries are the inputs of raw materials A and B, respectively, and two exits are the outputs of the final A-parts and B-parts, respectively. The machines 1 and 2 produce the same intermediate A-parts and machine 4 produces the intermediate B-parts. Machines 3 and 5 produce the final A-parts and B-parts, respectively.

Using above mentioned analogy the OPN-based model can be obtained. The graphical representation of the model is presented by means of the OPN given on Fig. 7. It can be knitted e.g. by means of the method presented

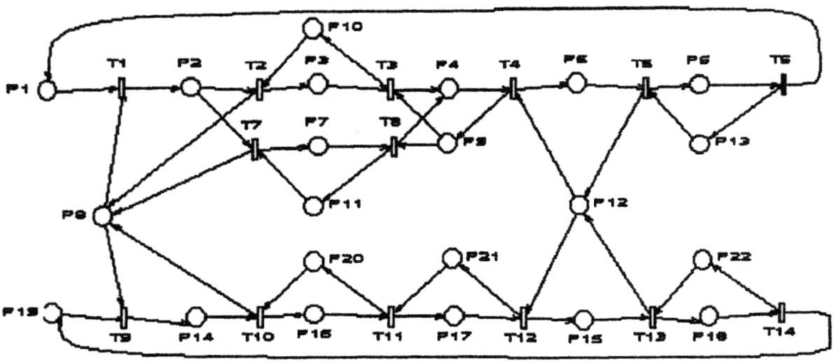

Fig. 7. Petri nets-based model of the FMS

in [10].

The meaning of the positions is: p1 - availability of A-raw material, p2 - loading by Robot 1 (R1), p3 - machining by Machine 1 (M1), p4 - delivering via AGV1, p5 - loading by R2, p6 - machining by M3, p7 - machining by M2, p8 - availability of R1, p9 - availability of AGV1, p10 - availability of M1, p11 - availability of M2, p12 - availability of R2, p13 - availability of M3, p14 - loading by R1, p15 - loading by R2, p16 - machining by M4, p17 - delivering via AGV2, p18 - machining by M5, p19 - availability of B-raw material, p20 - availability of M4, p21 - availability of AGV2, p22 - availability of M5. The trasitions t1 - t14 represent the starting or/and ending the corresponding operations i.e. the discrete events.

The structural matrices of the model are

$$
\mathbf{F} = \begin{pmatrix}
1\,0\,0\,0\,0\,0\,0\,0\,0\,0\,0\,0\,0\,0 \\
0\,1\,0\,0\,0\,0\,1\,0\,0\,0\,0\,0\,0\,0 \\
0\,0\,1\,0\,0\,0\,0\,0\,0\,0\,0\,0\,0\,0 \\
0\,0\,0\,1\,0\,0\,0\,0\,0\,0\,0\,0\,0\,0 \\
0\,0\,0\,0\,1\,0\,0\,0\,0\,0\,0\,0\,0\,0 \\
0\,0\,0\,0\,0\,1\,0\,0\,0\,0\,0\,0\,0\,0 \\
0\,0\,0\,0\,0\,0\,0\,1\,0\,0\,0\,0\,0\,0 \\
1\,0\,0\,0\,0\,0\,0\,0\,1\,0\,0\,0\,0\,0 \\
0\,0\,1\,0\,0\,0\,0\,1\,0\,0\,0\,0\,0\,0 \\
0\,1\,0\,0\,0\,0\,0\,0\,0\,0\,0\,0\,0\,0 \\
0\,0\,0\,0\,0\,0\,1\,0\,0\,0\,0\,0\,0\,0 \\
0\,0\,0\,1\,0\,0\,0\,0\,0\,0\,0\,1\,0\,0 \\
0\,0\,0\,0\,1\,0\,0\,0\,0\,0\,0\,0\,0\,0 \\
0\,0\,0\,0\,0\,0\,0\,0\,0\,1\,0\,0\,0\,0 \\
0\,0\,0\,0\,0\,0\,0\,0\,0\,0\,0\,0\,1\,0 \\
0\,0\,0\,0\,0\,0\,0\,0\,0\,0\,1\,0\,0\,0 \\
0\,0\,0\,0\,0\,0\,0\,0\,0\,0\,0\,1\,0\,0 \\
0\,0\,0\,0\,0\,0\,0\,0\,0\,0\,0\,0\,0\,1 \\
0\,0\,0\,0\,0\,0\,0\,0\,1\,0\,0\,0\,0\,0 \\
0\,0\,0\,0\,0\,0\,0\,0\,0\,1\,0\,0\,0\,0 \\
0\,0\,0\,0\,0\,0\,0\,0\,0\,0\,1\,0\,0\,0 \\
0\,0\,0\,0\,0\,0\,0\,0\,0\,0\,0\,0\,1\,0
\end{pmatrix}
; \mathbf{G}^T = \begin{pmatrix}
0\,0\,0\,0\,0\,1\,0\,0\,0\,0\,0\,0\,0\,0 \\
1\,0\,0\,0\,0\,0\,0\,0\,0\,0\,0\,0\,0\,0 \\
0\,1\,0\,0\,0\,0\,0\,0\,0\,0\,0\,0\,0\,0 \\
0\,0\,1\,0\,0\,0\,0\,1\,0\,0\,0\,0\,0\,0 \\
0\,0\,0\,1\,0\,0\,0\,0\,0\,0\,0\,0\,0\,0 \\
0\,0\,0\,0\,1\,0\,0\,0\,0\,0\,0\,0\,0\,0 \\
0\,0\,0\,0\,0\,0\,1\,0\,0\,0\,0\,0\,0\,0 \\
0\,1\,0\,0\,0\,0\,1\,0\,0\,1\,0\,0\,0\,0 \\
0\,0\,0\,1\,0\,0\,0\,0\,0\,0\,0\,0\,0\,0 \\
0\,0\,1\,0\,0\,0\,0\,0\,0\,0\,0\,0\,0\,0 \\
0\,0\,0\,0\,0\,0\,0\,1\,0\,0\,0\,0\,0\,0 \\
0\,0\,0\,0\,1\,0\,0\,0\,0\,0\,0\,0\,1\,0 \\
0\,0\,0\,0\,0\,1\,0\,0\,0\,0\,0\,0\,0\,0 \\
0\,0\,0\,0\,0\,0\,0\,0\,1\,0\,0\,0\,0\,0 \\
0\,0\,0\,0\,0\,0\,0\,0\,0\,0\,0\,1\,0\,0 \\
0\,0\,0\,0\,0\,0\,0\,0\,0\,1\,0\,0\,0\,0 \\
0\,0\,0\,0\,0\,0\,0\,0\,0\,0\,1\,0\,0\,0 \\
0\,0\,0\,0\,0\,0\,0\,0\,0\,0\,0\,0\,1\,0 \\
0\,0\,0\,0\,0\,0\,0\,0\,0\,0\,0\,0\,0\,1 \\
0\,0\,0\,0\,0\,0\,0\,0\,0\,0\,1\,0\,0\,0 \\
0\,0\,0\,0\,0\,0\,0\,0\,0\,0\,0\,1\,0\,0 \\
0\,0\,0\,0\,0\,0\,0\,0\,0\,0\,0\,0\,0\,1
\end{pmatrix}
$$

The initial state vector of the process is

$$\mathbf{x}_0 = (2\,0\,0\,0\,0\,0\,0\,1\,1\,1\,1\,1\,1\,0\,0\,0\,0\,0\,2\,1\,1\,1)^T$$

Consequently, the system dynamics development is the following

1. in the step $k = 0$ of the FMS dynamics development is

$$\mathbf{w}_0 = (1\,0\,0\,0\,0\,0\,0\,0\,1\,0\,0\,0\,0\,0)^T$$

Hence, the following control possibilities can be automatically generated

$$\mathbf{u}_0^1 = (1\,0\,0\,0\,0\,0\,0\,0\,0\,0\,0\,0\,0\,0)^T$$
$$\mathbf{u}_0^2 = (0\,0\,0\,0\,0\,0\,0\,0\,1\,0\,0\,0\,0\,0)^T$$
$$\mathbf{u}_0^3 = (1\,0\,0\,0\,0\,0\,0\,0\,1\,0\,0\,0\,0\,0)^T$$

The $\mathbf{u}_0^3$ does not satisfy the condition (5). Remaining two possibilities are admissible, however, not simultaneously. There is the conflict among them (i.e. among the enabled transitions t_1 and t_9). The model itself is not able to solve such a conflict. To solve it a rule has to be defined.

a) When the first possibility $\mathbf{u}_0^1$ is chosen the further development of the FMS dynamics is the following

$$\mathbf{x}_1 = (1\,1\,0\,0\,0\,0\,0\,1\,1\,1\,1\,1\,1\,0\,0\,0\,0\,0\,2\,1\,1\,1)^T$$

Consequently, the base for control synthesis in the step $k = 1$ of the FMS dynamics development is

$$\mathbf{w}_1 = (0\,1\,0\,0\,0\,0\,1\,0\,0\,0\,0\,0\,0\,0)^T$$

Hence, the following control possibilities can be automatically generated

$$\mathbf{u}_1^1 = (0\,1\,0\,0\,0\,0\,0\,0\,0\,0\,0\,0\,0\,0)^T$$
$$\mathbf{u}_1^2 = (0\,0\,0\,0\,0\,0\,1\,0\,0\,0\,0\,0\,0\,0)^T$$
$$\mathbf{u}_1^2 = (0\,1\,0\,0\,0\,0\,1\,0\,0\,0\,0\,0\,0\,0)^T$$

However, only the first two possibilities can be alternatively realized.

i. Consequently, when the first possibility $\mathbf{u}_1^1$ is chosen the further dynamics development is

$$\mathbf{x}_2 = (1\,0\,1\,0\,0\,0\,0\,1\,1\,0\,1\,1\,1\,0\,0\,0\,0\,0\,2\,1\,1\,1)^T$$

The base for control synthesis for the step $k = 2$ of the FMS dynamics development is

$$\mathbf{w}_2 = (1\,0\,1\,0\,0\,0\,0\,0\,1\,0\,0\,0\,0\,0)^T$$

Hence, the following control possibilities can be automatically generated

$$\mathbf{u}_2^1 = (1\,0\,0\,0\,0\,0\,0\,0\,0\,0\,0\,0\,0\,0)^T$$
$$\mathbf{u}_2^2 = (0\,0\,1\,0\,0\,0\,0\,0\,0\,0\,0\,0\,0\,0)^T$$
$$\mathbf{u}_2^3 = (0\,0\,0\,0\,0\,0\,0\,0\,1\,0\,0\,0\,0\,0)^T$$
$$\mathbf{u}_2^4 = (1\,0\,1\,0\,0\,0\,0\,0\,0\,0\,0\,0\,0\,0)^T$$
$$\mathbf{u}_2^5 = (1\,0\,0\,0\,0\,0\,0\,0\,1\,0\,0\,0\,0\,0)^T$$
$$\mathbf{u}_2^6 = (0\,0\,1\,0\,0\,0\,0\,0\,1\,0\,0\,0\,0\,0)^T$$
$$\mathbf{u}_2^7 = (1\,0\,1\,0\,0\,0\,0\,0\,1\,0\,0\,0\,0\,0)^T$$

However, only five possibilities are possible as to (5) - $\mathbf{u}_2^5$ and $\mathbf{u}_2^7$ are impossible. After choosing the suitable possibility the development could continue analogically into further steps.

ii. When the second possibility $\mathbf{u}_1^2$ is chosen the further dynamics development is

$$\mathbf{x}_2 = (1\,0\,0\,0\,0\,0\,1\,1\,1\,1\,0\,1\,1\,0\,0\,0\,0\,0\,2\,1\,1\,1)^T$$

The base for control synthesis for the step $k = 2$ of the FMS dynamics development is

$$\mathbf{w}_2 = (1\,0\,0\,0\,0\,0\,0\,1\,1\,0\,0\,0\,0\,0)^T$$

Hence, the following control possibilities can be automatically generated

$$\begin{aligned}
\mathbf{u}_2^1 &= (10000000000000)^T \\
\mathbf{u}_2^2 &= (00000001000000)^T \\
\mathbf{u}_2^3 &= (00000000100000)^T \\
\mathbf{u}_2^4 &= (10000001000000)^T \\
\mathbf{u}_2^5 &= (10000000100000)^T \\
\mathbf{u}_2^6 &= (00000001100000)^T \\
\mathbf{u}_2^7 &= (10000001100000)^T
\end{aligned}$$

However, only five possibilities are possible as to (5) - $\mathbf{u}_2^5$ and $\mathbf{u}_2^7$ are impossible. After choosing the suitable possibility the development could continue analogically into further steps.

b) When the second possibility $\mathbf{u}_0^2$ is chosen the further dynamics development is

$$\mathbf{x}_1 = (2000000111111000011111)^T$$

Consequently, the base for control synthesis in the step $k = 1$ of the FMS dynamics development is

$$\mathbf{w}_1 = (00000000010000)^T$$

Hence, only the following control possibility can be automatically generated

$$\mathbf{u}_1 = (00000000010000)^T$$

Consequently, when the further development of the FMS dynamics is the following

$$\mathbf{x}_2 = (2000000111111001001011)^T$$

Consequently, the base for control synthesis in the step $k = 1$ of the FMS dynamics development is

$$\mathbf{w}_2 = (10000000100010)^T$$

Hence, only the following control possibilities can be automatically generated

$$\begin{aligned}
\mathbf{u}_2^1 &= (10000000000000)^T \\
\mathbf{u}_2^2 &= (00000000100000)^T \\
\mathbf{u}_2^3 &= (00000000000010)^T \\
\mathbf{u}_2^4 &= (10000000100000)^T \\
\mathbf{u}_2^5 &= (10000000000010)^T \\
\mathbf{u}_2^6 &= (00000000100010)^T \\
\mathbf{u}_2^7 &= (10000000100010)^T
\end{aligned}$$

However, only five possibilities are possible as to (5) - $\mathbf{u}_2^5$ and $\mathbf{u}_2^7$ are impossible. After choosing the suitable possibility the development could continue analogically into further steps.

As we can see the process is branching very wide and it is practically impossible for an operator to pursue all possibilities. In situations when there are several possibilities of choosing the vector $\mathbf{u}_k$ in a step k a rule or several rules have to be defined. Generally, building the consistent KB for a complicated real system is not any easy process. The human expert from the domain in question is expected to do it. When the set of rules is defined, the above introduced analytical model of the KB can be utilized in the control synthesis process to automatize it - i.e. to select the most suitable control in each step.

How the rules can be formed we can see in the foolowing example.

6.2 A simple fragment of the FMS

Consider the simple fragment of the FMS on Fig. 8. Its PN-based model

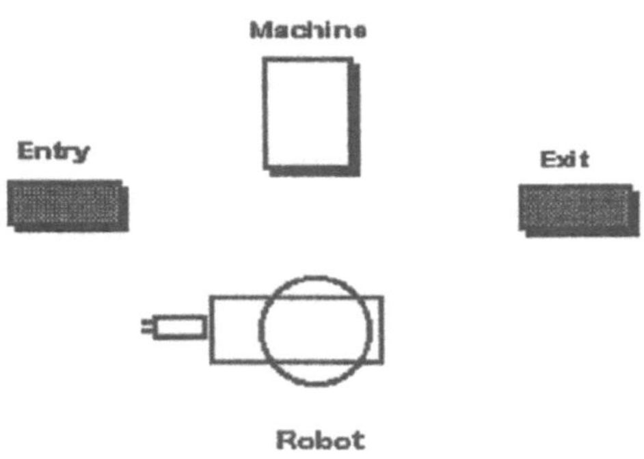

Fig. 8. The simple flexible manufacturing system

is given on Fig. 9 where $m = 16$ elementary positions model the following activities

p_1 - FMS is waiting to be started; p_2 - transport belt on the exit is switched on to enter the input parts; p_3 - robot is comming from the neutral position to NC-machine tool; p_4 - operation of grasping the input part; p_5 - robot is moving with the party to the NC-machine tool; p_6 - operation of putting the part into the machine; p_7 - robot is leaving the NC-machine tool to move into the neutral position; p_8 - robot is waiting in the neutral position; p_9 -

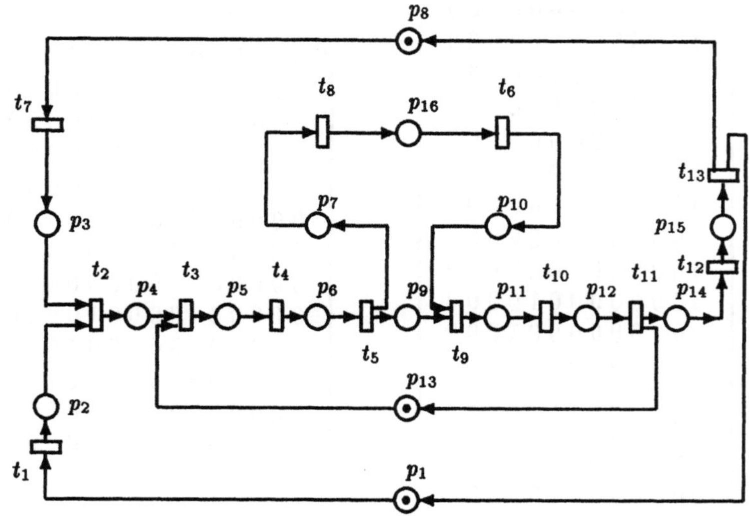

Fig. 9. The PNs-based model of the FMS

operation of tooling in the NC-machine tool; p_{10} - robot is comming to the NC-machine tool from the neutral position; p_{11} - operation of putting the finished part out of the NC-machine tool; p_{12} - robot is moving with the final part from the NC-machine tool to the exit; p_{13} - NC-machine is free; p_{14} - operation of putting the final part into the exit position (a container or pallet); p_{15} - robot is leaving the exit place and moving to the neutral position; p_{16} - robot is waiting in the neutral position.

The elementary discrete events occurring in the FMS during its operation can be expressed by means of $m = 13$ transitions as follows

t_1 - starting the working cycle; t_2 - starting the operation p_4; t_3 - starting the activity p_5; t_4 - starting the operation p_6; t_5 - ending the operation p_6 and starting the operations p_7, p_9; t_6 - ending the activity p_7; t_7 - starting the activity p_3; t_8 - starting the activity p_{10}; t_9 - starting the operation p_{11}; t_{10} - starting the activity p_{12}; t_{11} - ending the activity p_{12} and starting the operation p_{14}; t_{12} - ending the operation p_{14} and starting the activity p_{15}; t_{13} - ending the activity p_{15} and the working cycle in the whole.
The structural matrices of the FMS model are the following

$$
\mathbf{F} = \begin{pmatrix}
1&0&0&0&0&0&0&0&0&0&0&0&0\\
0&1&0&0&0&0&0&0&0&0&0&0&0\\
0&1&0&0&0&0&0&0&0&0&0&0&0\\
0&0&1&0&0&0&0&0&0&0&0&0&0\\
0&0&0&1&0&0&0&0&0&0&0&0&0\\
0&0&0&0&1&0&0&0&0&0&0&0&0\\
0&0&0&0&0&0&0&1&0&0&0&0&0\\
0&0&0&0&0&0&1&0&0&0&0&0&0\\
0&0&0&0&0&0&0&0&1&0&0&0&0\\
0&0&0&0&0&0&0&0&1&0&0&0&0\\
0&0&0&0&0&0&0&0&0&1&0&0&0\\
0&0&0&0&0&0&0&0&0&0&1&0&0\\
0&0&1&0&0&0&0&0&0&0&0&0&0\\
0&0&0&0&0&0&0&0&0&0&0&1&0\\
0&0&0&0&0&0&0&0&0&0&0&0&1\\
0&0&0&0&0&0&0&1&0&0&0&0&0
\end{pmatrix}
\quad
\mathbf{G} = \begin{pmatrix}
0&1&0&0&0&0&0&0&0&0&0&0&0&0&0&0\\
0&0&0&1&0&0&0&0&0&0&0&0&0&0&0&0\\
0&0&0&0&1&0&0&0&0&0&0&0&0&0&0&0\\
0&0&0&0&0&1&0&0&0&0&0&0&0&0&0&0\\
0&0&0&0&0&0&1&0&1&0&0&0&0&0&0&0\\
0&0&0&0&0&0&0&0&0&1&0&0&0&0&0&0\\
0&0&1&0&0&0&0&0&0&0&0&0&0&0&0&0\\
0&0&0&0&0&0&0&0&0&0&0&0&0&0&0&1\\
0&0&0&0&0&0&0&0&0&0&1&0&0&0&0&0\\
0&0&0&0&0&0&0&0&0&0&0&1&0&0&0&0\\
0&0&0&0&0&0&0&0&0&0&0&0&1&1&0&0\\
0&0&0&0&0&0&0&0&0&0&0&0&0&0&1&0\\
1&0&0&0&0&0&0&1&0&0&0&0&0&0&0&0
\end{pmatrix}
$$

Consider the initial state of the FMS to be

$$\mathbf{x}_0 = (3\,0\,0\,0\,0\,0\,0\,1\,0\,0\,0\,0\,1\,0\,0\,0)^T \tag{18}$$

It means that the FMS dynamics development is

$$\mathbf{v}_0 = (0\,1\,1\,1\,1\,1\,0\,1\,1\,1\,1\,1\,1)^T; \quad \mathbf{w}_0 = (1\,0\,0\,0\,0\,0\,1\,0\,0\,0\,0\,0\,0)^T$$

Hence, the following control possibilities can be automatically generated

$$\mathbf{u}_0^1 = (1\,0\,0\,0\,0\,0\,0\,0\,0\,0\,0\,0\,0)^T; \mathbf{u}_0^2 = (0\,0\,0\,0\,0\,0\,1\,0\,0\,0\,0\,0\,0)^T;$$

$$\mathbf{u}_0^3 = (1\,0\,0\,0\,0\,0\,1\,0\,0\,0\,0\,0\,0)^T$$

All of the control possibilities satisfy the condition (5). Consequently, they are admissible. The problem is that the model itself is not able to choose the most suitable one. To solve this problem a rule has to be defined. Suppose, that we want to use as high parallelism as possible - i.e. in the previous case we want to utilize the possibility $\mathbf{u}_0^3$. In order to form the KB in general, let us form the following statements:

S_1 - $n_{cvk} > 1$; S_2 - number of maximally possible parallel discrete events is $n_{max} = max\{n_{cv_1}, n_{cv_2}, ..., n_{cv_{n_{cvk}}}\}$; S_3 - the sequential number of the control vector with n_{max} nonzero elements is $n_{cv_{max}}$; S_4 - the choice of the control vector $u_{n_{cv_{max}}}$
where

n_{cvk} - the number of the realizable control vectors in the step k of the system dynamics development (in our case it is 3); n_{cv_i}; $(i = 1, 2, ..., n_{cvk})$ - the number of nonzero elements of the i-th control vector from the set of realizable ones (in our case it is the set {1, 1, 2}); n_{max} - the maximum from the $n_{cv_i}; i = 1, 2, ..., n_{cvk}$ (in our case it is 2); $n_{cv_{max}}$ - the sequential number of the control vector with the maximal number of nonzero elements (in our case it is 3). Consequently, the following rule can be formed to choose the most suitable control vector

R_1 : *IF* S_1 and S_2 and S_3 *THEN* S_4

6.3 A maze problem solving

The cat-and-mouse problem formulated by Ramadge and Wonham in [9] is solved here. Two "participants" - in [9] a cat and a mouse, however, they can be as well as e.g. two mobile robots, two induction cars in a transport system of a kind of the FMS, two trains or cars in a transport net etc. - are placed in a maze (however, it can also be e.g. complicated crossroads etc.) given on Fig. 10. The maze consists of five rooms denoted by numbers 0, 1,...,

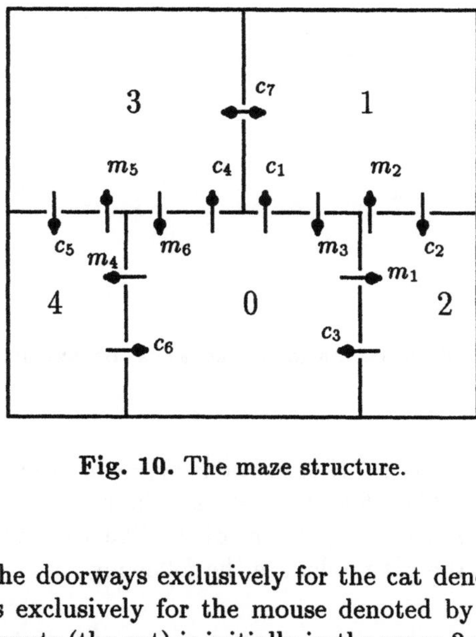

Fig. 10. The maze structure.

4 connecting by the doorways exclusively for the cat denoted by $c_i, i = 1,7$ and the doorways exclusively for the mouse denoted by $m_j, j = 1,6$. The first of the participants (the cat) is initially in the room 2 and the second one (the mouse) in room 4. Each doorway can be traversed only in the direction indicated. Each door (with the exception of the door c_7) can be opened or closed by means of control actions. The door c_7 is uncontrollable (or better, it is continuously opened in both directions). The controller to be synthetized observes only discrete events generated by sensors in the doors. They indicate that a participant ist just running through. The control problem is to find a feedback controller such that the following three constraints will be satisfied:

1. The participants never occupy the same room simultaneously.

2. It is always possible for both of them to return to their initial positions (the first one to the room 2 and the second one to the room 4).
3. The controller should enable the participants to behave as freely as possible with respect to the constraints imposed.

The behaviour of the animals can be expressed in the form of the abstract linear discrete dynamical systems (the PN-based models). A simple procedure of the control synthesis presented above can be utilized as well as the knowledge base expressing the predefined constraints imposed on the control task.

6.3.1 Modelling the system to be controlled. Consider the rooms of the maze to be the Petri nets positions (places) and the doorways of the animals to be the Petri nets transitions. The corresponding PN-based representation of the maze is given on Fig. 11. It can be seen that the uncontrolled

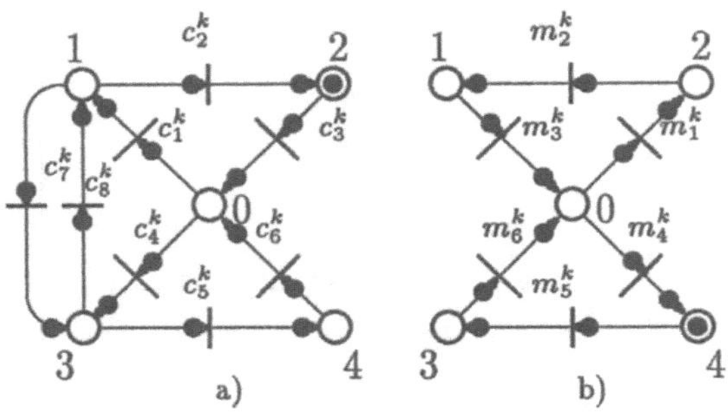

Fig. 11. The PN-based representation of the: a) cat behaviour; b) mouse behaviour

door c_7 is replaced by two PN transitions c_7, c_8 in order to avoid the problems specific for the PN. Both the cat and the mouse can be expressed, as to their dynamic behaviour, by means of the linear discrete dynamic model. The parameters of the cat model are the following

$n = 5,\ m_c = 8$

$$\mathbf{F}_c = \begin{pmatrix} 1 & 0 & 0 & 1 & 0 & 0 & 0 & 0 \\ 0 & 1 & 0 & 0 & 0 & 0 & 1 & 0 \\ 0 & 0 & 1 & 0 & 0 & 0 & 0 & 0 \\ 0 & 0 & 0 & 0 & 1 & 0 & 0 & 1 \\ 0 & 0 & 0 & 0 & 0 & 1 & 0 & 0 \end{pmatrix} \quad \mathbf{G}_c^T = \begin{pmatrix} 0 & 0 & 1 & 0 & 0 & 1 & 0 & 0 \\ 1 & 0 & 0 & 0 & 0 & 0 & 0 & 1 \\ 0 & 1 & 0 & 0 & 0 & 0 & 0 & 0 \\ 0 & 0 & 0 & 1 & 0 & 0 & 1 & 0 \\ 0 & 0 & 0 & 0 & 1 & 0 & 0 & 0 \end{pmatrix}$$

and the parameters of the mouse model are

$n = 5,\ m_m = 6$

$$\mathbf{F}_m = \begin{pmatrix} 1 & 0 & 0 & 1 & 0 & 0 \\ 0 & 0 & 1 & 0 & 0 & 0 \\ 0 & 1 & 0 & 0 & 0 & 0 \\ 0 & 0 & 0 & 0 & 0 & 1 \\ 0 & 0 & 0 & 0 & 1 & 0 \end{pmatrix} \quad \mathbf{G}_m^T = \begin{pmatrix} 0 & 0 & 1 & 0 & 0 & 1 \\ 0 & 1 & 0 & 0 & 0 & 0 \\ 1 & 0 & 0 & 0 & 0 & 0 \\ 0 & 0 & 0 & 0 & 1 & 0 \\ 0 & 0 & 0 & 1 & 0 & 0 \end{pmatrix}$$

The initial state vectors of the cat and the mouse are:

$$^{c}\mathbf{x}_0 = (0\,0\,1\,0\,0) \quad , \quad ^{m}x_0 = (0\,0\,0\,0\,1)^T$$

The structure of the control vectors of the animals is:

$$\begin{aligned} ^{c}\mathbf{u}_k &= (c_1^k, c_2^k, c_3^k, c_4^k, c_5^k, c_6^k, c_7^k, c_8^k)^T \ , \quad c_i^k &\in \{0,1\},\ i = 1,8 \\ ^{m}\mathbf{u}_k &= (m_1^k, m_2^k, m_3^k, m_4^k, m_5^k, m_6^k)^T \ , \quad m_i^k &\in \{0,1\},\ i = 1,6 \end{aligned}$$

6.3.2 The constraints acceptation and the knowledge base formulation. In order to accept three constraints defined in [9] and introduced above we have to express them in the form of suitable knowledgw base. An approach to this can be realized by means of the following statements $S_1 = (^{c}\mathbf{x}_{k+1} = {}^{m}\mathbf{x}_{k+1})$; $S_2 = (^{c}\mathbf{x}_{k+1} = {}^{c}\mathbf{x}_0)$; $S_3 = (^{m}\mathbf{x}_{k+1} = {}^{m}\mathbf{x}_0)$; $S_4 = (^{c}\mathbf{x}_{k+1} = {}^{m}\mathbf{x}_0)$; $S_5 = (^{m}\mathbf{x}_{k+1} = {}^{c}\mathbf{x}_0)$; $S_6 = (^{c}\mathbf{x}_{k+1} \neq {}^{m}\mathbf{x}_{k+1})$; $S_7 = (^{c}\mathbf{x}_{k+1} \neq {}^{c}\mathbf{x}_0)$; $S_8 = (^{m}\mathbf{x}_{k+1} \neq {}^{m}\mathbf{x}_0)$; $S_9 = (^{c}\mathbf{x}_{k+1} \neq {}^{m}\mathbf{x}_0)$; $S_{10} = (^{m}\mathbf{x}_{k+1} \neq {}^{c}\mathbf{x}_0)$; S_{11} = there is only one vector $^{c}\mathbf{u}_k$; S_{12} = there is only one vector $^{m}\mathbf{u}_k$; S_{13} = there are several vectors $^{c}\mathbf{u}_k$; S_{14} = there are several vectors $^{m}\mathbf{u}_k$); S_{15} = accept $^{c}\mathbf{u}_k$; S_{16} = accept $^{m}\mathbf{u}_k$; S_{17} = elimine $^{c}\mathbf{u}_k$; S_{18} = elimine $^{m}\mathbf{u}_k$); S_{19} = solution does not exist; S_{20} = take another $^{c}\mathbf{u}_k$; S_{21} = take another $^{m}\mathbf{u}_k$); S_{22} = (accept $^{c}\mathbf{u}_k$ and elimine $^{m}\mathbf{u}_k$) or (accept $^{m}\mathbf{u}_k$ and elimine $^{c}\mathbf{u}_k$); S_{23} = I do not know

and the following rules ($\circ$ symbolizes the <u>and</u> operator)
R_1: *IF* $(S_1 \circ S_7 \circ S_{12} \circ S_{13})$ *THEN* $(S_{16} \circ S_{17} \circ S_{20})$; R_2: *IF* $(S_1 \circ S_8 \circ S_{11} \circ S_{14})$ *THEN* $(S_{15} \circ S_{18} \circ S_{21})$
R_3: *IF* S_6 *THEN* $(S_{15} \circ S_{16})$; R_4: *IF* $(S_1 \circ S_2 \circ S_{14})$ *THEN* $(S_{15} \circ S_{18} \circ S_{21})$
R_5: *IF* $(S_1 \circ S_3 \circ S_{13})$ *THEN* $(S_{16} \circ S_{17} \circ S_{20})$; R_6: *IF* $(S_1 \circ S_5 \circ S_{14})$ *THEN* $(S_{15} \circ S_{18} \circ S_{21})$
R_7: *IF* $(S_1 \circ S_4 \circ S_{13})$ *THEN* $(S_{16} \circ S_{17} \circ S_{20})$; R_8: *IF* $(S_4 \circ S_6 \circ S_{12} \circ S_{13})$ *THEN* $(S_{16} \circ S_{17} \circ S_{20})$
R_9: *IF* $(S_5 \circ S_6 \circ S_{11} \circ S_{14})$ *THEN*$(S_{15} \circ S_{18} \circ S_{21})$; R_{10}: *IF* $(S_2 \circ S_6)$ *THEN* $(S_{15} \circ S_{16})$
R_{11}: *IF* $(S_3 \circ S_6)$ *THEN* $(S_{15} \circ S_{16})$; R_{12}: *IF* $(S_1 \circ S_2 \circ S_{12})$ *THEN* S_{19}
R_{13}: *IF* $(S_1 \circ S_3 \circ S_{11})$ *THEN* S_{19}; R_{14}: *IF* $(S_1 \circ S_2 \circ S_5 \circ S_{11} \circ S_{12})$ *THEN* S_{19}
R_{15}: *IF* $(S_1 \circ S_3 \circ S_4 \circ S_{11} \circ S_{12})$ *THEN* S_{19}; R_{16}: *IF* $(S_1 \circ S_{11} \circ S_{12})$ *THEN* S_{19}

R_{17}: *IF* $(S_1 \circ S_7 \circ S_8 \circ S_{13} \circ S_{14})$ *THEN* S_{22}; R_{18}: *IF* $(S_4 \circ S_6 \circ S_{13})$ *THEN* S_{20}
R_{19}: *IF* $(S_5 \circ S_6 \circ S_{14})$ *THEN* S_{21}
R_{20}: *IF* $(S_1 \circ S_7 \circ S_8 \circ S_9 \circ S_{10} \circ S_{13} \circ S_{14})$ *THEN* S_{23}
Consequently, we can obtain the structural matrices $\boldsymbol{\Psi}$, $\boldsymbol{\Gamma}$ (n = 23, m = 20) of the analytical model of the PN-based knowledge representation. The KB obtained on this way helps to find automatically the solution of the control task - i.e. to obtain the automatic synthesis of the control actions. The work of the KB can be illustrated as follows

$$\boldsymbol{\Phi}_0 = (11001001100011000000000)^T \tag{19}$$

$$\boldsymbol{\Omega}_0 = (00010100000000000000)^T \tag{20}$$

$$\boldsymbol{\Phi}_1 = (11001001100011100100100)^T \tag{21}$$

It means that when the statements $S_1, S_2, S_5, S_8, S_9, S_{13}, S_{14}$ are true the rules R_4, R_6 can be evaluated and, consequently, the statements S_{15}, S_{18}, S_{21} can be also true.

The permanently open doors c_7, c_8 must be always accepted in the control synthesis task. Consequently, the process of returning the cat into the room 2 can be infinite.

6.3.3 The step-by-step problem solving. As a result of the control system synthesis with respect to both the system model and the knowledge base mentioned above we have the behaviour of the animals given on Fig. 12 as well as in the Tab. 1. As we can see from the previous table, the uncontrollable doors c_7, c_8 become evident in the step $k = 2$. We cannot prohibit to the cat using those doors and, consequently, the doors must be respected. What we can only do in order to avoid a crash of the animals in the next step is, that we can support opening the door c_2 if the cat is in room 1 (this door should be opened because of the possibility of its returning to the initial state - i.e to the room 2). In addition to this fact, opening the door c_2 has priority to opening the mouse door m_1 leading the mouse to the room 2 what is undesirable. We prohibit this mouse door, in order to avoid a possible crash, and support the door m_4 leading the mouse to their initial state - the room 4. The other way round, opening the door m_4 has priority to opening the cat door c_5 leading the cat to the mouse initial state (the room 4). Hence, in the step $k = 2$ we prohibit opening the doors m_1, c_5 (in the table they are struck out) and support opening the doors m_4 and c_2 (in the table they are underlined). The permanently open doors c_7, c_8 must be accepted in the control synthesis task. Consequently, the process of returning the cat into the room 2 can be infinite. The solution of the task can be seen better on Fig. 12.

Table 1. The results of the problem solving process

PROCESS DEVELOPMENT		
Step	**Cat Behaviour**	**Mouse Behaviour**
$k=0$	${}^{c}\mathbf{x}_0=(0\,0\,1\,0\,0)^T$ ${}^{c}\overline{\mathbf{x}}_0=(1\,1\,0\,1\,1)^T$ ${}^{c}\mathbf{w}_0=(0\,0\,1\,0\,0\,0\,0\,0)^T$ ${}^{c}\mathbf{u}_0={}^{c}\mathbf{w}_0$ ${}^{c}\mathbf{x}_1={}^{c}\mathbf{x}_0+\mathbf{B}_c.{}^{c}\mathbf{u}_0$ ${}^{c}\mathbf{x}_1=(1\,0\,0\,0\,0)^T$	${}^{m}\mathbf{x}_0=(0\,0\,0\,0\,1)^T$ ${}^{m}\overline{\mathbf{x}}_0=(1\,1\,1\,1\,0)^T$ ${}^{m}\mathbf{w}_0=(0\,0\,0\,0\,1\,0)^T$ ${}^{m}\mathbf{u}_0={}^{m}\mathbf{w}_0$ ${}^{m}\mathbf{x}_1={}^{m}\mathbf{x}_0+\mathbf{B}_m.{}^{m}\mathbf{u}_0$ ${}^{m}\mathbf{x}_1=(0\,0\,0\,1\,0)^T$
	${}^{c}\mathbf{x}_1 \neq {}^{m}\mathbf{x}_1$	
$k=1$	${}^{c}\mathbf{w}_1=(1\,0\,0\,1\,0\,0\,0\,0)^T$	${}^{m}\mathbf{w}_1=(0\,0\,0\,0\,0\,1)^T$
	the control possibilities are: $\{\underline{c_1},\ \underline{c_4},\ \underline{m_6}\}$ all of them are possible	
	${}^{c}\mathbf{u}_1^1=(1\,0\,0\,0\,0\,0\,0\,0)^T$ ${}^{c}\mathbf{x}_2^1=(0\,1\,0\,0\,0)^T$ ${}^{c}\mathbf{u}_1^2=(0\,0\,0\,1\,0\,0\,0\,0)^T$ ${}^{c}\mathbf{x}_2^2=(0\,0\,0\,1\,0)^T$	${}^{m}\mathbf{u}_1={}^{m}\mathbf{w}_1$ ${}^{m}\mathbf{x}_2=(1\,0\,0\,0\,0)^T$ =n o n e= =n o n e=
	${}^{c}\mathbf{x}_2^1 \neq {}^{m}\mathbf{x}_2\,;\ {}^{c}\mathbf{x}_2^2 \neq {}^{m}\mathbf{x}_2$	
$k=2$	${}^{c}\mathbf{w}_2^1=(0\,1\,0\,0\,0\,0\,1\,0)^T$ ${}^{c}\mathbf{w}_2^2=(0\,0\,0\,0\,1\,0\,0\,1)^T$	${}^{m}\mathbf{w}_2=(1\,0\,0\,1\,0\,0)^T$ =n o n e=
	the control possibilities are: $\{\underline{c_2},\ c_7,\ \not{m}_1,\ \underline{m_4}\}$; c_2 has priority to m_1 $\{\not{c}_5,\ c_8,\ \not{m}_1,\ \underline{m_4}\}$; m_4 has priority to c_5	
	${}^{c}\mathbf{u}_2^{11}=(0\,1\,0\,0\,0\,0\,0\,0)^T$ ${}^{c}\mathbf{u}_2^{12}=(0\,0\,0\,0\,0\,0\,1\,0)^T$ ${}^{c}\mathbf{u}_2^{2}=(0\,0\,0\,0\,0\,0\,0\,1)^T$ ${}^{c}\mathbf{x}_3^{11}=(0\,0\,1\,0\,0)^T={}^{c}\mathbf{x}_0$ ${}^{c}\mathbf{x}_3^{12}=(0\,0\,0\,1\,0)^T$ ${}^{c}\mathbf{x}_3^{2}=(0\,1\,0\,0\,0)^T$	${}^{m}\mathbf{u}_2=(0\,0\,0\,1\,0\,0)^T$ =n o n e= =n o n e= ${}^{m}\mathbf{x}_3=(0\,0\,0\,0\,1)^T={}^{m}\mathbf{x}_0$ =n o n e= =n o n e=

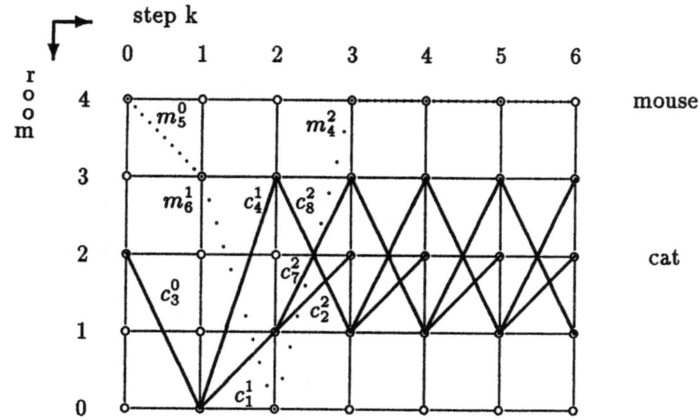

Fig. 12. The graphical expression of the task solution.

7. Knowledge inference and automatic reasoning

In order to explain the better the knowledge inference and the simple form of automatic reasoning. The KB "dynamics" development (i.e. the statements truth propagation) can be formally expressed - see [6] - as follows

$$\langle \Phi, \Omega, \delta_1, \boldsymbol{\Phi}_0, \mathcal{M} \rangle \quad ; \quad \Phi \cap \Omega = \emptyset \tag{22}$$

where

Φ $= \{\boldsymbol{\Phi}_0, ..., \boldsymbol{\Phi}_{N_1}\}$ is a set of the state vectors of the KB.

$\boldsymbol{\Phi}_K = (\phi^K_{S_1}, ..., \phi^K_{S_{n_1}})^T$; $K = 0, N_1$ is the state vector of the KB (the state of the statements truth propagation) in the step K.

K is the discrete step of the KB dynamics development

T symbolizes the vector or matrix transposition

N_1 is an integer representing the number of different situations during the KB dynamics development (i.e. during the statements truth propagation)

$\phi^K_{S_i}$, $i = 1, n_1$ is the state of the truth of the elementary statement S_i in the step K. It means that the statement is false (when 0), true (when 1) or that the statement is true with a fuzzy measure (when this parameter acquires its value from the real interval between these two boundary values). In other words, for each S_i the $\phi^K_{S_i}$, $K = 1, N_1$ are members of a fuzzy set with the corresponding membership functions $\mu_{\phi_{S_i}}(\phi^K_{S_i})$.

Ω $= \{\boldsymbol{\Omega}_0, ..., \boldsymbol{\Omega}_{N_1}\}$ is a set of the "control" vectors of the KB.

$\boldsymbol{\Omega}_K = (\omega^K_{R_1}, ..., \omega^K_{R_{m_1}})^T$; $K = 0, N_1$ is the "control" vector of the KB (i.e. the state of the rules evaluability) in the step K.

$\omega_{R_j}^K$, $j = 1, m_1$ is the state of the rule R_j evaluability in the step K. It means that the rule is not able to be evaluated (when 0), the rule is able to be evaluated (when 1) or that the rule is able to be evaluated with a fuzzy measure (when this parameter acquires its value from the interval between these two boundary values). It depends on the fuzzy values of the intput statements truth.

$\delta_1 \;: \Phi \times \Omega \longmapsto \Phi$ is a transition function of the KB.

$\boldsymbol{\Phi}_0$ is the initial state vector of the KB.

$\mathcal{M} = \{\mathbf{m}_1, ..., \mathbf{m}_{N_1}\}$ is the set of the vectors representing the rules truth values in the step K.

$\mathbf{m}_K = (m_{R_1}^K, ..., m_{R_{m_1}}^K)^T$; $K = 0, N_1$ is the vector of the rules truth values in the step K.

$m_{R_j}^K \in <0, 1>$, $j = 1, m_1$ expresses the fuzzy truth value of the rule R_j in the step K. In other words, for each R_j the $m_{R_j}^K$, $K = 1, N_1$ are members of a fuzzy set with the corresponding membership functions $\mu_{m_{R_j}}(m_{R_j}^K)$.

It is better for us to write $\mathbf{m}_K$ in the form of the (m_1 x m_1)-dimensional diagonal matrix $\mathbf{M}_K = diag\,\{m_{R_1}^K, ..., m_{R_{m_1}}^K\}$.

To imagine the introduced facts see Fig. 13. It can be said that the LPN are

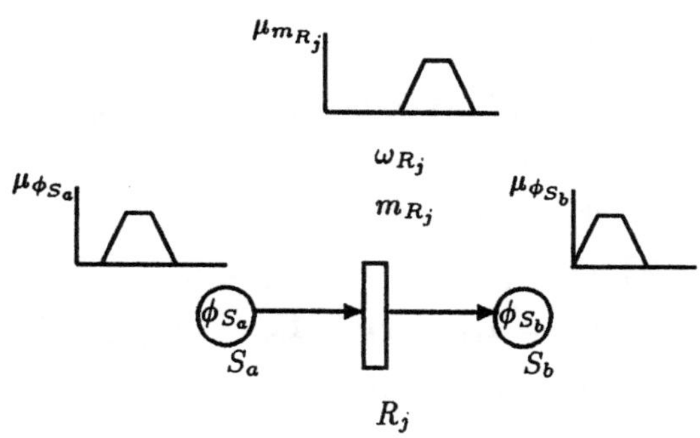

Fig. 13. The simple fuzzy rule R_j with corresponding membership functions

(using the fuzzy sets terminology) a crisp form of the FPN.

As to the membership functions of the statements truth, the rules truth and the fuzzy measures of existence of the causal interconnections are not analysed in details because they strongly depend on the actual application and they must be set by an expert from the actual domain.

The KB dynamics development (more precisely the transition function δ_1) can be expressed in analytical terms as follows

$$\boldsymbol{\Phi}_{K+1} = \boldsymbol{\Phi}_K \ \underline{or}\ \boldsymbol{\Delta}_K \ \underline{and}\ \boldsymbol{\Omega}_K \quad , \quad K = 0, N_1 \quad , \quad \boldsymbol{\Phi}_{K_{|K=0}} = \boldsymbol{\Phi}_0 \tag{23}$$

$$\boldsymbol{\Delta}_K = \boldsymbol{\Gamma}_K^T \ \underline{or}\ \boldsymbol{\Psi}_K \tag{24}$$

$$\boldsymbol{\Psi}_K \ \underline{and}\ \boldsymbol{\Omega}_K \leq \boldsymbol{\Phi}_K \tag{25}$$

where

$\underline{and}$ is the operator of logical multiplying in general. For both the bivalued logic and the fuzzy one it can be defined (for scalar operands) to be the minimum of its operands. For example the result of its application on the scalar operands a, b is a scalar c which can be obtained as follows: $a \ \underline{and}\ b = c = min\,\{a, b\}$.

$\underline{or}$ is the operator of logical additioning in general. For both the bivalued logic and the fuzzy one it can be defined (for scalar operands) to be the maximum of its operands. For example the result of its application on the scalar operands a, b is a scalar c which can be obtained as follows: $a \ \underline{or}\ b = c = max\,\{a, b\}$.

To derive the knowledge inference suppose that the inference mechanism consists of two parts:

1. the mechanism of the statements truth propagation (something like a carrier-wave) - when the rules truth values are crisp (equal to 1).
2. the influence of the fuzzy values of the rules truth (something like a modulation wave).

7.1 The truth propagation

The automatic mechanism of the statements truth propagation can be analytically described as follows

$$\overline{\boldsymbol{\Phi}_K} = \underline{neg}\,\boldsymbol{\Phi}_K = \mathbf{1}_{n_1} - \boldsymbol{\Phi}_K \tag{26}$$

$$\mathbf{v}_K = \boldsymbol{\Psi}_K^T \underline{and}\ \overline{\boldsymbol{\Phi}_K} \tag{27}$$

$$\mathbf{w}_K = \underline{neg}\,\mathbf{v}_K = \mathbf{1}_{m_1} - \mathbf{v}_K = \mathbf{1}_{m_1} - (\boldsymbol{\Psi}_K^T \underline{and}\,(\mathbf{1}_{n_1} - \boldsymbol{\Phi}_K)) \tag{28}$$

$$= \underline{neg}(\boldsymbol{\Psi}_K^T \ \underline{and}(\underline{neg}\,\boldsymbol{\Phi}_K)) \tag{29}$$

$$\boldsymbol{\Omega}_K = \mathbf{w}_K \tag{30}$$

where the meaning of quantities is the following

$\mathbf{v}_K$ is a m_1-dimensional auxiliary vector pointing out (by its nonzero elements) the rules that cannot be evaluated, because there is at least one false (of course in the LPN analogy) statement among its input statements

$\mathbf{w}_K$ is a m_1-dimensional "control" vector pointing out the rules that have all their input statements true and, consequently, they can be evaluated in the step K of the KB dynamics development. This vector is a base of the inference, because it contains information about the rules that can contribute to obtaining the new knowledge - i.e. to transfer the KB from the state $\boldsymbol{\Phi}_K$ of the truth propagation into another state $\boldsymbol{\Phi}_{K+1}$. These rules correspond to the nonzero elements of the vector $\mathbf{w}_K$.

$\underline{neg}$ is the operator of logical negation in general. For both the bivalued logic and the fuzzy one it can be defined (for scalar operands) to be the complement of its operand. For example the result of its application on the scalar operands a is a scalar b which can be obtained as follows: $\underline{neg}\, a = b = 1 - a$.

After imbedding (30) into equation (23) we have

$$\boldsymbol{\Phi}_{K+1} = \boldsymbol{\Phi}_K \;\underline{or}\; \boldsymbol{\Delta}_K \;\underline{and}\,(\underline{neg}(\boldsymbol{\Psi}^T_K \;\underline{and}\,(\underline{neg}\, \boldsymbol{\Phi}_K))) \tag{31}$$

7.2 The knowledge inference

The automatic mechanism of the knowledge inference can be described as follows

$$\overline{\boldsymbol{\Phi}_K} = \underline{neg}\, \boldsymbol{\Phi}_K = \mathbf{1}_{n_1} - \boldsymbol{\Phi}_K \tag{32}$$

$$\mathbf{v}_K = \boldsymbol{\Psi}^T_K \underline{and}\, \overline{\boldsymbol{\Phi}_K} \tag{33}$$

$$\mathbf{w}_K = \underline{neg}\, \mathbf{v}_K = \mathbf{1}_{m_1} - \mathbf{v}_K = \mathbf{1}_{m_1} - (\boldsymbol{\Psi}^T_K \underline{and}\, (\mathbf{1}_{n_1} - \boldsymbol{\Phi}_K)) \tag{34}$$

$$= \underline{neg}(\boldsymbol{\Psi}^T_K \;\underline{and}\,(\underline{neg}\, \boldsymbol{\Phi}_K)) \tag{35}$$

$$\boldsymbol{\Omega}_K = \mathbf{M}_K \,.\, \mathbf{w}_K = \mathbf{M}_K \,.\, (\underline{neg}(\boldsymbol{\Psi}^T_K \;\underline{and}\,(\underline{neg}\, \boldsymbol{\Phi}_K))) \tag{36}$$

After imbedding (36) into equation (23) we have

$$\boldsymbol{\Phi}_{K+1} = \boldsymbol{\Phi}_K \;\underline{or}\; \boldsymbol{\Delta}_K \;\underline{and}\,(\mathbf{M}_K \,.\, (\underline{neg}(\boldsymbol{\Psi}^T_K \;\underline{and}\,(\underline{neg}\, \boldsymbol{\Phi}_K)))) \tag{37}$$

8. An illustrative example

Consider a set of the following statements:

$S_1 = "A"$ $S_2 = "B"$ $S_3 = "C"$ $S_4 = "D"$

connected by the following systems of rules:

$R_1: \; IF(S_1 \;\underline{and}\; S_2)\, THEN(S_3)$ $R_2: IF(S_2 \;\underline{and}\; S_3)\, THEN(S_4)$

$R_3: \; IF S_4 \, THEN S_1$ $R_4: IF S_4 \, THEN S_3$

The situation is illustrated in Fig. 14. Let us consider the case of the KB where the structure is given as crisp.

$$\boldsymbol{\Psi} = \begin{pmatrix} 1 & 0 & 0 & 0 \\ 1 & 1 & 0 & 0 \\ 0 & 1 & 0 & 0 \\ 0 & 0 & 1 & 1 \end{pmatrix} \quad \boldsymbol{\Gamma}^T = \begin{pmatrix} 0 & 0 & 1 & 0 \\ 0 & 0 & 0 & 0 \\ 1 & 0 & 0 & 1 \\ 0 & 1 & 0 & 0 \end{pmatrix}$$

$$\boldsymbol{\Delta} = \begin{pmatrix} 1 & 0 & 1 & 0 \\ 1 & 1 & 0 & 0 \\ 1 & 1 & 0 & 1 \\ 0 & 1 & 1 & 1 \end{pmatrix}$$

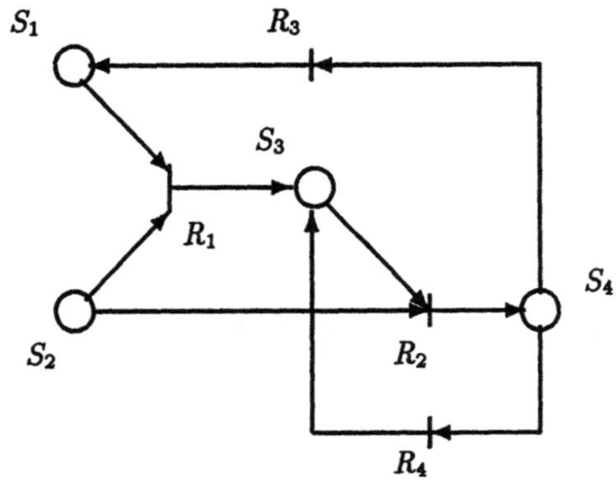

Fig. 14. The PN-based representation of the KB

However, both the statements truth and the rules truth values are fuzzy. Consider the initial state of the statements truth in the form:
"A" is true with fuzzy measure 0.3 (i.e. $\phi^0_{S_1} = 0.3$); *"B" is true with the fuzzy measure* 0.5 (i.e. $\phi^0_{S_2} = 0.5$).
Hence,

$$\boldsymbol{\Phi}_0 = (0.3, 0.5, 0.0, 0.0)^T \; ; \quad \underline{neg}\, \boldsymbol{\Phi}_0 = (0.7, 0.5, 1.0, 1.0)^T$$
$$\mathbf{v}_0 = (0.7, 1.0, 1.0, 1.0)^T \; ; \quad \mathbf{w}_0 = (0.3, 0.0, 0.0, 0.0)^T$$

When we consider the rules truth values $\mathbf{m}_K = (0.8, 0.7, 0.9, 1.0)^T$; $K = 0, N_1$

$$\boldsymbol{\Omega}_0 = (0.24, 0.0, 0.0, 0.0)^T \; ; \quad \boldsymbol{\Phi}_1 = (0.3, 0.5, 0.24, 0.0)^T$$
$$\underline{neg}\, \boldsymbol{\Phi}_1 = (0.7, 0.5, 0.76, 1.0)^T \; ; \quad \mathbf{v}_1 = (0.7, 0.76, 1.0, 1.0)^T)$$
$$\mathbf{w}_1 = (0.3, 0.24, 0.0, 0.0)^T \; ; \quad \boldsymbol{\Omega}_1 = (0.24, 0.168, 0.0, 0.0)^T$$
$$\boldsymbol{\Phi}_2 = (0.3, 0.5, 0.24, 0.168)^T$$

....

It means that (as a consequence of the state of the statements truth: *"A" is true with the fuzzy measure* 0.3 and *"B" is true with the fuzzy measure* 0.5) the statements *"C" will be true with the fuzzy measure* 0.24 and *"D" will be true with the fuzzy measure* 0.168 (i.e. $\phi^0_{S_3} = 0.24$ and $\phi^0_{S_4} = 0.168$) .

9. Conclusions

This chapter presents both the mathematical model of the DEDS to be controlled and the mathematical model of the knowledge base. Both of them are based on the Petri nets and they commonly used in the control synthesis procedure. The system model plays two principial roles. It makes the simulation of the process dynamics development possible and the results of simulation give us the principle image how the production rules of the knowledge base have to be built. On the other hand it generates principial information about the system behaviour which is necessary for the knowledge base operating. The knowledge base helps the control synthesis procedure to choose the most suitable control possibility in each step of the system dynamics development. Consequently, the process of the control synthesis can be automatic.

References

1. Čapkovič, F.: Petri nets-based computer aided synthesis of control systems for discrete event dynamic systems. In: Barker, H.A. (Ed.): Computer Aided Design in Control Systems. Preprints of the 5-th IFAC Symp., Swansea, U.K., July 1991, Pergamon Press, pp. 409-414.

2. Čapkovič, F.: Computer-aided design of intelligent control systems for discrete event dynamic systems. In: Mattson, S.E., J.O. Gray and F.F. Cellier (Eds.): Preprints of the IEEE/IFAC Joint Symposium on Computer-Aided Control System Design - CACSD'94, Tucson, Arizona, USA, March 1994. IEEE Press, Piscataway, pp. 55-60.

3. Čapkovič, F.: A knowledge-based approach to synthesis of intelligent control of DEDS. In: Aamodt, A., J. Komorowski (Eds.): SCAI-95, Fifth Scandinavian Conference on Artificial Intelligence. IOS Press Ohmsha, 1995, Amsterdam-Oxford-Washington DC-Tokyo, pp. 9-18.

4. Čapkovič, F.: Petri nets-based modelling and control of discrete event dynamic systems. In: Doležal, J.,J. Fiedler (Eds.): Preprints of the 17th IFIP TC7 Conference onSystem Modelling and Optimization, Prague, Czech Republic, June 10-14, 1995, pp. 199-202.

5. Čapkovič, F.: A Petri nets-based representation of rule-based knowledge for DEDS control purposes. In: Proceedings of the TAINN '95 Conference, Gebze, Turkey, June 1995, pp. 293-303.

6. Čapkovič, F.: Using Fuzzy Logic for Knowledge Representation at Control Synthesis. *BUSEFAL*, **63**, 1995, pp. 4-9.

7. Čapkovič, F.: Petri net-based approach to intelligent control synthesis of FMS. In: Proceedings of the 1995 INRIA/IEEE Symposium on Emerging Technologies and Factory Automation - ETFA'95, Paris, France, October 1995, IEEE Computer Society Press, Los Alamitos, USA, Vol. 1, pp. 293-303.

8. Looney, C.G.: Fuzzy Petri nets for rule-based decisionmaking. *IEEE Trans. Syst. Man Cybern.*, **SMC-18**, No 1, 1988, pp.178-183
9. Wonham, W.M., Ramadge, P.J.: On the supremal controllable sublanguage of a given language. *SIAM J. Control and Optimization*, **25**, No 3, May 1987, pp. 637-659.

10. Chao, D.Y., M.C. Zhou and D.T. Wang: Extending the knitting technique to Petri net Synthesis of automated manufacturing systems. *The Computer Journal*, **37**, No 1, 1994, pp. 67-76.

13 New Methods and Tools for Commissioning of Manufacturing Lines with Robots

Rolf Bernhardt

1 Introduction

Since manufacturing lines with robots are widely used in the automotive industry, aspects of this industrial sector are addressed. Car makers including their suppliers are confronted with a tremendous world-wide competition, even extended by additional competitors emerging in the last years. To meet this challenge, some key figures in car production must be dramatically changed. These figures are related to the time to market, production loss for new models, reliability and launching flexibility. Additionally, these changes must lead to an improvement in quality. Eventually an overall cost reduction of 30%-50%, a 50% reduction in the number of work hours per year accomplished by a company related to the number of cars produced, and a decrease in market lead time from 6-8 years to at least 3-4 years should be reached. A further important point is to decrease the commissioning time of manufacturing systems from 6-8 weeks to 48 hours, which in fact means one weekend. This last point saves not only cost, but tremendously increases the launching flexibility, i.e. bringing a new car on the market when the market requires it and not only after the factory vacations. This contribution mainly deals with this last aspect from a technical point of view. All necessary tasks for changing the manufacturing system for new car models should be done in parallel to the running production. Thereby, one important factor is the programming of manufacturing lines with robots. Programming of robots in parallel to the running production can only be done if simulation and off-line programming techniques and systems (SOP) are used.

SOP have reached a status of maturity enabling their application in industry. Especially the automotive companies are playing a leading role in using such systems. But also other industries (ship yards, airplane builders) as well as suppliers to the automotive industry are aware of the potential of these techniques: to reach a much better competitive position by increasing productivity, improving the quality of the products and at the same time reducing the cost for planning, set-up and operation of manufacturing systems. The aim of all these activities, undertaken in parallel in many companies, is to create task programs (robot programs) in the planning phase (off-line) and to run them at the shop floor, if possible, without any alteration. Even though there are many company internal activities as well as publicly funded and industrial projects under way, the overall goal is not yet reached today. There are still gaps which need R&D efforts and/or standardization activities. Concerning SOP, this is mainly related to the availability of accurate simulation models of manufacturing components and their exchangeability between SOP systems from different vendors (standardized interfaces).

The effective use of SOP systems requires the knowledge of tolerances of the manufacturing components in order to enable a realistic planning as possible. This especially concerns the availability of robots with a tenfold improvement in absolute positioning accuracy (from 10-20mm to 1-2mm). This considerable change of the end-user requirements related to the absolute positioning accuracy is a big challenge for the robot manufacturers. On the other hand, there is a huge market volume for robots and robotic systems for the near future.

Based on estimations by the UN Economic Commission (UN/EC) and the International Federation of Robotics (IFR) /1/, the number of installed robots world-wide will grow from 653,000 today to 831,000 by the end of 1997 (fig.1). Taking only the robots into account, this equals a market volume of 10-20 billion US$. Since only 5%-10% of the robots are sold as stand alone systems and 90%-95% as turn key applications, the market volume is at least threefold to fivefold. In order to realize this market, the requirements of the end-users, especially from the automotive industry, need to be fulfilled. However, this requires a close cooperation between end-users, robot and controller maufacturers, calibration system developers and, last but not least, simulation and off-line programming (SOP) system developers.

A first step in this direction was the initiation of the so-called "Realistic Robot Simulation (RRS)" project. In the RRS-project end-users from the automotive industry cooperate with robot/control manufacturers and SOP system developers in order to improve the accuracy of simulation. More details about these activities are given in chapter 3.1.

The demanded absolute accuracy, not only for robots but also for other components (tools, workpiece, fixtures, etc.) can only be reached, without increasing the costs, if adequate measuring systems and powerful tools for mathematical modeling, parameter identification and off-line and/or on-line error compensation are available. These aspects, which can be summarized under the term calibration, are presented in chapter 3.2.

A further important area is the optimization of robot programs related to cycle times and/or mechanical stress. These topics are dealt with in chapter 3.3.

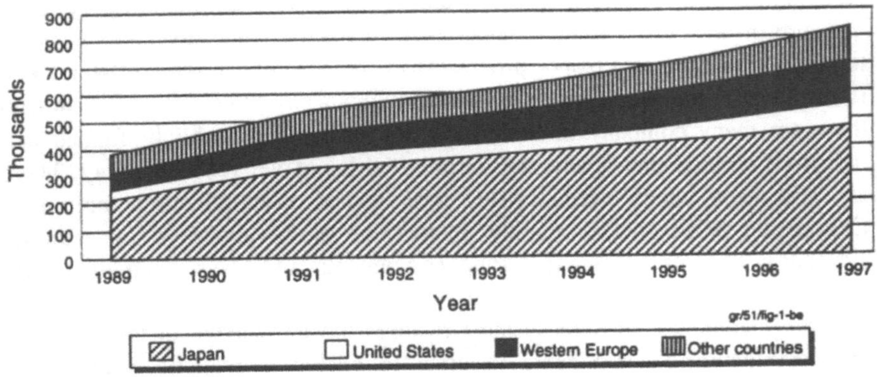

Fig. 1 Total stock of robots at year end, 1989-1993 and forecasts 1994-1997 (source: UN/ECE and IFR secretariats)

In all of these fields, much R&D has been undertaken in companies and research institutions. What is missing is a bundling of all these activities and a collection of the results already attained. Therefore it should be aimed at incorporating all these efforts, i.e. providing accurate simulation models and standardized interfaces for their vendor independent integration. Additionally optimization tools which are linked to or integrated in commercial SOP systems should be available. Furthermore powerful methods and tools for error compensation of manufacturing components and systems are needed in order to improve the product quality, reduce the commissioning time and cost and increase the launching flexibility.

2 State-of-the-art in Robot Planning and Future Needs

The majority of robot users still use the manual teach-in programming method for program generation, even though commercial SOP systems have been available for a number of years. Currently, the SOP systems are hardly economically viable even for companies using a large number of robots in a flexible manner. If SOP systems are used, the generated programs require manual on-line modifications at the shop floor. The manual effort to re-teach off-line generated programs still results in tremendous costs compared with those of the off-line programming effort itself, mainly due to the production loss during teaching. A typical welding line with 30 robots and 40 welding spots per robot takes about 400 hours only for robot teaching. Therefore, it should be aimed at avoiding these time consuming and costly efforts, i.e.

doing the robot programming in the planning office in parallel with the running production.

A first step in in overcoming these problems is the availability of the Realistic Robot Simulation (RRS) specification |2|, i.e. the integration of original robot control software parts into simulation (fig.2). It has been shown that RRS greatly improves the simulation accuracy related to the robot's motion behavior (deviations between simulation and reality with RRS: joint angles $< 0{,}06$ degree, cycle time of task programs: $<3\%$). A second step is related to robot calibration and error compensation |3,4|.

From this recognized need, several SOP developers have recently introduced calibration packages of limited capabilities. Each SOP calibration package requires a variety of measurements to be made in the production line, in order to supply the necessary data to their calibration routines. Neither the tools for measuring these data, especially in the running production, nor the methods for their collection, evaluation, administration and integration in factory information systems are defined, respectively sufficiently tested, in an industrial environment.

A further weak area is the deviation between the simulated and the real work cell. Results of measurements made by several end-users show that these differences are for example: robot location >10 mm, articulated and elastic structure $\approx$10-20mm, robot's end effector (welding gun) <10mm, and location of the workpiece $\approx$10mm.

It can be stated that there are software tools available for calibration and off-line compensation, but applicable methods and tools have to be developed for on-line compensation of robot errors (e.g. within the robot controller) at the shop floor. Many discussions with end-users, especially from the automotive industry, pointed out that only on-line compensation will be accepted. Off-line compensation would mean many additional efforts concerning the administration of robot programs. A general statement of the end-users is therefore to purchase only robots with a defined absolute positioning accuracy in the future.

Furthermore, the tools available for optimization of robot task planning are very poor. This is true for their functionality and their integration/linkage to commercially available SOP systems.

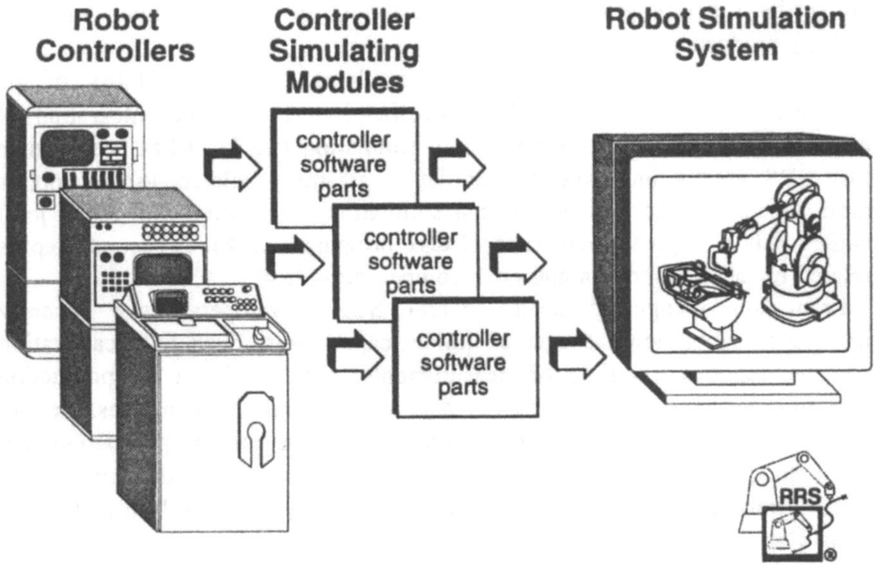

Fig. 2. Realistic robot simulation via integration of original controller software into simulation

Finally it can be stated that corresponding methods and tools for the integration of such procedures and systems in an end-user environment are almost completely missing. This also concerns aspects such as easy to use systems and the information flow (consistency, interfaces) from planning level to the shop floor level.

In order to reach the already mentioned goal of reducing commissioning time of manufacturing systems from 6-8 weeks to 48 hours from the point of view of robot programming, the following is required:

- The availability and exchangeability of accurate simulation models of manufacturing components delivered by different vendors. This is also very much associated with standardization of related interfaces.
- The availability of advanced robot systems providing on-line error compensation capabilities (i.e. high accuracy at low price).
- The availability of optimization tools for robot planning and programming in order to reach minimal cycle times, minimal stress of the robot links or even an optimum of both at the same time. These tools should be directly connected to or integrated into SOP systems.
- The availability of adequate information systems enabling a continuous information flow from the planning level to the shop floor level and vice versa. This also includes feed-back loops between the different levels and information consistency (information integration).

- Procedures and tools for the integration of all the different components and systems ensuring their ease of use and operation.

In the subsequent chapter three items from the above list will be dealt with in more detail. This concerns the aspects of robot accuracy, the so-called realistic robot simulation (RRS) and trajectory optimization.

3 Methods and Tools for Commissioning Time Reduction

3.1 Realistic Robot Simulation

Off-line generation of robot programs only make sense if they can really be used at the shop floor. This means that the correction efforts (re-teaching) are minimized or better superfluous. This requires a simulation environment which ensures the testing of their executability and correctness. Therefore, the simulation must contain accurate models of the real equipment and systems. One problem thereby is the availability of a model of the robot controller of which the algorithms defining the robot's behavior are not publicly available. To overcome this problem a consortium of automotive companies initiated the so-called Realistic Robot Simulation (RRS) project. It aimed at integrating original controller software (black box) into SOP systems via the specification of an adequate interface. To do so, the integration of robot, controller and SOP system developers was a must. The IPK-Berlin has been selected in a competitive procedure for the project management.

In the beginning of the project, it was decided to concentrate on the motion and kinematic software of controllers. To be able to reasonably operate these controller parts in a SOP system, additional functionalities had to be included (e.g. machine data, tracking, condition handling). Furthermore the interface had to be uniform for all controllers, but also had to cover a wide range of different concepts for motion handling, parametrization and performance, while being handable and easy to survey. In the result, the interface provides a set of principal services. Without these services no reasonable operation of a robot controller simulation (RCS)-module is possible. In addition, several groups of further services are defined, from which services relevant for a specific controller can be selected.

The operation concept of the RRS-Interface is subsequently described by the operation of the principle services (fig. 3).

The RRS-Interface supports the simulation of any number of robot controllers by one RCS-Module. An instance of a robot controller is generated by the service INITIALIZE. INITIALIZE returns a unique identifier for this instance. The identifier has to be passed to each further call of a service for this instance. The service TERMINATE ends the being of an instance, the identifier is then invalid.

General Operation:

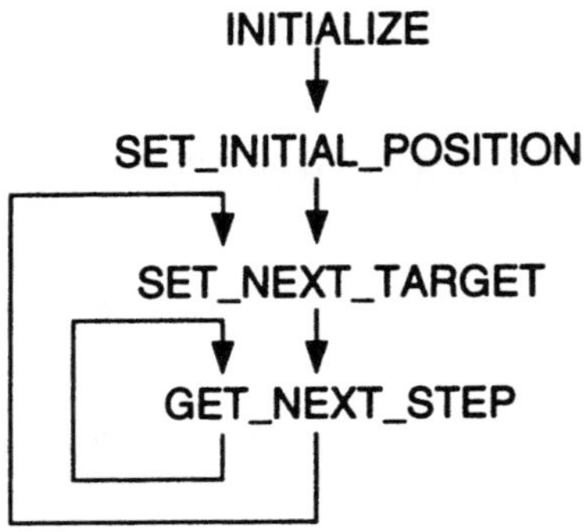

Machine Data:

MODIFY_RCS_DATA
GET_RCS_DATA

Fig. 3 The Principal Services

After INITIALIZE the position of the robot is still undefined. The service SET_INITIAL_POSITION sets the robot position, thereby defining the first starting position for interpolation. From now on target positions for motion can be passed to the instance by the service SET_NEXT_TARGET. Subsequently, the interpolated intermediate positions may be sampled by the service GET_NEXT_STEP.

GET_NEXT_STEP also returns control information to the simulation system. This concerns error and informational messages, occurred events and a status that controls e.g. requests for new targets. As long as the status indicates 'The service is successful', GET_NEXT_STEP returns robot positions. If the status indicates 'Need more data' further target positions may be passed with GET_NEXT_TARGET. In the case of circular motion or fly-by motion GET_NEXT_STEP may return 'Need more data' several times. If no further target positions are supplied, GET_NEXT_STEP may continue to return interpolated positions. If the last supplied target is reached GET_NEXT_STEP returns in the status 'Final step'.

The integration concept of RRS had to address the fact that the different programming systems as well as the controller software are written in several programming languages (C, Fortran, Pascal), run on different operating systems (MVS, Unix, VM, VMS, special developments) and on a variety of different hardware platforms (DEC, Hewlett Packard, IBM, Intel, Motorola,

Siclicon Graphics, SUN). Beyond these portability requirements, the interface had to cope with the problem of partial implementation (not all controllers require all services) and the interface had to be expandable. Both the portability and the compatibility requirements led to the development of the RRS-Calling Conventions (fig. 4).

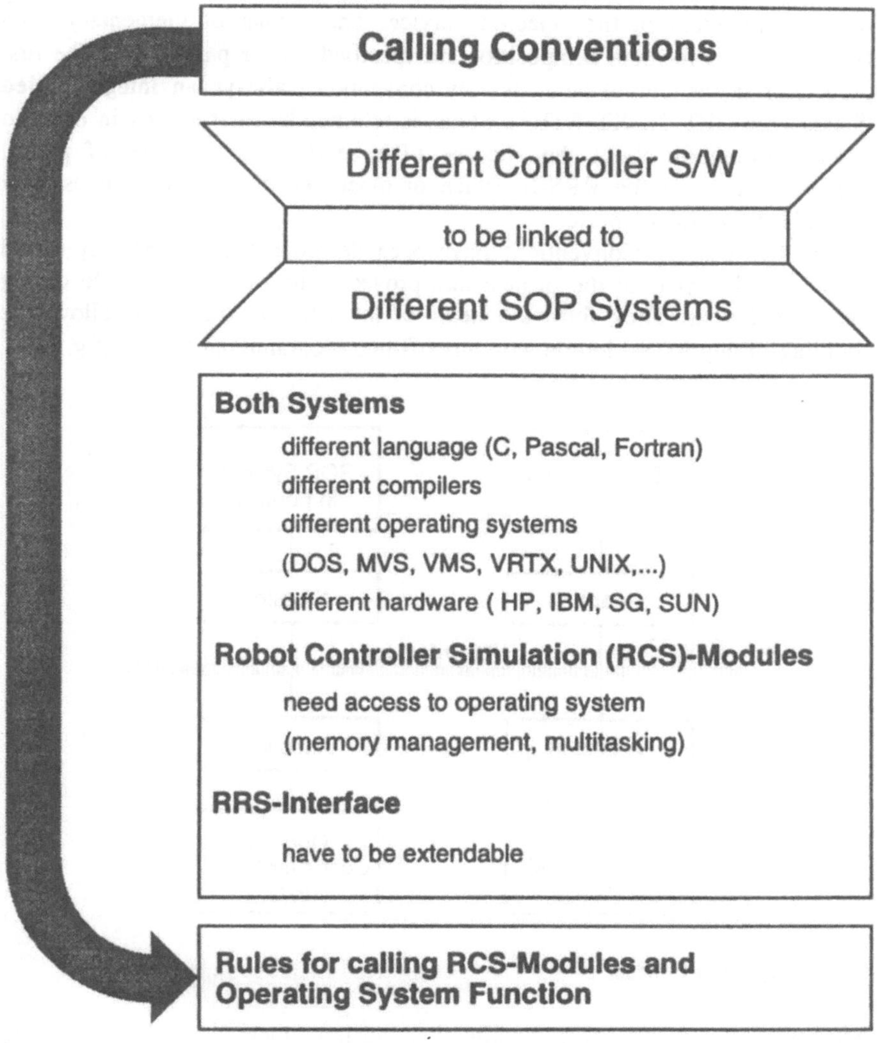

Fig. 4. RRS Calling Conventions

The basic concept of the RRS-Calling Conventions is to supply only one address per RCS-Module as a main entry point. This entry point is an ANSI-C function with two pointers as parameters. Such a function can be called and provided by most programming languages. The first parameter points to a data block for input parameters and the second one points to a data block for output parameters. Each of the blocks starts with a header containing information for data security. The header of the input block additionally contains an opcode that selects the called service.

The header of the input and output blocks are followed by the input and output parameters of the selected service. The format of elementary data types and rules for data composition are specified for the parameters. The first parameter in the output block is - by convention - always an integer called 'Status' (see GET_NEXT_STEP). Status is returned by all services in order to report information about the success of a service. In the case of partial implementation of the RRS-Interface or older versions, RCS-Modules may report 'Service not supported'.

The RRS-Calling Conventions have been developed and thoroughly tested during the first year of the project and proved to be robust and stable during the second year. They form the heart of the RRS-Interface and allow the coupling of any RCS-Module with any off-line programming system (fig.5).

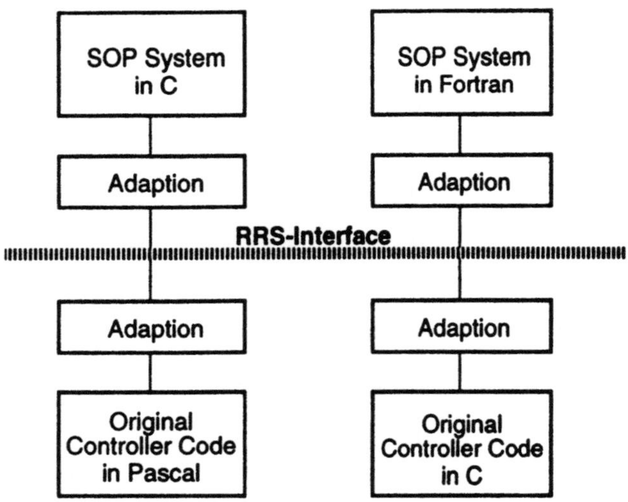

Fig. 5 Integration Concept

After two years of intensive specification and implementation, the concept of RRS proves to be extremely successful. It was possible to easily integrate the five developed RCS-Modules into the four off-line programming systems of the participants. Up to nine different robots from four different manufacturers have operated in parallel with satisfying response time. The desired simulation accuracy of 0.001 radians for joint values and 3% for cycle time was achieved in all cases and surpassed to up to 0.00005 radians and 1% cycle time. Today RCS-Modules are in industrial application.

For the future, the obtained success has to be maintained and extended. The RRS-Interface Specification has to grow with future extensions of the controllers in order to follow the state of the art. The deferred requirement for inclusion of the language/interpreter system into RCS-Modules remains an important issue. For this purpose, the project team will in the future be responsible for the maintenance of the RRS-Interface Specification and maintenance meetings including further companies will be held. The IPK-Berlin will stay available for contacts and for distribution of the RRS-Interface Specification.

Additionally, the IPK-Berlin has developed an RRS-Interface Shell Package (RISP) which adapts the RRS-Calling Conventions to ANSI-C-function calls. RISP reduces implementation efforts up to 50% and, more importantly, guarantees the formal correctness of the interfacing and facilitates debugging and integration (fig. 6).

The complete software, which is delivered in source code, includes the following packages:

- The computer-aided robotic (CAR)-shell which calls RRS-services via corresponding ANSI-C functions used for SOP-system openings and RCS-module testing.
- The RCS-shell which provides an RCS call via ANSI-C functions corresponding to RRS services used for RCS module implementation and SOP-system testing.
- A dummy RCS functionality which generates a complete RCS-module in combination with the RCS-shell. This function returns test values for SOP-system testing.
- An empty RCS functionality for step by step building of RCS modules.
- A test procedure which calls each RRS service via the CAR-shell. Since executable programs can be linked to the components of RISP, SOP-system opening or RCS-modules can be built step by step simply by replacing components of the RISP package.

In summary it can be stated that the required programming efforts for RRS-Calling Convention's implementation can be avoided by using RISP.

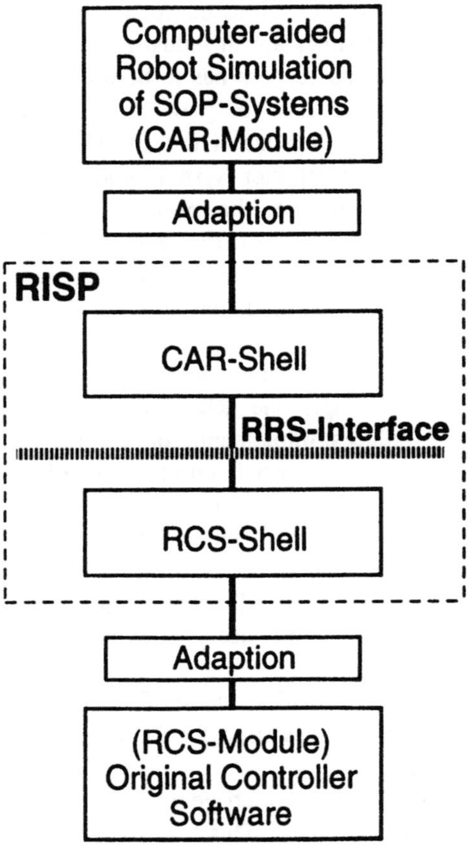

Fig. 6. RRS Interface Shell Package (RISP)

3.2 Robot Calibration

In contrast to teach-in programming of a robot which requires only a high repeat accuracy, the absolute accuracy in the positioning of the robot plays the dominant role for off-line programming. Absolute positioning accuracy depends on the manufacturing tolerances of the robot and the accuracy of the robot model used for motion control within the robot controller. These tolerances result in positioning errors. Instead of avoiding assembly and manufacturing tolerances, a model deviating from the nominal model can be established for each real system. The determination of a more complete and precise model is called calibration. The goal of calibration is to improve the absolute positioning accuracy to the repeatability level without any constructional modifications. Furthermore, calibration results offer versatile insights on the quality of the manufacturing process of robot systems and machine tools. Calibration results are a set of identified robot model parameters which can be used by the

- robot manufacturer, as a check on the quality of robot production,
- robot user, to improve the robot's absolute positioning accuracy, i.e. to use these data for the compensation of off-line generated robot programs, or
- the robot controller manufacturer, to implement algorithms compensating these errors on-line in the controller.

R&D in static calibration of robots have resulted in various methods for identifying actual internal features of a robot such as joint-axis geometries, actuator mechanisms and system compliances which are not accurately accounted for in the robot's nominal controller model. All these features influence the static positioning accuracy of the robot. In fig. 7 values of positioning errors of a typical spot welding robot with and without error compensation are given.

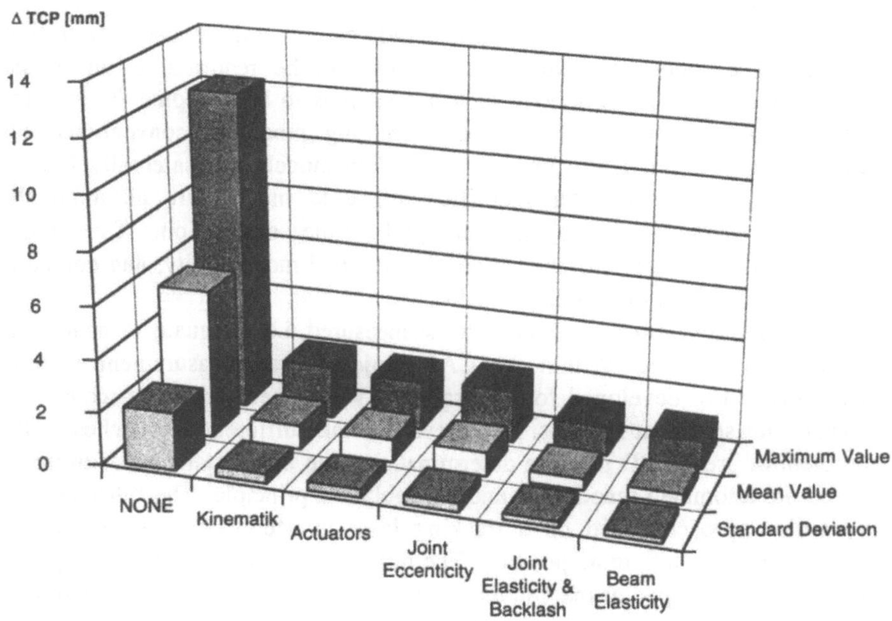

Fig. 7. Reduction of positioning error ΔTCP when identifying additional parameter classes from none to full model

An industrial robot calibration system which consists of automated procedures and equipment which do not require a technical knowledge of calibration theory for their use, and its application for quality control in robot production were developed by an R&D consortium within the ESPRIT project CAR-5220. The following methods, tools and equipment have been developed:

- Appropriate robot modeling techniques
- Reliable parameter identification methods
- Accurate measurement equipment
- Data acquisition and management system

Experience showed that not only kinematic but also mechanic features have a significant impact on accurate robot identification. The kinematic geometry of the robot model used is therefore extended to include effects of elastic deformations and actuator parameters. In order to include elasticity, the reaction forces and torques induced by the robot's payload and its own body mass must be computed.

The robot model is defined as an integration of four models; the actuator model, defining the mechanical relationship between robot motors and joints; the kinematic model, describing the robot's overall movements; the deformation model, characterizing the compliance in the robot joints and links; and the measurement target model, specifying the target-center-point (TCP) with respect to the robot flange.

Though modeling is a most important and complex task of robot calibration, a user of a robot-calibration system must not be required to understand modeling-convention theories if such a system is to be accepted by industry. This requirement is avoided by programming modeling conventions into calibration software. Thus, for each of the four models, the internally (i.e. for computation and parameter identification) used models are automatically generated from a simple and comprehensible description. The model generation procedure also checks consistency and modelability, and computes the nominal values of model parameters.

For calibration robot poses have to be measured, i.e. adequate measurement systems are required. Within the CAR project, three measurement systems were adapted or developed for calibration. Two of them are large-scale, non-contact measurement systems. These systems take different approaches to the problem of precise 3D position determination. One offers static measurements based on automated theodolites and triangulation principle. The other takes a dynamic approach based on a tracking laser interferometer, using distance and direction to determine position. The third system is a low-cost equipment which was developed at the University of Amsterdam. The system is based on a single camera in the robot hand, plus a specially designed, passive, flat target positioned in the robot work space. The system is self-calibrating, i.e. camera, lens and CCD deviations are identified and compensated during the evaluation of a set of measurements. Experience with this system shows that the camera parameters need to be recalibrated on every run, and preferably be monitored throughout the run. A prototype version of the measuring system was implemented using a simple off-line camera. The target plate consists of a blank aluminum plate with a black pattern of rings anodized onto it [5].

As a third step in the calibration procedure a numerical identification of parameter differences to the nominal model, so that position deviations between model function and measurement are minimized. Besides the identification of entire model, it is possible to determine partial models by an arbitrary combination of the parameter classes. The principle calibration procedure is illustrated in fig. 8.

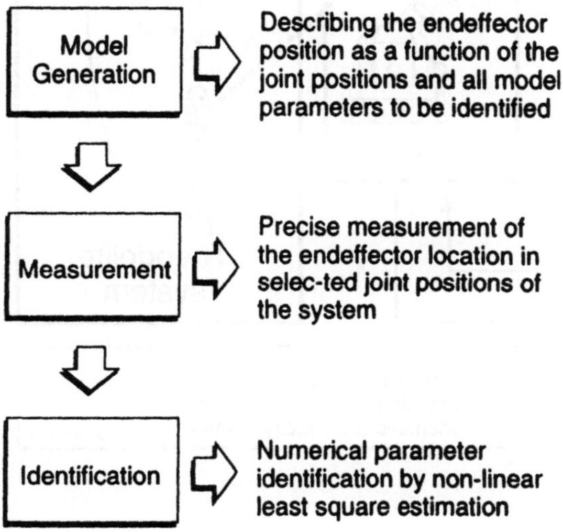

Fig. 8. Principle Calibration Procedure

The calibration system developed within the CAR project consists of four automated procedures:

- Generation of an internally used robot
- Generation of measurement positions for calibration
- Robot measurement
- Identification, verification and evaluation of robot-model parameters.
- Additionally, it offers the following opportunities:
- Partial recalibration after robot repair and for installation through selecting arbitrary model parameter subsets for identification
- Application of low-cost measurement systems and automatic evaluation of calibration results.

In fig.9 the calibration procedure is illustrated and the typical results are given when calibration data are used for error compensation.

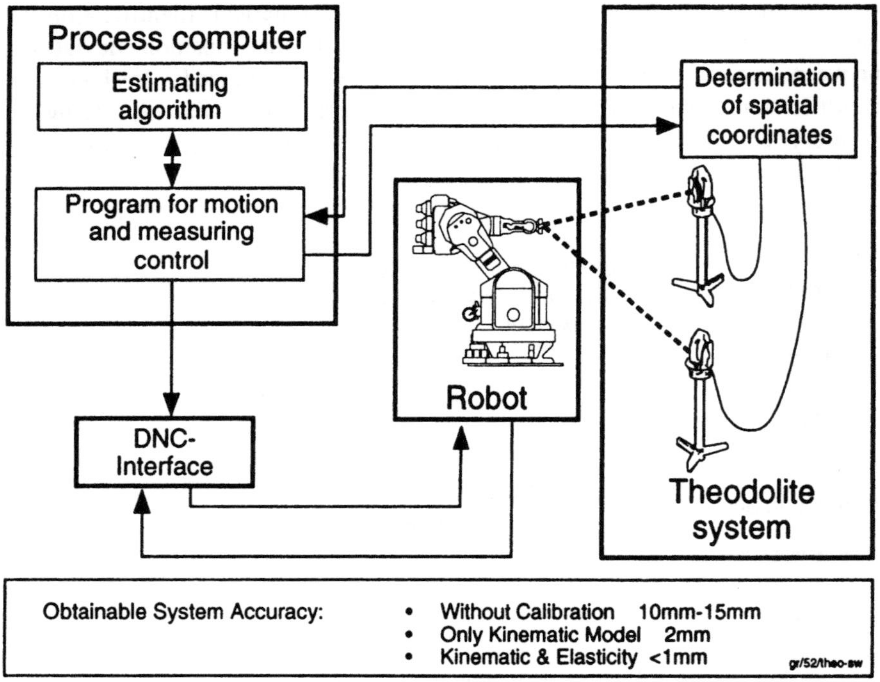

Fig. 9. Calibration procedure and results

Finally it may be emphasized that the robot is only one link in the manufacturing chain. All other elements such as welding guns and the car body also have tolerances and have therefore been taken into account. In fig. 10 these error sources are marked by coordinate systems. To deal with these errors additional means for measuring and compensation are required.

Based on the R&D work described above the IPK has developed a software package called RoboCal. It is composed of the components modeling, identification, evaluation and error compensation (fig. 11). Flexible interfaces enable the exchange of all required data with external systems and equipment.

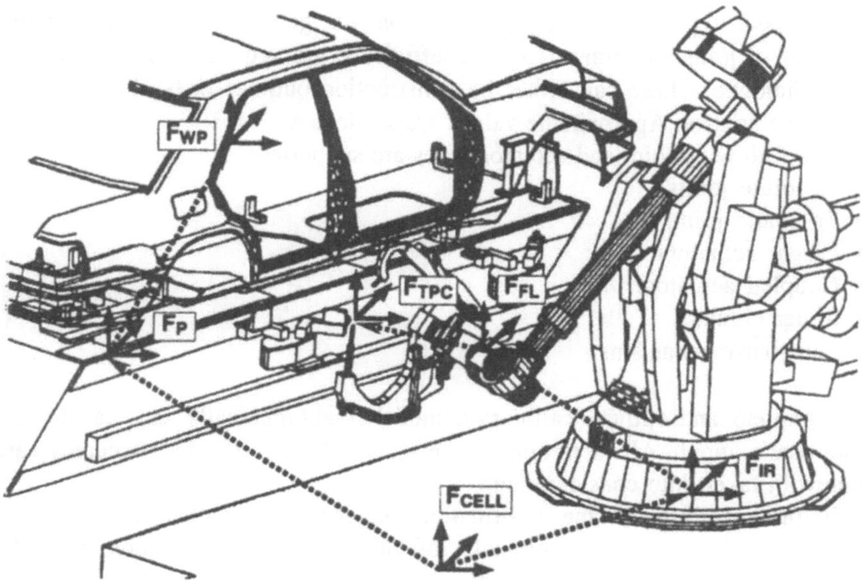

Fig. 10 Error model for off-line programming (source: Mercedes-Benz AG)

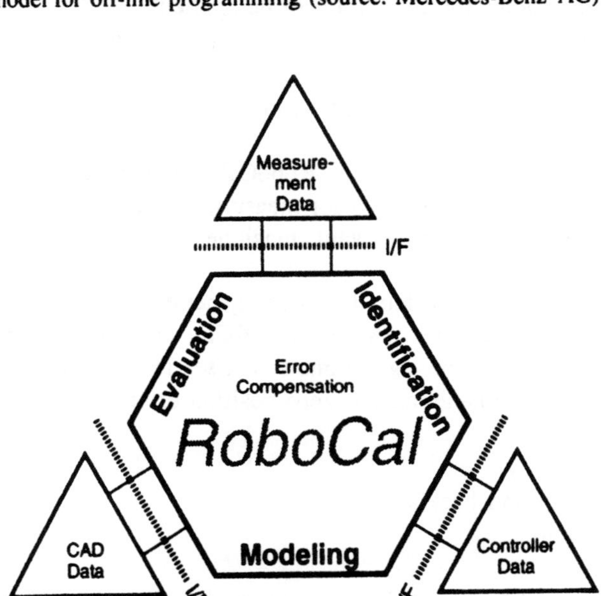

Fig. 11 Structure of the RoboCal Software Package

Model generation is largely automated and must be executed only once for each robot type. Necessary information on geometry, axes orientation, number of joints, or transmission coefficients are requested by RoboCal in user dialogue. These can either be construction outlines or be directly taken from the control. Approximate values are sufficient.

The following joint and actuator types are supported:

- Revolute joints
- Prismatic joints
- Linear actuators
- Coupled actuators
- Slider cranks
- Four-bar-mechanisms

A complete and optimal calibration model is automatically calculated from the user's input. All deterministic effects which influence accuracy are herewith taken into consideration.

In order to receive the information on the real system necessary for calibration, precise measurements of the end-effector position and the according motor encoder values must be executed for a number of selected robot positions.

RoboCal does not have an integrated measurement procedure. Instead a flexible interface has been implemented, which enables various types of measurement systems for calibration to be applied:

- Any type of position and orientation measurement system (1D - 6D),
- distance measurement system
- light beam connected to the control, etc.

For the system to be examined RoboCal automatically creates sets of maximal different joint positions for measuring. A general validity of the calibration results for the relevant machine operating space is herewith guaranteed.

Furthermore, RoboCal offers the direct generation of an appropriate robot program in the languages IRDATA, SRCL and IRL. Furthermore, the identification of the following parameter classes is enabled:

- Kinematik (axes mis-alignment, zero position errors, link lengths),
- actuators,
- joint eccentricity,
- joint elasticity and backlash,
- beam elasticity

Besides the identification of the entire model, it is possible to determine partial models by an arbitrary combination of the parameter classes.

The software package is available on all platforms with the operating systems OS/2 (DOS), Unix and VMS and offers various possibilities to use

the calibration results. As far as possible, calibration parameters can be directly used to correct the control model. For that RoboCal enables the output of identified models in various mathematical representations.
Programs generated off-line are modified using the calibrated model before transferring to the real system, so that the original positions created off-line will be reached when the inaccurate control model is used.

3.3 Optimal Trajectory Planning Tools

Programming of industrial robots generally takes place on the basis of the operator's experience, regardless of the system dynamics or relevant optimization criteria. The technical system is limited to testing the results of the human planning for compatibility with the limiting technical conditions of the manufacturing systems and technology. The attainable quality of user programs (working speed, mechanical stress on robot elements, motion flexibility, etc.) as well as the planning expenditure needed for such depend therefore strongly on the individual qualifications and motivation of the operator. Due to the complexity of industrial robots and manufacturing systems, even highly qualified operators can only reach a limited level of efficiency. The better exploitation of performance limits of robot integrated manufacturing systems can only be achieved using computer aided planning systems, whereby suitable planning and optimization procedures are necessary [7].

The application of robot integrated flexible manufacturing systems in production technology often implies for the execution of certain tasks a number of basic alternate solutions for robot motion behavior. For the optimal robot motion planning, the goal is to choose the optimal trajectory from among the range of allowable trajectories. Here, optimal is understood to mean choosing the best possible solution in the sense of a performance measure while taking into account the physical, technological, and equipment constraints. Numerous research works have been done in the area of optimal robot motion planning which can be formulated as a constrained optimization problem. Solutions using classic methods such as the variational method and the Pontrjagin´s minimum principle lead to a two-point boundary-value problem [8]. Such an approach can be solved with limited numeric expenditure only in the case of linear systems or simple two-jointed Scara robots [9]. A number of the proposed solutions are based on the simplified optimal solution such as assumption of the number of switches or the linearized dynamic models to obtain a sub-optimal solution [10]. An efficient method for the special case of the velocity optimization along a pregiven path has been developed in [11,12]. This method was extended to find the time-optimal path as well as the corresponding velocity [13,14]. Most of the proposed solutions are based on approximated constraints and neglect the realization conditions on the control systems of industrial robots. In addition, they consider only the case of time-minimization. Other important performance criteria such as mechanical stress or energy consumption are not taken into account. In SOP systemsoff-line programming systems presently used in the industry, optimization procedures for robot motion planning are still not very powerful in the sense of minimum cycle time or minimum mechanical stress under consideration of all realistic constraints imposed by the robot and technology.

However, the considerable progress made in off-line programming of industrial robots offers new possibilities for the integration of optimization procedures into SOP systems. Increasing productivity remains a primary impetus for applying robot integrated manufacturing systems. High working speed and high reliability represent often contradictory requirements which can only be solved by suitable optimization procedures. The IPK developed a practical optimization procedure for robot motion planning in terms of minimum cycle time and/or minimum mechanical stress under consideration of all realistic constraints imposed by the robots and technology. The developed procedure has been linked to an off-line programming system for the generation of optimal user programs.

The optimization criterion can be chosen from among the minimum cycle time and the minimum mechanical stress. The whole cycle time of a production cell is generally determined by certain time-critical motion sections which should be optimized with respect to the minimum time. For repetitive motions with given cycle times, in contrast, the minimum mechanical stress should be emphasized, leading to more favorable dynamic behavior and therefore extended robot reliability.

According to the two basic types of robot motions, the following tasks towards optimal robot motion planning can be formulated: For continuous-path control (CP), the geometric path (position and orientation) is given, while the velocity profile is to be optimized, as is the case for cutting, painting and arc welding. This task can therefore be formulated as follows: Set the velocity profile for the given path so that the criterion function such as the minimum cycle time or the minimum mechanical stress is minimized. For point-to-point (PTP) control, in contrast, only the beginning and the end positions of the robot are given, while the collision-free connecting trajectory between these positions is to be determined, as in the case of spot welding, pick-and-place-operations, as well as simple handling operations. Both optimal solutions have to fulfill the following robot specific and technological constraints:

- the maximum driving torques,
- the maximum joint velocities,
- the admissible velocity and acceleration of the robot hand,
- the joint angle´s limitations,
- avoidance of obstacles in the working space.

Both problems represent a non-linear, dynamic optimization task with complex constraints and can, in terms of system theory, be expressed as the following basic optimization task: Desired is the optimal trajectory, which minimizes the cycle time or the mechanical stress under consideration of all constraints imposed by the robot and technology.

The IPKsolution is based on the projection of the optimization problem from an infinitely dimensional functional space into a finite dimensional functional space. The key idea is that an arbitrary desired function can be approximated by a linear combination of base functions of a suitable functional space, i.e. φ_1, φ_2, φ_3, ... are a complete system of functions in the definition space D. Each function f(x) in *D* can be approximated by a linear combination of the functions φ_i with sufficient accuracy.

If the desired functions are approximated by the attachment function, i.e. $\Sigma\alpha_i\ \varphi_i$, the optimization problem is then put down to determining the unknown coefficients (α_1, α_2, ..., α_n). There are different possibilities when applying this principle, such as parametrization of the control functions or the state functions using a finite number of base functions of suitable functional space. Just as well, different base functions can be used, such as trigonometric functions, polynomial and Tschebyscheff-polynomial, etc.. The prerequisite for suitable base functions is that they build a complete function system, with that an arbitrary function can be approximated by a linear combination of these. In the selection of appropriate attachment functions to describe the optimal trajectories of industrial robots, above all, the following three aspects have to be taken into account, the approximation property, the realizability in the control systems and the required computation expenditure for on-line trajectory generation. Based on these requirements, cubic interpolation splines are chosen as attachment functions.

The solution is based on the parametrization of the state functions, i.e. the path and velocity profiles, using the cubic interpolation splines. The differential constraint equations turn then into nonlinear algebraic equations and inequations, and the criterion function into a function of a finite number of variables. The criterion function has to be minimized with respect to these variables. In this way, the original dynamic constrained optimization problem is reduced to a parameter optimization problem with nonlinear algebraic equations and inequations as constraints.

The developed optimization software package (OTP) for robot motion planning has been linked to an available off-line programming system (fig. 12). Using this system, the desired paths or poses can be defined by means of the robot programming language, derived from SOP systems, or set in teach-in mode. The constraints imposed by the robot and technology can also be textually given or derived using the information available in SOP systems. With the given robot and task-specific characteristic data the optimization module lays down the entire trajectory, i.e. the parameters for the definitive geometric and dynamic description of the optimal trajectory and translates it into a user program in the appropriate robot programming language. The user programs are then transferred by DNC interface from the programming system to the real robot control system. In practical tests, up to 40% cycle time reduction could be attained.

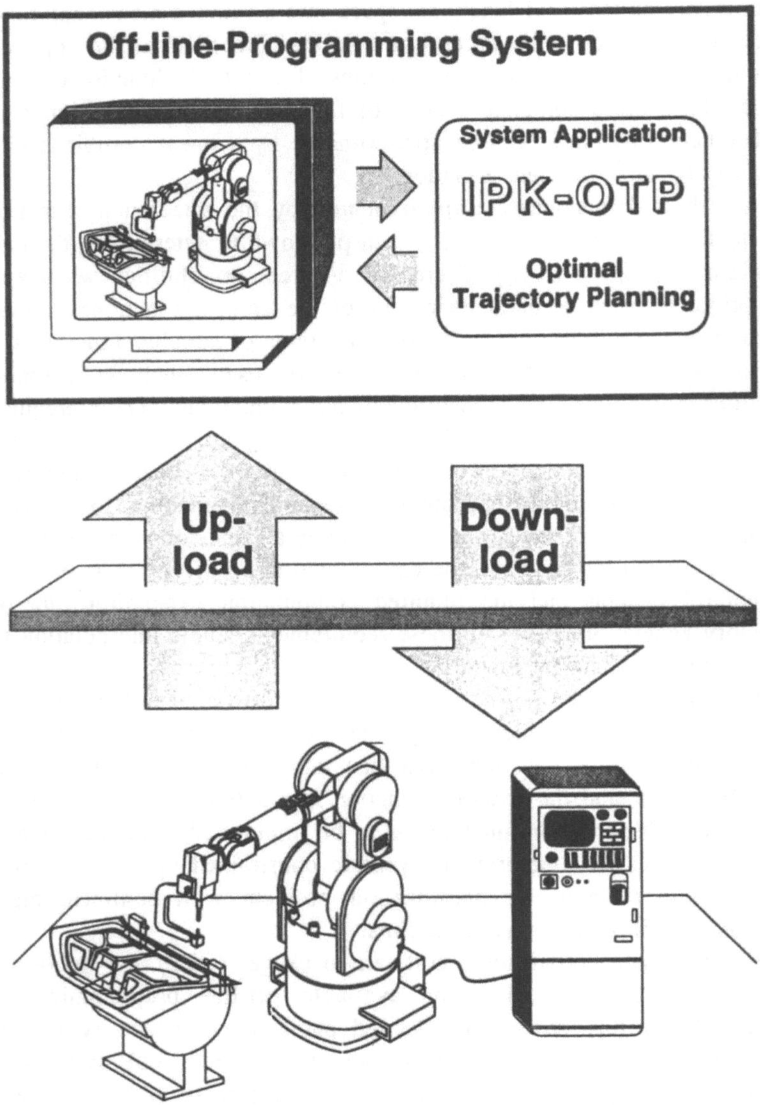

Fig. 12. Optimal Trajectory Planning (OTP) Software Package linked to a SOP system

4 Conclusion

The RRS project led to a tremendous improvement in simulation accuracy. The project approach is based on a cooperation of robot manufacturers and SOP system suppliers. It was initiated and driven by the users of robotic systems in the automotive industry. By a commonly defined interface, the RRS-Interface, robot control simulation modules provided by robot manufactures (so called RCS-Modules) can be integrated in SOP systems. The technical goals defined by the users were reached in the given time frame of two years. The RRS-Interface has been implemented in different SOP systems available on the market and first RCS-modules are integrated and successfully applied in industry.

Regarding robot calibration, it has been shown that these procedures are useful in several areas. It is a tool for improving robot accuracy which in turn enables off-line generated programs to be successfully implemented on the shop floor. Its results can be used to adjust or compensate existing programs after robot repair or replacement, and thus reduce production loss. Additionally these methods can lead to mechanical design improvements or manufacturing cost reductions. Finally, for the robot user, calibration is a monitoring and diagnostic tool for detecting robot component wear.

The optimal trajectory planning software package (OTP) allows a minimization of cycle times or mechanical stress. Additionally, in a mixed mode the minimum of both can be reached at the same time. Practical tests cases showed that cycle time reductions up to 40% are possible. Due to its open interfaces the OTP package can be easily linked to commercially available SOP systems, i.e. to have the full functionality of the SOP system beside the optimization functions.

For all areas related software tools are offered by the IPK-Berlin. To reach the overall goal, i.e. efficient off-line generation of robot programs and their operation at the shop-floor without corrections, additional R&D work has to be done. This concerns the analysis and the definition of application oriented "accuracy cycles" including more accurate robot and controller behavior, manufacturing environment accuracies. After this analysis appropriate measures concerning necessary specification, standardization and R&D work have to be taken.

5 References

[1] World Industrial Robots 1994
Statistics 1983-1993 and Forecasts to 1997. United Nations, New York and Geneva, 1994 (ISBN 92-11 100686-4

[2] Realistic Robot Simulation Specification, Version 1.0, 1994, IPK-Berlin, Pascalstr. 8-9, D-10587 Berlin

[3] R. Bernhardt, "Reduction of Set-up Times for Manufacturing Systems with Robots", Proceedings of the International Body Engineering Conference IBEC '95 (Body Assembly and Manufacturing), Detroit, 1995

[4] R. Bernhardt, S. L. Albright (Eds.), "Robot Calibration", Chapman & Hall, 1993 (ISBN 0 412 491 40 0)

[5] K. Schröer, "Calibration Applied to Quality Control in Robot Production (CAR)", final report of the ESPRIT project No. 5220

[6] D. G. van Albada, J.M. Lagerberg, Wei Z. Zhang, "Portable Calibration System for Robots", (chapter in [4])

[7] Y. Zhang, H. Münch, "Optimal Motion Programming for Industrial Robots", Proceedings of the 25th ISIR Conference, Hannover, 1994

[8] Kirk, D. E.: Optimal Control Theory. Printice-Hall, Inc. Englewood Cliffs, New Jersey, 1970

[9] Geering, H.P.; Guzzella, L.; Hepner, S.A.R.; Onder, C.H.: "Time-optimal motions of robots in assembly tasks," in IEEE Trans. Autom. Contr., Vol. AC-31, pp. 512-519, June 1986

[10] Wen, J. T., Desrochers, A. A.: "Sub-optimal control strategies for robotic manipulators," in Proc. IEEE Int. Conf. on Robotics and Automation, 1986, pp. 402-406.

[11] Dubowsky, S.; Shiller, Z.: "On the optimal control of robotic manipulators with actuator and end-effector constraints," in IEEE Int. Conf. on Robotics and Auto., pp. 614-620, 1985

[12] Shin, K.G.; McKay, N.D.: "Minimum-time control of a robotic manipulator with geometric path constraints," in IEEE Trans. Autom. Contr., vol AC-30 (June 1985), 531-541.

[13] Bobrow, J.E.: "Optimal robot path planning using the minimum time criterion," in IEEE J. Robotics and Automation. 1988, 4(4): 443-449

[14] Shiller, Z.; Dubowsky, S.: "Robot path planning with obstacles, actuator, gripper, and payload constraints," Int. Journal of Robotics Research, Vol. 8, No. 6, pp. 3-18, Dec. 1989

6 Acknowledgments

The article is based on results and discussions in many national and international projects and cooperations. Due to its importance, two of these projects are subsequently mentioned.

The CAR Project (Calibration Applied to Quality Control and Maintenance in Robot Production, EP 5220) was partially financed in the frame of the European ESPRIT program. Partners in this project were IPK, Germany; KUKA, Germany; LEICA, United Kingdom; University of Amsterdam, the Netherlands; TGT, Eire; Volkswagen, Germany. The project was coordinated by IPK-Berlin.

The RRS-project is totally financed by the involved companies: Opel, Germany; BMW, Germany; Audi, Germany; P.S.A., France; Mercedes-Benz, Germany; Volkswagen, Germany; Renault, France; Volvo, Sweden; Ford, Germany; ABB, Norway; Comau, Italy; Fanuc, Germany; Renault Automation, France; KUKA, Germany; Siemens, Germany; Dassault Systemes, France; Deneb, Germany, USA; Silma, Germany, France, USA; Tecnomatix, Belgium, Germany. In the initial phase of the RRS project (first two years) the IPK project management efforts were financed by the automotive companies. The management efforts of the RRS specification maintenance phase, running since 1994 is financed by all partners. Also for this phase the author of this chapter is the project manager.

14 Balanced Automation

L. M. Camarinha-Matos, R. Rabelo and L. Osório

1. Introduction

The concept. The concept of Balanced Automation tries to capture the idea of appropriate level of automation when considering the current challenges for flexibility, quality improvement and productivity, but taking into account the socio-economic context as well as the local traditions and skills. Most international efforts in search for new manufacturing paradigms have put too much emphasis on the technical aspects of automation, i.e., follow a technocentric approach. As a reaction, but still with lesser impact, the concept of Anthropocentric Manufacturing focuses the human-centered aspects. Both approaches are limited. A Balanced Automation System (BAS) aims the combination with adequate weights of, not only these two approaches, but considering also the economic aspects, the integration / migration of legacy systems and the full heritage of the enterprise, namely in terms of established know how and traditional skills.

A BAS is not an invariable construction as various factors may change along the time such as, for example, the cost of manual labor compared to the introduction of automation technology.

Although it is a recent concept, it can be considered a synthesis or development and integration of models and approaches that have been evolving in various sectors: social sciences, economics, university-industry experiences, enterprise modeling, holistic production management, human-machine interfacing, enterprise reference models, etc..

Pressing factors. Manufacturing companies worldwide are facing tough challenges as a consequence of the globalization of the economy, openness of markets, changes in the market structure and push of new technology and standards. The formation of economic blocks such as the European Union with its variable geometry extensions, NAFTA, and MERCOSUR, and the global agreements such as GATT, in addition to their obvious economic and social consequences, provoke strong paradigm shifts in the way the manufacturing systems are designed and operate.

As a result, two opposing tendencies can be observed. On one side there is a growing independence of the manufacturing sites regarding the geographical location. Many components are nowadays produced in different parts of the world basically under the same technological conditions. This tendency for the "uniformization" of the manufacturing conditions facilitates the re-location phenomena, tempered by the economic and political conditions. On the other hand, local societies and components / services suppliers try to become less vulnerable to such re-location raising the importance of their traditional skills and know how. Many regions are trying hard to identify, in some cases with obvious success, their local competencies and other unique assets. A Balanced Automation approach might give these "regional" companies the appropriate re-engineering tool for survival in a fast changing economic order.

Brief history. The concept of Balanced Automation emerged in the framework of the pilot collaboration research activities between the European Union and the Latin America, namely within the ECLA projects CIMIS.net and FlexSys. These projects were launched in 1993 by the Commission of the European Union as exploratory actions. As a pilot program, the intention was to evaluate the feasibility of such long distance cooperation and to investigate the most effective way of its implementation.

The CIMIS.net project was focused on Distributed Information Systems for Computer Integrated Manufacturing. The CIMIS.net consortium comprised academic and research centers from The Netherlands (University of Amsterdam), Portugal (New University of Lisbon, project leader), and Spain (Polytechnic University of Madrid) from EU, and Brazil (Federal Universities of Espirito Santo and Santa Catarina) and Costa Rica (University of Costa Rica) from Latin America. Covered areas include:

- -Integrating Infrastructures for the distributed industrial systems - the paradigms of federated architectures and virtual enterprises.
- -Distributed object-oriented information management systems - distributed access, interoperability, persistency, consistency, and concurrency.
- -STEP / Express based modeling of manufacturing systems.
- -Experiments on modeling and representation constructs for engineering applications.
- -Identification of information management requirements for concurrent engineering, and specification of the supporting functionalities.
- -Distributed control systems.

The FlexSys project addressed the Design Methods for FMS / FAS Systems in SMEs. This project aimed at promoting the collaboration in research, development and evaluation of methods for designing Intelligent Manufacturing Systems. The project involved partners from Portugal (Uninova Institute, project leader) and Germany (IPK, Berlin) from EU, and Argentina (National University of San Juan), Brazil (University of São Paulo) and Mexico (Cinvestav Institute) from the Latin America. The areas covered include:

- -Manufacturing systems control architectures - distributed approaches.
- -Electronic catalog of CIM components and interactive planning of IMS.
- -Modeling and evaluation of FMS/FAS systems.
- -Systems monitoring - multisensorial information acquisition and interpretation, diagnosis and error recovery.
- -Distributed factory information and communication systems.
- -Design and integration approaches.
- -Integration of legacy systems.

Although based on research groups, the two consortia tried, from the beginning, to establish various interactions with local industry. These interactions offered a rich understanding of the challenges industry is facing in both continents. The geographical distribution of the research centers involved in these two projects provided a rich variety of diversified contexts for research on balanced automation; namely the developed and developing countries, different directions and traditions in research and technology applied to industry, and the diverse cultural and economic situations. The two projects presented a unique opportunity for collaborative research, in spite of the rather limited financial investment. Within this context, through joint research, exchange of scientists, technical meetings and workshops, and the case study missions towards some industries, an adequate environment was created for the genesis of the Balanced Automation concept.

A major step was the decision to start the series of International Conferences on Architectures and Design Methods for Balanced Automation Systems (BASYS) whose first occurrences were held in Vitoria, Brazil (1995) and Lisbon, Portugal (1996). The concept of Balanced Automation is now accepted and applied by many research groups outside the CIMIS.net and FlexSys consortia and various other events and has been a subject for discussion in publications.

2 Difficulties and Contributions

2.1 Obstacles

In spite of all challenges manufacturing companies are facing, and the large number of international projects on advanced manufacturing systems, changes in the industry do not happen at the desired speed. A number of obstacles that justify this situation can be identified. Among them:

Size of enterprises. In most countries, as is the case in Europe, Small and Medium size Enterprises (SME) or even micro-enterprises represent the most important part of the industrial sector and also the most active in job development. But the engineering power of such enterprises, with the exception of a few high tech companies, is very limited, if any. In many cases the only engineer the company has is totally busy with the everyday survival activities.

On the other hand, many research activities address an industrial scenario only available on large (multinational) companies. The paradigms and solutions found for this context are hardly portable to the SME context. Although it is becoming fashionable to mention the SMEs needs, the reality is that the "large company scenario" still dominates in most research programs.

The European Union, through its R&D programs, namely the information technology initiative (ESPRIT) or the research on materials and processes (Brite / Euram), has launched some specific initiatives to facilitate the involvement of SMEs in cooperative projects. But the "bureaucracy" associated to the project proposals preparation and management is still too large for a micro or small enterprise. In most cases these companies only accept getting involved in a cooperative project when convinced by a research organization who volunteers to do most of the work related to the "project logistics". On the other hand, even though there is some political willingness towards SMEs, there is still a rather limited participation of representatives from the "SME culture" in the evaluation committees.

Engineering education. It is nowadays widely accepted that traditional engineering curricula, designed to prepare people who knows everything about their own field and nothing else, is not adequate to the contemporary manufacturing world. In various parts of the world promising approaches towards the education of people to be involved in team-oriented multidisciplinary industrial environments are taking place [1], [2]. The involvement of industry leaders and regional development experts together with academics in developing new curricula and teaching methods as well as assessing the results seems the right way to go.

An example of such trends can be found in a report from the American Society for Engineering Education [3] which includes, among others, the following recommendations as aspects to be incorporated in engineering education: a) team skills, including collaborative, active learning; b) communication skills; c) leadership; d) a systems perspective; e) an appreciation of different cultures and business practices, and the understanding that the practice of engineering is now global; f) integration of knowledge throughout the curriculum; g) a multi-disciplinary perspective; h) a commitment to quality, timeliness, and continuous improvement; i) an understanding of the societal, economic and environmental impacts of engineering decisions; j) ethics; etc..

Although the university - industry interactions have increased significantly in various parts of the world, the global situation is still not so optimistic. In many cases engineering curricula are developed with little interaction with the "outside world" and keep clear separations between departments / disciplines.

Thinking in terms of the SME environment, a real re-engineering of the engineering education is urgent. An SME cannot afford to hire engineers with a narrow view of the world.

Divorce industry - university. This is an old topic for many round-tables and symposia since long ago. In many cases the situation is having a positive evolution, but there are still many difficulties in various regions of the world.

One important aspect identified by various studies [1] is the faculty reward systems. In various places these reward systems are almost exclusively based on the "publish or perish" rule. As a consequence, any time spent working with industry on real problems is frequently a "waste" of time according to the rewarding policy. Therefore, some departments degenerate in a situation in which most of their engineering full professors never visited a shop floor. The obvious consequence is that instead of educating engineers, they prepare research assistants. This situation can even create a "cultural environment" in which any attempt to do applied research is considered a "lesser" activity.

It is not surprising to verify that in countries with a more developed industrial capability, like Germany, the criteria for appointing an engineering technology faculty member requires industrial experience. On the contrary, countries like Portugal, still privilege a "purely academic" career. Therefore a strong pressure from the society is necessary in order to force the faculty reward systems being assessed and modified to ensure they support the requirements of the new manufacturing reality.

On the other hand, many companies have a distorted idea about what they can expect from the universities. Some companies just look to the universities as a low cost subcontractor from which they can receive goods and services at a more competitive price compared to other suppliers. For instance, they do not understand that universities do not have the aim, skills and resources to develop and maintain products.

In Europe, due to the eligibility criteria for research projects under the European Community programs which strongly recommends, or even forces, the interaction university - industry, a new culture has emerged. Even in less developed regions the cooperation between universities and enterprises is becoming more natural.

Complexity: multiple interactions and variety of processes. Deciding about the appropriate level of automation for a given enterprise is really a difficult task. The number and variety of business processes and the interactions among all activities and subsystems in a manufacturing enterprise are quite large and complex. Besides a rich multidisciplinary background it is also necessary a great deal of practical experience that only a few senior engineers have.

For young researchers, the major workforce of many R&D projects, it is much easier to concentrate on a narrower subject (graduated students lack a general view). Most senior people become too engaged in management activities. Therefore there is really a shortage of people, worldwide, able to master the complexity of designing balanced automation systems.

2.2 Anthropocentric approaches

The concept of Anthropocentric Production System (APS) which can be defined as a manufacturing system which improves skills, participation in the decision making processes and the quality of working life [4], appeared as a reaction to the purely technocentric approaches. Under this approach new technologies should be designed to increase the participation of people on decision-making and control of the manufacturing processes. Therefore human centered systems are envisaged.

Observing the role of humans in a manufacturing environment an evolution can be noticed, from the early days in which the tasks of the human operator were to manually load, operate and unload machines, to a more decision-oriented activity as in systems' supervision. At some stage, with the technological evolution, there has been the goal of progressing towards the unmanned factory. Replacing skilled operators turned out to be too difficult and too expensive to achieve. In particular all decision making processes associated to error handling are difficult to automate. The expression "CIM-ruin" represents the idea of expensive installations that don't work due to small errors [5]. Therefore the role of human beings in the manufacturing environment has become more appreciated. Social pressure due to unemployment problems also contributes to this new view.

Although aiming a coherent set of technological and organizational innovations to improve quality and flexibility [4], APS are a little "biased" by socioeconomic and organizational factors and do not discuss the technical perspectives of automation. Some essential components of an APS are: i) Flexible automation, supporting human work and decision making; ii) Decentralized organization of work, with flat hierarchies and a strong delegation of power and responsibilities, especially at shop-floor level; iii) Reduced division of labor; iv) Continuous, product-oriented up-skilling of people at work; v) Product oriented integration within the broader production process.

It is interesting to notice, on this list, some influences of the "team work" area.

Although the emphasis is on the "human side" of the manufacturing environment, this area brings significant contributes to the BAS approach.

2.3 Enterprises evolution and reengineering

Traditionally, industrial enterprises were based on a vertical organization structure where all internal transformations were processed in-house in an independent and closed way, i.e., with very limited communications with the outside world (Fig1.1).

The first innovative step was basically supported by the introduction of numeric controlled machines leading to a production system with linear progression characteristics. In the sequence, new automation machines / robots were introduced.

Each time a new machine with new technological characteristics was introduced to modernize some sector, new factors contributed to change the productivity function. By the introduction of more advanced automation components the production and implicitly all associated costs changed from a linear function to a step-shaped function (structured by lots). The ideal lots are not any more arbitrary and the measure of the productivity / costs previously evaluated considering a linear function of the number of pieces, has now to be measured by lots of pieces. The production of one piece is similar to the production of one lot of 1000, for instance. The measurement of productivity started to be evaluated considering the "set up" time associated to each lot of pieces.

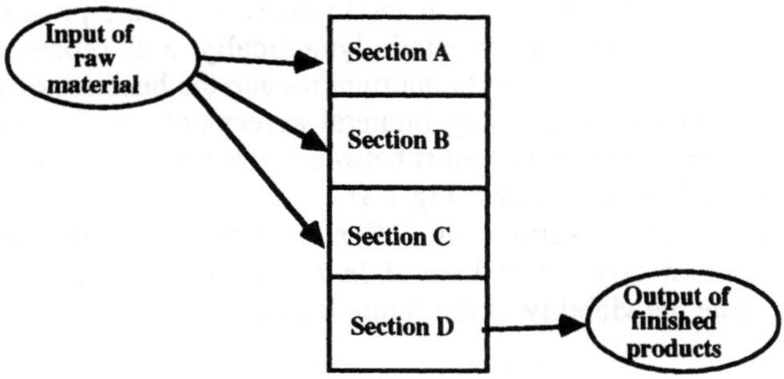

Fig. 1.1 Traditional enterprise structure

In parallel, and due to market changes, the size of orders changed from big to small quantities of each model. One consequence of this transformation was the introduction of new management strategies. Furthermore, if we take into account the new perturbation factor by which the production function loses its autonomy and closed character and becomes an integral part of the product life cycle, starting from the client orders until the product delivery or even product recycling, the consequences were tremendous and can be illustrated by the following example:

The introduction of a robot in a production line, increasing the production capability, may introduce bottlenecks in other zones, or may imply a big re-structuring effort in order to make it profitable. For instance, moving this sector out of the company and letting it behave as a separate company .

This situation has thus created a new scenario. The technological innovation, in parallel with the commercial relationship changes, increased the total production costs which can, in some cases and in some sectors, reach uncontrollable levels. To answer this new situation, some sectors like the automotive and aerospace, started to offer their services to other enterprises to generate new business opportunities and to maximize the return on investment in capital equipment and the skilled workforce (Fig 1.2).

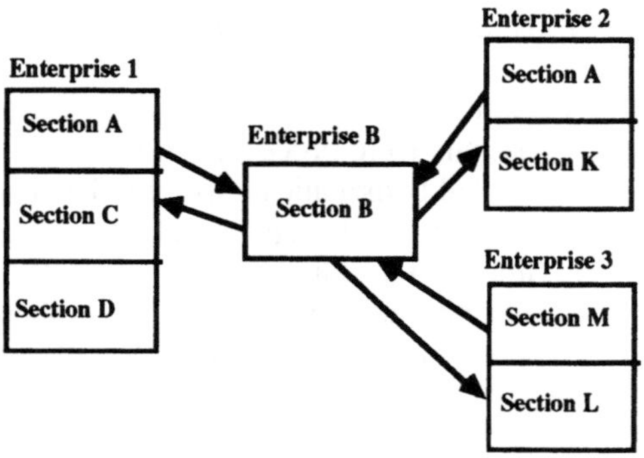

Fig. 1.2 Autonomization of sectors in an enterprise

This "emancipation" movement enabled sections within an organization to meet the needs of the internal requirements and the outside market place by being considered as an autonomous production unit driven by market forces and profitable business rules. This situation has been observed in many industry sectors, resulting in focused areas of specialization and expertise and, therefore, has transformed the way companies co-operate. This process has led to many inter-enterprise relationships. Now, enterprise networks are created dynamically and represent company associations required to meet market opportunities and the business objectives of all concerned. As a consequence, many bilateral agreements are in existence. This network associations create virtual paths between different enterprises and establish sets of virtual manufacturing chains (Fig 1.3).
This way, the old seventies vertical and inflexible enterprise is disappearing and in the same place a new concept of *Virtual Industrial Enterprise* is becoming the solution to promote the viability of the future industry.

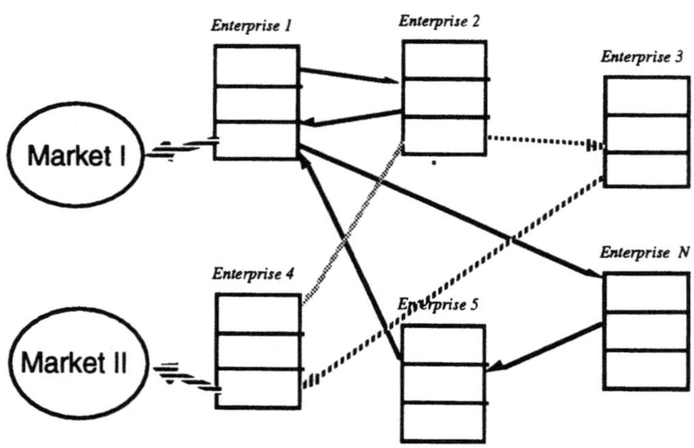

Fig. 1.3 Dynamic networks of enterprises

A good example of a virtual industrial enterprise is the partnership relationships which exist between the major European automotive manufacturers and their network of component suppliers. In fact, the manufacturing process is not any more carried on by a single enterprise, but each enterprise is just a node that adds some value (a step in the manufacturing chain). In virtual enterprises, manufacturers do not produce complete products in isolated facilities. They operate as nodes in a network of suppliers, customers, engineers, and other specialized service functions. Virtual enterprises are materialized by selecting skills and assets from different firms and synthesizing them into a single business entity.

The effort being put on the implantation of high speed networks (digital high ways), supporting multimedia information, opens new opportunities for team work in multi-enterprise - multi-site networks. This new scenario brings new requirements in terms of exchanging, sharing, managing and controlling information related to the products being produced. One key area is control; for example, access rights to product information, scheduling of information access, control of interactions. These requirements impact the design, availability and cost of appropriate distributed information logistics infrastructures.

There is here a major opportunity for a balanced approach. Although IT infrastructures and decision support tools are vital support to a virtual enterprise, it is important to find out the appropriate level of human participation. Even from the technological infrastructures point of view, migration from legacy systems and heterogeneity are key factors.

On the other hand, classical industrial organizations are suffering the pressure to join a process in which the classical circuit Raw materials => Producer => Transformer => Distributor => Consumer, works more and more under a truly Just In Time philosophy. In a traditional manufacturing cycle, there is a sequence of steps and some tolerance gap (in time) is left between each two steps in order to accommodate unforeseen situations. The tendency in last years has been to reduce the manufacturing cycle time, not only by eliminating the gaps between phases, but also by forcing some overlapping of phases.

On one side, this led to the concept of Concurrent Engineering. This concept is the result of the recognition of the need to integrate diverse expertise and to improve the flow of information among all "areas" involved in the product life cycle. The principle of concurrent engineering is therefore to create a team of experts from different fields and make the team responsible for the design, engineering manufacturing, and, marketing of a single product. For instance, production engineers in the team must be able to look at the early design, just as well as sales people. This is radically different from the traditionally approach known as "finish the design and throw it over the wall" to the production engineer and hardly interact at all. Team work, based on concurrent or simultaneous activities, potentially leads to a substantial reduction in the design - production cycle time, if compared to the traditional sequentially "throw it over the wall" approach.

But this tendency has consequences at a more global level, affecting the interaction between enterprises in a network. The need to shorten the production cycle imposes this parallelism of activities being carried out by different companies and the network must accommodate (or recover from) unexpected problems (delays) in any node; therefore, there is a need for an efficient flow of information between nodes and, especially, a much more effective way of managing orders in this network of clients - suppliers.

But to develop this new concept of a virtual enterprise a revision of the framework of experiences and models that are available for organizational design methods is needed. To cope with all these changes many companies have started

Business Process Reengineering (BPR) projects. Although several success cases can be found, various difficulties are also being detected. Among them: "implementation difficulties due to communication barriers between company sub-units, the unexpected size of the required BPR effort, its disruption to business operations, reluctance of top managers to commit the necessary funds for the project, making business mistakes under the pressure to produce quick results, etc." [6]. BPR is mostly a philosophy, still lacking formal methods (only in part being offered by modern systems analysis tools). On the other end, the necessary reengineering process should not be confined to the business processes. A shop floor reengineering is also necessary.

2.4 Systems modeling and analysis

An important requirement for the design of a Balanced Automation System is a modeling and analysis tool. In order to decide on the appropriate level of automation in a particular context it is important to have some sort of *what-if* analysis tool to apply on the model of the intended system. There is a large collection of modeling tools and analysis methods being applied to manufacturing systems. However, in most cases, these tools only model particular facets of the system and the integration of various modeling formalisms is, from a practical point of view, quite complex and time consuming. For instance, just to mention some of the most used formalisms:

-Functional modeling: SADT/IDEF0 [7], [10], BP/EA [8], CASE tools.

-Information modeling: EER, Express[9], NIAM [11], OO Languages, etc.

-Behavior modeling: Petri Nets (various dialects) [11], Stochastic models, Rules, OOP, Factory simulation tools (discrete events)

These formalisms / tools provide "separate" views of a system from which it is quite hard to identify correspondences between them. An "integration" effort in order to combine the various views is still needed. Other more subjective aspects - like quality, confidence, cost, etc. - are not clearly represented within some of these formalisms. The use of rules, OR (Operational Research) techniques, etc. might be a reasonable approach.

Another aspect to be considered is the acceptance of models by the target community. As a manufacturing system design is an interactive process involving engineers, managers, perhaps sociologists, it is very important to have a modeling framework from which easily understandable representations can be derived. This is however still a challenge.

An important tool for a manufacturing system modeller is a reference architecture that guides the design process. Major efforts in global enterprise reference models are: CIM-OSA [8], PERA [12], GRAI-GIM [7]. Of special mention is the IFIP/IFAC joint initiative on reference architectures, GERAM [13].

2.5 Legacy and migration

In our opinion, more important than the design of new systems is the transformation of existing systems in order to cope with new requirements. In this way, an important part of the problem is the accommodation of and migration from legacy systems. This is a hard engineering work which might involve:

- Tools wrapping: building adaptation layers around existing software applications and controllers;

- Face lifting: renovating the human-machine interface of old software;
- Machine retrofitting: changing parts of the control and mechatronic parts of manufacturing resources, adding sensors, etc..

On the other hand, the interoperation between different technologic generations, different data management systems and decision support tools is necessary. One can expect that the dissemination of standards will ease this migration process, but as standardization efforts do not proceed at the same speed as technology evolution, it is expectable that similar problems will remain.

2.6 Application areas

The concept of Balanced Automation can be applied or will have implications at all levels of a manufacturing system. In next chapters some examples illustrate some of the challenges a balanced approach introduces in the way systems are designed. The selected examples include shop floor systems, concurrent engineering, scheduling and support for virtual enterprises. This is not an exhaustive list but its scope is wide enough to allow a discussion of the main problems and to illustrate some solutions.

3. BAS and Shop Floor

A natural application area for a balanced automation approach is the design / re-design of manufacturing systems at shop floor level. In order to make a well founded decision on the level of automation and integration of human beings, it is important to define a good model of the system. Based on this model, a what-if analysis can be made. This is important, not only for the design of new systems but also, and perhaps the most frequent case, for the transformation of existing systems (shop floor reengineering).

Another aspect to be considered at this level is the design of ergonomic workplaces to allow for the improvement of working conditions.

Besides design and resources selection, a major engineering effort is necessary on the migration from existing technology / components. "Wrapping" (or "agentification") of controllers and retrofitting of machines is normally a hard but necessary task in order to allow the integration of manufacturing resources in a hybrid and highly heterogeneous context.

3.1 Modeling of manufacturing systems

One of the most "popular" tools for modeling the behavior of manufacturing systems is Petri nets. A number of researchers have applied PN to model FMS/FAS (see, for instance, [14], [19], [20], [21]), due to the following reasons:

- PNs capture the precedence relations and structural interactions of concurrent and asynchronous events. In addition, their graphical nature helps to visualize such complex systems.
- Deadlock situations, conflicts, and buffer sizes can be easily and efficiently modeled.
- PNs models represent a hierarchical modeling tool with a well-developed mathematical and graphical foundation.

- PNs allow the use of different abstraction level models and the use of powerful modeling methodologies: refinement and modular composition, each of them with its corresponding validation analysis.

Among the various extensions to the basic formalism, Temporized Petri Nets and Colored Petri Nets have been particularly used in the FlexSys project [15], [16]. Another example that has been used in various activities of this project [15] is the Mark Flow Graph - MFG [17] which is a technique derived from Petri nets and particularly suited to represent manufacturing systems, devices and control. A MFG is composed of the following structural elements (see Fig. 3.1.a):
a) Transitions that indicates the occurrence of events; b) Boxes that represents the pre' and pos' conditions, operations or tasks associated to devices; c) Directed Arcs that define the causal relationships between events and conditions; d) Permissive and Inhibitor Gates that enable the occurrence of events; e) Marks that indicate the maintenance of a condition. The dynamic behavior, as in normal PN, is indicated by the evolution of marks in the graph. Boxes have a capacity limited to 1 mark. To represent the interface between the system model and the real system, there are two other elements: f) Output signal arcs to send signals to real system, and g) External Gates to receive signals from the real system. Figure 1.b shows an example of a MFG model.

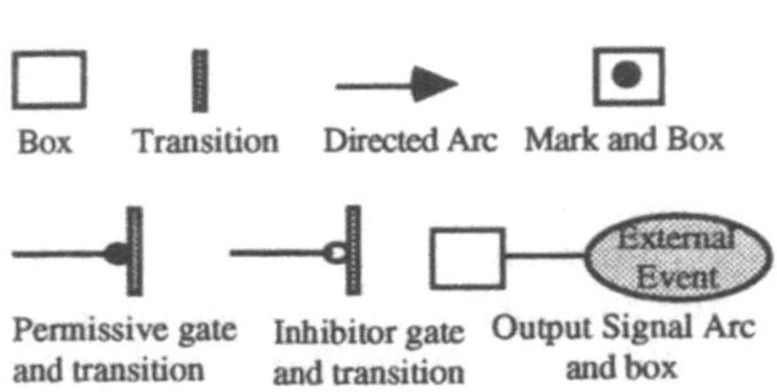

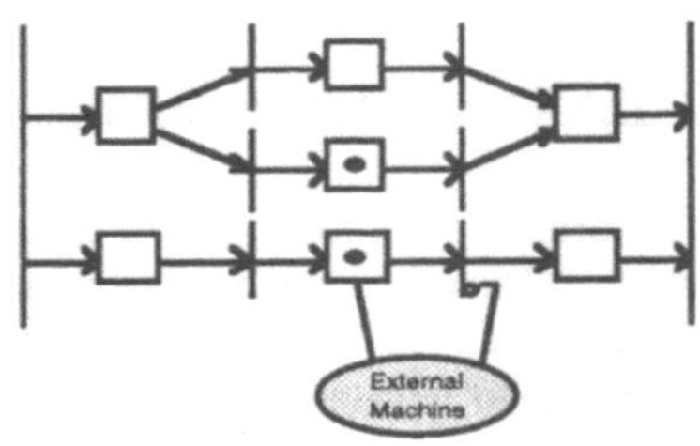

Fig. 3.1.a. Structural elements of MFG. 3.1.b Example of a MFG.

Like in colored PN, in the enhanced version of MFG (E-MFG) [18], marks may have distinct "colors", represented by attribute vectors: <a1, a2, a3, a4>, where a1= part type, a2= batch, a3= origin, a4= destination.

These attributes can be modified by "conditional changes" or "selective filtering", which makes the E-MFG similar to a Predicate Transition Net. Conditional changes are executed by another functional box named "controller box" according to if-then rules.

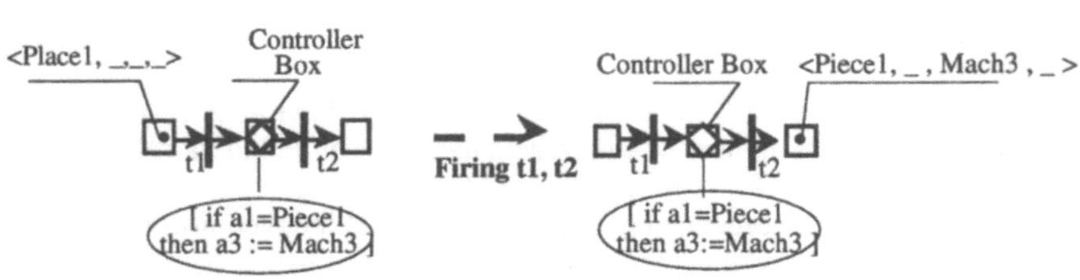

Fig. 3.2. Example of controller box changing the state of an individual mark.

The *selective filtering* establishes fields corresponding to attributes that must remain in the mark after the firing of a transition. This specification is accomplished through inscriptions in the directed arcs. It is important to note that the inscriptions do not represent restrictions on firing of the transition connected by the directed arc. The inscriptions limit the attribute fields to be transmitted. Other attributes are not transmitted.

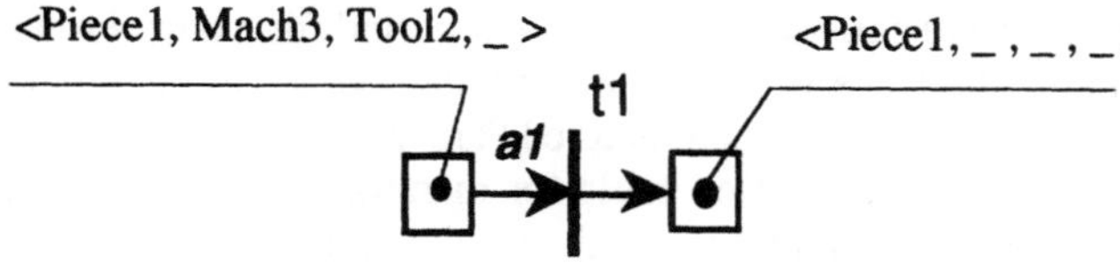

Fig. 3.3. Example of a selective filtering process.

The modeling phase has to be, in most cases, an interactive process involving engineers, managers and perhaps sociologists. To guarantee that the model is an accurate representation of the actual or intended system, this model must be understandable by those people. Although PN based models may lead to a good representation of a system, they are not very clear to non specialists. In order to facilitate the discussion of alternatives by a team, an "iconic" representation as available in various factory simulation packages may be more adequate. A combination of PN and an iconic representation ("graphical" labels) is a good compromise.

3.2 Analysis of manufacturing systems

Once a model is built, the analysis of the modeled system according to a given list of performance indicators is the next phase.

For a qualitative validation, it is possible to use some of the methods which are extensively described in the literature of Petri Nets [19], [20], [21] . For instance, the SimRdP package developed by one FlexSys partner [16], which includes the Coverability Graph method by means of simulation of the Petri Net token-game and Structural Analysis for obtaining the p- and t-invariant relationships of the net [22]. SIMRdP enables to work with the following kinds of Petri Net models: ordinary Petri Nets, generalized Petri Nets, Petri Nets with predicates associated to their transitions and temporized Petri Nets with time associated to their transitions. Similar simulators have been developed for the MFG variant [23].

After performing the simulation of the net evolution, and obtaining the invariants relationships, very important conclusions on the net behavior, concerning the dynamics of the modeled system can be stated, such as: boundedness, liveness, and reversibility of the net; reachability of some special state -marking- of the net; mutual exclusion among marking of places; capacity of storage zones and other modeled resources; the development of managing strategies for the optimal firing of transitions -optimal sequence of modeled operations- and refining of the modeled control processes, by adding, for instance, a new resource.

Once a Petri Net model with good behavior properties is achieved, the designer can proceed to the next stage of the design methodology, using a temporized Petri Net model. This new stage corresponds to the performance validation of the modeled system. Key outputs of SIMRdP are the system throughput, a work-in-process inventory, the cycle time, and machine utilization. By examining the above results, the designers are able to identify system bottlenecks, and to suggest alternatives in order to alleviate the problems that may arise.

Taking into account the statistical information obtained from the simulation of the temporized Petri Nets, new performance indexes can be defined. Nevertheless, this new information must be referred to the dynamical behavior of the modeled systems and their functioning specifications. For instance, in SimRdP the following indicators are considered: Production period for a part, Production period for a batch, Percentual use of a resource related to a part, Percentual use of a resource related to a batch, and Manufactured Parts per Time Units.

Other discrete event simulation tools for manufacturing systems also provide such kind of indicators. Difficult to include in such systems are aspects like:

-Quality of working conditions for human beings;
-Risk level of each technological solution;
-Costs of each solution;
-Local knowledge and traditions.

These factors could be more easily captured by a knowledge based system (KBS). Therefore, a hybrid decision support system combining a PN-based simulator / evaluator (or other simulator) and a KBS able to handle incomplete and imprecise knowledge would be a good design tool. The integration of this decision support tool with a layout planning component would enrich the shop floor modeling and analysis environment.

3.3 Integration and migration

Once a structure and layout have been agreed and resources have been selected, there is a need for the integration of all components and subsystems, taking into account legacy constraints.

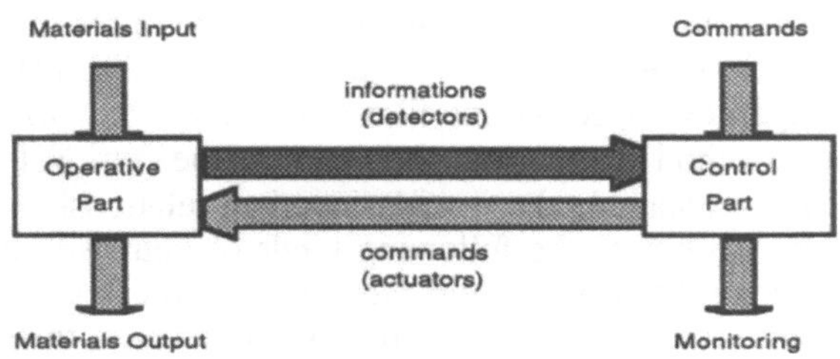

Fig. 3.4. Operative part and control part of production systems.

Manufacturing systems can be represented by two fundamental and interconnected subsystems that operate cooperatively. These subsystems represent the operative part and control part respectively, as illustrated in Figure 3.4. This approach is aimed at establishing a distinction between the execution sub-system (operative part) and the coordination sub-system (control part). The operative part sends information for the control part about its state and the control part processes this information to send commands to operative part.

The operative part can be seen by the control part as an abstract machine that supplies a set of services in the same sense an operating system provides services to the applications. In this way, this part will execute high level commands driven by the control part. A clear separation between the control and operative parts allows a transparent access, hiding the hardware peculiarities, the heterogeneity of the various local controllers driven by the operative part, and the manually performed tasks, from the control part. In this sense, the services exported by the operative part should be independent from the specific resources (hardware and humans), meaning that the variety of existing controllers and human interaction front-ends should be integrated by this platform. As in classical software engineering practice, this layered approach allows for future evolution of the level of automation with minimal disturbances on the upper levels.

The operative part services should not be the sum of all individual features that exist in each real controller or human front end. This is justifiable because: 1) available controllers show a big heterogeneity in terms of the level of abstraction they export; 2) the behavior of a particular controller has to be coordinated acording to its role in the integrated community of agents. Therefore, services offered by the integrating platform should be of higher level, using those low level services (existing in the real controllers).

This aspect can be clarified by an example of part of an assembly cell with a robot and an automatic tool exchange system. The operative services should not be only services like: open_gripper1, open_gripper2, move_point, set_acceleration, set_speed, which are just the sum of the services provided by the real robot and tool controllers. They should be instead: open_gripper, change_tool, transfer, place, ..., which are higher level services and require the knowledge of the cell structural aspects.

In order to develop such an operative part using the proposed requirements it is necessary to choose an adequate modeling paradigm. Two modeling requirements can be pointed out clearly: cell structural aspects and individual components modeling which includes both the modeling of their static and dynamic properties. The Object Oriented & Frames (OO&F) paradigm seems to be quite suitable for this purpose [16], due essentially to its mechanisms to model the operational aspects of the components (methods/demons) and to represent the structural aspects of the system (relations that can be user defined). Methods and demons associated to the component's model can hide the underlying manufacturing infrastructure. Another important aspect is the "relation" concept, which can provide a flexible way to describe inter-components' relationships. This concept has different semantic meanings in the Object and Frame paradigms, which makes modeling a little bit different in these two frameworks.

Based on these modeling / programming formalism, a good approach for migration from existing components towards an integrated manufacturing infrastructure is the "agentification" of manufacturing resources. In other words, the transformation of manufacturing resources into agents that can cooperate in a multi-agent community. This "agentification" process usually implies:

- -building a "wrapping" layer around the existing controllers in order to transform them into normalized servers; in the case of human resources, a human front end is necessary;
- -building an agent manager, i.e., a high level "mirror" of the resource's functionalities implemented in frames.

Building this "wrapper" can be a hard engineering task, as many manufacturing devices are not open enough. In some cases it might even imply some machine retrofitting, like introducing new sensors, new I/O functionalities, etc..

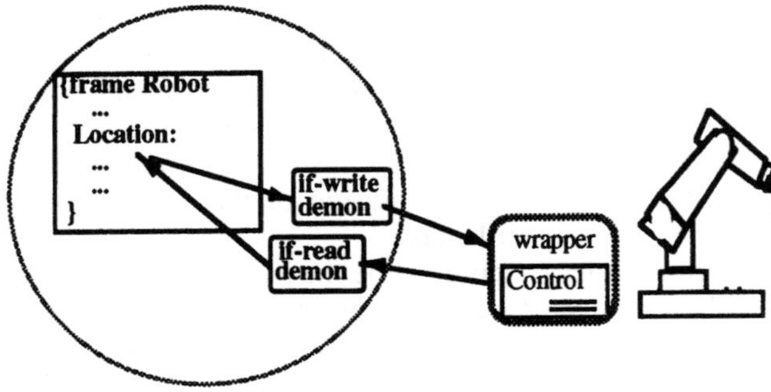

Fig. 3.5 "Agentification" of a robot controller

At this stage it is convenient to clarify the concepts of agent its relation with the components/manufacturing resources. **Components** are entities which participates in the productive process with a specific function and can be controlled by a computational entity. A component model is context independent descriptions of its static and dynamic characteristics. A robot component model, for instance, includes all the characteristics which completely characterize its structural and dynamic aspects. A robot **agent** (Figure 3.6) is a model of a robot and associated resources, like tools or auxiliary sensors, when inserted in a particular context. A robot can play different **roles** in different **contexts**. The (expected) behavior of a robot in an Assembly context is different from its behavior in a spot welding context.

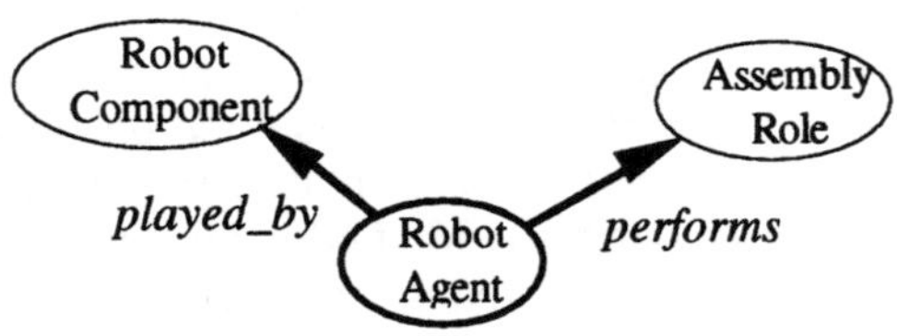

Fig. 3.6. Structure of a robot agent.

On the other hand, when a robot is performing a given role, it may resort to auxiliary resources, like tools, sensors, buffers, etc., that extend the robot functionality in order to fulfill the functionality required by this role. A robot agent is, therefore, a model of the robot when playing a particular role and extended by selected attributes inherited from its auxiliary resources.

The agent entity ASSEMBLY_ROBOT (Figure 3.7), for instance, is a structure which is supported by two relations: *performs* and *played_by*. The relation *performs* assures the inheritance of the role characteristics to the structure while *played_by* assures the inheritance of those agent relevant aspects, from the component. *Main_attributes* and *component_attrib* are attributes to be used by *played_by* and *performs* relations. The slots, *tools_domain* and *aux_res_domain* represent domain-knowledge that is important during configuration time. The slot *current_tool* is a relation that associates the main player of this role (robot component) to a particular tool. It could be defined as:

new_relation(current_tool,transitive,inclusion(tool_operations,attached_to)

where *attached_to* is the inverse relation. By the "inclusion" restriction, only the tool_operations attribute will be inherited by the ag_robot_assembly_role.

Assembly_device is an attribute describing where assembly operations are really done. *Fixture1* is an instance of a component specialized in holding parts.

FRAME ASSEMBLY_ROBOT
is-a: agent
performs: *ASSEMBLY_ROLE*
played by: *ROBOT_COMPONENT*
main_attributes:
force_sensor, current_tool,
available_tools,
available_resources
component_attrib:
Base_coord_system, Controlled_by,
Working_area, load, Current_position

FRAME AG_ROBOT_ASSEMBLY_ROLE
is-a: role
tools_domain:(gripper,screwdriver)
aux_res_domain: (buffers)
force sensor:
current tool:
available_tools: gr1, gr2, sd2
aux_resource: buf1, buf2
assembly_device: fixture1

RELATION PERFORMS
type: intransitive
inherit_slot: main_attributes
inverse_relation: performed_by

RELATION PLAYED_BY
is-a: relation
type: intransitive
inherit_slot: component_attributes
nverse_relation: play

Fig. 3.7. Definition of relations *performs* and *played_by*.

A cell is made of entities that are playing different roles. This modular approach to cell representation facilitates the creation of complex systems.

One important point in this discussion is the way a controller model is connected to the physical controller. From our experience a lot of effort has usually to be put in trying to "open" real controllers' architecture, and it implies the production of an interpreter ("wrapper") that runs on it. This interpreter accepts commands from the "image" that runs in the other world.

FRAME ROBOT_CTRL_COMPONENT
is-a: controller
move_wc: method move_wc_fn(x,y,z,q)
move_jc: method move_jc_fn(m1,m2,m3,m4)
hardhome: method hardhome_fn
acceleration: demon if write accel_dem
speed: demon if write speed_dem
input: byte demon if needed input_dem
output: byte demon if write output_dem

Fig. 3.8. Model of a component controller.

The methods of a Controller model (Fig 3.8) implement the actions that are needed to send the right commands to the real controller. The image of the real controller should be developed using a client-server approach. In this way, implementation methods can ask this server to perform the needed actions. These methods hide the underlying hardware structure from the application, i.e., any application using a robot component doesn't need to know anything about the real robot controller and its image or server. The applications only know what functionalities are provided by the robot component model. This approach is very suitable to integrate existing controllers, making the integration of legacy systems a less difficult task.

This approach has been successfully applied in various integration efforts [24], [16]. When the resource is a human, the human front-end module will perform the same role as the controller plus wrapper.

High level services supplied by the abstract machine (operative part) will be accessed by its client (controller part), by activating the existing methods in the frames that represent the system. The use of E-MFG or other PN to model the control object generates an explicit representation of local states reachable for each device and permits through its structural elements (external gates and output arcs) an effective exchange of signals between the controller and the control object.

3.4 Supervision

Supervision in manufacturing systems is an activity traditionally performed mostly by humans. The increase of complexity and flexibility at shop floor level has created a demand for Intelligent Supervision Systems. These systems are supposed to include, as basic functionalities:

-Dispatching and Global Coordination — The global coordination activities include: dispatching actions to the executing agents, driven by the scheduled task plan; synchronization of activities performed by different agents and synchronization with external events; world model update and information exchange, resorting to an information system or knowledge base. The dispatcher also coordinates the execution of the other modules of the Intelligent Supervisor as well as the interaction with the human operator.

-Monitoring — The monitoring function is used to detect non-nominal feedback in the system during the execution of task plans. Two monitoring modes are usually considered: discrete monitoring and continuous monitoring. Discrete monitoring is used to check preconditions before the execution and goal achievement after the execution of operations. Continuous monitoring is used to check sensory conditions during the execution of operations. In many contributions to the problem of execution monitoring, the sensory conditions to test in each situation are defined in a model, often coded as monitoring rules:

IF <situation> AND <sensory-condition> THEN <actions>.

However, very often it is not easy, even for an expert, to specify the sensory conditions that guarantee the success or the pre-conditions of an action and to identify the statistical significance of each situation.

-Failure Diagnosis — This diagnosis function will firstly check if there really is a failure (failure confirmation) and update the internal model. Then this function will try to classify and explain the failure. Considering a task described at various levels of abstraction (hierarchical plan), at each execution level, different levels of explanation for a detected failure may be generated, depending on the amount of information available [25].

-Failure Recovery — At each supervision level, the recovery function is called when the diagnosis function confirmed a failure and found an explanation. The recovery function will try to determine a recovery strategy to bring the execution to a nominal state. Since the detected error is some unexpected (abnormal) event and, therefore, the nominal plan is not to be altered, one basic question is "how to build recovery strategies?".

The complexity of the system, the increase of unexpected events as a result of the increase in flexibility, and the increase in quality, suggest requirements difficult to accomplish by humans. This justify the development of automated supervision systems. However, most of the required knowledge to support an effective supervision system is not available or is rather incomplete. One approach is to combine some partial automated supervision with human based decision making.

The problem of building the knowledge base, and in particular the models that the monitoring, diagnosis and recovery functions need, is not easily solved. Even the best domain expert will have difficulty in specifying the necessary mappings between the available sensors on one side and the monitoring conditions, failure classifications, failure explanations and recovery strategies on the other. Also, a few less common errors will be forgotten. Thus, two other functions can be included in the supervision system [25]: *training* and *learning* . The training module coordinates the interaction with the human operator in order to acquire new information about nominal execution of the assembly plans as well as descriptions of new error situations. The learning module compiles raw data generating classification knowledge, generalizes instances of target concepts, etc., in order to build the needed models.

According to the paradigm of Programming by Human Demonstration, complex systems are programmed by showing particular examples of their desired behavior and giving explanations for particular failure situations. For this objective, the interaction between the execution supervisor and the human operator is fundamental. The human will carry out an initial training phase for the nominal plan execution. The traces of all testable sensors will be collected during training in order to generate the corresponding monitoring knowledge.

An example of research following this approach and carried out in the framework of the Esprit B-LEARN II project can be found in [25], [26]. As natural in inductive processes, the experiences realized in B-LEARN II shown that the quality of the generated knowledge depends on the number of examples provided. In some cases, like in assembly, provoking a comprehensive set of error examples can be a very boring task for a human tutor. Therefore, in complement to an initial training phase, the system shall be designed to support incremental learning based on examples collected during the normal operation of the system. That is, in the beginning the system would strongly rely on the human supervisor. As the number of collected cases increases, less effort is required from the human operator. If a "learning observer" is plugged in in parallel with the hybrid supervision system, the human operator does not have to explicitly "teach" the systems. The learning function observes the operator during his normal operation of the system.

In some application domains, namely in continuous processes monitoring, there has been a large introduction of sensorial systems. The problem for humans is how not to be overloaded with information which could distract them and difficult the distinction between critical and non-critical situations. An "intelligent filtering" of the sensorial information might be a good complement to human supervision. On the other hand, this apparently excessive information might be useful for anticipating future deviations (prognosis) and enabling anticipated remedy actions. Again the use of machine learning is a promising approach to acquire such prognostic knowledge.

4. BAS AND SCHEDULING

A second example of application area that can benefit from a balanced automation approach is scheduling. An agile manufacturing system requires a scheduling system able to support a dynamic adaptation in the presence of unexpected events (such as machine breakdown, operation lateness, changes in orders' priority, etc.) and a flexible adaptation of the whole production structure according to given business process requirements (agile scheduling [27]). In despite of the technology applied upon the systems and the impacts of the globalization of the economy, industries keep demanding for custom-tailored scheduling systems, which besides being open, should support a flexible (re)configuration of themselves so that they can be adapted once the production methods, scheduling control procedures, layout of the production resources, etc., are changed.

The HOLOS framework [28] is a software system developed to derive particular agile scheduling systems based on a reference architecture. In this system, three main subjects are taken into account in terms of balanced automation : i) an anthropocentric approach in the design of the scheduling system, ii) an open framework both of the generic reference architecture and the particular system derived, iii) and the integration approach for legacy systems.

4.1 An anthropocentric approach for the scheduling system design

The development of such 'flexible custom-tailored' agile scheduling system is still a challenge. However, there are already some important considerations to be taken into account, such as : i) the scheduling domain is intrinsically distributed; ii) the scheduling domain requires a joining of different expertises to be executed; iii) the desired scheduling agility may be improved by augmenting and decentralization of the decisions related to a scheduling; iv) the desired scheduling agility needs a framework capable to support an efficient reaction to unexpected events; v) the life cycle of a scheduling system may be increased if its supporting framework is open enough to allow different programming paradigms be used by different / existing scheduling / production methods.

The Multiagent Systems (MAS) paradigm [29] was adopted in HOLOS since it is able to support those considerations. HOLOS is a reactive scheduling system based on a MAS approach, which means that a scheduling is made and dynamically adapted and supervised / executed by a set of distributed agents. These agents may be located in different geographical places (computers), depending on the global system architecture specification, which in turn may consider various criteria for (shopfloor layout, topological relations, enterprise model, etc.). A HOLOS agent can be defined as a semi-autonomous software component with :

- some decision making capabilities and knowledge on how to execute a task;
- some knowledge about the other agents;
- the capability to communicate with the other agents;
- self perception capabilities;
- a cooperative behavior.

In fact, the underlying objective of HOLOS aims at reaching a "state of harmony" within the enterprise via a complementary and cooperative relationship between the agents involved in the scheduling. Nevertheless, this state of harmony in an agile context is a tough task to be reached. The manufacturing environment is dynamic, normally over-constrained and unpredictable to some extent. Besides that, current constraints are commonly conflicting to each other so that a trade-off or requirements relaxation of a certain task (an "enterprise activity") has to be *negotiated*. The HOLOS makes use of the Contract-net protocol / Negotiation paradigm [30] for supporting the desired flexibility in conflicts resolution during a schedule generation and execution. Figure 4.1 illustrates such Negotiation process in HOLOS approach for scheduling.

In the HOLOS framework, a *task* to be executed is modeled as a hierarchy of "business processes" (BP), "enterprise activities" (EA) and "procedural rules sets" (PRS), concepts borrowed from the CIM-OSA methodology [8]. A BP corresponds to an abstract task description (one *production order* in the case) which can be decomposed either into a set of EAs or BPs. An EA corresponds to a BP's operation description (one process plan's operation), and a PRS basically defines the precedence relations between the BPs and/or EAs. A production plan is therefore viewed as a logical, temporary and hierarchical grouping of these concepts, and it represents the way all planned BPs must be accomplished. In this sense, in the negotiation process the agents exchange information with other agents about a given BP / EA's requirements (phase 1) until a production-resource agent (a robot, a CNC machine, etc.) is selected to execute it (phase 2).

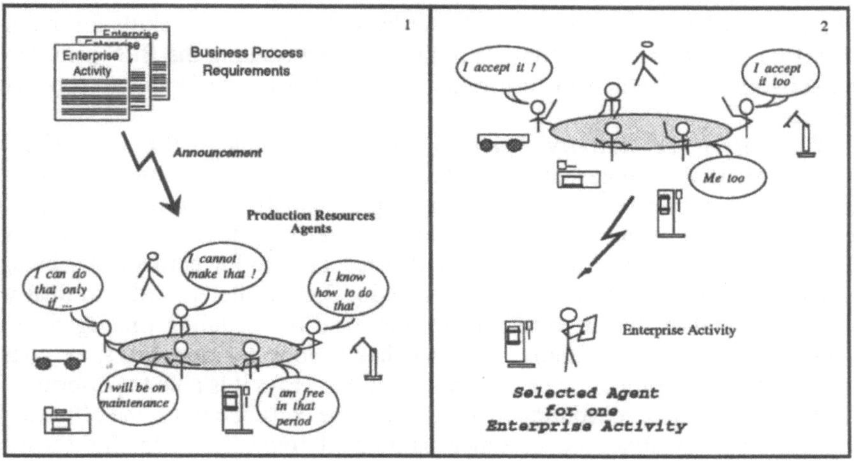

Fig. 4.1 Negotiation process in scheduling

4.2 The HOLOS agents

HOLOS agents are heterogeneous and functionally hierarchical, and they are designed according to the four problem solving phases in Distributed Artificial Intelligence: problem decomposition, sub-problem distribution, sub-problem solution and answer synthesis [31]. There are four "classes" of HOLOS agents :

The *Scheduling Supervisor* agent (SS) is an semi-hierarchical agent created to control the execution of schedules. In fact, the SS agent makes a *loose* control since the other agents in the architecture have some degree of responsibility / autonomy in what respects the control of execution of the schedules.

The prime and general function of the SS agent is, once a BP has arrived, to guarantee that a schedule will be generated and then executed. Thus, the principle utilized consists in spreading an announcement about a BP to the other agents in order to select the best group of production resources to execute it. Because an enterprise has in general to process several BPs, there will be as many BPs to be announced as many arriving BPs. The SS agent is also responsible for evaluating the quality of a schedule according to the current objectives and based on some production parameters, like flow time, utilization time, work in process, costs, lateness, etc.. Furthermore, because the other agents are distributed and can be responsible to solve only part of a BP (i.e. they do not have a global and complete view neither about the whole BP nor about the other BPs being executed in the shop by other agents), the SS agent can intervene (once it is requested by one of the agents) in order to help the agent which is in difficulties, specially in rescheduling situations.

The *Local Spreading Center* agent (LSC) corresponds to a more abstract view of a physical organization of the production resources, like FMS, cells, workstations, etc. In fact, it represents a process of avoiding that a BP announcement (BPN) is broadcasted to all agents. Since a BPN has a basic specification about the general type of agent required for the BP execution (or part of it), the strategy consists in sending it only to those agents which "match" the specification. For example, it does not make sense to send a BPN related to a milling operation to agents of type robot, simply because robot agents do not have the appropriate skills for this task. Therefore, the control flow consists in the SS agent sending the BPN primarily to the suitable LSC, which in turn re-sends the BPN to the "known" production resources.

The *Enterprise Activity Agent* (EAA) corresponds to the agent (i.e. the virtualization) that represents a production resource's local controller (i.e., an "agentified" controller). An EAA has three essential functions: *i.* from the scheduling point of view it is responsible for executing tasks (a whole BP or an EA); *ii.* from the control point of view it is responsible for permanently being in contact with the real production resource's local controller in order to deal with some local failure, and if so to perform the local rescheduling (managed by the Consortium agent); *iii.* finally, the EAA has to look for new tasks, i.e., has to participate in the negotiation process related to BP / EAs it is able to perform.

Because dynamic scheduling involves real time considerations, the efficiency in performing those functions assumes paramount importance. In this sense, an innovative solution has been adopted : an EAA is not modeled as a monolithic agent, but as a composition of two types of interacting processes, *Manager* and *Server*. A Server is responsible for the task execution itself and for reporting its operation status to the Manager. The Manager controls the task execution getting attentive to some possible abnormal occurrences in a Server. Besides that, it tries to maximize the Server utilization. One advantage it provides is more efficiency in task execution : the parallelism between the Server and Manager functionalities allows the Manager to look for more tasks to the Server, to reason on the need for rescheduling as well as on possible optimization on the sequence of planned tasks to be executed, and it allows the Server to keep executing the planned tasks with no interruption unless its manager orders so or some failure occurs. This architecture can also support a hybrid decision making process. The human intervention can take place at the server (operator) or at the manager (local supervisor) level.

The *Consortium* agent represents a *logical* group of EAAs specially created to execute a certain BP. From the organizational point of view, an analogy could be made with the concepts of "virtual cell" [32],[33] and "logical cell" [8], for instance. Unlike the other agent types, a Consortium agent is not persistent. It is dynamically created when a new BP arrives and a schedule for it has to be made. As the existence of a BP within the manufacturing environment is naturally temporary, the "life" of a Consortium is so temporary; i.e. once the schedule for a BP has been executed, the Consortium agent kills itself. Due to the temporary life cycle of a Consortium, an EAA will belong to several Consortia along the time, i.e. as soon as it finishes the execution of its EAs for a certain Consortium, the EAA can start the execution of other planned EAs which belong to other Consortium.

There are some criteria for this composition, such as minimum path, avoiding bottlenecked EAAs, fewer number of EAAs involved, etc., which at last try to

consider the current shop status; therefore we could say that each Consortium represents the *best* team of EAAs. In spite of the difficulties it represents, that kind of logical grouping is able to face three drawbacks of rigid organizations : routing flexibility, machine replacement flexibility and idle time [34].
Beyond the organizational improvement represented by the use of the Consortium concept, from the scheduling control point of view other improvements are introduced. Due to its autonomy, a Consortium has the responsibility to execute a BP *locally*. It means that when a problem occurs, for example when a machine breaks down, the Consortium primarily tries to find a substitute and to make the local rescheduling. If it cannot find an equivalent substitute according to the current constraints, then it calls the SS agent for "help".

4.3 The HOLOS Derivation Framework

When the development of a particular agile scheduling system is envisaged, the number of parameters and criteria to be considered which are related to each particular manufacturing context is so large that makes unrealistic to have a single general purpose scheduling system. The attempt to develop a "scheduler derivator" system, i.e., a decision support system that helps human experts in deriving a scheduling system tuned to a particular context seems to be a more rewarding approach [35]. Thus, instead of trying to develop a very large scheduling system, the approach is to follow a derivation process that, from a general reference model, derives a particular scheduler. The HOLOS Framework is the software system used for deriving a particular scheduling system. This framework is constituted by three components:

- *HOLOS Generic Architecture* - HOLOS-GA [36]: represented by the four HOLOS agents classes (their knowledge), which includes the explicit definition of the communication flows between them, the communication support infrastructure, and the high level protocol for agents intercommunication;
- *HOLOS System Generator* - HOLOS-SG [37]: to assist an user to derive a particular (multiagent) agile scheduling system based on the HOLOS-GA;
- *HOLOS Methodology* [27]: the underlying methodology which guides the HOLOS-SG to derive instances of HOLOS agents classes.

CIM-OSA provides a general methodology related to derivation of CIM architectures from an abstract and general reference model. The HOLOS methodology is more restricted in scope than CIM-OSA since it specifically addresses scheduling systems derivation as well as it is tightly biased by the multiagent approach.

The Derivation of an instance of the HOLOS-GA is the interactive creation of instances of the HOLOS agents classes, according to the enterprise characteristics and requirements. A particular scheduling system is a collection of distributed agents organized for a concrete structure and configuration of production resources (humans, machines, robots, transport systems, cells, etc.), and that can perform a schedule supported by negotiation. As mentioned before, as the Consortium agent is dynamically and temporary created, it is not generated during the derivation process; its creation will occur when there is a BP to be scheduled and executed.

A scheduling system derivation process is normally too complex. It comprises lots of parameters and information about production, engineering and scheduling control, which in turn may be combined to each other. A customized solution for that scenario requires an exhaustive discussion between all people engaged in, as well as the evaluation of its impacts on existing technology and human resources management. Such discussion can involve not only technical points (like production system, integration, heterogeneity of existing systems, production resources layout, solution costs, etc.), but also the enterprise's organizational culture and its work organizational methods (like team work, decentralization levels and autonomy for decision making, human resources qualifications, their functions (re)definition, training policies, etc.), and some national singularities (like work shifts related to special holidays or local tradition, efficiency criteria specifications, etc.). The adequate consideration of all these aspects may determine the success of the system and of its implantation [38].

4.4 The HOLOS Methodology

Due to the complexity, implications to the enterprise, and the enormous quantity of aspects to be taken into account, the process of deriving (an instance-of) a particular scheduling system from the HOLOS-GA is based on a support methodology : The HOLOS Methodology. It corresponds to a sequence of interdependent general procedures and considerations which have to be followed by a "human deriver" towards the implantation of an agile scheduling system for a particular enterprise. On the other hand, although all those aspects are pointed out as crucial, they are in part out of the scope of the HOLOS methodology itself. The HOLOS methodology assumes the whole phase mentioned above is done <u>before</u> starting a derivation process. In other words, it is assumed that a preliminary analysis and a rough system evaluation is made and their results are incorporated in the respective and suitable information models.

In the HOLOS approach, a derivation process basically consists in interactively creating instances of HOLOS agents classes and filling their *skeleton* through the methodology steps. Basically they are:

I) Enterprise Characterization. This first step aims at describing the enterprise (for which the particular system is to be derived) in terms of information about engineering, production and scheduling control (EPS). This information comprises the specification / selection of policies for stocking, the production system (MRP, JIT, OPT), scheduling visualization, criteria for performance and schedule evaluation, scheduling strategy, type of scheduling system (backward, forward, etc.), among many other aspects. This selection may be a difficult task, specially when the information have to be combined to each other. A wrong specification and/or combination can provoke a situation of domain inconsistency. The indication can be made manually, semi-automatically (as a choice between options), and automatically. In fact, associated to each criterion or to a group of them, there might be a computer procedure (a small program), which represents part of the scheduling knowledge. It means that while the user and the system specify the EPS aspects, this knowledge is transparently included in the specific agent instances. However, depending on the EPS information (and combination) and complexity for its specification, a decision support system may be utilized. Therefore, this step is strongly based on a hybrid decision making, in which is necessary to find a balance in terms of human based decision making and automatic functionalities.

II) Agents Integration. In the HOLOS methodology, the migration from legacy systems is the underlying approach applied for systems integration. A reactive/agile scheduling system has to interface the shop floor controllers (in order to have run time feedback). An *agentification* process for such controllers has to be devised (as presented in 4.2) in order to transform them in cooperating agents for the MAS scheduling framework. There are three integration layers to be made :

ii1. Local controller - EAA: it aims at allowing a given machine controller to be represented within the community of the HOLOS agents (or other "external application"). Monitoring, diagnosis and error recovery functions are being added to these controllers [39], which increase their decision making capabilities and degree of autonomy. On the other hand, the client-server paradigm is "invading" the area, which facilitates the integration task. However, in spite of standardization efforts, as exemplified by MAP/MMS and fieldbus, the situation is still of a large heterogeneity. Different generations of production resources, like robots, NC machines, sensors, etc., based on particular hardware and control languages, have to considered. This imposes, in many cases, the development of encapsulating layers in order to transform an actual controller into an integratable server.

ii2. EAA - other HOLOS agents: it aims at allowing the EAA (the Manager) to make a conversation (in a "multiagent scheduling language") with the other agents of the architecture, and possibly with other subsystems. A high level protocol can be utilized for that, such as the KQML [40], and it can be supported by RPC, sockets, etc. In fact, considering implantation and legacy aspects, the Manager is just a more abstract control level to be added upon the Server's local controller, but in a detached way; i.e. resorting to the classical Client-Server mechanisms.

ii3. HOLOS agents - CIM-IS: it aims at allowing an agent EAA to have access to the CIM Information System services (the source of most of the information required by the scheduler). The paradigm client - server is also applied here. The communication between the agents and the CIM-IS may be supported by SDAI (STEP Data Access Interface) [41] at the application level, and RPC or sockets.

III) Agents Specification. This step is related to the knowledge to be incorporated into each agent instance (from the classes SS, LSC, Consortium and EAA) when they are created. Before starting the agents instantiation, the whole architecture of the particular scheduling system may be designed. This design consists in defining an "agents map" to help the user for having a global visualization of the system architecture. A map is a collection of "dummy" agents which will be further created/instantiated, which also describes the relationships between the agents.

Creating agents means to instantiate them, that is creating instances of HOLOS agents classes and fill in and/or choose some of their attributes and functionalities (i.e. definition of the agents' knowledge), regarding the specific target system. Similarly to CIM-OSA, these attributes and functionalities can be of type Generic (those which should exist in an agent independently of the particular site), Customized (those which need to be customized for a particular scheduling system, but still independently of the particular physical site), and Particular (those which should exist and/or have to be customized taking into account the particular physical site).

IV) Agents Reconfiguration. An open solution implies giving the user the possibility to modify a portion of the agents' knowledge created during step III. This can be done through modifications, extensions, and/or insertions of attributes and / or functionalities into some agents which have not been generated during the creation, or even modifications in some system configuration files. This step is designed to cope with the natural evolution of manufacturing components, namely in terms of level of automation.

V) Agents Launching. This step is related to the agents implantation. Launching agents means to put the logical agents instances created in execution into the system. After this phase they can be recognized in the system, can communicate with other agents and can execute actions. In this sense, once the logical agents instances are completely created, the communication channels are assigned to them, their functionalities and other programs are compiled and linked. The creation of libraries, the linking with the main information repositories, and the adjustment of graphical interfaces are other procedures done.

After this step, the derivation process is considered finished and therefore the particular scheduling system is raised.

VI) System Architecture Reconfiguration. During the life cycle of a derived particular scheduling system, its global architecture may change. For instance, new production resources may be bought, some production resources may leave from the shopfloor, or the shopfloor layout can be altered. Hence, the global system architecture can be reconfigured according to these changes, which means the possibility for incorporating new agents, discard existing agents, and altering some parameters or the relationships between existing agents.

VII) Generic Architecture Reconfiguration. Production domain and scheduling policies and their control procedures may change along the time, either due to its obsoleteness or due to some adaptation for a specific derivation. Thus, an open architecture has to contemplate the possibility of a user to change the agents' classes and the domain knowledge (for a future derivation).

4.5 The HOLOS System Generator

The HOLOS System Generator corresponds to a prototype implementation of the HOLOS methodology. It is a "computer aided derivation" tool. By means of a strong interaction with an "user-deriver", the HOLOS methodology steps are passed so that at the end a particular scheduling system is derived.

The Engineering-Tool-Kit [42] and OPIS [43] represent two other related works. In spite of the fact that they are older, apply different approaches and are already more established than HOLOS-SG, their general objective is equivalent : generating particular architectures and systems based on generic concepts. However, due to the HOLOS potentiality in terms of flexible modeling and control, system modularity and expandability, integration and MAS approach, a HOLOS-SG instance system appears to fit in a more suitable way with the requirements of the industries which have adopted a virtual manufacturing towards the extended enterprise paradigm.

The HOLOS framework offers a good match with most of the characteristics considered as trends and emerging concepts in scheduling [44] and utilizes some anthropocentric concepts in applying a decentralized control and in exploiting the autonomy and natural distribution of the production resources (agents), using the Negotiation and distributed decision making as the support technique for that.

However, it does not intend to cover all kinds of industries. This prototype assumes that the generic architecture is "good enough" to cover the most typical manufacturing situations and scenarios. Furthermore, it also assumes that the human deriver has all knowledge on how the particular scheduling system (the instance-of) has to operate. Nevertheless, the architecture can be modified along the time, and the derived instance can be adapted by the user after a derivation.

5. BAS and Concurrent Engineering

Concurrent Engineering (CE) is another area for the application of a balanced automation approach. On one side there is a large legacy of engineering tools that although not specifically designed for a cooperative work, have to be "kept", either due to economic reasons (previous investment, existing libraries, costs of porting, etc.) and to social reasons (trained human skills and experience). There is an increasing tendency for the adoption of supporting tools for the various engineering and other product related activities, but it is necessary to cope with a large heterogeneity of such tools. On the other hand design and complementary engineering activities require considerable amount of high level (human) decision making. Cooperation between members of a team is a complex process (social behavior) not totally understood yet which requires some kind of coordination supporting tools. There is, therefore, a need for a platform supporting hybrid decision making.

5.1 Supporting Platform

For all areas involved in the product life cycle there is a considerable number of computer-aided tools (CA) on the market that help the human experts in their tasks. At a particular enterprise level various of these tools may be available, together with some proprietary software developments. It might also happen that, depending on the size and other characteristics of the company, some of these activities are performed manually or by loosely integrated tools. For instance, due to subcontracting, some activities may be totally absent from a particular company. A platform that supports the integration of such tools (information and knowledge sharing) as well as the interaction among their users (team work) is a computational requirement for Concurrent Engineering. This platform has also to consider less automated situations and has to cope with the difficulties of the integration of legacy systems - a perspective of Balanced Automation.

The definition of a platform for Concurrent Engineering involves, in our opinion, three related sub-problems:

i) *Definition of common models.* This is a basic requirement in order to enable communication and co-operation among members of the engineering team. The adoption of common modeling formalisms is a first requirement. Formalisms like IDEF0, NIAM, Express/ Express-G, Petri nets are being widely used. The consolidation of STEP [45] may help in terms of product modeling, but many other aspects not covered by STEP have to be considered, like process and manufacturing resources modeling. Domain and Business Processes modeling, as proposed by CIMOSA [7,8], is also contributing to facilitate dialogue.

ii) *Engineering Information Management.* A second major requirement is an information integration and management infrastructure which allows that information from multiple disciplines can be meaningfully organized and accessed by team members. Definition of integrating infrastructures and information management systems able to cope with the distributed and heterogeneous nature of CIM, has been the subject of many research projects from which various approaches and prototypes have been proposed in last years [46], [47].

iii) *Co-ordination.* To build a platform that supports concurrent engineering it is not enough to guarantee that the various computer-aided tools used by a team are able to communicate and share information. In other words, it is not enough to provide an integrating infrastructure and to normalize information models. Even though these aspects are essential, there is also the problem of co-ordination. For this it is necessary to establish a supervision architecture that controls or moderates the way and time schedule under which computer-aided tools (team members) access the infrastructure and modify or provide shared information. Effective co-ordination is critical to the success of this Concurrent Engineering process since the involved distributed activities are typically highly interdependent [48]. In other words, it is necessary to model the various business processes and to implement a process interpreter or supervisor. Some contributions come from the workflow and scheduling areas. Workflow systems are basically interpreters of process models represented in some process modeling language. But current systems are not flexible enough to cope with a changing co-operation environment, not completely known a-priori.

The mentioned areas cover technical-oriented aspects. But there are other facets, like conflict resolution, negotiation, etc., that can be better understood if analyzed from a social perspective. Therefore, the anthropocentric manufacturing approach [4], [5] should also contribute to the definition of an effective infrastructure for Concurrent Engineering. In fact the coordination activity should be supported by an hybrid system combining human decision making and automatic functionalities. Achieving the right balance here is quite delicate as there are various agents, based on different background knowledge and supported by heterogeneous tools. Privacy and cooperation are competing facets in this scenario. This might be even more difficult in the context of a virtual enterprise, with the various team members located in different nodes of the network. The platform will have to support different levels of mutual trust in the team. The organizational aspects are also very important in order to understand the various levels of responsibility.

5.2 The CIM-FACE approach

The platform for integration and concurrent engineering -- CIM-FACE: Federated Architecture for Concurrent Engineering -- proposed by the Group of Robotic Systems and CIM of the New the University of Lisbon addresses various of these issues [49].

CIM-FACE is a **federated multi-agent system**. An agent in this context is an abstract construct representing the human expert and the CA tools this human is using. The concept of federation, by analogy with a federal country, combines central decision making (federal rules) with local autonomy (agents' decision making capabilities). The degree of autonomy varies depending on the specificities of the participating components.

From the engineering activity point of view, this infrastructure provides a kind of "virtual round table" allowing the various experts to cooperate as a team.

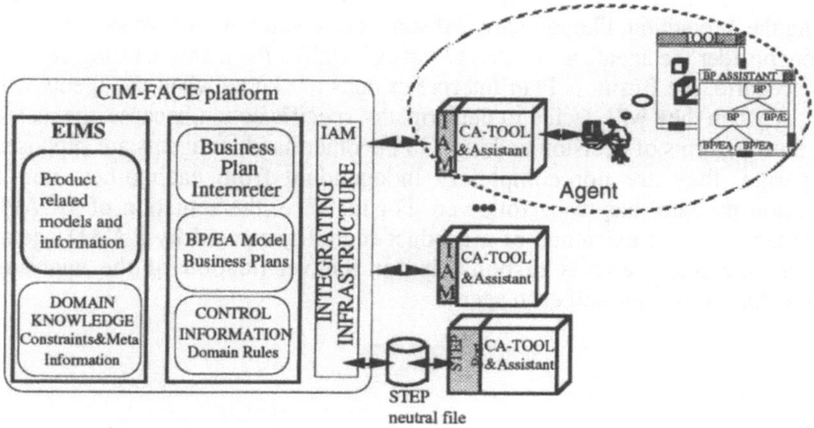

Fig. 5.1 CIM-FACE platform for CE

The Business Process / Enterprise Activity modeling constructs of CIM-OSA are used to model the functional view of the co-operating activities in terms of hierarchies of Business Processes (BA) and Enterprise Activities (EA)[8]. At each level of the hierarchy, extended Procedural Rule Sets (PRS) define precedence constraints between BP/EAs of that level, as well as their starting (firing) conditions [49]. This hierarchical structure is called business plan. The co-operative design activity is considered a business plan. Special tools - Business Plan Assistants - provide a front end to the human experts allowing them to take part in the execution of the business processes. The supervisor or Business Plan Interpreter is responsible for controlling the evolution ("execution") of the various BPs.

Each agent may have its own data models (partial views of the world) and its decision making capabilities. For instance, the actual time window within which a CA tool is active, i.e. interacting with the federation, and the kind of performed interactions are decided by the couple "CA tool - human user". This couple represents a hybrid decision making component (hybrid agent).

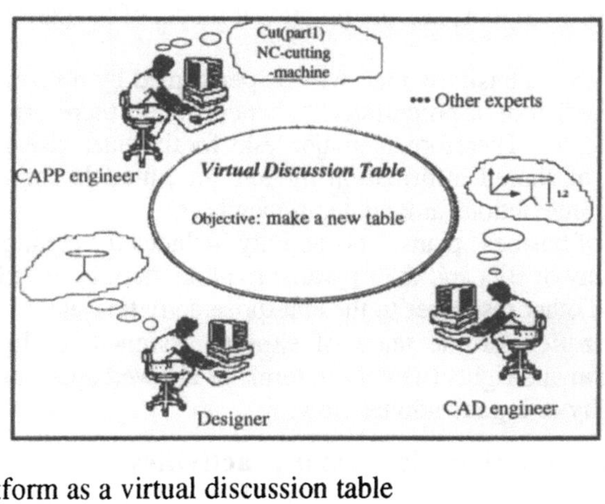

Fig. 5.2 CE platform as a virtual discussion table

A global supervision system -- "federal government" -- can impose some rules regarding the interaction, like refusing it if some pre-conditions are not satisfied, but it cannot consider the agents as obedient "slaves". Unlike the manufacturing devices control scenario, the Business Plan Interpreter does not know when the agents will "log in" or when they will decide to perform any specific action because agents are autonomous in terms of decision making. On the other hand, as agents are supposed to cooperate, they are not completely independent from each other, i.e., a coordination protocol has to be followed. For instance, the actuation of a CAPP agent depends on the existence of a product model generated by a CAD agent. Therefore, the Interpreter is responsible for the verification of the enabling conditions for the actuation of each agent.

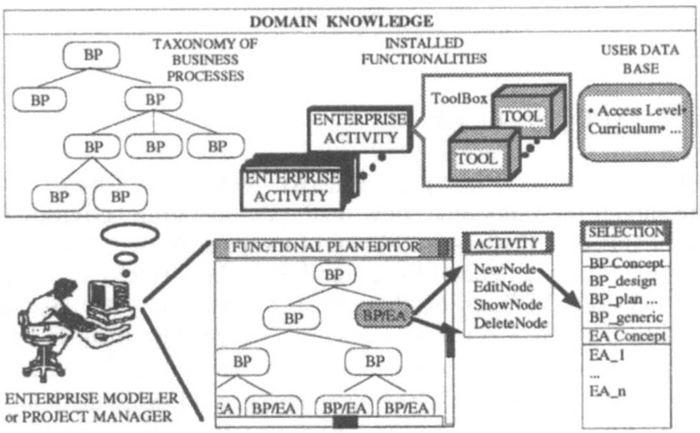

Fig. 5.3. Definition of the enterprise's business plans.

Building up business plans is a task for an enterprise modeler or project manager, Figure 5.3 [49]. This expert may resort to some domain knowledge, specific to the target industrial sector or to his / her company. Such knowledge includes a taxonomy of classes of Business Processes and a catalogue of installed functionalities (Enterprise Activities, i.e., the functional model of the particular enterprise).

Considering that a business plan will be performed by a team of experts (with the help of several tools), potentially in concurrency, there are other additional aspects to be modeled. Therefore the major tasks for the enterprise modeler are:

- Definition of shared information models, i.e., models common to, and that support the interactions among, team members.
- Definition of business plans. This activity includes the creation of instances of the taxonomy of BPs and their parametrization, definition of PRS, assignment of tools and other resources to the enterprise activities, etc..
- Characterization of the team of experts assigned to the business plan: identification and rights (namely in terms of allowed operations), definition of responsibility or organizational view.

5.3 Precedence relationships among activities

The Procedural Rule Set construct of CIM-OSA doesn't seem quite flexible to represent the temporal interdependencies among activities in a context of Concurrent

Engineering. For instance, lets consider the activities of "Product Design" and "Process Planning". Although they can proceed with some degree of concurrency (i.e., process planning can start once a first draft of the product is made), Process Planning cannot be finished before Product Design is complete. At least some details of the process plan definitely depend on the final commitment regarding the product model. Classical precedence relationships don't capture the full semantics of this temporal overlapping situations. Therefore CIM-FACE considers rules that combine some aspects of PRS with the Allen's temporal primitives. Examples:

i) Activity A1 must start before all other activities and should finish during execution of A2

ii) Activity A2 starts after A1 and finishes after A1

iii) Activity A3 must be done during A2

The specification of this kind of constraints is very important to drive the business plan interpreter. Such rules are specified by the enterprise modeler.

The model interpreter is responsible for the "execution" of the various enterprise's business plans. After the enterprise modeling and, in particular, after the definition of a business plan, a model interpreter is necessary to support the operational phase. It is responsible for:

- keeping track of the execution status of each business plan (and each node in the plan);
- supporting consistency maintenance (by verifying pre- and post-conditions for each BP/EA);
- helping or directing CA-tool selection for each EA, when more than one tool is available for an EA;
- providing a common access to the enterprise information models and supporting a multicontext framework;
- allowing communication among team members;
- providing a platform for progressive improvement of the control structure (addition of new rules).

Associated to this interpreter, a global project monitoring and evaluation subsystem can be thought of as a support tool for the project manager. This “monitor” should keep an audit trail of the decision making process and observed problems.

Although the emphasis is put here on CE, the proposed platform can support the "execution" of any business plan and not only the engineering related BPs.

5.4 Business plan assistant

An important part of the execution environment is the set of CA-tools that - in co-operation with the various team experts - actually implement the agents. A "protocol" is necessary to specify the interactions between each couple tool-user (agent) and the integrated federation. The "performer" of this protocol can be a layer separated from the applicative part of the tool. For legacy systems it is quite hard or nearly impossible to modify their control architectures. For new tools this "protocol performer" can be seen as a common script (library) that can be linked to the too (tool wrapper). Taking into account that:

- the "tasks" to be realized in a given business plan are hierarchically decomposed;
- "real" tools appear only associated to the leaves of this hierarchy (EAs);
- the human experts may interact with (inspect, start, finish, etc.) activities at various levels of the hierarchy;

This protocol performer is materialized by a separate (detached) process - business plan assistant [49] (see Figure 5.4).

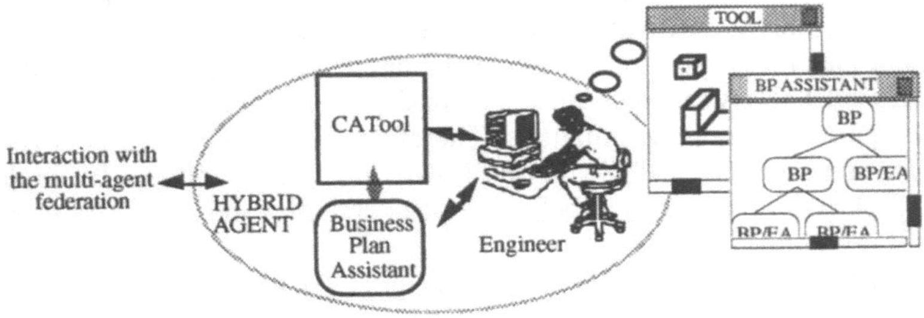

Fig. 5.4. Business Plan Assistant.

In computational terms, whenever a user "joins" the federation (logs in), a business plan assistant is launched. Through a graphical interface, the user has access to a specific business plan (or part of it, according to his/her access rights or responsibility level). For each node in the plan he/she can inspect its status, start or stop the BP/EA, create or change the default context, etc.. As a result of login, the user is positioned in the execution context of one or more activities of the selected business plan. From that plan, the human can select, for instance, "StartActivity" for an activity not yet being executed (if he/she has enough rights). Depending on the confidence or responsibility level assigned to this user, by the enterprise modeler, the interpreter can interdict the execution if some validation rule fails. When starting an EA, the associated application tools are launched. It shall be noticed that the selection of appropriate CA tools for each EA (Implemented Functional Operation, in the CIM-OSA terminology) was made by the enterprise modeler when he/she created the business plan or by the user in run time.

5.5 Contexts

As there might be many business plans running in parallel, it is important to provide a different context to each business plan. On the other hand, even inside one team / one business plan, some members (or subgroups) might need some "privacy" while they are exploring some rough ideas and before they decide to make their results public to the team. Therefore, CIM-FACE offers the possibility to create hierarchies of contexts. All objects created / modified in a given context won't be visible outside this context or its children contexts. When a business plan is started, a root context is associated to it. Offspring contexts may be created outside the current context provided the team member has enough rights. The new context is created as a subcontext of the current one and normal inheritance rules apply, i.e., all objects visible in a parent context are available to the child context. In this way, a team member (or subgroup) can explore his / her ideas without interfering with other members involved in the business plan. Once he / she is satisfied with his / her results, he / she can make them available to the others, i.e., export them to parent context.

5.6 Characterization of supervision knowledge

The human expert plays an important role in the evolution of the business plan execution. As the control events are, to a large extent, decided externally, by the human, how can the process interpreter be sure the decision was appropriate? For instance, lets suppose the current user of tool T_i informed the system that he finished the generation of a process plan for a given production task. Should the control system simply accept such information as a fact or should it be cautious and try to investigate the accuracy of the information? Therefore different "kinds" of control systems can be defined, ranging from a totally confident system to a cautious one. This is a key point for the achievement of a balanced decision making system. In order to support a flexible co-ordination infrastructure, a control knowledge base is necessary. From the concurrent engineering point of view, the execution of an activity should be flexible enough to allow engineers the exploration of different alternatives or even the redefinition of a previous approach regarding a new idea received from another co-operating team member.

It is necessary however to provide the system with mechanisms to express constraints on the execution of planned activities. Taking into account the fact that the execution environment is strongly based on a distributed infrastructure, maybe even located in different enterprises, various levels of validation can be considered. Not all of these levels are direct responsibility of the plan interpreter. We think that control knowledge may be defined at three levels:

1. Information System level - constraints may be embedded into the common information models. This can be represented by local and global rules of the EXPRESS language. The execution of such rules is the responsibility of the Information Management System.
2. Application level. Various control aspects, due to their specificity, are better defined inside specific application software (enterprise's proprietary tools).
3. Business plan level. Rules like precedence constraints and other specific pre- and post- conditions may be associated to each BP/EA node. The business plan interpreter has mainly to take this part of control into account.

Precedence constraints are only a part of the necessary coordination knowledge. Although a few projects started the analysis of the co-ordination control aspects in CE (see for instance [48]), there is not yet a sound foundation of co-ordination knowledge, which has to combine a model of the engineering process and a model of team work.

As a first step, in some cases and for some application domains, it will be possible to define a set of verification rules to test the validity of each access protocol action issued by business plan assistants. In other cases that might be difficult. Therefore, the approach in CIM-FACE is to have an architecture that can start from a level of total confidence and progress towards a more cautious system once verification rules are added to its control knowledge base.

An example for a minimal checking level is a rule that implements a CIM-OSA PRS rule:

IF <all previous activities are finished with success>
and <Self was requested to start>
THEN <enable execution of Self>

A more specific example can be associated to the BP_Design_P1 of type BP_design:

IF <MarketEvaluation_p1 of type MarketEvaluation was started> and
<BP_Design_P1 was requested to start>
THEN <enable execution of BP_Design_P1>

For instance the following rule could be attached to the EA124:

IF <RoughDesign.verified of EA123 Output is true> and
<EA123 is inactive> and <Self was requested to start>
THEN <enable its execution>.

The "evolutive" approach here proposed allows for an incremental improvement of the control / coordination knowledge based on experimental evaluation cases. This evolutive approach is also appropriate to cope with technology evolution. The coordination knowledge has to change with the evolution of the CA tools and their level of autonomous decision making.

As CE implies a design culture that is still quite different from the traditional working conditions in most companies, not only a better understanding of the design process but also the understanding of the new collective behavior is necessary. If we consider the reality in most developing countries, the situation is more difficult as most companies do not even have a design culture at all as they manufacture goods designed outside. As the manufacturing process is becoming global, with no geographical or political borders, the challenge for these companies is considerable. But, at the same time, they have the chance to start anew, without bias from the past, provided they have access to effective support platforms.

A more careful analysis of the social implications of the CE paradigm and how to capture their impacts on the proposed platform is necessary. The implantation aspects represent also a difficult task, especially in the context of small and medium sized enterprises, in which it is necessary to change traditional working habits. Therefore, an implantation methodology and training plan have to be associated to any new platform for CE.

6. BAS and Virtual Enterprises

Finally lets consider the case of virtual enterprises. The rapidly changing manufacturing scenario in which companies are becoming nodes of a production chain, adding some value to the components received from their suppliers and delivering the transformed components to their clients, led to the concept of virtual or extended enterprise [50]. There is, therefore, the need to consider production management at the network level and not only at each enterprise level. This networked production management has been done by humans with little intervention of automated functions. However, the need to react quickly to changes on the network, to assure high quality at all levels of interaction with other nodes, is leading to the need of a computational infrastructure to support the network operation.

6.1 General requirements

First requirements for such infrastructure are traditionally related to improve the information flow between enterprises, namely in terms of orders (commercial information) and product technical specifications. Nevertheless other facets are rapidly emerging. A more complete list of requirements for the virtual enterprise infrastructure will include:

i. Production management aspects

Follow up of orders status. Fax or other non-electronic media lead to a non-effective monitoring of the orders' status. Delays in orders processing, temporary incapacity of a supplier, changes in not completed orders, the need to re-adjust delivery times, etc., are some factors that point to the need for a more flexible and reliable interchange of information. A client node wants to know the status of processing of its orders at the supplier's site in order to prevent any difficulties for itself. The implementation of a truly Just-In-Time philosophy requires an infrastructure that supports a stronger clients - suppliers interaction. However, the supporting infrastructure does not eliminate the need for a negotiation process between human interlocutors whom ultimately decide which information to make available.

Distributed scheduling and workload optimization. In the context of distributed manufacturing, there is a scheduling function at each node and a scheduling at the network level. Breakdown of key equipments, changes in priorities, etc., may require the transfer of work load from one node (enterprise) to another. Thus, there is a need for an infrastructure which supports negotiation between nodes for workload distribution/contract assignment. The contract-net paradigm of distributed artificial intelligence may be applied for the implementation of a decision support system that helps nodes (i.e. the local responsible humans) in making decisions based on an update knowledge of the current status of the network and historic information regarding the performance, quality and reliability of each node (Supplier). A rational balanced distribution of the workload by the network nodes is necessary to achieve a global optimization of the resources utilization and minimization of costs. It shall be noted also that sometimes the decision making systems can face social problems that indirectly can influence the workload optimization.

Incomplete and imprecise orders. A very challenging requirement for the networked production management system is the capability to support the flow - and the acceptance by suppliers - of incomplete and imprecise orders. The growing pressure to reduce the production cycle time is originating - and forcing enterprises to accept - orders that are not completely specified in their initial formulation. However, the supplier cannot wait for the order to be completely defined, because then it would not have time and resources to fulfill the order. Thus, there is a need for an information logistics and special scheduling system which helps to avoid "hard picks" in the production phases.

In some areas - like the garment industry - some aspects of the product are dependent on "fashion waves" and therefore not know until a date very close to the delivery date. In other areas there are "global orders" for a long period (one year, for instance), but the exact splits of these orders between the intermediate delivery dates are not known. These factors raise questions on the following issues: orders representation and management; decision making with incomplete information; determination of what can be started with the available information: this is related to the product model - which parts are invariant (relationship between imprecise / incomplete orders and the product structure); scheduling in a context of incomplete information; the impact of such order in the node's production management system; once a node accepted such an order, which "part" of uncertainty can be passed on to the supplier of this node.

ii. Architecture and information management

Typically, SMEs preexist before they decide to join in an information sharing and exchange network. Consequently, every SME is autonomous, developed independently of other SMEs and uses distinct information management and control strategies that serve its purposes best. However, for these enterprises to cooperate, on one hand they need to share and exchange a part of their information with the others and on the other hand every SME would like to preserve its local autonomy. Furthermore, distinct enterprises have distinct and often contradictory views and semantic associated with information they store.

Another information management problem among autonomous cooperating enterprises rises from the fact that every enterprise requires privacy and security for a part of its information. Namely a part of information that is managed by an enterprise may be classified information and therefore another enterprise should not be able to access it. For instance the administration of clients and orders should not be visible to other competing enterprises. On the other hand, every company is only interested in a part of information that another enterprise is willing to share and not all of it.

The classical approach to support the information management in a network of cooperating nodes is through defining a single shared global schema (structure of information) to be shared and used by all enterprises. It is however clear, that this approach does not support information management requirements in the cooperation of SMEs as described above. In this application, a specific architecture needs to be introduced to support different aspects of nodes' cooperation. For instance, there are three different categories of information that need to be managed by every node. Namely, (1) the information that this node wishes to share with others, (2) the information stored by other nodes that is of interest to this node, and (3) the information handled within a workspace area, that is created, accessed, and updated by two or more nodes , e.g. to handle the order management among nodes.

The scope of information related to a product is often considerable, even for a simple component which ends up as a part of a complex product. At present there is not one single mechanism for the communication of the total product data. Strategies and techniques are in place or are being developed for specific requirements between related teams; but there is not a single global schema in place. As a result of the work being undertaken in specific communicative areas in these networks, methodologies are emerging which could be exploited to provide the "glue" required to relate information and data to communication of total product data. Key steps in establishing a virtual enterprise communication scheme require an understanding of the business processes in the virtual enterprise, the information, control and resources needed to support these processes, and the scope of product data and information which must be shared or exchanged.

Taking into account all new communication capabilities, protocols and document formats, companies will have severe problems to manage and track these data flows, with the risk to loose all benefits coming from advanced communication systems. To prevent this risk, SMEs need solutions to cover:

- Commercial documents (by extensive use of EDI standard, instead of fax, to support the orders follow up),
- Technical product data (i.e. STEP),
- Scheduling and Workload optimization data,
- Quality (tracking and tracing) data,
- And the most important: Tools to manage and follow-up all data flows with a business approach and not via a technical point of view, as it is always the case.

These are aspects where automated functionalities are preferable to other more classic (manual) approaches.

6.2 SME independence

SMEs are often under pressure from large companies, trying to force sub-contractors to use their standards, sometimes with an open solution, but more often with a proprietary approach. The key difference between an open and a proprietary approach is that open solutions are reusable whereas the proprietary solutions are not. When forced to use proprietary solutions, SMEs are required to invest for each of these customers. This is a difficult situation for SMEs as they prefer easy-to-use and low-cost solutions without much importance placed on systems integration.

SMEs with multiple proprietary systems to support different customer demands have the overhead of management of different information systems. Files are copied and modified between different systems. This introduces the risk that the SME is unable to manage this data flow. Furthermore, when a customer change is required, this impacts the supplier environment more than the customer. This supplier / customer dependency created by using proprietary solutions significantly increases the cost base of the supplier who has to purchase different systems for different customers. Therefore, to be more competitive, by reducing costs and enabling flexibility as well as freedom to integrate different networks, SMEs need an open platform to support the information interchange needs as well as their contribution to the global decision making processes.

It shall be noted however that strategical decision making related to most negotiation processes at network level is performed by humans, although computer assisted functions might improve the efficiency of such processes. Thus, an extended / virtual enterprise model, leading to the implementation of team-based structures and new production concepts, should develop an anthropocentric production system strategy. The technocentric strategy would presuppose a single-oriented focus on technical and technological issues for the management of the decision process. In a virtual enterprise the focus is not on the equipment but on the way it could be used for a better result. And this means complex inter-relations between sub-units in the network of the extended enterprise. The geographical distribution of the network, even world wide, imposes dealing with different social situations, traditions and skills. It is therefore necessary to build a reference framework for the concepts of extended / virtual enterprise and for the balanced distributed production systems.

7. Conclusions

In this section the emerging concept of Balanced Automation was introduced as an attempt to achieve a right combination of automation and manual based operations, taking into account the economic and socio-organizational aspects as well as the legacy assets of the enterprise or region in which it is implanted. Specially for SMEs, the achievement of a right balance between automated components and manual and hybrid solutions, taking into account the economic and social aspects as well as the migration from legacy systems and other inherited assets like the established know how and skills, is a factor of competitiveness, if not of survival. The need for a "balanced" approach is not restricted to the shop floor aspects but can likely be found in all other areas, as illustrated for the areas of scheduling, concurrent engineering and virtual enterprises. Although many results from various disciplines - CIME, anthropocentric systems, human-machine interfaces, cooperative activity and cooperative decision-making, etc. - can be used as a supporting toolbox to the development of Balanced Automation Systems, a lot of work has still to be done. And, in our opinion, the development of hybrid solutions, coping with a variety of automation levels / manual approaches, is a more challenging research problem then the search for a purely automatic systems.

Acknowledgments. This work was supported in part by the European Commission (projects CIMIS.net, FlexSys and B-Learn II) and the Portuguese Board for Scientific and Technological Research (CIM-CASE project). We also thank the contributions of Luis Seabra Lopes and José Barata.

8. References

[1] Masi,C.G.-Re-Engineering Engineering Education,*IEEE Spectrum*,Sept 1995.

[2] Geppert, L. - Educating the Renaissance Engineer, *IEEE Spectrum*, Sept 95.

[3] ASEE - Engineering Education for a Changing World, *American Society for Engineering Education*, Oct 94.

[4] Kovacs, I.; Brandão Moniz, A. - Issues on the anthropocentric production systems, *Balanced Automation Systems: Architectures and Design Methods*, L.M. Camarinha-Matos, H. Afsarmanesh (Eds.), Chapman & Hall, 1995.

[5] Roshardt, R.; Uhrhan, C.; Waefler, T.; Weik, S. -A complementary Approach to Flexible Automation, in *Balanced Automation Systems: Architectures and Design Methods*, L.M. Camarinha-Matos, H. Afsarmanesh (Eds.), Chapman & Hall, 1995.

[6] T. Guimaraes - Reengineering Manufacturing Organizations: Benefits, Problems and Success Factors, *Communication to 5th Portuguese Conference on Computer Aided Design, Planning and Production*, Guimaraes, Portugal, May 1995.

[7] CEN TC310 WG1 - An evaluation of CIM modelling constructs - Evaluation report for views according to ENV 40 003, *Computers in Industry*, Vol. 24, N. 2-3, Sep 94.

[8] Esprit Consortium AMICE (1989) Open System Architecture for CIM, Springer-Verlag.

[9] Schenck, D.; Wilson, P. (1994), Information Modeling the EXPRESS Way, Oxforf University Press, New York.

[10] Ross, D.T. -Structured Analysis (SA): A Language for communicating ideas, *IEEE Trans. on Software Engineering*, vol SE 3, N.1, Jan 77.

[11] Peterson, J. - Petri net theory and the modelling of systems,Prentice-Hall, 81.

[12] Williams, T. - The Purdue Enterprise Reference Architecture, *Computers in Industry*, Vol. 24, N. 2-3, Sep 94.

[13] Bernus, P., Nemes, L. - A framework to define a generic enterprise reference architecture and methodology, CSIRO, Australia, 1995.

[14] Colombo, A.W.; Carelli, R. - Petri Nets for Designing Manufacturing Systems, a chapter in this book.

[15] Miyagi, P.; Camarinha-Matos, L.M.; Santos, D.; Barata, J.; Arakaki, J.- The application of enhanced Mark Flow Graph in Real Time Control Systems, *Proc. IFAC Conf. on Low Cost Automation*, Buenos Aires, Sep 95.

[16] Barata, J.; Camarinha-Matos, L.M.; Colombo, W.; Carelli, R.- Integration of OOP and Petri Nets for Discrete Event Supervision and Simulation in

Balanced Automation Systems: Architectures and Design Methods, L.M. Camarinha-Matos, H. Afsarmanesh (Eds.), Chapman & Hall, 1995.

[17] Hasegawa, K.; et al., 1984, "Proposal of Mark Flow Graph for Discrete System Control", *Transactions of the SICE; SICE*, Vol 20 N 2, Japan.

[18] Santos Fº, D.J. -Proposal of Enhanced Mark Flow Graph for Modeling and Control of Integrated Manufacturing Systems, Master Thesis, University of São Paulo - Escola Politécnica, São Paulo, SP, Brazil (in Portuguese).

[19] Silva, M. ; R. Vallete, *Petri Nets and Flexible Manufacturing,* in *Advances in Petri Nets.* 1989, p. 374-417.

[20] Zhou, M., F. DiCesare, and A. Desrochers, *A Hybrid Methodology for Synthesis of Petri Net Modells for Manufacturing Systems,* in *IEEE Transactions on Robotics and Automation*1992, p. 350-361.

[21] David, R. and H. Alla, *Petri Nets for Modelling of Dynamic Systems - A Survey,* in *Automatica - IFAC*1994, p. 175-202.

[22] Murata, T., *Petri Nets: Properties, Analysis and Applications,* in *Proceedings of the IEEE*1989, p. 541-580.

[23] Camarinha-Matos, L.M. ; Osório, A.L. - Monitoring and Error Recovery in Assembly Tasks, 23rd ISATA - *Int. Symp. on Automotive Technology and Automation*, Vienna, Austria, Nov. 90.

[24] Barata, J. and L.M. Camarinha-Matos. *Dynamic Behaviour Objects in Modelling Manufacturing Processes.* in *CAPE'95 - The Fifth International Conference on Computer Applications in Production and Engineering*. 1995. Beijing - China.

[25] Seabra Lopes,L.; Camarinha-Matos, L.M. - Inductive generation of diagnostic knowledge for autonomous assembly, Proc. *IEEE Robotics and Automation*, Nagoya, Japan 21-27 May 95.

[26] Seabra Lopes,L.; Camarinha-Matos, L.M. - A Machine Learning Approach to Error Detection and Recovery in Assembly , *IROS'95: IEEE/RSJ International Conference on Intelligent Robots and Systems*, Pittsburgh, USA, August 1995.

[27] Rabelo, R.; Camarinha-Matos,L.M. - "HOLOS: a methodology for deriving scheduling systems", Balanced Automation Systems: Architectures and Design Methods, Eds. L.M.Camarinha-Matos and H. Afsarmanesh, Chapman & Hall, 1995.

[28] Rabelo, R.; Camarinha-Matos, L.M. - "Deriving Particular Agile Scheduling Systems using the HOLOS Methodology", International Journal in Informatics and Control, Vol 5 N 2, Jun 1996.

[29] Huhns, M., editor - "Distributed Artificial Inteligence", Pitman Publishing / Morgan Kaufmann Publishers, San Mateo, CA, 1987.

[30] Davis, Randall; Smith, Reid - "Negotiation as a Metaphor for Distributed Problem Solving", Artificial Intelligence, N 20, pp.63-109, 1983.

[31] Ge, Qing; Findler, N. - "Distributed Planning and Control for Manufacturing Operations", Lecture Notes in Artificial Intelligence, Eds. C. J. Barter & M. Brooks, Springer-Verlag, 1988.

[32] McLean , C. R.; Bloom, H.M; Hopp, T.H. - "The Virtual Manufacturing Cell", Proceedings 4th IFAC/IFIP Conference on Information Control Problems in Manufacturing Technology, Maryland, USA, Oct 1982.

[33] Drolet, J; Montreuil, B.; Moodie, C. - "Decision Architecture for Scheduling Virtual Cellular Manufacturing Systems", IFAC/IFIP Int. Workshop on Decisional Structures in Automated Manufacturing, Italy, 1989.

[34] Chryssolouris, George - Manufacturing Systems: Theory and Practice, Springer-Verlag, NY, 1992.

[35] Dietrich, B.- "Automation in Manufacturing, Control vs Chaos", Advances in Agile Manufacturing, Ed. P.T. Kidd and W. Karwowski, IOS Press, 1994.

[36] Rabelo, R.; Camarinha-Matos, L.M. - "A Holistic Control Architecture Infrastructure for Dynamic Scheduling", Artificial Intelligence in Reactive Scheduling, Eds. Roger Kerr and Elizabeth Szelke, Chapman & Hall, 1995.

[37] Rabelo, R.; Camarinha-Matos, L.M. - "Generation of Multi-Agent Infrastructures for Dynamic Scheduling and Control Architectures", *Proceedings 27th ISATA / Conference on Lean/Agile Manufacturing in the Automotive Industries*, Germany, 1994.

[38] Jones, B. - "Essential Cultural Aspects, Strategies and Techniques - A Comparative View of Work Technology and Flexible Production", in *Flexible Manufacturing Systems and Work Reorganization* [in portuguese], Ed. by Ilona Kovács at all, Portugal, 1992.

[39] Rabelo, R.; Camarinha-Matos, L.M. - "Negotiation in Multiagent Based Dynamic Scheduling", *Journal on Robotics and Computer Integrated Manufacturing*, Vol 11 N 4, Dec 1994, Pergamon.

[40] KQML - Knowledge Query and Manipulation Language, ARPA Knowledge Sharing Effort Project, USA.

[41] Fowler, J. - "Proposal for the STEP Data Access Interface Specification", STEP Implementation Specifications Committee, NIST, Jan 1992.

[42] Hirsh, B., Kuhlmann,T., Marciniak, Z. - "Engineering Tool-Kit for Implementation of Shop Floor Control Systems", *Proc. IFIP WG5.7 Conf. on Eval. of Product. Manag. Methods*, Gramado, Brazil, Mar 1994.

[43] Smith, S. - "Configurable Systems for Reactive Production Management", Proceedings KBRS'93 - *Knowledge-Based Reactive Scheduling IFIP Workshop*, Eds. E. Szelke, R. Kerr, IFIP North-Holland, 1994.

[44] Szelke, E.; Kerr, R. - "Knowledge-Based Reactive Scheduling", *Int. Journal of Production Planning & Control*, Vol 5 N 2, Taylor & Francis, 1994.

[45] STEP, ISO - Reference Manual, ISO/TC 184 /SC4.

[46] Camarinha-Matos, L.; Sastron , F. - Information Integration for CIM planning tools, *CAPE'91 - 4th IFIP Conference on Computer Applications in Production and Engineering*, Bordeaux, 10-12 Sep, 1991.

[47] Camarinha-Matos, L.M.; Osório, A. L. - CIM Information Management System: An Express-based integration platform, *IFAC Workshop on CIM in Processes and Manufacturing Industries*, Espoo, Finland - published by Pergamon Press.

[48] Klein, M. - iDCSS: Integrating Workflow, Conflict and Rationale-Based Concurrent Engineering Coordination Technologies, Boeing Computer Services, *Journal of Concurrent Engineering Research and Applications*, vol. 3, N. 1, Jan 95.

[49] Osório, A.L.; Camarinha-Matos, L.M - Support for Concurrent Engineering in CIM-FACE,*Balanced Automation Systems: Architectures and Design Methods*,L.M.Camarinha-Matos,H.Afsarmanesh(Eds.),Chapman&Hall, 1995.

[50] Brown, J. - The extended enterprise - manufacturing and the value chain, in *Balanced Automation Systems: Architectures and Design Methods*, L.M. Camarinha-Matos, H. Afsarmanesh (Eds.), Chapman & Hall, 1995.

15 Factory Principles Applied to Analytical Chemistry : An Integrated Laboratory Management System

J. N. Thompson

1. Introduction

Equipping a laboratory with state-of-the-art analytical instrumentation and allocating routine tasks to preparation and analytical robots is a laudable aim which has tangible business and technical benefits. To be totally effective, however, there exists a need for brains to combine the brawn of these components into a cohesive functional unit. The brains behind a laboratory operation have, in the past, been laboratory information management systems (LIMS). The nomenclature is significant.

Traditionally, LIMS have been directed towards manual operations with a prime requirement to generate an analytical schedule (worklist) and to provide a mechanism for recording the results of analysis against that schedule. Some data manipulation and validation functions may have been provided to verify the results and to track result trends. Such systems, were, as the name suggests, information management systems with no elements of instrument control.

The developments at North West Water (NWW) mark a significant achievement on the water supply industry world stage where the aim has been to apply factory principles to analytical chemistry. The parallel between the factory and the laboratory is an interesting one. An examination of the evolution of factory computing puts the development of LIMS into context.

In the late 1970s, factory control systems emerged which communicated directly with the components of automation (cranes, conveyors, etc.) via low speed serial (RS-232) links to on-board programmable logic controllers (PLCs). This required the control systems to handle the communication protocol to the PLC directly with a consequential processing overhead. Such single-layered systems typically communicated upwards to a management information system, again via RS-232 communications.

In the 1980s, when system integration began to emerge as the preferred approach, factory control systems took on a more diverse role. Instead of the discrete control systems of yesteryear, targeted on a single process, the control systems for discrete processes were integrated via a supervisory scheduling system which communicated upwards to corporate systems. The control systems continued to communicate to PLCs directly.

In the late 1980s and into the 1990s, PLC manufacturers have performed integration at their level of control. Thus, instead of control systems communicating directly with one or more PLCs, they now communicate with a local area controller which integrates the PLCs associated with a particular process. In this way, the communication protocol is now isolated from the control system and more sophisticated and specialist techniques reside within the local area controllers.

Coincident with the move to integration at a lower level in the factory control system hierarchy was the development of networking capabilities. The impact on factory control systems has been as profound as in other application areas. Now, instead of direct connections from the scheduling systems to the control systems and then onto the local area controllers and their attendant PLCs, all components of the integrated systems can be connected with one another, removing the need for the hierarchy of factory control systems outlined above. Add to this the increased processing power of modern computers, reduced hardware costs, the development of open systems and increased manpower costs and an industrial revolution beckons.

In the context of factory control system evolution, therefore, the traditional LIMS aligns with the factory management systems of the 1970s which were essentially data capture and manipulation systems, with no components of automation and control. Against this background, the achievement of NWW in the development of a new central laboratory has brought the capabilities of their LIMS from the factory control systems of the 1970s to those of the 1990s in one giant leap. The term LIMS is not appropriate to NWW's new system as it incorporates elements of information management and process control, giving rise to a true integrated laboratory management system (ILMS).

2. Business Motives

The business context in which NWW undertook a review of their water quality service was the need to significantly improve their efficiency as a result of an increase in workload following privatisation, brought upon them by regulators and the EU. These pressures apply to the entire business of water supply and wastewater treatment. The laboratory would be the key to the success of the business because it had to provide both regulatory compliance and the process control information without which the business could not operate. Therefore, the key business goals which NWW Laboratory Services had to meet were to reduce the unit cost of analysis, increase the speed of analysis and reporting and surpass the existing level of compliance to meet higher external regulatory standards.

The conclusion reached by NWW Laboratory Services was that the most cost-effective solution was to build a new, centralised, highly automated laboratory to replace the 28 laboratories scattered throughout the region. This approach demanded the realisation of the integration advancements envisaged by Linder[1]. However, automation could not be implemented wholesale, it being neither cost-justifiable nor, in some cases, feasible to automate analyses which demand the application of complex analytical techniques. Manual analysis was therefore retained, leaving the routine, non-specialist work to robots.

Full integration of the laboratory with the rest of the business was necessary to enable it to react rapidly to changes in circumstances, be they pollution incidents in the supply network or alterations to the EU drinking water regulations. Such a laboratory could not work without a fully-integrated laboratory management system unifying the various components of laboratory operation with each other and with the rest of NWW's operation to support the water supply, treatment and disposal networks.

3. Business Integration

Laboratory integration is not solely concerned with the co-ordination of sample preparation and analysis units since it must integrate the laboratory with the business processes of analytical scheduling and the reporting of analytical results. The key is the speed with which analytically validated and approved results are made available to the corporate business, as it is here that operational decisions can be taken.

Figure 13.1 depicts the work flow through the laboratory and emphasises the criticality of the ILMS to the operation of the water quality business since it alone is aware of the status of all the work which the laboratory is required to perform (hence all work flow arrows pass through the laboratory management system). Each of the components is described below, with the emphasis placed on business-level integration. A detailed account of the ILMS support of laboratory operation is provided in section 4.

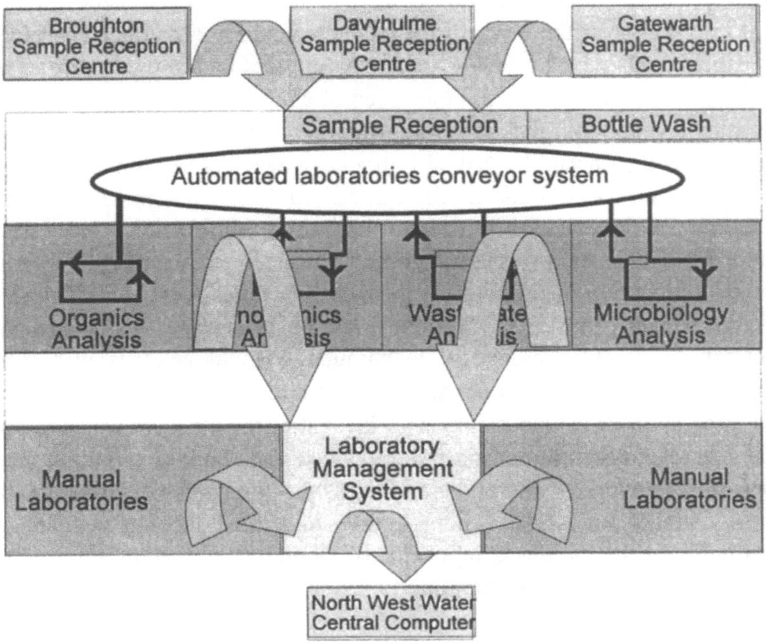

Figure 13.1: Operational Work Flow

3.1 Out in the Field

3.1.1 Sampling Requirements

NWW carry out sample analysis (at the rate of 2,000 samples per day, over 3 million measurements per annum) in order to ensure that they conform to the requirements of the regulatory bodies to whom they are accountable. These bodies monitor drinking, bathing and river water quality and the disposal of trade and other effluents. Each defines performance levels which must be met and must be seen to be met. As a consequence, NWW are required to take regulatory samples at a specific frequency in prescribed locations so that statutory measurements can be provided to these external bodies.

In addition, operational sample analyses must be performed to provide control information to the water and wastewater treatment plants throughout the region and to monitor, internally, NWW's performance against regulatory standards. A further requirement is to take water samples in response to specific customer requests, where a customer reports a problem with the quality of water being supplied or reports some abnormal situation such as a pollution incident.

3.1.2 Planned Samples

All these requirements are converted into work plans for the sample collection personnel. A sample scheduling system, running on an IBM RS/6000 computer, generates a schedule of samples and field tests for each water quality officer (WQO) to perform on the following day. This schedule identifies the sampling location, the type of bottle in which the sample is to be placed and any field measurements which need to be taken. These planned sample details are transferred to the ILMS to enable the samples to be registered, in advance, on the ILMS database. This occurs overnight, when the system load is at its lightest.

3.1.3 Taken Samples

As the WQOs take the samples and perform tests in the field, they record details against their schedule and, having completed their daily round, they deliver bottles of samples to a regional sample reception centre. Here, the taken sample details and any field test measurements are registered on the sample scheduling system, bottles of the same type are loaded into crates and the sample taken information, which includes the identities of the crates in which the sample containers have been placed, is sent to the ILMS. This takes place from about 11 a.m. to 4 p.m. when the ILMS system load is heavy. For this reason, an ILMS cloning facility was established which enables the sample registration process to replicate itself, up to a configurable maximum number of instances, when certain workload thresholds (held as configuration parameters) are exceeded.

Naturally, there are occasions when samples cannot be taken or field tests cannot be performed. The ILMS is notified of such cancellations and the sample scheduling system re-schedules the taking of the sample and any required field tests.

The crated samples accumulated at the three regional sample reception centres are taken by courier to NWW's central laboratory in Warrington.

3.2 Laboratory Processing

At the business integration level, the central laboratory is effectively an analytical black box into which samples and work schedules are fed and from which results of analysis are received.

3.3 Regulatory Reporting

When the validation level of a test reaches a certain value, as dictated by the test procedure, it triggers the execution of an ILMS process to transmit the results to the corporate systems resident on the NWW central computer. This business level integration provides the benefit of making analytical results available to the business at the earliest opportunity so that problems in the treatment process or the supply and disposal networks can be identified quickly and remedial action instigated.

4. Laboratory Integration

4.1 Laboratory Classification

Before considering the operational cycle within the laboratory, it is necessary to understand the three classifications into which the discrete laboratory areas within the central laboratory fall.

Firstly, there are fully-automated laboratory areas in which both sample preparation and sample analysis are performed by robotic cells under the control of a laboratory area controller (LAC). The second type of area is termed a semi-automated area in which sample preparation is performed by robotic cells under the supervision of a dedicated LAC but sample analysis is performed by an analytical instrument which may be connected to the ILMS or operated in stand-alone mode. Finally, there are manual laboratory areas in which sample analysis is performed by analytical staff, such analyses tending to be either low volume or involving methods that require application of complex analytical techniques.

The layout of NWW's central laboratory is depicted in Figure 13.2 and a description of the operation of the individual laboratories supported by the ILMS follows.

4.2 Crate Routing

4.2.1 Sample Reception

When couriers arrive at the central laboratory, they drive to the unloading bay and place the crates on the input spur of the sample transport system (STS). The crates proceed into the laboratory passing a barcode reader which scans the crate's barcoded label and the STS informs the ILMS of the crate's identity. This information should have been received from the sample scheduling system as part of the sample taken data.

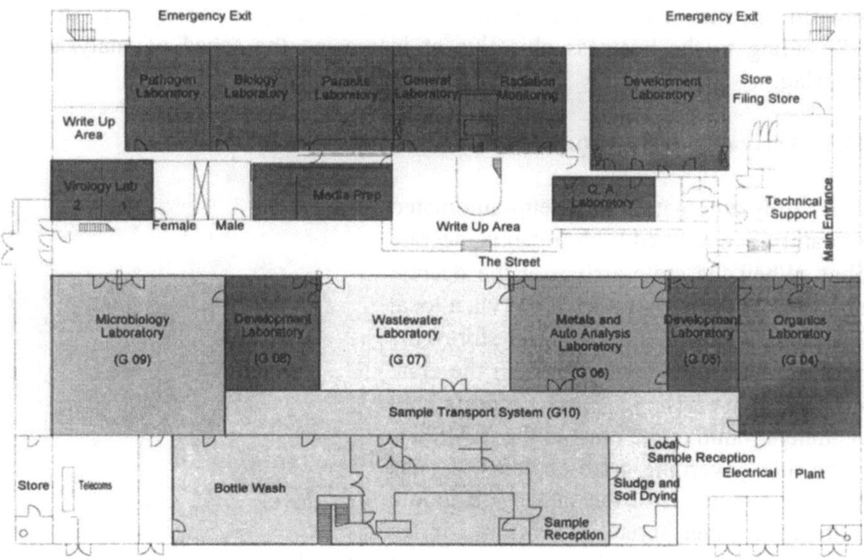

Figure 13.2: Laboratory Layout

Provided that the crate is known to the ILMS, it instructs the STS to transport the crate to its destination laboratory as denoted by the crate identity. The ILMS also instructs the STS to deliver unexpected crates to a reject spur in the sample reception area of the laboratory for manual intervention. Should the STS be unable to read a crate identity barcode then it automatically directs the crate to this reject spur.

4.2.2 Sample Transportation

While the crate is in transit to a fully-automated or semi-automated area, the ILMS schedules the work to be performed on the containers in the crate. It sends worksheets down to the LAC when the crate arrives in the laboratory, so that the potential exists for work to commence immediately, although the LAC actually schedules the analyses according to crate priority and sample age.

For crates in transit to a semi-automated laboratory area, the ILMS also allocates an instrument to perform the analysis, as dictated by the test procedure. Where more than one instrument of the same type exists, the ILMS balances the workload between instruments to achieve the maximum sample throughput, thereby contributing to the business objective of increasing the speed of analysis and reporting.

4.2.3 Arrival in Laboratory Area.

Each fully-automated or semi-automated laboratory is equipped with a gantry decrate robot. When the crate arrives at the decrate robotic cell in the laboratory area (via a local STS loop serving as a crate buffer store), the robot removes each container from the crate and its supervisory LAC reports all containers found in the crate to the ILMS.

The rule governing crate contents dictates that only containers of the same type should exist in a given crate. Therefore, if an unexpected container of the correct type is found in the crate then the preparation or analysis of the sample in that container is scheduled by the ILMS. Detection of an unexpected container of the wrong type results in cancellation of the worksheet previously sent to the LAC, although the sample itself remains available for analysis (in another area).

4.3 Analytical Processing

4.3.1 Fully-automated Laboratory Areas.

To support operation of the fully-automated wastewater and sludge laboratory areas, the ILMS generates worksheets identifying all the methods to be performed on each container in the laboratory area, for use by the robots, and downloads these to the LAC which co-ordinates the shop-floor units.

Results of the preparation stage are sent by the LAC to the ILMS and are stored on the database. When sample analysis is complete, the results of the analysis are transmitted by the LAC to the ILMS.

4.3.2 Semi-automated Laboratory Areas

The organics, microbiology and metals and autoanalysis laboratories are semi-automated areas. Here, preparation robots are used to carry out the routine sample preparation stage, again dictated by the test procedure. The ILMS sends a preparation worksheet to the LAC which, in turn, downloads the data to the robotic cell. Once sample preparation is completed, results are sent to the ILMS by the LAC.

Based upon the results received from the robotic cell which performed the sample preparation, the ILMS generates an analytical worklist which, for instruments connected to the ILMS, it transfers automatically to the instrument control system. Some degree of data manipulation is required at this stage to match the results elicited from the robotic cell with the instrument sequence file requirements of the individual control system. Based on the contents of the worklist, the instrument runs the prescribed analytical method and returns a results file and, in some cases, a status file to the ILMS.

4.3.3 Manual Laboratory Areas

The remaining areas of the laboratory are devoted to purely manual sample analysis. Although sample preparation for manual analysis may be performed by robotic cells in an automated laboratory thence transported by hand to a manual area, the majority of crates of samples for these areas are directed to a manual spur in the development laboratory (G.05) where they are collected by hand. The ILMS provides automated interfaces to a limited number of instrument control systems resident in manual laboratory areas. Analytical instruments operated in stand-alone mode are supported by ILMS manual functions to generate worklists and record results of analysis.

4.4 Result Validation and Approval

Results of analysis received from LACs, instrument control systems and manual results entry instigate the mapping of results onto the ILMS database against the test procedure. The resulting data changes can be used by the test procedure to trigger the execution of a number of ILMS functions, logically grouped together as a result validation tool-kit, to perform automated result validation and approval at the test, sample and sample container level, as well as AQC validation. Additional functions within the tool-kit enable an instrument to be taken off-line (to prevent its allocation to further sample analyses) and the initiation of a re-test of a sample.

The AQC validation function performs statistical analysis on the test results for standards within a worklist followed by trend analysis based upon previous (historical) results for the specific instrument and test method combination. When a measurement fails AQC validation, all results for the worklist (including unknowns) are deemed to have failed and a local alarm is raised. Following investigation, the results can be manually validated or cancelled.

4.5 Container Disposal

One of the result validation processes checks that all analytical work on the contents of a container has been completed and when all the containers in a crate reach this disposal status the ILMS instructs the STS to send the crate to either the bottle wash area or to sample reception. Crates will only be sent to sample reception when a problem with the crate contents has been identified which means that one or more containers in the crate still need to undergo analysis. Upon arrival, such containers are placed in another crate, the contents of the crate logged with the ILMS and the analytical cycle repeated.

Containers despatched in crates to the bottle wash area are cleaned and sterilised and then re-registered with the ILMS prior to return to the sample reception centres in readiness for collection by WQOs for the next day's sampling.

5. Anatomy of the ILMS

Having explained the operational context, both at the business and laboratory component level, a description of the computer system necessary to support the integrated laboratory is considered appropriate. The various hardware and software layers which make up NWW's ILMS are depicted in Figure 13.3 and the remainder of this section describes the components involved.

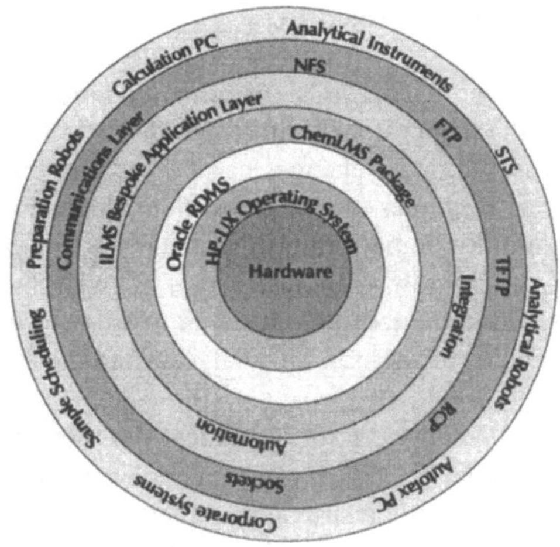

Figure 13.3: NWW ILMS Component Layers

The significance of this model is that each layer is reliant on all the underlying layers so that a change to one layer potentially affects all layers above it. The architecture of the ILMS is consistent with that described by Mattes and McDowall[2].

5.1 Technical Platform

Figure 13.4 identifies the connectivity of the various hardware components which comprise the integrated laboratory, each of which is described below.

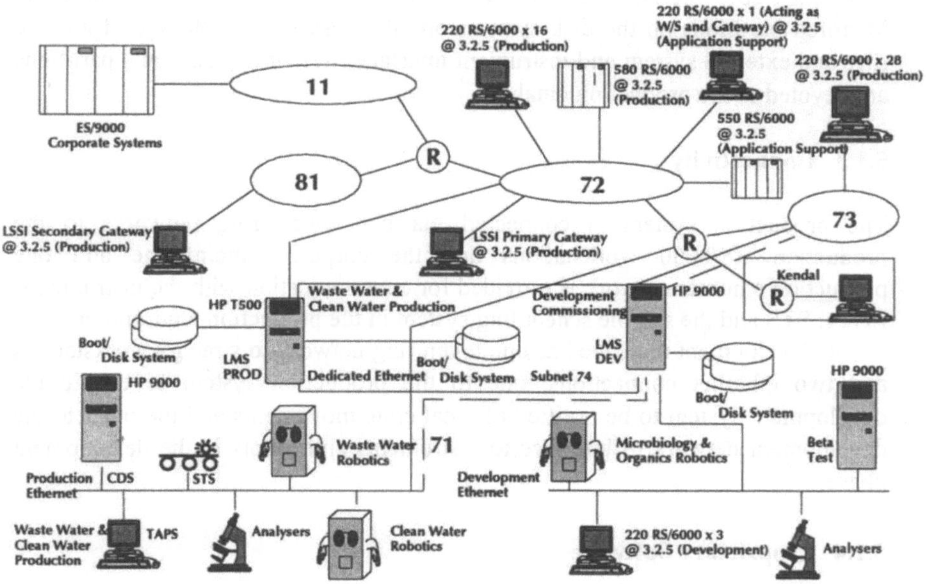

Figure 13.4: NWW ILMS Technical Architecture

5.1.1 Hardware Platform

The NWW ILMS is based on two stand-alone HP computer systems. The production system, upon which the operation of the live laboratory is supported, comprises a T500 multi-processor system with mirrored disk systems, housed in NWW's Data Centre some 400 metres from the laboratory building. The multi-processor system offers considerable resilience since the service can be recovered following failure of a single component processor by simply re-booting the T500 computer. Mirrored disk systems provide the data security necessary to support continuous system operation.

The development system is an HP9000 series computer equipped with a local disk system and is used to support the development of additional application software and new analytical methods and procedures. It is also used to commission newly acquired items of equipment in isolation from the production environment.

5.1.2 System Architecture

The system architecture is a client-server configuration with the production system acting as the NWW ILMS server and the RS/6000 workstations acting as the client. Mirrored partitions on the production system disks are used to hold the proprietary software and the NWW bespoke ILMS source and executable code. Mirrored partitions on the disk systems are also reserved for storage of archive files and external system and instrument interface files while additional partitions are devoted to the production database.

5.1.3 Connectivity

The production system is connected via two token ring networks to the production RS/6000 workstations and the corporate mainframe and one production ethernet network is provided for communication with the instruments, LACs, STS and the sample scheduling system in the production environment.

The development machine has one token ring network to strategy workstations and two ethernet connections, one to the production system to enable the development system to be notified of local crate movements and the other to the development network to interface to instruments and robots in the development phase.

5.1.4 Proprietary Software

This term is used to identify the system and package software environment under which the ILMS application code went through its development and production life-cycle. There are three elements to this software, namely the HP-UX (Unix) operating system software, the Oracle relational database management system (RDMS) and the core ChemLMS package.

The significant factor here was the ability to migrate the ILMS application software through the various proprietary software upgrades without any impact on the functionality provided. NWW were provided with a user confidence test script for this purpose, which provides a typical profile of the ILMS facilities to be exercised on the development system in order to prove successful application migration. This was used by NWW as the basis for the decision to install proprietary software upgrades in the production environment.

ChemLMS is Hewlett-Packard's laboratory management package which provides a tool-kit to enable customisation of the package to meet the operational and procedural needs of both laboratory users and the business. It provides suites of native commands to enable the progress of a sample through the various stages of processing to be registered in a logical way. These are essentially manual functions for recording sample analyses but with the *capability* to automate sample log-in, instrument interfaces and result validation and approval.

5.2 Bespoke Application Layer

The ILMS bespoke application layer concentrates on migrating the capability of ChemLMS from a manual system to that of the event-driven system necessary to support automated laboratory operation and to react to external events. In addition, a number of manual facilities are provided to complement the automated components and to apply common processes to handle manual analyses.

ILMS facilities providing business-level integration with the sample scheduling system and corporate systems and operational integration with the STS, LACs and analytical instruments have been described in earlier sections. It therefore only remains to describe the less obvious features of the ILMS which are required to support automated analysis.

5.2.1 Event Management

The event management suite of processes provides the capability to handle all events which are detected by the ILMS centrally. All processes in the ILMS generate events as they occur and route them to a dedicated ILMS process which places them on an event log. Events are data driven in that their definition includes the event status (event or alarm), the users who are to be informed of the event occurrence (by electronic mail), an option to specify the name of a macro to be executed, along with run-time parameters, when the event has occurred and an indication as to whether the event is to be routed to the central error log, to computer operations staff or locally.

Other processes in the suite enable suitably privileged users to view events, clear alarms and produce event log reports. An ILMS process is automatically run each night to generate an event log report and to clear down all events, outstanding alarms being preserved for operational integrity.

5.2.2 Equipment Monitoring

Two groups of ILMS processes are provided which monitor the status of all laboratory equipment to which the ILMS interfaces. This is an important component of the automation process in that it monitors the status of robotic cells, LACs, the STS and instrument control systems, takes equipment in and out of service as the result of user requests and automatically sets the availability status of items of equipment in response to laboratory events. In this way, ILMS processes which allocate sample analysis to items of automated equipment will only select from those pieces of equipment which are both in service and available.

An important quality requirement is that only items of equipment which are not due a maintenance service are used for analytical work. To achieve this, the ILMS records the next service date for each component of an item of equipment. The ILMS will set the equipment to be unavailable when any one of its components is due a service, as part of the end of day processing. Where necessary, this status can be overridden by suitably privileged users to allow analysis to proceed on an instrument which requires a maintenance service.

5.2.3 Statistics Gathering

A business requirement of the ILMS was the ability to monitor the performance of the laboratory against regulatory targets and the availability of instruments and the conveyor and robotic systems. A suite of batch processes is provided by the ILMS which captures the information in real-time and updates such statistics overnight when the load on the system is at its lightest. Various statistical reports can be selected by authorised users.

5.2.4 Archive and Recovery

The core ChemLMS package has native facilities to archive to tape details of analytically completed samples which have been on the system for a configurable number of days and then delete them from the ILMS database as a housekeeping function. A recovery facility is also provided to restore previously archived sample information to the ILMS database.

The problem for the ILMS was that external Oracle database tables were created and ChemLMS objects used to hold additional information associated with a sample as demanded by the application. Consequently, it was necessary to develop processes which augment the ChemLMS archive and recovery functions to automatically include these tables and objects in the process.

5.3 Communications Layer

The ILMS uses a number of industry-standard communication protocols for both internal messaging and to implement interfaces to the STS, LACs, instrument control systems and external systems.

A common network management protocol used by the ILMS to communicate with instrument control systems is network file server (NFS) which permits files on a remote machine to be accessed as if they are held locally by making part of the directory tree on the remote machine appear as part of the local tree.

Communication to external systems was also achieved using Unix remote copy protocol (RCP) and Unix file transfer protocol (FTP). Both provide limited facilities for guaranteed transfer of files between two machines, access to the remote system being controlled via a user identity and password. A third protocol, trivial file transfer protocol (TFTP), was used to communicate with one type of instrument control system. Whereas RCP and FTP provide two-way communications, TFTP provides a one-way service, the instrument control system being switched between client and server mode to receive and transmit data, respectively.

Communication between the ILMS and the STS and LACs was implemented using TCP/IP sockets, a transport level service which runs over both ethernet and token ring networks. The same mechanism was used to provide internal messaging between ILMS processes to implement the event-driven responsiveness described earlier.

5.4 External Systems Layer

5.4.1 STS and LACs.

Dialogue between the ILMS and both the STS and each of the 5 LACs (one per automated laboratory area) is achieved by a strictly defined message protocol in which the ILMS is the master and the receiving system the slave. The ILMS regularly polls the slave system to establish its state of health and the slave LAC indicates in its reply the status of both itself and its components and any events of which the ILMS should be aware, such as result availability, batch cut-off requests and crate contents. The STS informs the ILMS of crate arrival and delivery events as well as its state of health.

The batch cut-off request is a particularly good example of the relationship between the ILMS and the LACs. The LAC accumulates batches of samples for preparation or analysis but does not have sufficient information to enable it to commit to processing the samples.

Therefore, it asks the ILMS whether it can proceed to process the batch of samples. The ILMS is aware of samples that are in transit to the central laboratory and their destination laboratory area. Upon receipt of the batch cut-off message from the LAC, the ILMS determines whether any further samples are destined for the LAC and will either give the LAC permission to process the batch of samples or instruct it to await further samples.

5.4.2 LSSI Gateway

The interface to the corporate systems running on the IBM ES/9000 mainframe is via a laboratory system strategy interface (LSSI) gateway resident on an IBM RS/6000 computer under the AIX operating system. The interface is used by the ILMS to transfer results to the corporate systems and to send alarms to a remote, central error log.

Result Export. The principle behind the implementation of this interface was that the RS/6000 gateway houses only the Unix operating system and strategy middleware, minimal application software being resident so that a replacement RS/6000 can be used in the event of failure. The RS/6000 gateway is transparent to the interface in that confirmation or failure of receipt of data by the corporate system is sent to the ILMS. The ILMS uses a configurable time-out value to detect failure of the corporate system to send a return code in response to a data transfer and, in such an event, stores result files locally pending recovery of the corporate system. In recovery mode, the ILMS transmits data in time sequence so that, for instance, a create transaction precedes an update transaction for the same data entity.

The ILMS issues events to its local event management suite to record transmission failures.

Central Error Logging. The ILMS is required to send alarms generated by the HP-UX operating system to a central point for operational monitoring. As a result of the open systems approach adopted by both HP and IBM, the transfer of messages to the LSSI is achieved using the same native Unix commands on both the ILMS and LSSI gateway, a configuration file on the ILMS acting as the sole filter of messages for central error logging.

5.4.3 Instrument Control Systems

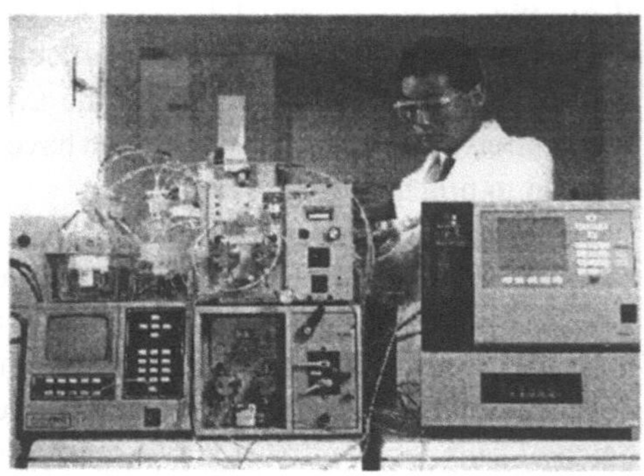

An ILMS process generates an analytical worklist to a pre-defined format specific to the instrument type and transfers the file via a script which is dedicated to the chosen communication protocol between the ILMS and the instrument control system.

The ILMS sets the status of an instrument to unavailable when a result or status file (when produced) is not received within a configurable time period. To achieve this, the ILMS starts a timer when the worklist files have been successfully transferred to the instrument control system and cancels the timer when a result file or status file is received from the instrument. If a time-out occurs then the instrument is set unavailable and an internal event is raised.

NWW analysts defined the components of the result files which had to be mapped into ChemLMS test procedures, based upon fixed-format ASCII report files, and the data manipulation rules to be applied. For instruments which generate status files, NWW analysts also advised on the critical values within the status file which should trigger the ILMS to report an internal event, such values typically relating to temperature, pressure and component status.

The directory or logical drive in which the worklist file was to be placed, the location of the result and status files on the instrument control system and file naming conventions were agreed with each instrument supplier. Instrument method files were also required to be stored on the ILMS to take advantage of the configuration management facilities of the ChemLMS package. The actual mechanism of method file transfer varied from instrument to instrument with some control systems being the active partner in transfer while others were passive, with the ILMS downloading and uploading the method files to/from specific directories or logical drives of the control system with an agreed file name format.

Specific tailoring of instrument control systems was required to align with standard operating procedures (SOPs) defined by NWW analytical staff, requiring the development of macros and scripts to support the operational requirements and to ensure guaranteed data transfer to the appropriate interface directory on the ILMS when analytical verification had been established.

5.4.4 Autofax PC

Certain, user-defined, sample types which are outside specific limits are reported immediately to a PC-based system which automatically transmits a facsimile message to the relevant party (such as a water treatment works manager) so that remedial action can be instantly taken.

5.4.5 Calculation PC

The LACs are connected to a number of instruments which provide raw readings of colour and turbidity, in milli-volts, rather than the absolute determination values in hazen and NTU units. The solution to this problem was to provide a calculation PC which applies a quadratic least squares fit to the data and calculates determination results for both standards and unknowns.

To implement the requirement, the ILMS treats the LAC as a preparation robot, maps the raw results and uses them to generate an instrument sequence file (worklist) which is sent to the calculation PC. The contents of the results file created by the PC are mapped by the ILMS against the test procedures, as if the PC was an analytical instrument.

This adaptation of the standard instrument interface mechanism emphasises the flexibility of the ILMS and the underlying ChemLMS package.

6. The Integration Process

6.1 Business Context

The developments in Laboratory Services were part of a much wider strategy to increase the efficiency and competitiveness of NWW as a business. This business strategy required an IT strategy to support it and a number of partners were contracted to implement specific components. Following selection by NWW as the partner for the new laboratory system, a jointly-staffed team prepared a detailed understanding of the scope of the project and the business benefits which it provided. The outcome of this work, a project definition, was used as a basis upon which to select the components of integration.

Constructing the ILMS as a bespoke system was rejected from the outset on the grounds that packages were available which would provide a significant proportion of the functionality which NWW required. NWW considered that the benefits of a package-based approach far outweighed those of a custom solution, provided that the package offered sufficient flexibility. NWW were provided with an extensive list of candidate packages and undertook a detailed process of selection over a period of 3 months.

NWW selected the ChemLMS product, the primary reasons being its flexibility, integration capability and extended audit trail facilities. At the same time as selecting an ILMS kernel package, NWW were identifying the equipment for the laboratory both in terms of procurement of new analytical instruments and selection of the supplier of the robotic and sample transport systems. However, the key to the success of the whole project remained the ILMS and systems integration.

Production of the ILMS for NWW was a two-phase project, the first of which was the development of the software to agreed functional and technical specifications. The second phase primarily involved driving the commissioning process but also required tailoring of the ILMS to integrate with all laboratory components and to implement agreed changes to bring the system in line with the SOPs identified by NWW analytical staff. The major stages in the production of the ILMS are summarised in Figure 13.5.

6.2 The Development Phase

The ILMS went through a traditional application build life-cycle which is shown in Figure 13.6 and both the technical definition and implementation demanded that the project team possessed a wide range of skills, from knowledge of the fundamental principles of the ChemLMS package, Oracle and Unix to experience in process automation and control.

While a description of the various stages in the development of the ILMS is not considered appropriate, a number of approaches are worthy of note.

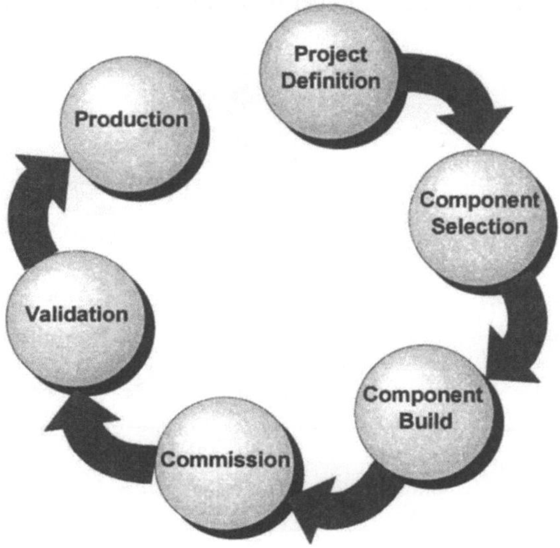

Figure 13.5: ILMS Project Cycle

6.2.1 Interface Specifications

Important areas for definition were the interfaces to the STS, LACs, instruments and the external sample scheduling system and corporate systems. The conceived approach was to write a separate specification for each interface, including each instrument type, so that the various suppliers could agree the contents of the specification and then develop or amend their systems accordingly.

6.2.2 Analytical Method Specification

To enable NWW analytical staff to develop ChemLMS analytical procedure specifications, the project team combined forces with the package supplier to provide a training course on the subject and then provided guidance in the development of test methods, procedures and reports to project standards. The underlying principle throughout the project was the transfer of knowledge to laboratory staff so that they would ultimately be self-sufficient in supporting the ILMS.

6.2.3 Design Philosophy

The design philosophy applied to the development of the ILMS was to make maximum use of the native facilities of the core ChemLMS package and to minimise development using the C programming language in order to reduce the total development effort and sustain flexibility. To reduce the memory overhead and contention for resources, the number of processes was kept to the minimum.

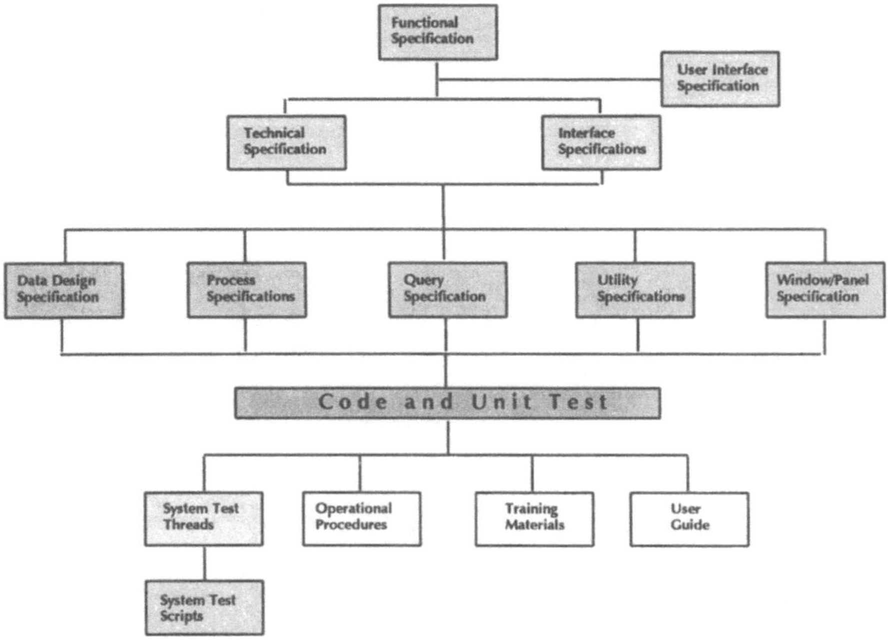

Figure 13.6: ILMS Application Build

The number of processes accessing database tables was minimised by localising data items such that they are only updated by one process which retains complete control of that data, thereby avoiding locking contention and increasing overall throughput. Most of the data was designed to be stored within the ChemLMS database to make full use of the package's security system, configuration management and archiving functions. The only data items to be stored in tables external to the ChemLMS database were those that either did not require value changes to be audit trailed or were transient and short-term.

To protect applications from the low-level operating system and database calls, specific utility functions were developed to isolate performance critical processing. Time-critical processes handling large volumes of data or communicating with external systems were designed to store data in memory to improve performance. Configuration parameters, such as instrument time-out values, were assigned to specific system global variables to enable their maintenance centrally.

6.2.4 Capacity Modelling

In most commercial projects, the sizing of the hardware is performed as a one-off exercise against known data volumes (with contingency for expansion) as part of proposal construction. This is an acceptable approach for mature packages but represents a risk when a new software product is a major component of the business solution.

In this instance, the youth of the ChemLMS package was compounded by a need to stretch the product's capabilities to sustain an automated laboratory operation, a purpose for which it was not specifically intended. The requirement emerged, therefore, to construct a model of the system's performance to ensure adequate processor power to support live data volumes and throughput requirements.

The first stage in this process was to obtain very accurate timings of the system performance by pushing high data volumes through the application and recording processor, disk and elapsed timings to obtain exact transaction throughput rates. At the same time, laboratory managers identified the required sample throughput rates on an hourly basis, since there was an uneven distribution of demand during the operating day due to the effective batching of samples during transportation.

Armed with this information, it was possible to accurately model the performance of the system. The next stage was to identify an acceptable catch-up time following system failure. The load required to recover from failure during the peak hour (the hour of most intensive system demand) was then derived and the processor power required to meet this demand calculated.

As revisions to both package and bespoke software were made to improve system performance, so timing readings were re-taken and fed into the model. As a result of the design approach to isolate database calls, it was possible to significantly improve bespoke software performance in a number of application areas by altering a single utility process.

Continued capacity modelling and the adoption of a system design philosophy which readily lends itself to adaptation are important contributors to the delivery of quality (fit for purpose) manufacturing and control systems. A full understanding of the process to be controlled must be attained before a system capable of managing that process can be designed and built. If the process is new, as in this case, then approaches can be taken, such as those cited, to deliver a system capable of meeting a variable (and potentially increasing) workload.

6.3 Commissioning Groundwork

In parallel with the latter stages of the ILMS development phase, two senior members of the team were contracted by NWW Laboratory Services to provide consultancy on the commissioning phase of the entire project, identifying the procedures and staffing requirements to carry out the commissioning of the integrated laboratory and defining the tests to be performed on the ILMS and the components of integration, including services and resources.

The basis for identifying the tests which were to be performed during the commissioning phase was the delivery of component systems which had already undergone system test, in some cases a factory acceptance test and, in all cases, a site acceptance test. It follows, therefore, that the assumption could be made that all internal conditions within the components of integration had been proven prior to the migration of the components into the commissioning phase.

Consequently, the scope of commissioning was confined to the interaction between the discrete and proven components of laboratory integration, together with the support of manual analyses. In addition to integration tests, tests were also devised to ensure that all SOPs were in place, to address NAMAS accreditation issues, and to confirm that all resources necessary to conduct the various types of sample analysis were available. In this way, the completion of commissioning, by definition, was the signal that the laboratory could smoothly progress into production.

The logical grouping of tests within commissioning threads was their relation to manual, semi-automated and fully-automated facilities in each of the laboratory areas. This approach guaranteed a degree of flexibility, as the successful completion of all tests within a thread denoted the availability of the laboratory area to migrate into production for manual, fully-automated or semi-automated analysis.

Scripts for all identified tests were produced, initially by the commissioning consultants but latterly by laboratory support staff. In all cases, the scripts were reviewed by NWW analysts who also identified the data and resource requirements of the tests which were then documented before the laboratory managers approved the scripts. Only when so approved were test scripts scheduled for execution.

6.4 The Commissioning Phase

When system test had been completed, the ILMS equipment and project team moved to the central laboratory to perform site acceptance test against specifications and scripts produced by the project team and authorised by NWW Laboratory Services. This was carried out throughout September 1993 and signed off by laboratory personnel.

This marked a significant change in the relationship with NWW from that of customer and supplier to one of partners sharing the goal of commissioning what is believed to be the most technologically advanced laboratory in the field of clean and waste water analysis. Under the newly founded agreement, the team (which included NWW staff) implemented a number of major functional changes to the ILMS to support the reality of the laboratory in late 1993, rather than the theoretical laboratory of early 1992. Changes included the integration of new analytical equipment and functional relocation from corporate systems to the sample scheduling system.

While these changes were being implemented, user documentation, including a user guide and operational procedures, was produced and laboratory personnel were provided with a training course and training material to enable them to train NWW analysts. In parallel, the team assisted NWW to define the requirements of strategy integration testing which was successfully carried out in January 1994.

Thereafter, the team concentrated on commissioning the entire laboratory complex, initially focusing on the business flow from the sample scheduling system, through the ILMS and ending with transmission of analytical results to the corporate systems. That completed, the team went on to integrate the ILMS with the LACs, STS and the automated analytical instruments, concentrating on those laboratories which would provide NWW with the greatest business benefit in reducing operating costs, namely the wastewater laboratory, the metals and autoanalysis laboratory and the sludge room. Upon completion of the commissioning of fully-automated laboratory areas, NWW staff undertook analytical validation of the robotic cells, prior to their migration into the production environment, using the ILMS to drive the process.

Commissioning highlighted the need for significant change in both functionality and mode of implementation, with the requirement for detailed planning to schedule changes to the ILMS software to meet timescales agreed with NWW. Following release of the software into the production environment, planning retained its importance to schedule releases of software to support production.

The reason for such a degree of change was that only when the instruments, automated equipment and the ILMS were resident in the laboratory could communication testing be carried out. No instrument simulators were available during the ILMS development stage, although LAC and STS simulators were used. However, the use of simulators realised little benefit.

An important concept, borne out of the ILMS system test phase and carried through to the commissioning phase was the stringent documentation of all anomalies identified in any component of laboratory integration, including its technical documentation, user documentation and the SOPs associated with the component. Such anomalies were not scheduled for implementation until signed by the designated authority and their implementation was then fed into a detailed production plan to provide an estimated delivery date.

7. The Experience

Commissioning this system, not to mention the building of the laboratory and discrete systems themselves, has been a very large, complex project in its own right. Many lessons have been learned from the experience and the following advice is offered to those brave enough to contemplate an undertaking of similar magnitude.

7.1 The Crucial Role of Systems Integration

In general, laboratories do not have the knowledge in-house and must rely on a systems integrator to advise the best testing approach, staffing structure and procedures to adopt. In examining the role of the systems integrator, the following key attributes can be identified.

Business understanding. A systems integrator who does not understand the business will not be able to deliver the business benefits nor advise on the best approach to meeting the business objectives.

Wide-ranging integration skills. The key to laboratory automation is the ability of the systems integrator to knit together all component systems of the laboratory to provide a seamless unit. Major cost savings result from a systems integrator's ability to identify the most suitable components for implementing a given functional requirement.

Proven delivery capability. It is essential to have confidence in the systems integrator's ability to deliver. Since the role of systems integrator is crucial to successful implementation, it is important to build up a trust that they have the interests of the laboratory at heart, understand the goals and are committed to realising those objectives.

Application expertise. Along with business knowledge, the systems integrator must fully understand how the operation functions and apply that knowledge to integrate the core package with the components of laboratory integration and the business.

Co-operation. It is crucial to select a systems integrator with whom laboratory staff can work in a partnership at all levels, marrying in-house analytical skills with their technological skills to jointly achieve the business objectives.

Adaptability. The systems integrator must be able to dip into the technical detail yet retain sight of the business drivers. As development moves into production, the systems integrator must show the flexibility to provide cost-effective solutions in as timely a manner as befits the business needs.

Integrity. Cultivating the open and honest relationship which is fundamental to success requires the ability to trust the systems integrator and take full advantage of their experience in the field of laboratory automation.

7.2 Management and Control

Although the systems integrator is the key to the success of the project, it is of paramount importance that laboratory management own the project and direct the decision-making process, particularly where more than one supplier is involved. Critical areas for management attention are listed below.

Planning is Crucial. First of all, establish a plan identifying the work to be done, deliverables upon which the plan is based and milestone dates. Publish the plan and get the commitment of all parties to the plan. Next, monitor progress *closely and regularly* against that plan and implement changes to the plan to meet the business objectives. Keep all parties informed of revised targets.

Retain Control. It is important that laboratory management drive the process, own the plan, direct decision-making while at the same time keeping an eye on the end objective. Be prepared to spend considerable effort on commissioning the installation and do not be over-anxious to cut corners.

Keep the Business Objectives in Mind. Remain aware of the end objectives, consider the alternatives should the current situation change and always question whether there is a better way of achieving a requirement.

Risk Management. Identify and manage the risks to the success of the project. Maintain a list of the risks and document how the risks are being addressed (managed). Always review the implications of a risk becoming reality.

Be Flexible. An essential requirement is the ability to react to changing requirements by having contingency plans in place should delivery of instruments or systems be delayed.

Change Management. No project of this scale is bereft of change. It is important to resist the need for change unless there is a strong business case for its implementation. If possible, delay change until the initial goal (a live system) has been achieved and then carefully control the implementation of change with thorough testing. Keep asking the question 'Is it needed?' and, if so, 'When?'. Be careful not to aim for perfection at the expense of practicality and (more importantly) budget.

Decision Making. All decisions, be they reached in a meeting or informally, need to be documented and counter-signed by affected parties. This reduces the risk of having to pay for changes which arise from misinformation or inconsistent implementation. Demand that suppliers document their solutions; it makes them feel more comfortable too.

Training is Vital. A good approach is to arrange for internal training staff to be trained by the systems integrator who will also be able to prepare and provide training materials. Cost savings can then be realised by in-house trainers giving courses to analytical staff. Invest in training and make it available to all staff so that they get involved in the overall commissioning scheme. Review supplied documentation to ensure it meets requirements; remember that the laboratory will have to become self-sufficient *before* the systems integrator leaves site. Staff training is an invariable requirement of any laboratory accreditation scheme and should be given the attention it rightly deserves.

Communication. The highest risk in any major project is that uncertainty and fear are bred into staff should they be kept in the dark, the one set of people ultimately responsible for the successful completion of the undertaking. Technology tends to take centre stage, key deliverables are the focus of attention and the end users tend to be forgotten. Keep them informed, tell them of the problems as well as the successes; they may have solutions. People do not like change. It makes them feel uncomfortable and threatened which is why it is important that they know what is going on, why it is being done and how they can contribute.

Acknowledgements

Figures 13.1 and 13.2, which are based on original diagrams produced by NWW Laboratory Services, and all photographs are reproduced by kind permission of North West Water Limited.

HP, T500, HP9000, HP-UX and ChemLMS are trademarks of the Hewlett-Packard Company.

RS/6000 and AIX are trademarks of the IBM Corporation.

Oracle RDMS is a trademark of the Oracle Corporation.

References

[1]Linder M. *Laboratory Automation and Robotics - Quo Vadis?* Scientific Computing and Automation (Europe) 1990. Elsevier Science Publishers B.V., Amsterdam.

[2]Mattes D.C. and McDowall R.D. *A Universal LIMS Architecture*. Scientific Computing and Automation (Europe) 1990. Elsevier Science Publishers B.V., Amsterdam.

16 The Generalized Network Model : Algorithms and Application to Manufacturing Operations

Spyros Tzafestas and George Kapsiotis

1 Introduction

The growth of computer industry has had a profound influence on many areas; Management and Manufacturing Sciences are undoubtedly two of them. The exploitation of the available computational power of the new technology has rendered possible the development and solution of realistic models, which capture many of the intricacies and take into account the plethora of data involved in most of the real-life problems. *Network programming* is one of the most extensively studied areas, and there has been a lot of work carried out on network formulation and computer implementation techniques. The major reasons that network modelling is such a popular tool among decision-makers are: i) many of the problems coming up in real-life applications have a network structure; the shortest path, assignment, scheduling, transportation and transshipment problems are the most common ones, ii) the pictorial nature of the model permits the easy statement of the problem, the conceptualzation, interpretation and verification of it and of the optimal solution, and also, the communication of relating ideas to non-scientific staff, iii) finally, the enhancement of the *pure* network model to the so-called *Generalized Network (GN)* model makes the model applicable to a great variety of problems ranging from integrated production and distribution planning to such exotic areas like file reduction and plastic-limits analysis.

This chapter covers the *GN optimal flow* (otherwise *Transshipment)* problem; specifically it gives: i) the mathematical and pictorial formulation of it, ii) a survey of the major algorithms developed for the solution of the problem by modern computer technology, iii) areas of application of the model, and finally iv) a new decision support tool for the solution and analysis of the problem.

2 Network formulation and Mathematical Problem Statement

According to Jewell, the **GN optimal flow problem** can be stated as follows:
"A network is constructed of oriented branches that can pass a limited amount of flow from one node to another; for every branch-transfer the outcoming flow is the incoming flow modified by a branch-gain. However, flow conservation laws are assumed to hold at all nodes, except where some required amount of flow is supplied to and withdrawn from the network. Given that the cost (or disutility) of flow in each branch is directly proportional to the amount of flow, what flow pattern will satisfy restrictions at minimum total cost?"

A **network** is defined by a set V of **vertices/nodes** i and a set B of **directed branches** (i,j) which is a subset of V¥V. With every branch we associate a nonnegative **flow** X_{ij} running through it, the **accepted interval** (L_{ij},M_{ij}) for the flow values, the **cost** C_{ij} for every unit of flow transfered through the branch, and a **gain/multiplier** K_{ij} which modifies the incoming flow X_{ij} to the outgoing $K_{ij}\cdot X_{ij}$. Nodes are classified as i) **supply nodes**, where positive flow S_i is inputed into the network, ii) **demand nodes**, where flow D_i leaves the network, and iii) **transshipment nodes**, where branch-flow running into the node balances the branch-flow running out of it.

A pictorial representation of the above notions has been established in order to facilitate network modelling, the **NETFORM** presentation. A typical branch in this representation looks like that of Fig. 1. In this figure, the representation of nodal supply and demand are also given.

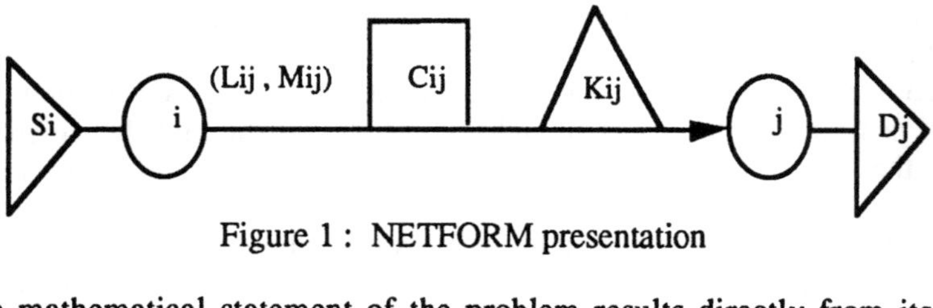

Figure 1 : NETFORM presentation

The mathematical statement of the problem results directly from its verbal expression and the parameter definitions given above. It reduces to the following *Linear Programming (LP)* problem:

$$\min c = \Sigma_{(i,j)\in B} C_{ij}X_{ij} \quad (1a)$$

$$\text{sbj. to } \Sigma_{j,(i,j)\in B}(X_{ij}-K_{ji}X_{ji}) = S_i-D_i \text{ , } i\in V \quad (1b)$$

$$L_{ij}\leq X_{ij}\leq M_{ij} \text{ , } (i,j)\in B \quad (1c)$$

The A-matrix of this linear problem has the special property that at most only two entries in each column are non-zero; in fact, this remark constitutes the principal criterion for the identification of a GN structure. By scaling and/or complementing to variables' upper bounds the matrix and right-hand-side coefficients, it is always possible to make one of the non-zero entries of each column equal to -1. Taking the corresponding node to be the branch start, the other coefficient is equal to the branch-gain. Usually, the constraint $L_{ij} \leq X_{ij}$ in formulation (1) is handled by setting $X'_{ij}=X_{ij}-L_{ij}$ and transforming problem (1) to another one in variables X'_{ij}, $0 \leq X'_{ij} \leq M_{ij}-L_{ij}$ and with the right-hand-side vector appropriately modified. In what follows, we assume that this transformation has taken place although we use unprimed symbols.

3 Solution Algorithms

Since problem (1) is a LP program, it can be treated by the Simplex method in any of its variations. This method is established by *the fundamental theorem of Linear Programming*, and constitutes a systematic searching among the set of *basic solutions* (i.e. solutions in which the number of variables with a non-zero value is equal to the rank of the A-matrix) for the optimal one (i.e. the one that satisfies (1a)). However, the special structure of the A-matrix gives rise to the following theorem concerning the basis topology:

Theorem: *Any subset B_b of B spanning all vertices V constitutes a basis if and only if it consists of a number of components of connectivity with each component having only one loop in it and the loop-gain is different from unity.*

The **loop-gain** is related to a direction of transversion and it is defined as the product of gains of the branches transversed in the direct sense divided by the product of gains of the branches transversed in the reverse sense. A loop with gains equal to unity is said to be *degenerate*. A pictorial representation of the basis topology is given in Fig. 2.

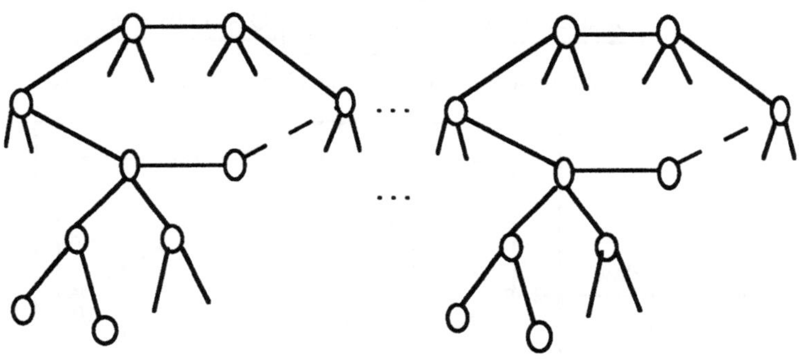

Figure 2 : GN Basis topology

The conceptualization of the problem structure and its primary features has led to the development of specialized algorithms for its solution. According to these approaches, data structures are suggested by the GN concepts and underlying topologies, while computations make explicit use of these concepts and are related directly to their schemes. Pure and applied computer science and software engineering have played a critical role to the success of this effort since they have offered the algorithmic and programming tools necessary for the efficient development and implementation of these algorithms; linked lists and tree representation and processing algorithms, as well as recursion and dynamic memory allocation are some of the most prominent features of the algorithms under consideration. As a result, *storage space economy* (by circumvention of the A-matrix and its sparsity, and also, by dynamic memory allocation), and *improved execution rates* (by exploitation of the independence of the basic connectivity components) have been achieved.

All the same, eventhough the developed algorithms are tailored to the GN structure, they depart from the Simplex method and make extensive use of the relevant theory for their validation. Thus, they can be classified as *Primal, Dual,* and *Primal-Dual*, in a similar fashion. Primal and Primal-Dual algorithms appear to be the most efficient ones, while it is established that duality and dual methods are most appropriate when it comes to post-optimality aspects. During the past thirty years, a variety of algorithms belonging to the above three categories has been developed. Some of them treat the problem to its general formulation, while some others are adapted for the solution of specific GN configurations. In [1],[8-9],[11-12],[15],[18] citation of the algorithms and codes developed in the literature is given. In what follows, a Primal-Dual algorithm developed by Jewell is firstly presented, while in continuation, a dual algorithm developed by the authors is discussed; finally, a more recent approach, which has been developed by Bertsekas and Tseng at MIT and is classified as a *relaxation method*, is briefly discribed. The overview is not intented to be complete by no means since the related literature is vast. It is rather intended to underline the different approaches that have been used to tackle the problem.

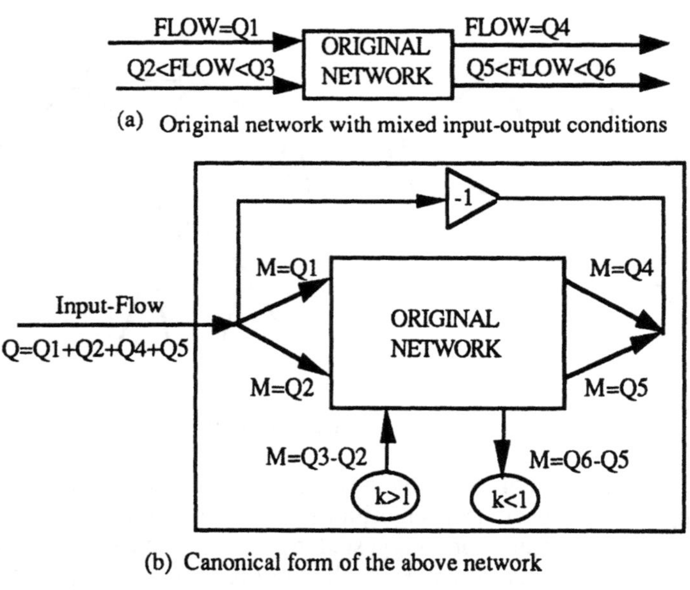

(b) Canonical form of the above network

Figure 3 : Reformulation of the network into the canonical form

3.1 Jewell's algorithm

In relation to the problem (1), which is characterized as the **Primal** problem, the **Dual** problem is defined, in variables Uij, Vi (corresponding to the network branches and nodes) as follows:

$$\max\ c = \Sigma_{i\in V}(S_i - D_i)Vi - \Sigma_{(i,j)\in B} M_{ij}U_{ij}$$
$$\text{sbj. to}\quad V_i - K_{ij}V_j - U_{ij} \le C_{ij} \qquad (2)$$
$$U_{ij} \ge 0$$
$$V_i \text{ unrestricted}$$

To solve the problem by the algorithm considered, it must be transformed into the **canonical form.** The transformation is a two-step process: first, the least-flow constraints are incorporated into the nodal boundary conditions, and in continuation, artificial nodes and branches appropriately capacitated are added, so that there remains only one boundary node (condition). The transformation technique is presented in Fig. 3.

The transformed LP is:

$$\min\quad c = \Sigma_{(i,j)\in B}\, C_{ij}X_{ij}$$
$$\text{sbj. to}\quad \Sigma_{j\in V}\,(X_{ij} - K_{ji}X_{ji}) = \begin{cases} Q, & i=0 \\ 0, & i\neq 0 \end{cases} \qquad (3)$$
$$O \le X_{ij} \le M'_{ij}$$

Generalizing the Ford and Fulkerson techniques for pure networks, a **restricted Primal** problem is formulated and solved at every cycle of the algorithm. Specifically, by stipulating non-negative costs, the algorithm is initiated with a zero dual feasible solution and zero network-flow; in this way, all of the constraints in (3) are satisfied except for the balance equation at node 0. The solution of the restricted Primal problem at every iteration results either in the reduction of the non-feasibility of the running network-flow, or in the resolution of the problem's infeasibility.

The formulation of the restricted Primal has as follows:

$$\max\ F_o$$
$$\text{sbj. to}\quad \Sigma_{j\in V}\,(X_{ij} - K_{ji}X_{ji}) = \begin{cases} Fo, & i=0 \\ 0, & i\neq 0 \end{cases} \qquad (4)$$
$$0 \le X_{ij} \le M'_{ij} \quad \text{for} \quad V_i - K_{ij}V_j = C_{ij}$$
$$0 = X_{ij} \quad \text{for} \quad V_i - K_{ij}V_j < C_{ij}$$
$$X_{ij} = M'_{ij} \quad \text{for} \quad V_i - K_{ij}V_j > C_{ij}$$

It is obvious that (4) constitutes a *maximal-flow* problem for an appropriately specified part of the original network. To solve this problem, the notions of the *generating* and *absorbing* loops are exploited; it can be shown that a loop with loop-gain greater than unity can operate as a source of flow for the entire network, while a loop with loop-gain less than unity operates as a sink. Thus, the maximal-flow routine reduces to the identification of the absorbing structures of the restricted network, and the determination of the maximal-flow they can absorb, in an incremental mode. This is done through a labeling and relabeling procedure which uses the *spanning-tree* concept.

At every iteration, once the restricted Primal has been solved without reaching optimality, an improved dual solution is computed, which in turn leads to a new restricted Primal. As the inspirator of the algorithm remarks, its efficiency is partly due to the fact that most of the network-type problems are loosely constrained, so that cost consideration (duality concepts) while trying to establish a feasible network-flow, is the dominant factor to obtain optimality. An outline of the algorithm is given in the flowchart of Fig. 4.

3.2 A Dual Algorithm

The relation of the primal and dual problem formulations, presented in (1) and (2) respectively, is given by the *Duality theorem* which states that *if there exists a finite optimum for one of the problems (1) and (2), then there exists a finite optimum for the other, and these two optima are equal; if one of them is found to have an unbounded objective, then the other is infeasible*. Moreover, if **X** and **(V,U)** are the primal and dual solutions (in case of bounded problems), the *weak theorem of Complementary Slackness* demands that:

$$
\begin{aligned}
&V_i-K_{ij}V_j<C_{ij}\ (U_{ij}=0) \Rightarrow X_{ij}=0 && \text{(a)}\\
&0<X_{ij}<M_{ij}\ \text{fi}\ V_i-K_{ij}V_j=C_{ij} \text{ and } U_{ij}=0 && \text{(b)}\\
&U_{ij}>0\ (V_i-K_{ij}V_j=C_{ij}+U_{ij}) \Rightarrow X_{ij}=M_{ij} && \text{(c)}
\end{aligned}
\qquad (5)
$$

A basic dual feasible solution is a set of V_i, U_{ij} for which there are exactly $|V|$ (V's cardinal number) branches satisfying the relation:

$$V_i-K_{ij}V_j=C_{ij}\ ,\ U_{ij}=0 \qquad (6)$$

Given a basic dual feasible solution, the set of reals X_{ij} satisfying (5a,c) and the continuity principle (1b) is called the **network pseudoflow** related to that basic dual feasible solution. Furthermore, **branch coflow** δ_{ij} is defined to be the quantity:

$$\delta_{ij}=V_i-K_{ij}V_j-C_{ij} \qquad (7)$$

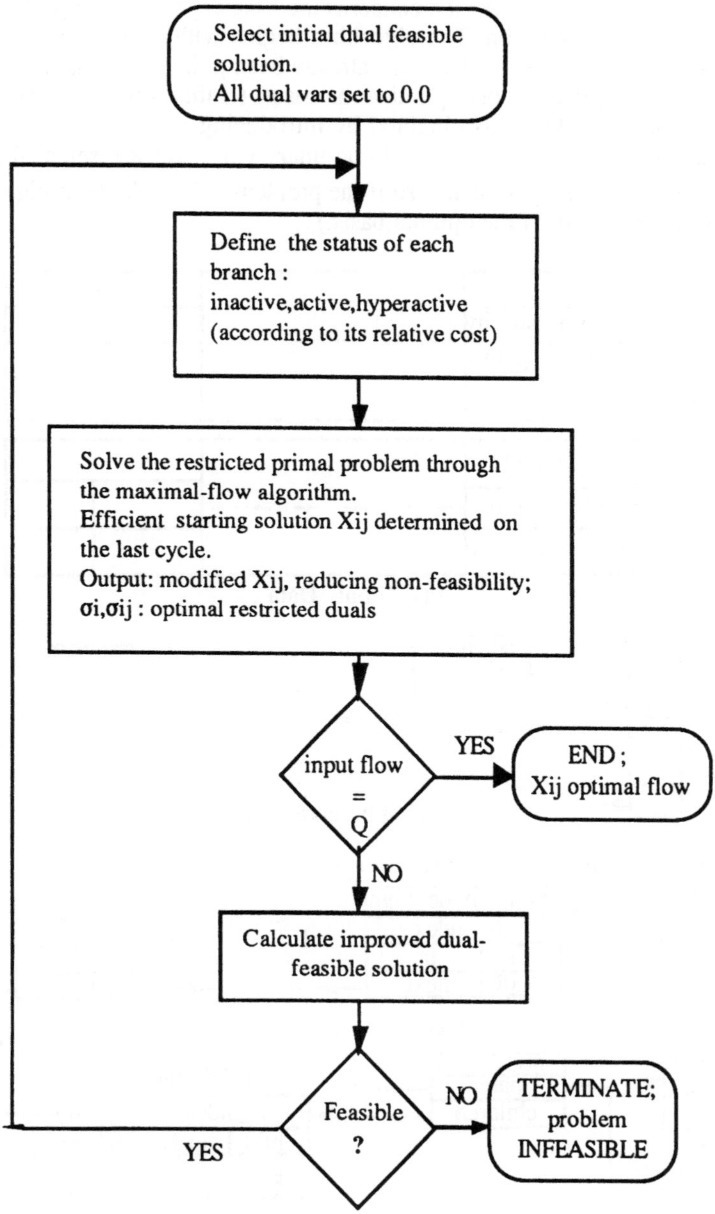

Figure 4 : Jewell's algorithm flowchart

The data structures that are used in the algorithm are given in Fig. 5, and stem from the basis topology theorem and network concepts. Data that describe the network topology is static, while that which concerns the basis representation is dynamically reorganized in each iteration. The algorithm begins with a dual feasible solution which is assumed to be available; it has already been pointed out that the method is efficient when appplied to post-optimization cases. (If this is not the case, an initial basic dual feasible solution is obtained by introducing artificial loop-branches at every node, with a very high cost and non-unary gain. V_i's are selected such that loop-branch coflows are priced at zero.If the problem is feasible, their high cost will keep these branches out of the optimal basis.)

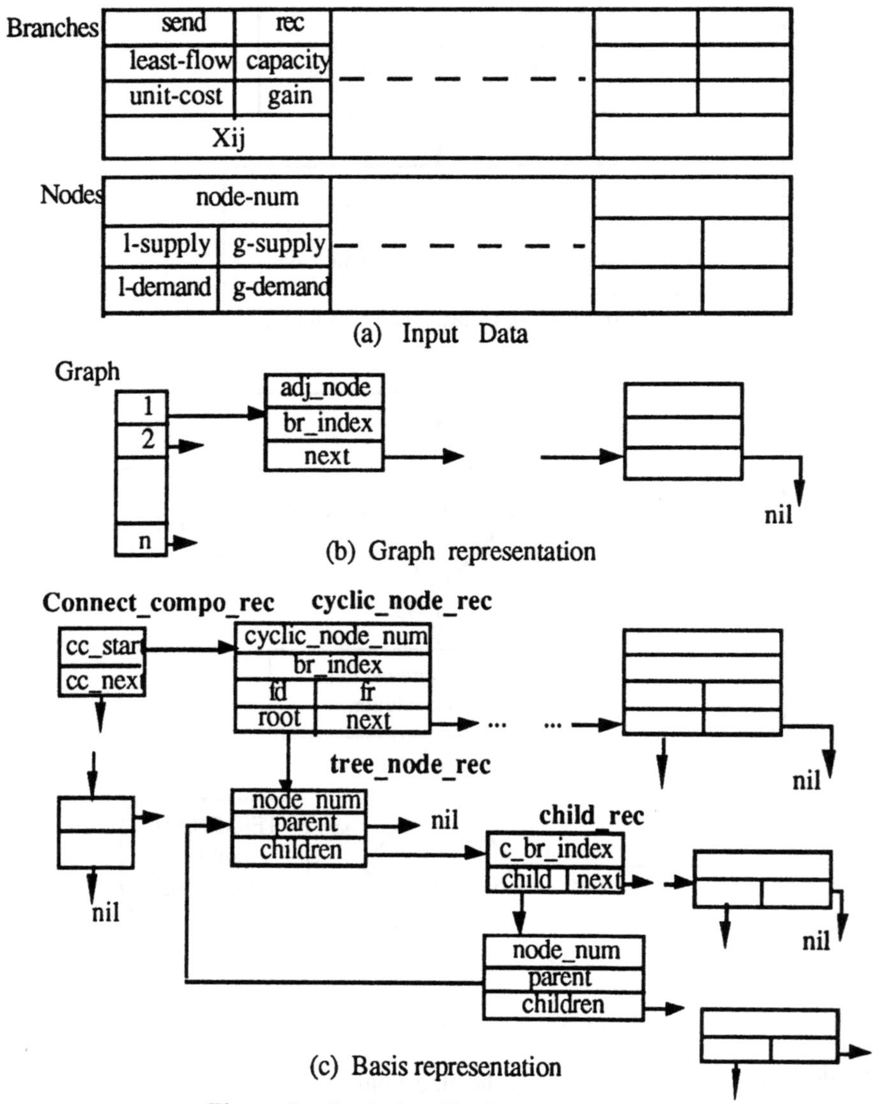

Figure 5 : Dual algorithm's data structures

Once an initial basic dual feasible solution is available, pseudoflows of the network branches are calculated by making use of the complementary slackness.

Specifically, branches with positive coflow are assigned a pseudoflow equal to their capacity, while those with negative coflow are set to zero. The pseudoflows of the remaining basic branches are calculated such that they satisfy the continuity principle. The basis topology allows for an efficient organization of the computations involved: in each basic connectivity component, the calculation of the branch pseudoflows of every basic subtree hanging from a cyclic node can be performed starting from its end-nodes and progressing to the root, demanding that the continuity principle is satisfied at every visited node; the application of the same principle at all of the loop-nodes of the component leads to a well-defined system of equations, in unknowns the loop-branch pseudoflows. This procedure, combined with the non-degeneracy of the basic loops, implies that for a given basic dual feasible solution, the set of related pseudoflows is single-valued. If the resulting basic flows are found to be within their boundaries, the Duality Theorem guarantees the optimality of the solution, and the algorithm terminates. Otherwise, a new basis has to be produced.

The branch (i_o,j_o) to leave the basis is the one that violates most the capacity constraints. The determination of the branch to enter the basis relies on a mechanism that ensures dual feasibility of the resulting basis. Initially, the non-basic branch coflow variations to a hypothetical *unary* variation of the coflow of the outgoing branch are calculated, with the rest of the basic branch-coflows remaining at zero. Taking into consideration the relation (7), $\Delta\delta_{ij}=\Delta V_i-K_{ij}\Delta V_j$, and the calculation previously mentioned, the problem reduces to the solution of the system of equations:

$$\begin{aligned} &\Delta V_i-K_{ij}\Delta V_j = 0 \qquad\qquad (i,j)\in B_b-\{(i_o,j_o)\} \\ &\Delta V_{i(o)}-K_{i(o)j(o)}\Delta V_{j(o)}=1 \end{aligned} \tag{8}$$

in ΔV_i. It is a well-defined system, and the basic property of the branches involved implies the uniqueness of its solution. A pattern similar to that described for the computation of the pseudoflows can be used for a fast solution of this system. Afterwards, the maximal variation that sets one (or more in case of degeneracy) non-basic branch coflow to zero, leaving the others at their previous signs, is performed. The branch whose coflow is set to zero, enters the basis. Dual variables are assigned their new values, and data structures storing basis information are appropriately updated.

The efficiency and promptness of the algorithm results from the exploitation of the fact that only the connectivity components where the outgoing and entering branches belong to, are actively involved in the required computations and basis updating. At that point, a new iteration begins. A flowchart of the algorithm is given in Fig. 6. It can be shown that the dual objective function increases at every iteration of the algorithm (or it remains constant, in case of degeneracy). So, convergence to optimality is achieved in a finite number of iterations. Finishing the presentation of the dual method, it is worth-mentioning that by making use of the same concepts and data structures, the *Primal* Simplex method has been tailored, in a similar fashion, for the efficient solution of the GN optimal flow problem.

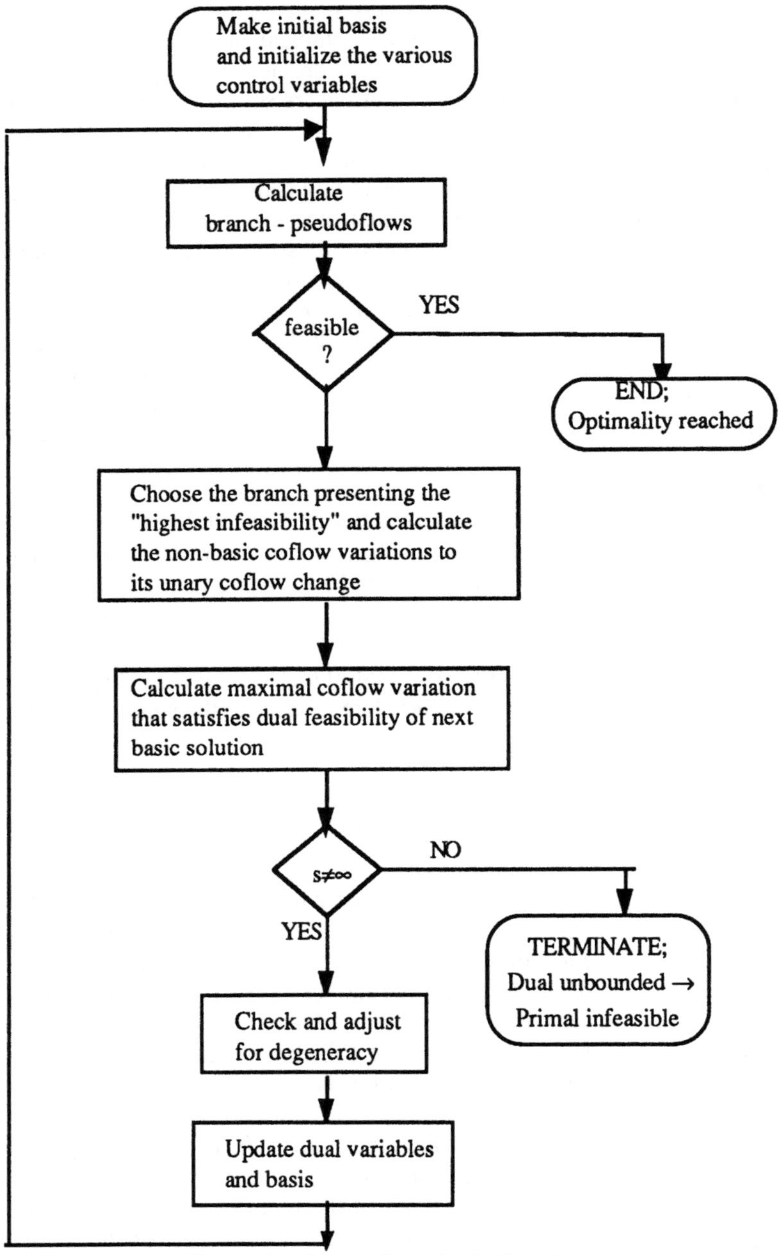

Figure 6 : Dual Algorithm's flowchart

Example. As an illustrating example consider the network of Fig. 7a. The network parameters are given in Table 1. Actually, this network constitutes the canonical form transformation of the network of Fig. 7c, given in NETFORM presentation. In Fig. 7a the numbers next to the nodes are the nodal prices in the initial dual basic solution, while the pairs along each branch are the branch coflow and pseudoflow, in that order. By cross-referencing Table 1, it can be found that the branch-pseudoflow with the greatest distance from its legitimate interval of values is that of branch (0,3). By performing the calculations suggested by the algorithm, the dual unary variation ($\Delta\mathbf{V}, \Delta\boldsymbol{\delta}$) is extracted and given in Tables (2.1, 2.2). The actual step is found to be s=+1.0, resulting to the new nodal prices **V'** of table 2.3 and the branch coflows $\boldsymbol{\delta}'$ of Table 2.4. From this Table it is gathered that the new branch to enter the basis is (0,1). Attention should also be paid to the fact that the non-zero entries of the variation vector occur only for the nodal prices of the subtree hanging from the branch to leave the basis and for the coflows of the non-basic branches incident to those nodes; this is consistent with the remark made at the last paragraph of the previous section. At this point the first iteration is completed. The algorithm reaches optimality after four iterations; the primal and dual optimal solutions are given in Fig. 7b. It is worth-noting that the rank of basis connectivity has increased from 1 to 2.

Table 1: Canonical form data

s/n	Branch	least-flow	capacity	gain	unit cost
1	(1,2)	0	∞	1.2	5
2	(1,3)	0	12	0.9	2
3	(4,3)	0	∞	1.0	8
4	(4,5)	0	15	1.0	5
5	(2,3)	0	∞	1.0	20
6	(2,6)	0	15	1.5	8
7	(2,5)	0	∞	1.0	6
8	(3,6)	0	15	1.0	10
9	(3,5)	0	∞	1.0	7
10	(3,7)	0	10	1.0	2
11	(8,9)	0	10	1.0	1
12	(0,1)	0	12	1.0	0
13	(2,10)	0	5	1.0	0
14	(0,3)	0	7	1.0	0
15	(0,4)	0	26	1.0	0
16	(5,10)	0	19	1.0	0
17	(6,6)	0	2	1.0	0
18	(7,10)	0	7	1.0	0
19	(8,8)	0	1	11.0	0
20	(9,10)	0	5	1.0	0
21	(0,10)	0	∞	-1.0	0

Table 2.1

node	0	1	2	3	4	5	6	7	8	9	10
ΔV_i	1	0	0	0	0	0	0	0	0.8	-1	-1

Table 2.2

branch	1	2	3	4	5	6	7	8	9	10	11	12	13
$\Delta\delta_{ij}$	0	0	0	0	0	0	0	0	0	0	0	1	1

branch	14	15	16	17	18	19	20	21
$\Delta\delta_{ij}$	0	1	1	0	1	8	0	0

Table 2.3

node	0	1	2	3	4	5	6	7	8	9	10
V'_i	11	11	5	10	18	3	0	8	-7.8	-11	-11

Table 2.4

branch	1	2	3	4	5	6	7	8	9	10	11	12
δ'_{ij}	0	0	0	10	-25	-3	-4	0	0	0	0	0

branch	13	14	15	16	17	18	19	20	21
δ'_{ij}	16	1	-7	14	0	19	78	0	0

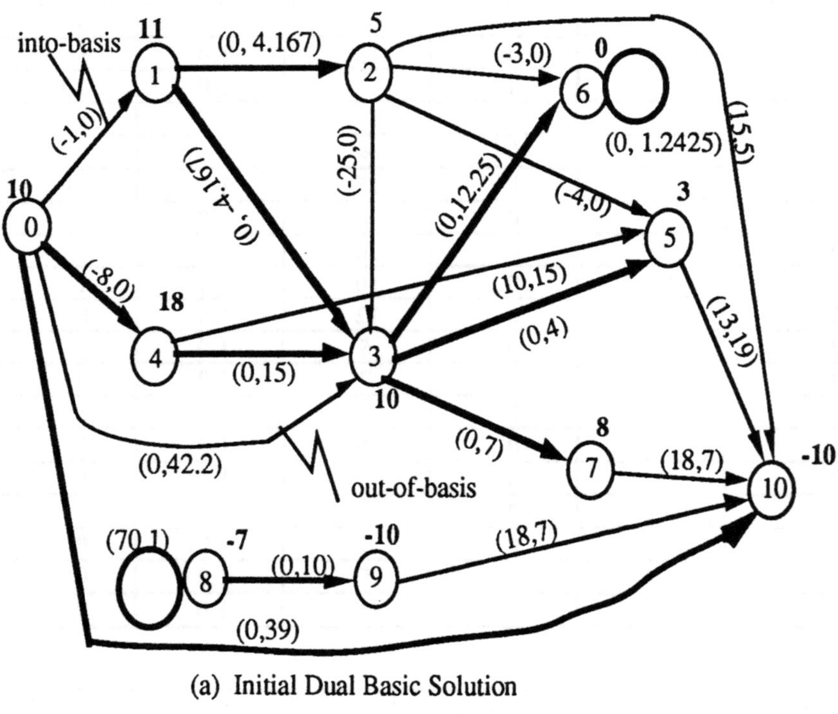

(a) Initial Dual Basic Solution

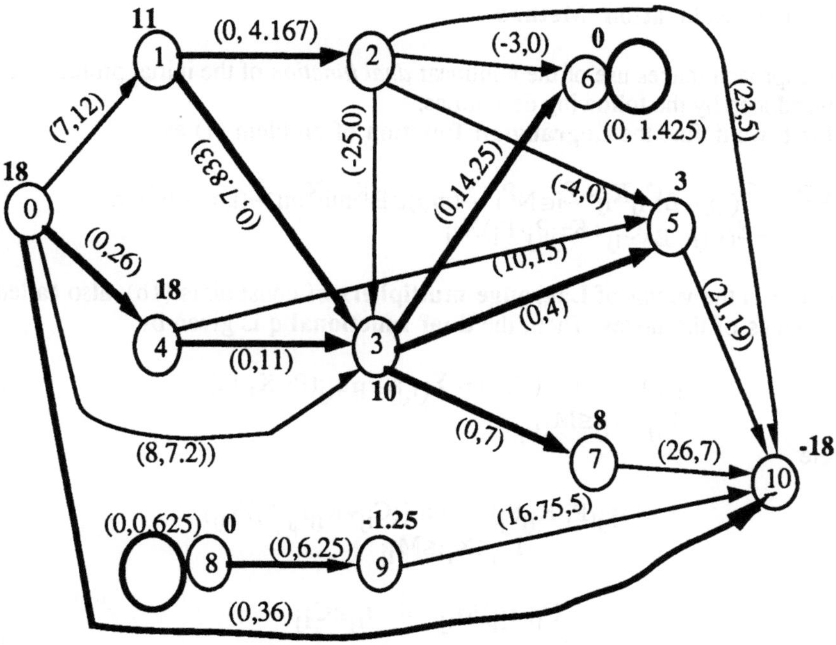

(b) Optimal Solution (Primal/Dual)

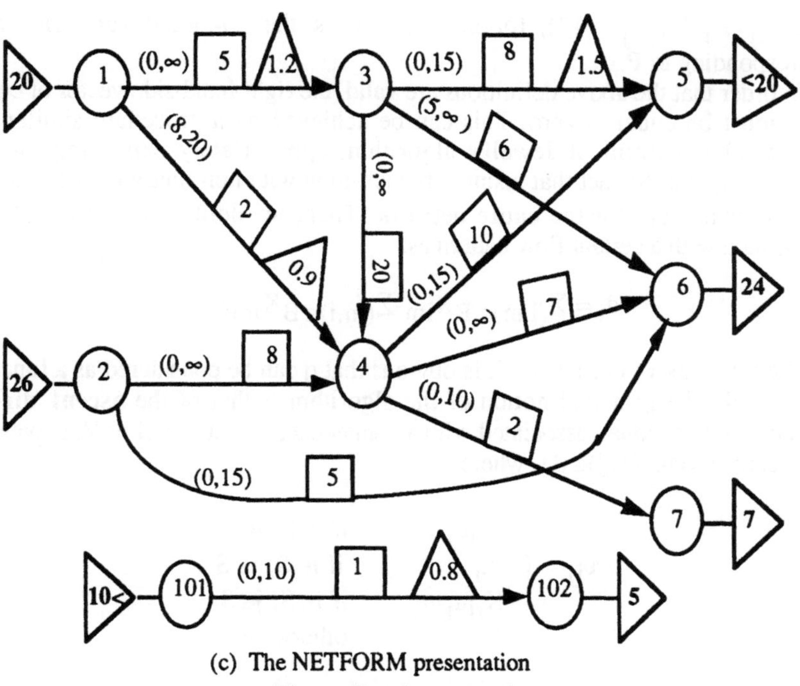

(c) The NETFORM presentation

Figure 7 : An example

3.3 The Relaxation Method

This approach makes use of the nonlinear *dual function* of the initial problem. This is introduced by the following definitions:

First, we define the **Lagrangian function** of problem (1) as:

$$L(\mathbf{X},\mathbf{P}) = \Sigma_{(i,j)\in B}C_{ij}X_{ij}+\Sigma_{i\in N}P_i\cdot(\Sigma_{(m,i)\in B}K_{mi}X_{mi}-\Sigma_{(i,m)\in B}X_{im})$$
$$=\Sigma_{(i,j)\in B}(C_{ij}+K_{ij}P_j-P_i)X_{ij} \qquad (9)$$

where **P** is the vector of **Lagrange multipliers** of constraints (1b), also called the **prices** of the nodes. Then, the **dual functional** q is given by

$$q(\mathbf{P}) = \min_{L_{ij}\leq X_{ij}\leq M_{ij}} L(\mathbf{X},\mathbf{P}) = \Sigma_{(i,j)\in B}q_{ij}(P_i-K_{ij}P_j) \qquad (10)$$

where

$$q_{ij}(P_i-K_{ij}P_j) = \min_{L_{ij}\leq X_{ij}\leq M_{ij}} \{(C_{ij}+K_{ij}P_j-P_i)X_{ij}\}$$

$$= \begin{cases} C_{ij}-t_{ij})M_{ij} & \text{if } t_{ij}\geq C_{ij} \\ C_{ij}-t_{ij})L_{ij} & \text{if } t_{ij}\leq C_{ij} \end{cases} \qquad (11)$$

and $t_{ij}=P_i-K_{ij}P_j$ (12), for all (i,j)∈ B, is the so-called **tension vector** corresponding to **P**.

In order that the above definitions are valid, the right-hand-side vector of equation (1b) must be equal to zero. This can be achieved by a procedure similar to the canonical transform of Jewell's algorithm, appropriately exploiting the lower bounds L_{ij} and the fact that a single branch-loop with non-unary gain operates as a source or a drain for the entire network. Then, we define the **deficit** of node i associated with a certain flow pattern as:

$$d_i = \Sigma_{(i,m)\in B}X_{im}-\Sigma_{(m,i)\in B}K_{mi}X_{mi} \qquad (12)$$

From the above definitions it is obvious that q can be considered as a function in **t**, as well. The principal notion of the algorithm is that of the **ascent direction** of the dual functional, associated with a connected strict subset S of V, expressed by the vector $\boldsymbol{\upsilon}=\{\upsilon_{ij} / (i,j)\in B\}$, where:

$$\upsilon_{ij} = \begin{cases} K_{ij}u_j & \text{if } i\notin S,\ j\in S \\ -u_i & \text{if } i\in S,\ j\notin S \\ K_{ij}u_j-u_i & \text{if } i\in S,\ j\in S \\ 0 & \text{otherwise} \end{cases} \qquad (13)$$

In relation (13), u_i's are a set of positive numbers which are calculated for all nodes of subset S, with the aid of a spanning tree T of it, such that:

$$u_i - K_{ij} u_j = 0 \qquad \text{for all } (i,j) \in T \tag{14}$$

The **directional derivative** of the dual function at point **t**, in direction **v**, is:

$$\begin{aligned} C(\mathbf{v},\mathbf{t}) &= \Sigma_{(i,j)\in B} \lim_{\alpha \to 0+} [q_{ij}(t_{ij}+\alpha v_{ij}) - q_{ij}(t_{ij})]/\alpha \\ &= \Sigma_{i\in S} u_i d_i - \Sigma_{\substack{i\in S, j\notin S \\ (i,j):\ \text{active}}} (X_{ij}-L_{ij}) u_i \\ &\quad - \Sigma_{\substack{i\notin S, j\in S \\ (i,j):\ \text{active}}} (M_{ij}-X_{ij}) K_{ij} u_j \\ &\quad - \Sigma_{\substack{i\in S, j\in S \\ (i,j):\ \text{active},\ u_i > K_{ij}u_j}} (X_{ij}-L_{ij})(u_i - K_{ij}u_j) \\ &\quad - \Sigma_{\substack{i\in S, j\in S \\ (i,j):\ \text{active},\ u_i < K_{ij}u_j}} (M_{ij}-X_{ij})(K_{ij}u_j - u_i) \end{aligned} \tag{15}$$

The *weak duality theorem of Lagrangian Relaxation* postulates that at optimality

$$q(\mathbf{P_0}) = \max_{\mathbf{P}} q(\mathbf{P}) = c \tag{16}$$

The algorithm initiates with a set of dual prices **P** and flows **X** satisfying complementary slackness, but not the constraints (1b). At every iteration the algorithm proceeds towards optimality either by a *redistribution of flow* which reduces nodal deficits, or by a *revision of the nodal prices*, so that the dual functional increases, whichever comes first. Specifically, at every iteration a non-balanced node is spotted, and in continuation, a scanning mechanism is put to work in order to determine one of the above-mentioned possibilities of improvement. A deficit reduction can be achieved any time a route to a node with deficit of opposite sign to that of the set-out node is established or an appropriate cyclic structure is developed; a dual functional increase is possible if the directional derivative associated with the scanned subset of nodes S is positive. It must be mentioned that primal-dual algorithms operate in a similar mode but the direction of correction is that of the maximal rate of ascent. However, according to the algorithm's inspirators, its efficiency stems from the fact that in many iterations the direction subset S may consist of the initial node only, which is attributed by the characterization *coordinate ascent algorithm*. In that case, it is shown that the negative of the nodal deficit, $-d_s$, is a *subgradient* of the dual functional at $\mathbf{P_s}$ in the sth coordinate direction. A final remark concerning the algorithm's convergence, is that the algorithm is proven to be finite if *ε-optimal* solutions are searched, with $\hat{A}$ defined arbitrarily small. At every iteration, vectors **P**, **X** must satisfy a modified version of complementary slackness, the so-called *Â-complementary slackness*, defined by

$$\begin{array}{ll} \varepsilon\text{-inactive} & \text{if } t_{ij}<C_{ij}-\varepsilon \\ \varepsilon\text{-active} & \text{if } C_{ij}-\hat{A}\leq t_{ij}\leq C_{ij}+\varepsilon \\ \varepsilon\text{-hyperactive} & \text{if } C_{ij}+\varepsilon<t_{ij} \end{array} \qquad (17)$$

Then, the produced solution will be within a radius of $\varepsilon\cdot\Sigma_{(i,j)\in B}(M_{ij}-L_{ij})$ from the optimal.

RELAX-II, an implementation code of the algorithm, is reported to be an order of magnitude faster than the latest state-of-the-art codes, which implement Primal and Primal-Dual methods.

4 Applications

Generalized networks can be used to model numerous problems for which there are no pure network equivalents. Essentially there are two ways in which GN multipliers can function: they can simply act to modify the amount of flow traversing the branch, representing for example phenomena of evaporation, seepage, deterioration, breeding, etc. or they can transform flow from one type to another, modelling in this way processes of manufacturing, production, conversion of fuel to energy, etc. ([19]). The principal applications of GNs in various areas of Manufacturing and Operational Research (O.R.) are the following: ([5])

4.1 Scheduling and Planning

Scheduling and Planning models usually require several time periods for their basic structure. Observing that they possess a network structure, one can see that the GN model has a significant impact on their solution. For one thing, large-scale network applications can be solved easily, and the level of detail provided to the model can be expanded considerably. Secondly, planning tools with quite fast response time may be designed, so that they are executed interactively. Representative areas of this category where considerable work has been done are *Hydroelectric Scheduling* and *Air Traffic Control.*

4.2 Finance

Mathematical planning models are an established part of financial analysis. Work in this area has begun since 1950. Several authors have proposed network models in which branch multipliers play a primary role. Some suggest that multipliers can be used for translating currencies across countries (e.g. dollars to Deutsch marks); others have designed multiperiod models in which interest, dividends and loans are modeled by means of multipliers. *Cash flow management* provides an ideal application for this methodology.

4.3 Equilibrium Models

There are many instances of such models; some of them are summarized below.

- *Traffic equilibrium*, in which case branches represent transportation paths (e.g. highways) and nodes are the connections between paths.
- *Market equilibrium*, with nodes indicating spatially separated markets for products, and branches indicating the interdependence between them.
- *Demographic equilibrium*, with nodes indicating distinct demographic zones, and branches indicating possible migration patterns among the zones.
- *Water/gas pipeline distribution systems*, with branches representing the pipes, and nodes indicating the connection points.

4.4 Statistics and Large Data Bases

The timely collection of very large data bases has become increasingly important over the past two decades. Many government agencies and private companies routinely depend upon these files for maintaining their operations. One important class of data bases is known as *microdata* whereby the file consists of a large number of individual decision units - individuals, families, corporations, etc. Typically, microdata files range in size from one thousand to over one million observations. Once collected, the raw data must be processed before they can be served up to the computer user. There are numerous steps involved, several of which employ network optimization. Representative example of this category is *the estimation of Social Accounting Matrices* and *File Merging (SAM).*

4.5 Manufacturing and Logistics

Models of integrated production, inventory and distribution often are based on a network form. The flow of materials through the factory, and the transport of the final product through distribution centers onto the market present an appealing case where a large part of the problem can be depicted as a graph. The commensurate gain in efficiency allows the modeller room for including and modelling features such as more time periods, disaggregate customer zones, additional potential sites for warehouses, or even a more realistic cost structure.

Of course, there are more active areas of research in Engineering Design and O.R., in which GNs play an important role for problem solvability. An extented version of the GN's, the **0-1 GN problem**, broadens the scope of the problems to be treated so that classical *Integer Programming (IP)* and *Mixed IP (MIP)* problems are included. According to this formulation, certain branch-flows are obliged to take the values of only 1 and 0, and the incorporation in the network of structures similar to that of Fig. 8, can lead to the integrality of all of the variables, while only a small number of them are explicitly stated to be of integer type. In case that a *Branch and Bound* method is used for the solution of the problem, this fact results to a considerable reduction of the underlying *enumeration tree* structure and thus, to very fast solution times.

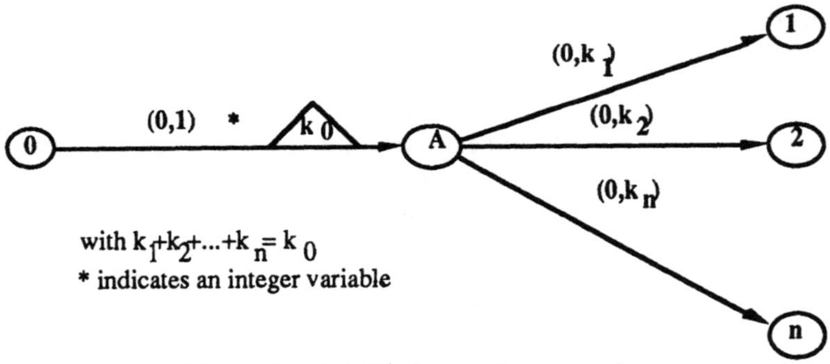

Figure 8 : 0-1 GN integrality structure

Taking the integer variables to be *decision variables*, a new group of problems can be faced as 0-1 GN's, including job assignment to computers, the design and operation of capacitated communication networks, financial capital allocation, plant location, energy models and physical distribution.

4.6 Optimal lot sizing and machine loading for multiple products.

The objective is to minimize the combined costs of production and inventory holding by determining optimal lot sizing and optimal assignment of production to machines. The principal characteristics of the problem are :

- The planning horizon is a single period, t weeks in length.
- The products are assigned to meet different needs, and they can not be substituted for one another. Production of each product is a single stage process.
- Lot sizes are selected from a predetermined set of possible lot sizes.
- All lots of any single product must be produced on the same machine.
- The machines work in parallel. They are similar in function, but they may differ in their rate and cost of operation. Some machines may be capable of producing several (all) of the products while others may be more specialized.
- The production capacities of all machines over the planning horizon are known constants. Each machine can produce only one product at a time.
- Demand for each product is assumed to occur continuously at a known constant rate.

The following variables are defined: X_{ijk} is a *binary* decision variable which takes the value 1 if the product j is produced on machine i in the kth possible lot size,.and 0 otherwise. C_{ijk} is the corresponding combined set-up, production and holding cost. r_{ijk} denotes the production capacity required on machine i to produce product j in the kth possible lot size. Finally, b_i denotes the aggregate production capacity of machine i over the t-week planning horizon. According to these notions, the mathematical model of the problem is:

$$\min \sum_{i}^{m}\sum_{j}^{n}\sum_{k}^{l} C_{ijk} X_{ijk} \qquad (18a)$$

$$\text{sbj.to} \quad \sum_{i}^{m}\sum_{k}^{l} X_{ijk} = 1, \quad j=1,\dots,n \tag{18b}$$

$$\sum_{j}^{n}\sum_{k}^{l} r_{ijk} X_{ijk} \le b_i, \quad i=1,\dots,m \tag{18c}$$

$$X_{ijk} = 0, 1 \qquad \forall\ i,j,k \tag{18d}$$

The model is designed only to load the machines. The management retains the prerogative of determining the precise sequence and timing for implementing the canditate assignment over the planning horizon. So, the model is rather a capacity planning tool, which combined with managers' experience, provides flexibility and responsiveness to the needs of the company/management, and further supports and evaluates different scenarios, based on alternative assumptions of demands and capacities. Of course, in order for this to be done efficiently, it is imperative for the model to be solved quickly. Excessive tests with existing 0-1 IP codes were reported to be unsuccesful because of the excessive amount of computation needed. An alternative approach can be developed based on the GN model, resulting on the graphical representation of Fig. 9.

As it can be seen, two sets of nodes are created: (i) *origin* nodes, associated with products, and (ii) *destination* nodes, associated with machines. Since each product is to be scheduled on exactly one machine (in one lot size exactly), the product nodes are given a supply of 1. Consequently an arc leading from a product to a machine will transmit a flow of 1, if its tail node is to be produced on the machine at its head node, while the lot size is specified by the multiplier of the arc. If the flow is not 1, then, because of the imposed discreteness, it must be 0. Thus, arc leading to machine i from product j for the kth possible lot size, corresponds to X_{ijk} variable defdined earlier. Cost and gain of each of these arcs correspond to the cost and resource consumption of the associated variable. The upper limit on the demand at each machine node implements the restriction on the aggregate production capacity of the machine.

4.7 Strategic Distribution Planning.

This is an application developed by the authors, and its importance as a case study stems from the fact that it highlights the decomposition of the initial problem to a sequence of MIGN subproblems, incorporating all of the most prominent features of GN technology. A description of the problem under consideration has as follows:

The Problem: In modern industry, production in many plants is strongly dependent on a wide range of raw materials and semi-processed products supplied to the firm by a large number of subcontractors. Quite often it is advisable, for reasons of better organization, process control and material-flow smoothing, that the whole volume of materials, supplied from a certain geographical area, is aggregated and despatched to the factory from a specially erected terminal. Many times, one of the primary subcontractors may operate as such a terminal. The question is to select among a set of candidate subcontractors the one that performs optimally on terms of warehouse setup and operational, as well as transportation costs. In a concise statement, the developed model is the following:

The Model: Given a *subcontractor network* S={N, B} and a *demand node* D, select the node to be operated as a *terminal* (depot) among a set of *candidate* nodes, so that the *cumulative* cost is minimized. The cumulative cost is analyzed to *transportation* cost, *terminal operational* cost and *terminal setup* cost. The terminal setup cost is a fixed quantity, while transportation and terminal operational costs are functions of the *life-cycle time* of the network (*linear* in a first approach; however, terminal operational cost may be any function of time). The *setup cost* and the *slope of the operational cost* for every candidate node consist input-data for the solution algorithm. The *slope of the cumulative transportation cost* for every candidate node has to be calculated by applying the GN optimal flow algorithms; so, we have a case where the Generalized Network comes up as a decomposition component of the entire model.

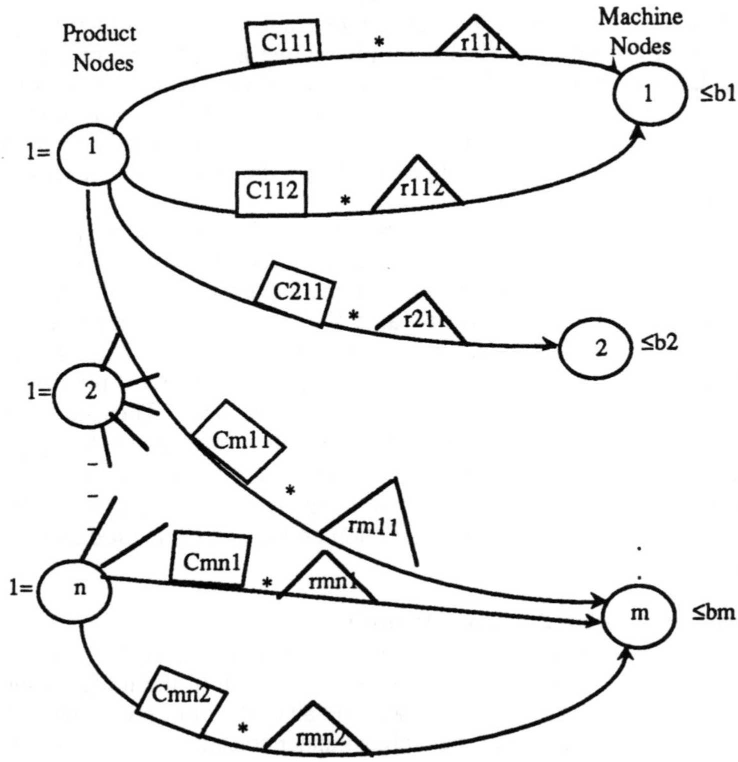

Figure 9 : Multiple product optimal lot sizing and machine loading NETFORM model

To highlight the analysis of the cumulative cost to its partial terms, and the influence that network life-cycle duration has on the selection of the optimal node, we give the diagram of Fig. 10.

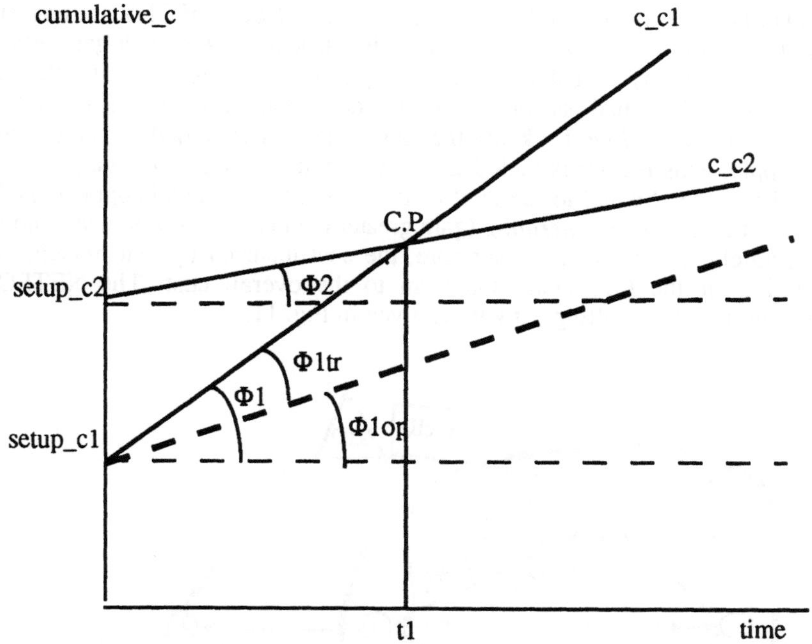

Figure 10 : Cumulative Cost Analysis

It can be easily seen in the diagram, that the choice of the optimal terminal node strongly depends on the life-cycle time of the network; i.e. if the latter is less than t_1, then c_1 must be selected as a terminal, otherwise c_2 is the optimal choice. To make the model more realistic, care has been taken of the fact that it is very likely that the exact duration of the network life-cycle is not known in advance (for one thing, the network parameters are liable to change). So, the model allows for a prediction of this parameter in terms of a *list of expected duaration times, accompanied by their probabilities (weights summing to unity)*; in this way, the model is converted to a *stochastic* one, and what is optimized, is the *expected cumulative* cost. The mathematical expression of the objective is

$$\min_{c\_n} \{\Sigma_i[p_i(t_i)\cdot t_i\cdot(op\_c_{c\_n}+tr\_c_{c\_n})]+st\_c_{c\_n}\} \tag{19}$$

where c_n is anyone of the candidate nodes, t_i, p_i the expected times and their associated probabilities, and $op\_c_{c\_n}$, $tr\_c_{c\_n}$, $st\_c_{c\_n}$ are the operational, transportation and setup costs in case that node c_n is selected for terminal.

In order to calculate the transportation cost, $tr\_c_{c\_n}$, the network is augmented with branches leading from the candidate nodes to the demand node. In the optimal solution, only one of those *artificial* branches must have a nonzero flow. A *supply* quantity (average) is attributed to every node of the subcontractor network, and *unit transportation costs* are associated with every branch, artificials included. *Capacity constraints* are imposed to every branch linking two subcontractors, to allow for the modelling of limited transportation media availability, and problems arising from natural or product conditions. Making use of the gain feature of GN's, the concept of *reliability factor* for every branch transfer is introduced. A branch-gain less than unity can model any probable *losses* during product transfers through that route (caused by stealing, bad packaging, product deterioration, etc.). A flow pattern is considered to be *feasible* if it keeps the total losses to a specified percentage of the total supply. This feature is included into the model through the assignment of a demand to node D equal to the total of the subcontractor nodal supplies, and the addition of an artificial *generating* (gain greater than unity) loop-branch at node D, appropriately capacitated. Furthermore, the cost incurred by that branch, is the expression of the *loss* cost as a term to the overall cost. The NETFORM presentation of the resulting network is given in Fig. 11.

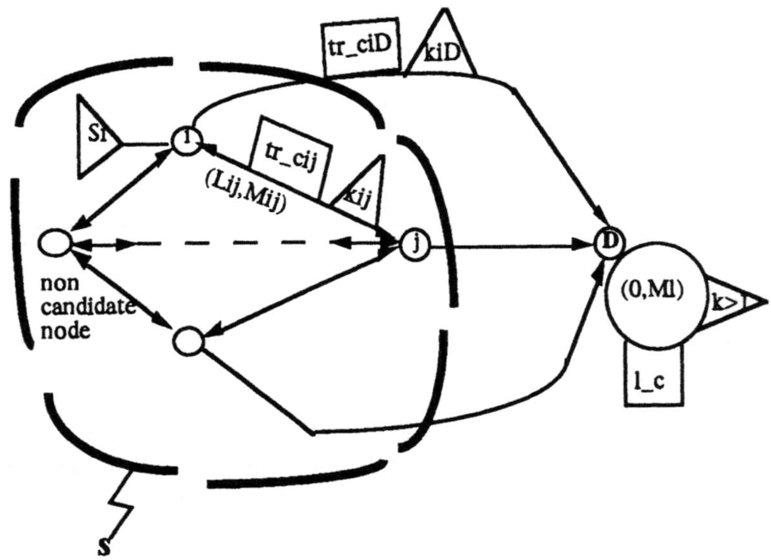

Figure 11 : Strategic Distribution Planning GN Model

The optimal terminal location and its associated material flow pattern is calculated by satisfying the objective (19) under the constraints:

$$\Sigma_j(X_{ij}-K_{ji}X_{ji}) = S_i-D_i \;\forall\; i\in N\cup\{D\} \qquad (a)$$
$$L_{ij}\leq X_{ij}\leq M_{ij} \qquad (b) \qquad (20)$$
only one branch (n_c,D) has non-zero flow (c)

Constraint (20a) ensures that there is no material loss at any node of the network, constraint (20b) takes care of the flow limitation on the branches, while constraint (20c) expresses the fact that one and only one node should be operated as terminal. This constraint suggests that we have to do with a *Specially Ordered Set of type 1 (SOS 1)* problem; these are sets of variables in which no more than one member is permitted to have a nonzero value. For the solution of this class of problems, appropriate BRANCH and BOUND methods have been developed. Practically, in each iteration of the method, a *relaxed* subproblem is solved, to get the optimal transportation cost in case that a whole subset of the candidate branches are active. The solution is characterized *acceptable* if only one branch of the subset has a nonzero value. All the same, a cumulative cost of an unacceptable solution is calculated, by taking the cumulative terminal cost (expected operational plus setup) to be equal to the minimum of the cumulative terminal costs of the branches belonging to the subset; this constitutes a *lower bound* of the cumulative cost in case that any single member of the subset operates as terminal. Each time an unacceptable solution with a cumulative cost lower than that of the best acceptable solution, found up to that point, is produced, the subset is *partitioned* to two new ones, and the possibility that the optimal solution lies in any of them is examined; if, on the other hand, the unacceptable solution presents a cumulative cost higher than that of the currently "optimal" acceptable solution (*incumbent cost*), or the resulting relaxation subproblem is infeasible, then there is no point in continuing the search in this subset, and thus, another subset is examined (*backtracking*). This prediction mechanism can reduce the number of GN optimal flow subproblems that have to be solved to get an optimal solution. It can be proven that in the best case the number of subproblems solved is 2, while in the worst, that number is $2\cdot(2^{[\log2(n)]+1}-1)$.

Another efficient feature of this approach is that since each subproblem is formulated by appropriately capacitating the branches leading from candidate nodes to the demand node; branch capacity of the non-active candidates is set to zero; post-optimization can be applied. So, the dual algorithm presented in comes into use. Employing decomposition schemes of Linear Programming, the model can be expanded so that it covers cases where various network parameters follow a regular pattern, for example, when the demand has a predetermined seasonal fluctuation.

Application Example. This example has been extracted from [4], and constitutes a typical case of 0-1 GN modelling. The problem is to choose the schedule of a fleet of ships among a variety of schedules, so that a given demand at numerous ports is served over a specified time period. In Fig. 12, the part of the model that applies to a single ship with exactly two schedules, A and B, is shown. Schedule A requires the ship to carry 10 tons of ore, which is distributed among the ports by dropping 3 tons at Port 1, 5 tons at Port 2, and 2 tons at Port 4. Schedule B requires the ship to carry 8 tons of ore, dropping 4 tons each at Ports 3 and 4.

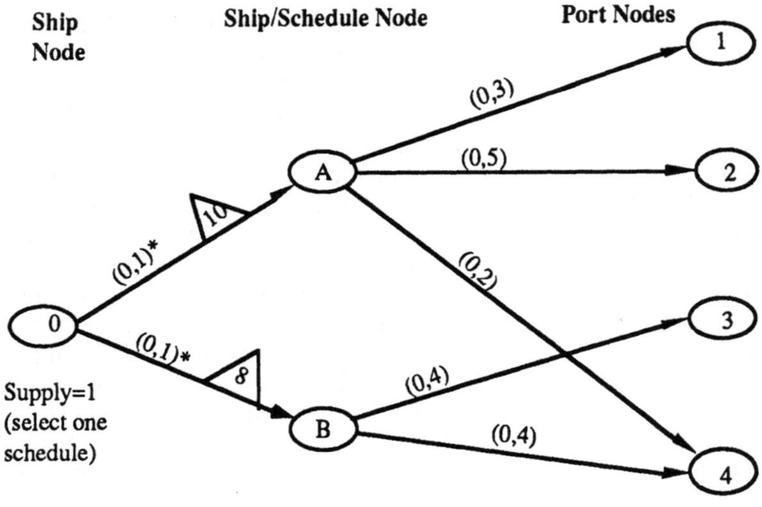

Figure 12 : Ship Scheduling

The supply of 1 at the Ship node, combined with the integrality of the departing branch flows, indicates that the ship can select precisely one of the available schedules. The multipliers of 10 and 8 convert the flows on the corresponding branches into the number of tons required in schedules A and B, respectively. For example, if the flow on the branch to Ship Schedule node A is 0, then $0 \cdot 10 = 0$, and therefore no flow gets transmitted to node A for distribution to the ports (by means of this schedule). But if the flow on this branch is 1, then the multiplier assures that 10 units of flow (tons of ore) are transmitted to A. Further, because the upper bounds on the branches from A to the Port nodes exactly sum to 10, each of these branches must carry a flow equal to its upper bound, thereby assuring that the desired amounts are delivered to each of the ports. Branches can be added emanating from the port nodes with lower and upper bounds limiting the total amount of ore received at each port during the time period from these and other ship schedules.

Since GN modelling is such a widespread and popular tool among decision-makers, the importance of the development of effective computer-based tools for the optimization and analysis of such models cannot be underestimated. In the next section, a PC-based tool for integrated GN modelling and analysis, developed by the authors, is briefly presented.

5 GENET-OPTIMIZER and GENETEXP: A new GN-modelling tool

Genet-Optimizer is a new tool for analyzing GN problems [16]. With this package, any real-life problem that has the mathematical formulation (1) can be modeled and analyzed. Actually, the problems to be treated with Genet-Optimizer are of a broader class than the one stated in (1), as it can deal with boundary condition constraints of inequality type, i.e. the nonzero supply for a ginen node can be specified either as a resource to be compulsory conveyed through the network, or as the available quantity of the resource considered. In the same way, the value of a demand can be considered either as a specific requirement to be met, or as a minimum of the required quantity.

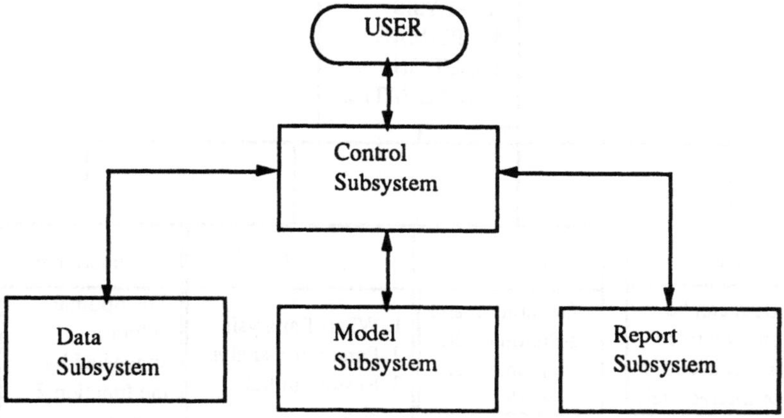

Figure 13 : System Components

Genet-Optimizer offers an integrated environment for the analysis of GNs. The main architecture of the package is that of Fig. 13; it is the typical architecture of packages of this class. The user controls the execution of the program interactively, by selecting among the "active" options of a hierarchically organized menu. The structure of the menu, accompanied by a brief description of each of its functions is given in Fig. 14. The routines implementing the mathematical algorithms necessary for the solution of the model are aggregated into the Model Subsystem. The role of the Data and Report subsystems is self-explanatory. Care has been taken so that the program is user-friendly and robust to any user's mistypings.

GENETEXP is the outcome of the integration of GENET OPTIMIZER (a previous pure numeric GN-DSS package) with an expert system shell developed at NTUA and used extensively for engine fault diagnosis (ENGEXP) and medical therapeutic treatment (BIOEXP). The package was developed in PASCAL under DOS on an IBM-PC compatible with sufficient memory to support the model's data and knowledge structure. A number of experiential rules have been accommodated to the knowledge-base for heuristic (expert based) post optimization analysis. Work is under development by the authors to incorporate in the knowledge base subsystem fully heuristic models accompanied by user-decision maker preferrential rules. This is done in cooperation with a European-level supply - distribution chain. The system will then be capable of evaluating the solutions and use the one that satisfies not only numerical but also other experiential preferential requirements.

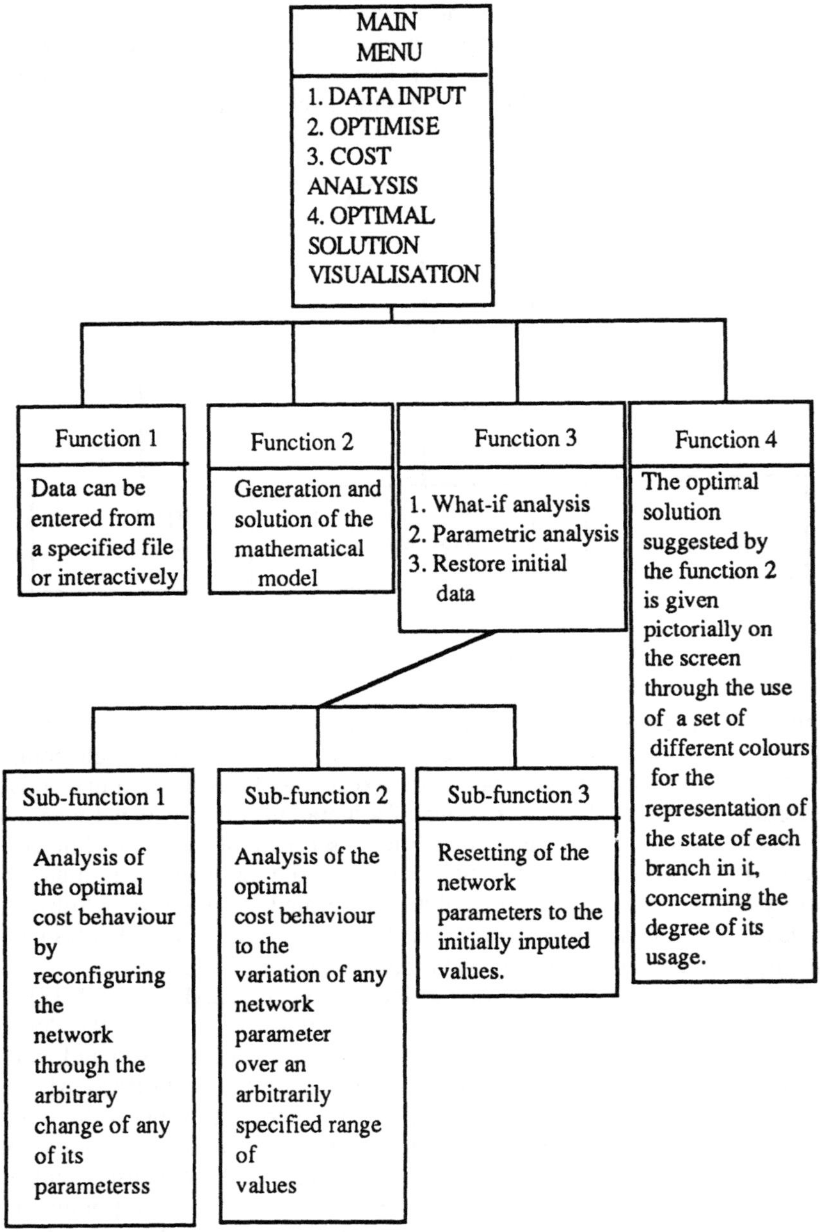

Figure 14 : Genet-Optimizer functions

6 Conclusions

Generalized network modelling is a powerful and popular tool among decision makers. In the past two decades extensive work has been carried out in the development of specialized algorithms for the solution of the GN optimal flow / transshipment problem. This paper has presented an overview of the GN model including the major solution algorithms. A new GN modelling tool was described, and a set of important applications of real life was discussed with emphasis on manufacturing applications. Currently, work is in progress by the authors to further enhance the GENETEXP tool particularly in the direction of increasing the capabilities of its knowledge-based component.

ACKNOWLEDGEMENT

Part of the work described in this chapter (viz. the GENET-Optimizer) was partially supported by ESPRIT CIM Project 2277 (CMSO).

REFERENCES

[1] Bertsekas P.D., ***Linear Network Optimization : Algorithms and codes***, MIT Press, 1992.

[2] Bertsekas P.D. and P. Tseng , “Relaxation Methods for Minimum Cost Ordinary and Generalized Network Flow Problems", **Oper. Res.**, Vol. 36, pp. 93-114, 1988.

[3] Brown J. and R. McBride, "Solving Generalized Networks", ***Management Sci.***, Vol. 30, pp.1497-1523, 1984.

[4] Dantzig B.G., ***Linear Programming and Extensions,*** Princeton University Press, Princeton, N.J., 1963.

[5] Dembo S.R., Mulvey M.J., and A.S. Zenios , "Large-Scale Nonlinear Network Models and their Applications", ***Operations Res.***, Vol. 37, pp. 353-372, 1989.

[6] Elam J., Glover F. and D. Klingman, "A strongly Convergent Primal Simplex Algorithm for Generalized Networks", ***Math. Oper. Res.***, Vol.4, pp. 39-59, 1979.

[7] Glover F., Jones J., Karns D., Klingman D. and J. Mote, "An Integrated Production, Distribution and Inventory Planning System", ***Interfaces*** , Vol. 9, pp. 21-35, 1979.

[8] Glover F., Hultz J., Klingman D., "Improved Computer-Based Planning Techniques, Part I", ***Interfaces***, Vol. 8, pp. 16-25, 1978.

[9] Glover F., Hultz J., Klingman D. and J. Stutz , "Generalized Networks: A Fundamental Computer-Based Planning Tool", ***Management Sci.***, Vol. 24, pp. 1209-1220, 1978.

[10] Glover F., Hultz J., Klingman D., "Improved Computer-Based Planning Techniques, Part II", ***Interfaces***, Vol. 9, pp. 12-20, 1979.

[11] Gongran M., Minoux M. and S. Vajda , ***Graphs and Algorithms***, John Wiley 1984.

[12] Jensen A.P. and G. Bhaumic, "A Flow Augmentation Approach to the Network with Gains Minimum Cost Flow Problem", ***Management Sci.***, Vol. 23, pp. 631-643, 1977.

[13] Jewell S.W., "Optimal Flow through Networks with Gains", **Operations Res.**, Vol. 10, pp. 476-499, 1962.

[14] Maurras J.F., "Optimization of the Flow through Networks with Gains", ***Mathematical Programming***, Vol. 3, pp. 135-144, 1972.

[15] McBride R. "Solving Embedded Generalized Network Problems", ***Eur. J. Opl. Res.***, Vol. 21, pp. 82-92, 1985.

[16] Reveliotis S. and G. Kapsiotis , "Genet-Optimiser", ***ESPRIT CMSO Report***, DECE-IRAL, NTUA, Athens, 1990.

[17] Tzafestas S., Kapsiotis G. and S. Reveliotis, "A Dual Algorithm for Post-Optimization of the Generalized Network Optimal Flow Problem", ***Foundation of Computing & Decision Sci.***, Vol. 16, No.1, pp.39-54,1991.

[18] Tzafestas S., Kapsiotis G. and S. Reveliotis, "The Generalized Network Approach to Optimized Decision Making and Planning", In : ***Systems and Simulation*** (A. Sydow Editor), North Holland 1992.

[19] Zahorik A., Thomas J. and W. Trigeiro, "Network Programming Models for Production Scheduling in Multi-stage, Multi-item Capacitated Systems", ***Management Sci.***, Vol. 30, pp. 308-325, 1984.

[20] Tzafestas S., Pimenidis T. and G. Kapsiotis, "Expert Decision Support Based on the Generalized Network Model : An Application to Transportation Planning", **EUROSIM '95 : European Simulation Congress**, Technical University of Vienna, Vienna, Austria (Sept. 11-15, 1995).

Biographies of the Contributors

Gabriella M. Acaccia : Doctor in Mathematics (1965) from the University of Genova; work experience in the Computation Centre of the University of Genova; professor of 'Modelling and Simulation of Mechanical Systems'; member of the Academic Senate of the University of Genova. Author or coauthor of some 200 papers in the areas of simulation, of motion of spacecrafts, design of mechanisms and robots, application of AI for intelligent manufacturing. Member of : IMACS (International Assosiation for Mathematics and Computers in Simulation), AIIA (Associazione Italiana di Intelligenza Artificiale), etc.

Eddy Bajic is Assosiate Professor at University Henri Poincare Nancy 1. After a Doctorate Thesis in 1988, he has obtained his Habilitation to be Research Director in 1995. He co-manages research activities at CRAN (Research Center for Automatic Control of Nancy) in the Quality Engineering research team. His research activities are focused on Quality Control and Management, Product Data Exchange and Specification, Distributed Information System Definition and Management with Nomadic Manufacturing Object Concept supported by escort memories and automatic identification techniques and methodologies. He participates in french-german research activities and a European research program with industrial partners on escort memories applications. He gives lectures and courses of Quality Management at University Henri Poincare, and mainly at the University Institute of Technology of Nancy Brabois.

Rolf Bernhardt studied Electrical Engineering at the Fachhochschule Coburg and the Technical University Berlin (TUB). From 1979 to 1984 he was a researcher at the TUB, Institute for Aeronautics and Astronautics. He began working at the Fraunhofer Institute for Production Systems and Design Technology in 1984. Since 1989 he has been the head of various departments in the fields of Automation, Robotics and Control Engineering. He is also a lecturer in Automation at the Technical University Berlin since 1993. Presently, he is head of the Department of Robot and System Control, and temporary director of the Control Engineering Division.

Massimo Callegari : Doctor in Mechanical Engineering (1986) from the University of Genova. Industrial work experiences at San Giorgio System Technologies, in the field of research and the development for factory automation. In 1990 he joined the Industrial Robot Design Research Group, where he is currently a researcher in the area of robotics and factory automation.Author of some 30 papers in the area of control of dynamic systems and robots, computer integrated manufacturing, and engineering applications of expert systems.

Luis M. Camarinha-Matos received his Computer Engineering degree and Ph.D. in Computer Science (Robotics and CIM) from the New University of Lisbon. Currently he is auxiliary professor (eq. associate professor) at the Electrical Engineering Department of the New University of Lisbon and leads the group of Robotic Systems and CIM of the UNINOVA's Center for Intelligent Robotics. His main research areas are : CIM systems integration , Intelligent Manufacturing Systems, Virtual Enterprises and Machine Learning in Robotics. He has been involved both as researcher and as project leader in several international projects in the following programmes. ESPRIT, TEMPUS, UNIDO, COMETT, EC-Latin America, LEONARDO, etc. and has published about 100 papers in national and international conferences and journals.

Frantisek Čapkovič received the engineering degree in Technical Cybernetics (Automation and Control) from the Electrotechnical Faculty of the Slovak Technical University in Bratislava. Since 1972 to 1991 he has been with the Institute of Technical Cybernetics of the Slovak Academy of Sciences. In 1980 he received the candidate of sciences (doctorate) degree (C.Sc.) in Technical Cybernetics (Automation and Control). Since 1991, when the institute was divided, he works at the successional Institute of Control Theory and Robotics of the Slovak Academy of Sciences. From 1991-1993 he was the head of the research project "Synthesis of control systems with ability of automatic reasoning" under a grant of the Slovak Grant Agency. Since 1994 he is the head of the research project "Modelling of DEDS and synthesis of their intelligent control by using artificial intelligence methods" under a grant of the same agency. His current research interests lie in the DEDS modelling area (especially using different kinds of Petri nets) and in the knowledge-based and intelligent control areas.

Ricardo Carelli received the Electrical Engineering degree (with honours) from the National University of San Juan (Argentina) in 1976 and the Ph.D. degree in Electrical Engineering from the National University of Mexico (UNAM) in 1989. He is presently professor of Automatic Control at the National University of San Juan and research-worker of the National Council for Scientific and Technical Research (CONICET). His current research interests are in the fields of robot control, manufacturing systems and artificial intelligence applied to systems control. He is a member of the IEEE and AADECA, the Argentine NMO of IFAC.

Ye-Hwa Chen was born in Taiwan, ROC. He received his B.S. in Chemical Engineering from the National Taiwan University in 1979 and his M.S. and Ph.D. in Mechanical Engineering from the University of California, Berkeley, in

1983 and 1985, respectively. From 1986 to 1988, he served as a faculty of the Department of Mechanical and Aerospace Engineering of Syracuse University. He then became a faculty of the School of Mechanical Engineering of Georgia Institute of Technology. His research interests include scheduling, control, fuzzy logic theory, and neural networks. Dr. Chen has published more than one hundred research articles in journals, conference proceedings, and monographs. He received the 1991 Sigma Xi Junior Faculty Award and 1994 Sigma Xi Best Research Paper Award. He is a member of Sigma Xi, ASME, and IEEE.

Armando Walter Colombo was born in Mendoza, Argentina in 1960. He graduated as an Electronic Engineer from the National Polytechnic University of Mendoza in 1990. He received the Master Degree in Control Engineering at the National University of San Juan in 1994. He is currently a Ph.D. candidate on the subject of Design of Flexible Manufacturing Systems and is developing a DAAD stage at the University of Erlangen-Nuremberg, Germany.

Juergen Fenchel is with a consulting firm, in Hamburg Germany. He received a Diploma in Mechanical Engineering from the University of Stuttgart, Germany, in 1991, and a Ph.D. in Mechanical Engineering from the Georgia Institute of Technology in 1995. His interests include flexible manufacturing systems, discrete event systems, scheduling, simulation, and system optimization.

George Kapsiotis was born in Volos, Greece,on November 5th,1964. He received his B.Sc. degree in Electrical Engineering from the National Technical University of Athens (NTUA) in 1988. From 1988, he joined the Intelligent Robotics and Automation Laboratory and worked in a number of EU funded research programs as a researcher (ESPRIT II, EUREKA, etc.). In 1993, he obtained the Ph.D. degree in Electrical Engineering. The subject of his dissertation was "Modelling, Optimization and Control of Computer Integrated Manufacturing Systems". He has published many journal and conference papers. His research interests include production planning control systems, computer integrated manufacturing systems, and applications of hierarchical optimization theory to the Strategic Business Planning. He is consultant of an R&D company.

George L. Kovács received his Diploma (Master) in Electrical Engineering from the Technical University of Budapest (1966). Since 1966 he is in the Computer and Automation Research Institute of the Hungarian Academy of Sciences and part-time lecturer at the Technical University of Budapest (El. Eng. , Mech. Eng.). He has been : one-year visiting researcher in the USA, two years in Russia, one year in Germany, and visiting professor in Mexico and Italy. He received the Doctoral Degree from the Technical University of Budapest (1976), the Degree of the Candidate of Sciences (Ph.D.) from the Hungarian Academy of Sciences (1987), and the Prof. and Habil. Degree from the Technical University of Budapest (1995). Since 1987 he is the Head of the CIM Department, and since

1991 he is the Head of the CIM Research Laboratory. His research interests include : FMS design, analysis, scheduling and control, application of AI in engineering, and intelligent manufacturing. He has more than 100 publications in journals and conference preprints and proceedings. He is a member of several Hungarian and international scientific organizations, and Vice-President of EEMIG (East-European MAP/TOP Interest Group).

Norbert Kratz studied Computer Science at RWTH Aachen and holds a Doctoral Degree in Computer Science from the University of Kaiserslautern. After his studies in Aachen he joined the research group of Prof. Dr. M.M.Richter at the University of Kaiserslautern and did some research work on knowledge-based systems supporting mechanical engineering design (Intelligent Design Assistant). In 1989 he started as project manager at FAW Ulm, where he was involved in a project enhancing CAD with knowledge-based modules in order to support design for manufacturability. Since 1991 he is as senior scientist at FAW heading the application area Business Process Management. His current research activities include business process modelling, as well as various applications of business engineering and knowledge-based systems in administration and production.

David Little is currently Professor of Manufacturing Systems at the University of Huddersfield. A Chartered Engineer and graduate of Loughborough University, he has worked in a number of companies which include GEC Telecoms, A Reyrolle and Molins Ltd before moving into higher education, initially with Huddersfield Polytechnic and then with the University of Liverpool where he had responsibilities as a Sub-Dean of the Faculty of Engineering, and Director of Studies of the Integrated Graduate Development Scheme which provides a modular route to a masters degree for graduates working in industry. He is a former BPICS President and has operated his own DTI listed manufacturing consultancy since 1979. His research interests include planning and scheduling systems, integrated performance measures, and the use of structured methods and tools in manufacturing.

Rinaldo C. Michelini : Doctor in Industrial Engineering (1960) from the University of Genova (Italy); work experiences in Switzerland: HSS-Genéve and ETH-Zurich: professor of Mechanical Engineering at the University of Genova, with international research/teaching experiences (visiting professor at the information System Laboratory, Stanford University, California USA). Leader of the Industrial Robot Design Research Group. Author or coauthor of more than 300 papers and 3 books in the areas of stability of motion, system identification and filtering theory, attidute control of spacecrafts, instrumental robotics, nonlinear control, industrial diagnostics, factory automation, CAD/CAM and CAT applications, etc. Member of IFIP WG. 5.3, ASME fellow, etc.

Martin Mišút is a Senior Lecturer in Information Systems in Production at the Department of Information Technologies and Automation of the Slovak Technical University. Prior to joining the faculty at the Slovak University in February 1987, he was a quality engineer at the Trnava Motor Company. He received his MSc. degree in production management with high distinction from the Slovak Technical University in 1985, and his Ph.D. Degree in production engineering at the same University in 1993. His current research interest include operation scheduling, production system management and the application of information technologies in production.

Rezia M. Molfino : Doctor in Electrotechnical Engineering (1996) from the University of Genova; work experiences as a member of research and development staff in CETENA (ship research centre) and in ELSAG (San Giorgio Electronics) of Genova; professor of 'Control of Mechanical Systems' and 'Robot Mechanics' at the University of Genova. Author or coauthor of some 200 papers in the areas of modelling and simulation of complex dynamic systems (weapon systems, satellites, robots); multivariable control synthesis; CAD/CAM programs for FMS and robot design with A.I. applications. Vice-president of SIRI (Associazione Italiana di Robotica), etc.

Oliver Moravčik is an associate Professor of Software Engineering at the Department of Information Technologies and Automation of the Slovak Technical University. He received his MSc. degree in Computer Science and Simulation from Ilmenau Technical University in Germany in 1976, and his Ph.D. in Automation of Production Systems at the same University in 1982. His current research areas include software engineering and the application of information technologies in production systems.

Luis Osorio : BSc. degree in Electronics and Telecommunications from the Instituto Superior de Engenharia de Lisboa, 1979, Licenciatura (5 years Portuguese degree) in Computer Science from the Faculdade de Ciências e Tecnologia da Universidade Nova de Lisboa, 1981. He is adjunct professor at the Instituto Superior de Engenharia de Lisboa (ISEL), teaching disciplines in the area of Systems Engineering, namely operating systems, distributed operating systems, logic programming and others. He is finishing his PhD in Information Systems and Systems Integration in CIM at the Universidade Nova of Lisboa, Electrotechnic Department and Intelligent Robotics Group, and is a researcher at the UNINOVA Institute. He is member of IEEE, APPIA and Portuguese Engineers Association.

Ricardo J. Rabelo received his degree in Computer Science from the Federal University of Santa Catarina (UFSC), Brazil, in 1984, and he has obtained a specialization in robotics at the New University of Lisbon (Portugal), UNINOVA in 1988. He worked as consultant for several large Brazilian companies in the area of CAM as a collaborator of GRUCON Research Group / UFSC. He is finishing his Ph.D. at the New University of Lisbon / UNINOVA in the area of agile scheduling. He has published several articles in international conferences journals, and books. His main interests are : agile scheduling, virtual manufacturing / enterprise, multiagent systems, negotiation in manufacturing, integration and enterprise modeling. He has been participating in several international projects which involved these areas.

Roberto P. Razzoli : Doctor in Mechanical Engineering from the University of Genova in 1991. Work experience in Consorzio Genova Ricerche in the field of real-time expert systems. In 1993 he joined the Industrial Robot Design Research Group where he is currently a researcher in the area of the Machine Design. His research subjects are in CAD, factory automation and robotics. Member of ADM (Associazione Nazionale Disegno di Macchine).

Jacques Richard is a Professor at the University Henri Poincaré Nancy 1. He leads the research activities of the Quality Engineering research team at CRAN-Research Center for Automatic Control of Nancy. He actually manages a research project, supported by the french research ministry, in collaboration with several national laboratories and industry, in the field of quality management methodology within a concurrent engineering-based project, with multi-sites constraints. He is responsible of the postgraduate DEA Diploma specialized in Automated and Integrated Manufacturing. He gives lectures and courses in Robotics and Quality Management at the Faculty of Sciences of the University Henri Poincaré, especially in the ESIAL University Engineers school.

Thomas Rose received his Diploma in Computer Science from the University of Dortmund in 1985, and the Doctoral degree in Computer Science from the University of Passau in 1991. Since 1993 he is a Senior Researcher at the FAW Ulm. He is heading the telematics group with projects in the context of enterprise integration , telematics applications for environment and health care, and electronic trade networks. From 1990 to1993 he has been Research Associate of Computer Science at the University of Toronto. He has been the leader of a project on the "Development of Intelligent Information Systems" which was funded by the Federal Networks of excellence programme through the Institute of Robotics and Intelligent Systems (IRIS). From 1986 to 1990 he worked at the University of Passau for the ESPRIT project DAIDA on a prototype environment for information systems development and maintenance. His current interests include software engineering and repositories, team work, enterprise engineering, conceptual modelling, and development of added-value services in telematics.

M. Szafarczyk received the M.Sc. (1956) and Ph.D. (1970) degrees both at the faculty of Mechanical Engineering of the Warsaw University of technology. Since 1983 he is Full Professor at the Warsaw University of Technology, and Head of the Department of Machine Tool Design and Automation , Institute of Mechanical Technology, Faculty of Production Engineering. He has been a consultant at the Central Institute of Machine Tool Design and Research, CBKO-Pruszkow (1970-now). His research interests are in machine tool design with computer support, automatic control of machine tools and manufacturing systems, automatic monitoring and supervision in manufacturing, and mechatronics in manufacturing equipment. He is a member of the International Institution for Production Engineering Research (CIRP), the International Society for Engineering education (IGIP), the Academy of Engineering in Poland (AIP), and the Polish Society of Mechanical Engineering (SIMP). He is a member of the editorial board of the International Journal of Advanced Manufacturing Technology and the Journal of Systems Engineering.

John Thompson has worked in the IT Industry for almost 20 years, initially following a technical discipline which culminated in the design of an automated storehouse control system for the M.o.D. This was one of a number of flagship sites with which he has been associated, delivering an advanced automated capability to NATO in the storage and retrieval of supplies. He was project manager of, and consultant to, the laboratory project described in this book and, as a result of his involvement, he presented a paper on the implementation of the ILMS at the 4th International Conference on Automation , Robotics and Artificial Intelligence applied to Analytical Chemistry and Laboratory Medicine held in Montreux, Switzerland in 1995. He is a charter member of the Assosiation for Laboratory Automation.

Spyros G. Tzafestas received the following degrees : B.Sc. in Physics and Electronics (Athens University, 1963); D.IC. and M.Sc. in Control (Imperial College, London,1966); Ph.D. (1969) and D.Sc. (1978) in Systems and Control (Southampton University). From 1969 to 1973 he was a research leader at the Nuclear Research Center "Demokritos" (Athens), and from 1973 to 1984 he was professor of systems and control engineering at Patras University, Greece. Since 1985 he is professor of robotics and control at the National Technical University of Athens, Greece. He has published 30 research books and over 500 technical papers. He is editor of the Journal of Intelligent and Robotic Systems, associate editor of ten journals, and editor of the Kluwer book series Microprocessor and Intelligent Systems Engineering. His research interests include intelligent systems, knowledge-based systems, robotics, CIM, control, and fault diagnosis. He is a fellow of IEEE and IEE, honorary president of the Society of Intelligent Robotics and Expert Systems (SIRES), and a member of ASME and the New York Academy of Sciences. His experience includes the leadership of several European R&D projects.

W. A. Vervoort is an engineer who started work at the Electrotechnical Department of the Twente University in the Netherlands. In the early days he helped to develop the field of computer science. From 1972 he works at the Computer Science Department of the same University. At first in the area of program verification and then in the field of real-time computer operating system kernels. Later on he moved to the topic of production automation, where he was involved in an Esprit project, together with the Department of Mechanics, which was the inspiration for his chapter of the book. Presently he is involved in the design of embeded real-time systems, especially autonomous mobile robotic systems and teams of autonomous mobile agents.

Index

MIX
Papier aus verantwortungsvollen Quellen
Paper from responsible sources
FSC® C105338

If you have any concerns about our products,
you can contact us on
ProductSafety@springernature.com

In case Publisher is established outside the EU,
the EU authorized representative is:
Springer Nature Customer Service Center GmbH
Europaplatz 3, 69115 Heidelberg, Germany

Printed by Libri Plureos GmbH
in Hamburg, Germany